Functional Coatings for Food Packaging Applications

Functional Coatings for Food Packaging Applications

Editors

Stefano Farris
Jari Vartiainen

MDPI • Basel • Beijing • Wuhan • Barcelona • Belgrade • Manchester • Tokyo • Cluj • Tianjin

Editors
Stefano Farris
University of Milan DeFENS
Italy

Jari Vartiainen
VTT Technical Research Centre of Finland Ltd.
Finland

Editorial Office
MDPI
St. Alban-Anlage 66
4052 Basel, Switzerland

This is a reprint of articles from the Special Issue published online in the open access journal *Coatings* (ISSN 2079-6412) (available at: https://www.mdpi.com/journal/coatings/special_issues/packag_coatings2017).

For citation purposes, cite each article independently as indicated on the article page online and as indicated below:

LastName, A.A.; LastName, B.B.; LastName, C.C. Article Title. *Journal Name* **Year**, *Article Number*, Page Range.

ISBN 978-3-03936-850-1 (Hbk)
ISBN 978-3-03936-851-8 (PDF)

© 2020 by the authors. Articles in this book are Open Access and distributed under the Creative Commons Attribution (CC BY) license, which allows users to download, copy and build upon published articles, as long as the author and publisher are properly credited, which ensures maximum dissemination and a wider impact of our publications.

The book as a whole is distributed by MDPI under the terms and conditions of the Creative Commons license CC BY-NC-ND.

Contents

About the Editors ix

Preface to "Functional Coatings for Food Packaging Applications" xi

Luis J. Bastarrachea, Dana E. Wong, Maxine J. Roman, Zhuangsheng Lin and Julie M. Goddard
Active Packaging Coatings
Reprinted from: *Coatings* 2015, 5, 771–791, doi:10.3390/coatings5040771 1

Lluís Palou, Silvia A. Valencia-Chamorro and María B. Pérez-Gago
Antifungal Edible Coatings for Fresh Citrus Fruit: A Review
Reprinted from: *Coatings* 2015, 5, 962–986, doi:10.3390/coatings5040962 19

Chuying Chen, Nan Cai, Jinyin Chen, Xuan Peng and Chunpeng Wan
Chitosan-Based Coating Enriched with Hairy Fig (*Ficus hirta* Vahl.) Fruit Extract for "Newhall" Navel Orange Preservation
Reprinted from: *Coatings* 2018, 8, 445, doi:10.3390/coatings8120445 41

Yaowen Liu, Shuyao Wang, Wenting Lan and Wen Qin
Fabrication and Testing of PVA/Chitosan Bilayer Films for Strawberry Packaging
Reprinted from: *Coatings* 2017, 7, 109, doi:10.3390/coatings7080109 53

Elena Fortunati, Geremia Giovanale, Francesca Luzi, Angelo Mazzaglia, Josè Maria Kenny, Luigi Torre and Giorgio Mariano Balestra
Effective Postharvest Preservation of Kiwifruit and Romaine Lettuce with a Chitosan Hydrochloride Coating
Reprinted from: *Coatings* 2017, 7, 196, doi:10.3390/coatings7110196 69

Monserrat Escamilla-García, María J. Rodríguez-Hernández, Hilda M. Hernández-Hernández, Luis F. Delgado-Sánchez, Blanca E. García-Almendárez, Aldo Amaro-Reyes and Carlos Regalado-González
Effect of an Edible Coating Based on Chitosan and Oxidized Starch on Shelf Life of *Carica papaya* L., and Its Physicochemical and Antimicrobial Properties
Reprinted from: *Coatings* 2018, 8, 318, doi:10.3390/coatings8090318 85

Manuela Rollini, Tim Nielsen, Alida Musatti, Sara Limbo, Luciano Piergiovanni, Pilar Hernandez Munoz and Rafael Gavara
Antimicrobial Performance of Two Different Packaging Materials on the Microbiological Quality of Fresh Salmon
Reprinted from: *Coatings* 2016, 6, 6, doi:10.3390/coatings6010006 99

Clara Silvestre, Donatella Duraccio, Antonella Marra, Valentina Strongone and Sossio Cimmino
Development of Antibacterial Composite Films Based on Isotactic Polypropylene and Coated ZnO Particles for Active Food Packaging
Reprinted from: *Coatings* 2016, 6, 4, doi:10.3390/coatings6010004 107

Gracia López-Carballo, Pilar Hernández-Muñoz and Rafael Gavara
Photoactivated Self-Sanitizing Chlorophyllin-Containing Coatings to Prevent Microbial Contamination in Packaged Food
Reprinted from: *Coatings* 2018, 8, 328, doi:10.3390/coatings8090328 121

Maria Rosaria Corbo, Daniela Campaniello, Barbara Speranza, Antonio Bevilacqua and Milena Sinigaglia
Non-Conventional Tools to Preserve and Prolong the Quality of Minimally-Processed Fruits and Vegetables
Reprinted from: *Coatings* 2015, 5, 931–961, doi:10.3390/coatings5040931 135

Marina Ramos, Arantzazu Valdés, Ana Beltrán and María Carmen Garrigós
Gelatin-Based Films and Coatings for Food Packaging Applications
Reprinted from: *Coatings* 2016, 6, 41, doi:10.3390/coatings6040041 161

Arantzazu Valdés, Nuria Burgos, Alfonso Jiménez and María Carmen Garrigós
Natural Pectin Polysaccharides as Edible Coatings
Reprinted from: *Coatings* 2015, 5, 865–886, doi:10.3390/coatings5040865 181

Cristina Costa, Annalisa Lucera, Amalia Conte, Angelo Vittorio Zambrini and Matteo Alessandro Del Nobile
Technological Strategies to Preserve Burrata Cheese Quality
Reprinted from: *Coatings* 2017, 7, 97, doi:10.3390/coatings7070097 199

Dan Xu, Jing Wang, Dan Ren and Xiyu Wu
Effects of Chitosan Coating Structure and Changes during Storage on Their Egg Preservation Performance
Reprinted from: *Coatings* 2018, 8, 317, doi:10.3390/coatings8090317 209

Juan Manuel Tirado-Gallegos, Paul Baruk Zamudio-Flores, José de Jesús Ornelas-Paz, Claudio Rios-Velasco, Guadalupe Isela Olivas Orozco, Miguel Espino-Díaz, Ramiro Baeza-Jiménez, José Juan Buenrostro-Figueroa, Miguel Angel Aguilar-González, Daniel Lardizábal-Gutiérrez, María Hernández-González, Francisco Hernández-Centeno and Haydee Yajaira López-De la Peña
Elaboration and Characterization of Active Apple Starch Films Incorporated with Ellagic Acid
Reprinted from: *Coatings* 2018, 8, 384, doi:10.3390/coatings8110384 221

Masaki Nakaya, Akira Uedono and Atsushi Hotta
Recent Progress in Gas Barrier Thin Film Coatings on PET Bottles in Food and Beverage Applications
Reprinted from: *Coatings* 2015, 5, 987–1001, doi:10.3390/coatings5040987 239

Samir Kopacic, Andrea Walzl, Armin Zankel, Erich Leitner and Wolfgang Bauer
Alginate and Chitosan as a Functional Barrier for Paper-Based Packaging Materials
Reprinted from: *Coatings* 2018, 8, 235, doi:10.3390/coatings8070235 253

Maria Pardo-Figuerez, Alex López-Córdoba, Sergio Torres-Giner and José M. Lagaron
Superhydrophobic Bio-Coating Made by Co-Continuous Electrospinning and Electrospraying on Polyethylene Terephthalate Films Proposed as Easy Emptying Transparent Food Packaging
Reprinted from: *Coatings* 2018, 8, 364, doi:10.3390/coatings8100364 269

Oisik Das, Thomas Aditya Loho, Antonio José Capezza, Ibrahim Lemrhari and Mikael S. Hedenqvist
A Novel Way of Adhering PET onto Protein (Wheat Gluten) Plastics to Impart Water Resistance
Reprinted from: *Coatings* 2018, 8, 388, doi:10.3390/coatings8110388 283

Pieter Samyn
Raman Microscopy for Classification and Chemical Surface Mapping of Barrier Coatings on Paper with Oil-Filled Organic Nanoparticles
Reprinted from: *Coatings* 2018, 8, 154, doi:10.3390/coatings8050154 299

Christoph Metzger, Solange Sanahuja, Lisa Behrends, Sven Sängerlaub, Martina Lindner and Heiko Briesen
Efficiently Extracted Cellulose Nanocrystals and Starch Nanoparticles and Techno-Functional Properties of Films Made Thereof
Reprinted from: *Coatings* **2018**, *8*, 142, doi:10.3390/coatings8040142 . . . . **317**

About the Editors

Stefano Farris earned his Ph.D. degree in Food and Microbial Biotechnology in 2007. From 2007 to 2008, he was a postdoctoral fellow at Rutgers University (NJ) with the Food Packaging laboratory led by Prof. K. Yam, where he worked on the use of hydrocolloids as biopolymer coatings. In 2011, he joined Prof. M. Hedenqvist's group at KTH in the Department of Fibre and Polymer Technology, where he worked on the development of hybrid materials. At present, he is an Associate Professor in the Department of Food, Environmental and Nutritional Sciences (DeFENS) of the University of Milan, where he coordinates a team of approximately 10 people. His research activities mainly focus on the design and development of new high-performance packaging materials using different approaches such as soft chemistry and nanotechnology. In addition, he is actively involved in the exploitation of green strategies (e.g., enzymatic routes for waste recovery) for the generation of new materials. He is the author of more than 70 papers in peer-reviewed journals, 5 international patents, and several book chapters.

Jari Vartiainen is the Senior Scientist witht he VTT Technical Research Centre of Finland.

Preface to "Functional Coatings for Food Packaging Applications"

The food packaging industry is experiencing one of the most relevant revolutions associated with the transition from fossil-based polymers to new materials of renewable origin. However, high production costs, low performance, and ethical issues still hinder the market penetration of bioplastics. Recently, coating technology was proposed as an additional strategy for achieving a more rational use of the materials used within the food packaging sector. According to the packaging optimization concept, the use of multifunctional thin layers would enable the replacement of multi-layer and heavy structures, thus reducing the upstream amount of packaging materials while maintaining (or even improving) the functional properties of the final package to pursue the goal of overall shelf life extension. Concurrently, the increasing requirements among consumers for convenience, smaller package sizes, and for minimally processed, fresh, and healthy foods have necessitated the design of highly sophisticated and engineered coatings. To this end, new chemical pathways, new raw materials (e.g., biopolymers), and non-conventional deposition technologies have been used. Nanotechnology, in particular, paved the way for the development of new architectures and never-before-seen patterns that eventually yielded nanostructured and nanocomposite coatings with outstanding performance. This book covers the most recent advances in the coating technology applied to the food packaging sector, with special emphasis on active coatings and barrier coatings intended for the shelf life extension of perishable foods.

Stefano Farris, Jari Vartiainen
Editors

Review

Active Packaging Coatings

Luis J. Bastarrachea, Dana E. Wong, Maxine J. Roman, Zhuangsheng Lin and Julie M. Goddard *

Department of Food Science, University of Massachusetts Amherst, 102 Holdsworth Way, Amherst, MA 01003, USA; lbastarr@foodsci.umass.edu (L.J.B.); dewong@foodsci.umass.edu (D.E.W.); mjroman@foodsci.umass.edu (M.J.R.); zhuangshengl@foodsci.umass.edu (Z.L.)

* Author to whom correspondence should be addressed; goddard@foodsci.umass.edu; Tel.: +1-413-545-2275; Fax: +1-413-545-1262.

Academic Editor: Stefano Farris

Received: 25 September 2015; Accepted: 30 October 2015; Published: 6 November 2015

Abstract: Active food packaging involves the packaging of foods with materials that provide an enhanced functionality, such as antimicrobial, antioxidant or biocatalytic functions. This can be achieved through the incorporation of active compounds into the matrix of the commonly used packaging materials, or by the application of coatings with the corresponding functionality through surface modification. The latter option offers the advantage of preserving the packaging materials' bulk properties nearly intact. Herein, different coating technologies like embedding for controlled release, immobilization, layer-by-layer deposition, and photografting are explained and their potential application for active food packaging is explored and discussed.

Keywords: active food packaging; antimicrobial; antioxidant; biocatalytic; surface modification

1. Introduction

Active packaging, in which the packaging material performs an additional function beyond containment and basic protection, remains an area of active research with great potential for commercial applications. Active packaging has application in packaging of food, pharmaceutical, and consumer goods products, with a common goal of improving shelf life, safety, or quality of packaged goods. A number of excellent reviews have been written on targeted applications of active packaging materials with less focus on material synthesis techniques [1–5]. Synthesis can be achieved by incorporating an active agent (e.g., antioxidant, enzyme, antimicrobial, oxygen scavenger) within or at the product contact surface of a packaging material. Positioning the active agent at the product contact side (*versus* bulk incorporation) has several benefits, including retention of bulk material properties and minimizing the amount (and therefore cost) of active agent required to impart efficacy. Understanding the technologies and challenges associated with various coating methods for preparation of active packaging materials will support effective technology transfer to commercial applications. The goal of this review is to describe key technologies for preparing active packaging coatings, including embedding, layer-by-layer deposition, and photografting, with a discussion of the difference between covalent (non-migratory) and non-covalent (migratory) immobilization chemistries. We then survey the current literature for active packaging technologies using these coating methods to impart antimicrobial, antioxidant, and biocatalytic activity. Smart/intelligent packaging [6,7], in which indicating devices are incorporated into the packaging structure, and sachet-based technologies [8] are well covered in the current literature and are outside the scope of this review. We conclude with a discussion of the challenges that remain in achieving commercial translation for active packaging coatings.

2. Overview of Coating Technologies

In the following subsections, the main approaches employed to prepare coatings for active packaging that have been reported in the literature are explained. These are the most frequent techniques found in the field of coatings for active packaging. Figure 1 shows a graphic summary of them.

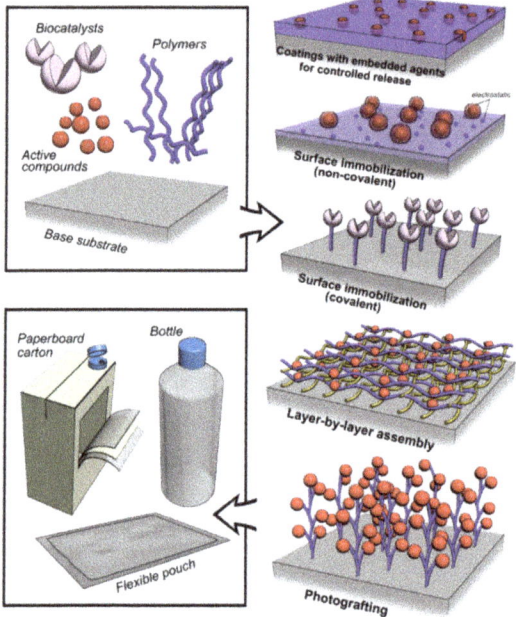

Figure 1. Summary of coating technologies.

2.1. Coatings with Embedded Agents for Controlled Release

As illustrated in Figure 1, active agents can be incorporated into coatings for active packaging by a number of technologies, including those in which the active agent is intended to migrate to the packaged good (embedding, non-covalent immobilization, some layer-by-layer deposition techniques) and non-migratory technologies in which the active agent is intended to remain stable in the packaging matrix (covalent immobilization, some layer-by-layer deposition techniques, photografting). Controlled release coatings are those in which active agents have been incorporated into the matrices of polymeric materials, from which they are expected to migrate and exert their specific function within a packaged good, whether antimicrobial, antioxidant, biocatalytic, or neutraceutical [6]. The main reported mechanisms through which active compounds have been incorporated into polymeric materials have been extrusion/blending and solution casting [9,10]. Solution casting consists of dissolving the polymer intended for packaging in a suitable solvent and simultaneously incorporating the active compound of interest, followed by pouring the solubilized polymer and active agent onto a surface for the solvent to evaporate, resulting in formation of plastic film with targeted functionality (antimicrobial, pharmaceutical, biocatalytic, *etc.*). In extrusion, the active compound is incorporated with polymeric material melted by heat transfer forming a blend from which films can be formed. The first method, although extensively used at a laboratory scale, exhibits some limitations for its practical and commercial application as most of the polymers intended for food packaging can only be dissolved at high temperatures with organic solvents [11], which would compromise the stability and effectiveness of many active compounds of interest. This is probably the main reason why incorporation of active compounds through solution casting has been studied more frequently with biodegradable polymers that can be dissolved at milder temperatures, for example poly(lactic acid), poly(butylene adipate-*co*-terephtalate), and cellulose derivatives [9]. Extrusion and blending, although industrially scalable, can present serious limitations due to the intrinsic lack of thermal stability of many active compounds, which can be lost through degradation and evaporation during the heat transfer these unit operations involve [3,9]. Ensuring uniform distribution of the active compound and retaining thermomechanical film properties are additional challenges to preparation of extruded active packaging materials. Controlled release can rely on physical and chemical phenomena, like the degree of affinity between the active compound and the matrix of the packaging material, as well as its morphology and porosity, which if low could allow a slower release [3,9,12]. In addition, controlled release can be provided by a multilaminated system in which the layer that harbors the active compound is covered by an adjacent layer that could serve as a barrier that controls the rate of release of such compound [9,13]. Temperature also plays a relevant role in the release of active compounds embedded in polymer packaging materials. It has been observed extensively that diffusion increases with temperature, which could represent an advantage specifically for the case of antimicrobial packaging as microorganisms also reproduce faster with an increase in temperature [9,13]. However, even if after incorporation of the compound of interest its activity and effectiveness is demonstrated at a laboratory scale, it is well known that the incorporation of a foreign compound into the matrix of a plastic material (even at small concentrations, at which no activity may be observed) can substantially affect its properties relevant to processing, production and machinability. These include the tensile and mechanical properties (tensile strength, elastic modulus, elongation at break, *etc.*), the thermal properties (melting point, glass transition temperature, crystallization temperature, heat capacity, *etc.*), and the gas barrier properties (water vapor permeability, O_2 and CO_2 permeability, *etc.*) [9,14]. As the goal of active packaging is to enhance material performance, losing desirable material properties as a result of active agent incorporation limits commercial applicability. There is therefore an interest in the application of thin coatings rather than bulk modifications, as thin coatings are not expected to affect relevant physical and chemical properties [15]. Nevertheless, a major benefit of embedding active agents into a packaging film by coating or coextrusion is their use of currently available converting equipment.

2.2. Surface Immobilization

In addition to embedding, active packaging materials can be prepared by immobilization of an active agent on the surface of a bulk packaging material. Although less studied as a material for active packaging, paperboard can be functionalized to immobilize active compounds by modifying the chemistry of its main component cellulose, with the formation of reactive groups like aldehydes, epoxy, carboxylic acids, *etc.* [16]. More commonly reported is the surface modification of polymer materials for use in active packaging. Polymer packaging materials are typically inert, and require an initial functionalization to enable immobilization of an active agent. Surface activation techniques of plastic polymer substrates can be divided into physical and wet methods. Physical methods include flame, corona discharge, UV radiation, and plasma [5,15,17]. Wet methods involve the use of corrosive liquids to which the polymer substrates are directly exposed, like piranha solution (dissolved hydrogen peroxide and sulfuric acid), combined sodium hydroxide and sulfuric acid, chromic acid, potassium permanganate, and nitric acid [18–22]. The main effect of these surface activation methods is the formation of reactive oxygenated species on the polymer surface like carbonyl, hydroxyl and carboxylic acid groups [21,22]. Some methods of industrial relevance to activate the surface of polymeric packaging materials are flame, corona, and atmospheric or vacuum plasma [23], which remove contaminants from the surface of polymer films, increase their surface energy and wettability (in order to apply coatings of different types, like inks for printing or metallic coatings), and increase the level of oxygen on their surface. In contrast to the wet methods, the physical methods are more widely used due to their industrial scalability, given the fact that no liquid reagents of any kind are needed in their application, avoiding accumulation and handling of harmful waste [22]. Once a polymeric surface has been functionalized and the mentioned oxygen-containing moieties haven been incorporated on it, compounds of different functionality can be applied through either covalent or non-covalent immobilization. Non-covalent binding relies mainly on electrostatic interactions and affinity [22]. In the first case, the polymer surface possesses a certain net charge which can attract molecules with an opposite net charge. The second case refers to specific ligand–receptor interactions, like in the case of biotin–avidin [24]. Covalent immobilization as its name implies involves the covalent coupling of the active compound of interest onto the polymer surface. A potential benefit of covalent immobilization is that the active compound is not expected to migrate to the packaged product matrix, which could otherwise compromise its commercial application [22]. Active agents can be covalently linked directly to the polymer surface or by use of a crosslinker, which may either share a permanent covalent bond with both the polymeric substrate and the bioactive compound, or may promote covalent bond formation between the activated substrate and the bioactive compound without forming part of that link ("zero-length" crosslinkers) [22,25]. The covalent attachment between a polymeric surface and an active compound relies mainly on the formation of amide, ether, ester, and thioether bonds, created between the hydroxyl, amine, imine, carboxylic acid, and thiol groups the active compounds of interest may possess intrinsically (or are incorporated in their structure) and the functional groups created on the substrate [22,25,26]. A benefit to immobilization technologies is the potential versatility: once functional groups are introduced to the polymer surface, a range of bioactive agents (e.g., enzymes, peptides) can be immobilized through standard bioconjugation techniques.

2.3. Layer-by-Layer Assembly

Layer-by-layer assembly is a versatile method of surface modification that relies on the deposition and mutual attraction of alternating polyelectrolytes with opposite net charges onto a solid support [27,28]. A polymeric substrate can be made reactive through the previously explained methods of surface activation, imparting a certain charge on it, or a polyelectrolyte can be covalently attached onto its surface to further apply the alternate polyelectrolyte layers. The polyelectrolytes can be polymers of any kind (proteins, polysaccharides, synthetic polymers, *etc.*) that harbor a net charge, and they can be modified by the incorporation of functional groups to facilitate their deposition [27–29]. Deposition can be accomplished either by submersion of the substrate into polyelectrolyte solutions or by spraying of solutions onto the substrate; both techniques have potential for scalability as high throughput coating technologies. The deposition of the polyelectrolytes can be optimized by adjusting the pH of their solutions to a point of full protonation or deprotonation, to maximize the presence of charge [30]. The deposition can rely exclusively on electrostatic interactions [29,31], or covalent bonds can be built between the alternating layers through the use of crosslinkers [25,29,32,33]. Although theoretically an indefinite number of polyelectrolyte layers can be applied on a substrate [15], a state of saturation can be reached in which no more polyelectrolytes will deposit [32]. There seems to be a positive association between molecular weight of the polyelectrolytes and the thickness and stability of the system [32,34], although the type of bonds that hold together the layers (when covalently bound) and their likelihood to undergo hydrolysis under different environmental conditions also dictates stability [32]. Layer-by-layer deposition can be used to prepare active packaging coatings by the incorporation of active agents either between layers or within the structure of an individual polyelectrolyte. Pilot scale layer-by-layer deposition tools have been developed, suggesting the potential for scalability to high through put production of active packaging coatings via layer-by-layer deposition [35].

2.4. Photografting

In photografting, polymer chains are grafted from a surface by exposure to UV light in the 315–400 nm range in the presence of photoinitiators and monomers [36]. UV irradiation generates free radicals on the polymeric surface after which photoinitiators, reactive compounds that generate unpaired electrons upon UV light exposure, initiate polymerization of monomers [28]. Photografting may take place through three types of mechanisms: hydrogen abstraction, electron transfer, and cleavage [37]. In hydrogen abstraction, the energy provided by the UV light removes a hydrogen atom from a substrate (either from a hydrocarbon, an alcohol, or an ether), which leaves unpaired electrons on its surface. At the same time, the ketone group of a photoinitiator undergoes reduction by coupling with the removed hydrogen from the substrate and coupling its unpaired electron to the unpaired electrons on the substrate, which is followed by polymerization of any monomer of interest. In electron transfer, the carbonyl group of the photoinitiator is also reduced in an electron acceptor-donor system, in which an aminated compound (normally a tertiary amine, but secondary and primary amines can also take place in the reaction) provides the hydrogen from the α carbon of its structure. In the cleavage mechanism, the photoinitiator undergoes homolytic scission upon UV light exposure, which generates free radical species able to initiate polymerization between vinyl monomers. Some common photoinitiators are benzophenone [38], anthraquinone [37], thioxantone [37], and phenyl azide [39]. Polymeric photoinitiators have also been studied [37], as well as naturally derived photoinitiators like curcumin [40,41]. Photografting can be used to prepare active packaging coatings either by direct incorporation of the active agent during photografting, or by subsequent immobilization of the active compound after grafting of a polymer chain with reactive functional groups (e.g., acrylic acid).

3. Applications

There are a number of active agents that can be incorporated into or onto coatings for active packaging using the technologies described previously, as summarized in Table 1. Major classes of active agents include antimicrobials, antioxidants, and enzymes, with applications in active packaging coatings ranging from controlling microbial growth, inhibiting oxidative degradation reactions, and targeted biocatalysis. The nature of the bulk packaging material onto which the active coating is deposited varies, including materials such as polyethylene (PE), low density polyethylene (LDPE), polypropylene (PP), polyethylene terephthalate (PET), ethylene vinyl alcohol (EVOH), poly(lactic acid) (PLA), polyamide (PA), polystyrene (PS), polyvinyl alcohol (PVOH), *etc*. Much of the published and commercialized research on active packaging coatings has targeted food packaging applications, largely for maintaining safety and quality and controlling spoilage. While the focus of the applications surveyed below is largely on food applications, these technologies have relevance to packaging of consumer products (e.g., health and beauty) as well as pharmaceuticals and biomedical devices. Smart packaging intended for monitoring of temperature, oxygen, pH, moisture, *etc.* [42,43] of packaged products represents a promising area for active packaging coatings, but falls outside the scope of this review.

3.1. Antimicrobial

Probably the most studied application of active packaging technologies has been antimicrobial packaging to control growth of pathogenic and/or spoilage microorganisms in packaged products. Antimicrobial agents such as essential oils, organic acids, peptides, enzymes, and biopolymers have been introduced into antimicrobial active packaging by a range of coating technologies. Regarding controlled release, a European patent describes a liquid formulation composed of essential oils, adhesion promotors (e.g., acrylic or vinyl resins or nitrocellulose), and fixatives to modulate the release of the essential oils for application on common packaging materials [44]. A recent study by Manso *et al.* demonstrated the anti-fungal character of the patented coating [45], in which a 30 μm thick coating containing cinnamon essential oil (coating grammage of 2.5 g·m^{-2}, with essential oil concentrations of 2%, 4% and 6%) was applied to PP and evaluated against different species of *Aspergillus* and *Penicillium*. Seeded agar was exposed to coated films at a distance normally given between the agar surfaces and Petri dish lids, and it was observed that total inhibition was obtained at 4% and 6% concentrations for all the fungi evaluated regardless of temperature and pH. In another work also involving controlled release of thyme and oregano essential oils [46], Valderrama Solano *et al.* reported coating onto corona treated LDPE films at concentrations of 1% and 4% (based on the weight of LDPE). Antimicrobial activity was demonstrated by zone of inhibition assays against *Escherichia coli* 0157:H7, *Salmonella* Typhimurium and *Listeria monocytogen*es, showing substantial inhibition at a concentration of 4% for all the pathogens, and no inhibition at 1% concentration. The authors further demonstrated that the reported coating method had not significant changes in the film mechanical properties (tensile strength and elongation at break) and although the changes in barrier properties (oxygen and water vapor transmission rates) were significant, they were probably not relevant enough to compromise their usage since a reduction in both parameters was observed. Another approach for controlled release involves the application of the sol-gel technique, which consists of the condensation of hydroxylated monomers and polymers into a network that can harbor active substances [47]. Lantano *et al.* recently reported on the preparation of a sol-gel coating in which the antifungal agent natamycin was embedded in a tetraethyl orthosilicate/EVOH gel and applied to plasma treated PLA films [48]. The natamycin loaded PLA films demonstrated release rates that were in accordance to European standards and were able to inhibit mold growth on cheese stored for 30 days at 4 °C. In another study, Zhu *et al.* showed the potential to apply antimicrobial photocatalysts (e.g., TiO_2) using the sol-gel technique [49].

Table 1. Summary of technologies and applications in active packaging coatings.

Coating Technology	Application		
	Antimicrobial	Antioxidant	Biocatalytic
Controlled release	Essential oils [45,46] Natamycin [48] TiO_2 [49] Cinnamaldehyde [58] Sorbic acid and lauric arginate ester [59] Lauric arginate ester [60] Nisin [61]	Citrus oil [50–52] Rosemary extract [54] α-tocopherol [56] – – – –	Lactase [53] Laccase [55] Oxalate oxidase [57] – – – –
Immobilization	3-aminopropyltrimethoxysilane [62] Lysozyme [66] (3-bromopropyl)triphenylphosphonium [69] SO_2 [73]	Gallic acid [63] Aluminum oxide [67] – –	Lactase [64,65] Catalase [68] Naringinase [70–72] –
Layer-by-layer assembly	Chitosan [74,75] Lysozyme [78]	Tannic acid [76] –	Lactase [77] –
Photografting	– – –	Caffeic acid [79] Acrylic acid [83–85] Hydroxamic acid [87,88]	Trypsin [80–82] Urease [86] –

There have been several reports in which controlled release coating technologies have been combined with surface immobilization to prepare antimicrobial coatings. In a recent study [58], Makwana *et al.*, covalently attached polydiacetylene liposomes containing cinnamaldehyde onto an amine-functionalized silane monolayer on piranha treated glass. The authors also explored immobilizing the liposomes on amine-functionalized PLA films. Approximately three logarithmic reductions were obtained against *Escherichia coli* after 50 min of exposure to the liposome-encapsulated cinnamaldehyde (for both glass and poly(lactic acid)). For the case of *Bacillus cereus*, PLA was more effective (~4 logarithmic reductions) than glass (~3 logarithmic reductions). What may be remarkable about this study is that it suggests the possibility of applying an approach for controlled release of antimicrobial substances on both organic and inorganic substrates. In another study, Guo *et al.* reported the direct coating of PLA with a mixture of chitosan (a widely studied polycation with known antimicrobial activity), sorbic acid, sodium lactate, and lauric arginate ester [59].

Antimicrobial effectiveness was confirmed both in bacterial suspensions (to an undetectable level for all microorganisms) and on ready to eat meat against *Listeria innocua* (2–3 logarithmic reductions), *Listeria monocytogenes* (2–3 logarithmic reductions) and *Salmonella* Tiphymurium (1–1.5 logarithmic reductions) under storage at 10 °C. In other work [60], Theinsathid *et al.* spray coated corona treated PLA with solutions having varying concentrations of lauric arginate ester (0%–10%). Between two and three logarithmic reductions were reached against *Listeria monocytogenes* and *Salmonella* Tiphymurium in wrapped ham samples after seven days of storage at 4 °C under the lowest antimicrobial concentration tested (0.07% based on the weight of PLA). The authors demonstrated that the coating treatment did not affect the mechanical properties of poly(lactic acid); however, the gas barrier properties measured as CO_2 permeation were adversely affected. In another work, the antimicrobial peptide nisin was entrapped in polyethylene oxide brushes grown on silane modified silicon wafers [61]. It was hypothesized that the polyethylene oxide brushes would protect nisin from being eluted by foreign proteins and subsequently losing its activity. Antimicrobial activity of the modified silicon wafers was confirmed through zone of inhibition assay against the Gram positive bacterium *Pediococcus pentosaceous* over a period of seven days. Retention of antimicrobial activity even after introduction of bovine serum albumin to the agar media supported their hypothesis that the polyethylene oxide brushes protected nisin. Although this coating was tested on an inorganic substrate, the authors proposed its potential for application on polymer packaging films.

Preparation of antimicrobial active coatings via covalent immobilization of antimicrobial agents is probably the least studied approach explored for antimicrobial packaging applications, likely because the most commonly studied antimicrobial agents (e.g., nisin, essential oils) must migrate from the packaging material to be effective. Cationic polymers and some antimicrobial enzymes, however, may retain efficacy after covalent immobilization onto a solid support. The cationic property of amines was tested in a study by Fernandes *et al.*, in which pullalan powder was rendered cationic by reaction with an amine terminated silane, after which films were formed by solution casting [62]. The cationic pullalan films demonstrated more than three logarithmic reductions *Staphylococcus aureus* and *Escherichia coli*. In addition, the cationic pullulan films showed improved mechanical and thermal properties as compared to unmodified pullulan. In recent work by Muriel-Galet *et al.*, the antimicrobial enzyme lysozyme was covalently attached onto UV-ozone treated EVOH films via carbodiimide chemistry and reported more than one logarithmic reduction of *Listeria monocytogenes* in bacterial suspension [66]. In another work, Anthierens *et al.* tested the possibility of functionalizing the flexible polymer poly(butylene adipate-*co*-terephtalate) with a quaternary phosphonium compound, (3-bromopropyl)triphenylphosphonium [69]. For that purpose, the biodegradable polymer was modified to introduce alkyne bonds to enable an azide-alkyne "click" reaction with the quaternary compound. The modified polyester was challenged against *Escherichia coli* both in bacterial suspension and under direct contact, reaching ~4 logarithmic reductions after 1 h in bacterial suspension and after 24 h under direct contact. In another recent study by Mackiw *et al.* [73], a multilayered film made of PA and PE was subjected to atmospheric plasma treatment (Ar, Na_2O and SO_2) on the PA side of the films. Na_2O was applied either alone with Ar or combined with Ar and SO_2. The temperature for Na_2O sublimation in the reactor was varied from 300–640 °C. It was observed that the highest inhibitions were obtained from films possessing the highest concentrations of SO_2 (the antimicrobial agent) against *Escherichia coli* (82%), *Staphylococcus aureus* (86%), *Listeria monocytogenes* (63%), *Bacillus subtilis* (79%), and *Candida albicans* (75%). Even though less than one logarithmic reduction was reached for every microorganism, this type of approach seems promising due to its potential industrial scalability, as plasma treatment is a widely used method of surface activation for food packaging films [23].

The possibility of using the layer-by-layer approach has been explored through several studies. In one recent work published by Pinheiro *et al.*, κ-carrageenan (polyanion) and chitosan (polycation) were applied onto aminated PET resulting in improved gas barrier properties (reduced oxygen and water vapor transmission rates) [74]. In other work presented by Carneiro-da-Cunha *et al.* [75], alginate was employed as the polyanion, suggesting it as an attractive alternative as there was not significant difference in water vapor permeability between the aminated PET film used and its coated counterpart. Finally, Medeiros *et al.*, have reported on layer-by-layer deposition of three layers of κ-carrageenan alternated with two layers of the antimicrobial enzyme lysozyme on amine-functionalized PET films [78]. The modified PET exhibited improved values of oxygen and water vapor permeability as compared to aminated PET. To date, layer-by-layer deposition technologies have largely focused on improving barrier properties, but those described above incorporate components with antimicrobial character, suggesting their potential application in antimicrobial active packaging coatings. While antimicrobial active packaging technologies have focused largely on inhibiting microbial growth or inactivating viable organisms, there may be opportunity in adapting food processing equipment coatings technologies which seek to inhibit microbial adhesion.

3.2. Antioxidant

Oxidative degradation is a key limiting factor in overall shelf life of many food and consumer products, as well as nutritional supplements. Many packaged products undergo quality deterioration during transport and storage due to oxidative reactions that cause lipid rancidity (e.g., mono and polyunsaturated fatty acids), color loss (e.g., carotenoids, chlorophyll, anthocyanins), and vitamin degradation (e.g., Vitamin A and its precursors, Vitamin C, Vitamin D, Vitamin E). There are several preservation strategies for packaged goods that contain oxidation-sensitive ingredients based on direct addition of antioxidants (e.g., free radical scavengers, metal chelators, singlet oxygen quenchers, oxygen scavengers) and design of appropriate packaging systems. One such preservation strategy is antioxidant active packaging, wherein an antioxidant has been incorporated into a packaging material to enhance food preservation [14]. The most widely used commercially available antioxidant active packaging technologies include oxygen scavengers, manufactured as sachets or labels [89]. However, there is emerging research in the development of migratory and non-migratory antioxidant coatings for active packaging, as surveyed in this section.

Migratory antioxidant active coatings are designed for controlled release of an antioxidant over the course of product shelf life. The majority of research for antioxidant active packaging that is designed for controlled release has been focused on blending of antioxidants with polymers followed by extrusion [4,90]. However, inclusion of antioxidants in packaging may have a negative impact on bulk material properties that may limit applications. Research in the design of active coatings for controlled release of antioxidant is promising for overcoming this challenge. Corlini *et al.* prepared antioxidant packaging materials by spray deposition of citrus oil in methanol onto the surface of PET trays and observed that plasma pretreatment enhanced the adhesion of citrus oil to the surface of PET trays, which demonstrated higher overall antioxidant activity [50]. This coating exhibited antioxidant activity with cooked turkey meat and retained activity after six months of storage [51,52]. Bolumar *et al.* coated rosemary extract directly onto LDPE plastic wrap (0.45 mg rosemary cm^{-2}) and demonstrated significant delay in the onset of surface lipid oxidation of wrapped chicken patties. Controlled released of migratory antioxidant coatings may be enhanced by using a polymeric carrier for antioxidants rather than direct surface application [54]. For example, Lee *et al.* applied α-tocopherol to paperboard using a vinyl acetate-ethylene copolymer as a carrier for controlled release (D = 2.91 × 10^{-11} m$^2\cdot$s^{-1}) and demonstrated its ability to retard oxidation in an oil-in-water emulsion [56].

Non-migratory antioxidant coatings are designed to scavenge prooxidants, such as free radicals, oxygen and transition metals, to extend the shelf life of oxidation-sensitive foods. Garces *et al.* patented an antioxidant coating composed of a polymer blend that contains plant extracts that scavenge free radicals [91]. Tovar *et al.* demonstrated that this antioxidant coating may be classified as non-migratory as it demonstrated migration that was 20 times less than the legal limits for the European Union [92]. Furthermore, this antioxidant active packaging coating derives activity from free radical scavenging in the packaging headspace, thus not requiring direct contact with the food [93–95]. Shutava *et al.* imparted free radical scavenging properties on glass slides via layer-by-layer deposition of tannic acid and poly(allylamine hydrochloride) and found that 2,2′-azino-bis(3-ethylbenzothiazoline-6-sulphonic acid) (ABTS) radical scavenging activity per mole immobilized tannic acid decreased with increasing number of bilayers, most likely due to inability of ABTS to diffuse through bilayers [76]. However, increasing the number of bilayers did result in a linear increase in overall scavenging activity (mol ABTS cm^{-2}). There has also been significant research on metal oxide coatings that have been applied to packaging materials to reduce oxygen diffusion into packaging to extend shelf life [67,96].

Non-migratory antioxidant coatings may also be applied by covalent immobilization of functional groups to the surface of packaging materials. Gallic acid was immobilized to chitosan by carbodiimide assisted reaction that resulted in an active packaging material that demonstrated significant reduction in oxidation of ground peanuts [63]. In other work, Arrua *et al.* utilized photoinitiated graft polymerization of a polymer functionalized with caffeic acid to coat polypropylene packaging materials and demonstrated its ability to prevent oxidative degradation of ascorbic acid in orange juice [79]. Photoinitiated graft polymerization has also been used to fabricate metal chelating active packaging coatings that extended the lag phase of lipid oxidation in oil-in-water emulsions by chelating transition metals, which are the most influential prooxidants in food emulsions [83,87]. Further research in this area demonstrated that active packaging coatings composed of photografted polyhydroxamate chelators retain activity in a wide range of pH values (3–5), viscosity (~1–10^4 CP) and competing ion (Na, Mg, Ca) conditions typically found in food and consumer products [97]. In addition to potential "clean label" regulatory benefits of non-migratory antioxidant coatings, such technologies have product quality advantages over migratory antioxidant packaging coatings, as migratory antioxidant coating technologies typically use active agents which must be used at a concentration which alters sensory perception (flavor, color, viscosity) of the packaged product. Nevertheless, technical hurdles remain in adapting non-migratory active coatings for commercial application.

3.3. Biocatalytic

The partnership of enzymes and solid support materials provides many opportunities for biocatalytic coatings for active packaging. Enzymes are proteins with enhanced specificity to a substrate, which catalyze reactions by lowering their activation energy to create products. Enzymes are employed in numerous processes used in the food, pharmaceutical, and diagnostic industries. In food processing, biocatalysts are used in production of ingredients, enhancement of product quality, and breakdown of undesirable components that may be harmful or may decrease product quality [98]. As the working conditions for biocatalytic materials can be very specific and variable with the immobilization method and material, there is much research on the immobilization of enzymes onto and into solid supports in order extend their thermostability, pH optima, and solvent stability. Integrating biocatalysts into packaging materials facilitates their use as active packaging coatings [99] which may enable so-called "in-package processing" where food constituents undergo changes to improve quality or shelf-life while in transport and storage.

The goal of many biocatalytic active packaging technologies is to control the growth of spoilage and/or pathogenic microorganisms. Lysozyme is a commonly utilized enzyme for such antimicrobial active packaging. Such antimicrobial enzymes have been incorporated into active packaging coatings via blending, non-covalent binding for controlled release, and covalent immobilization [100–102]. For additional information, refer to the comprehensive section on antimicrobial coatings of this review. Herein outlines methods for the incorporation of enzymes into biocatalytic coatings for active packaging, and a few works which have the potential for application in active packaging coatings.

Incorporation of enzymes into active packaging coatings by embedding and blending requires compatibility between the enzyme and the coating matrix. Compatibility can be achieved by surface functionalization, altering bulk material composition, and enzyme modification. Embedding and blending may simplify commercialization and scale-up as one-pot preparation methods become more available [16]; however, these techniques typically result in the enzyme becoming a part of the packaged product, reducing catalytic activity of the coating over time [103]. A targeted application for several biocatalytic active packaging coatings technologies involves the removal of oxygen for increased food shelf stability. Efforts towards incorporation of glucose oxidase in a low-density polyethylene and paper board multilaminate have included details for scaled-up production. Various compositions of LDPE, glucose oxidase, and catalase were produced in Tetra Pak's pilot plant, and showed up to 97% activity could initially be achieved even after exposure to 325 °C during production. Control of production parameters was key to maintaining the package's oxygen scavenging capability [104]. Johansson et al. worked to improve embedding the glucose oxidase and catalase oxygen scavenging pair by varying combinations of LDPE, polypropylene, and polylactic acid [55]. Variations of these embedding methods for oxygen scavenging by laccase and oxalate oxidase have shown similar results [55,57]. Recent work by Talbert et al. demonstrates that enzymes may be modified by hydrophobic ion pairing to be soluble and retain activity in solvents used in ink formulations, enabling the preparation of biocatalytic active packaging coatings using existing printing technology [105].

Much of the work on biocatalytic coatings for food packaging involves covalent immobilization. Covalent immobilization allows for biocatalytic coatings to interact with packaged products without being incorporated into the food matrix, enabling their regulation as contact materials rather than direct additives. Non-migratory biocatalytic coatings therefore offer a potential regulatory benefit, as there is growing demand for "clean label" products. For example, Soares and Hotchkiss developed non-migratory packaging films to de-bitter fruit juices by covalently immobilizing fungal naringinase [70,71]. Naringinase activity was maintained for 15 days storage after preparation of the biocatalytic active packaging coating, and k_M values were lower when compared to free enzyme. Nunes et al., achieved similar success by cross-linking naringinase to polyvinyl alcohol and alginate using boric acid for de-bittering [72]. In other work, β-galactosidase was bound to amine-functionalized polyethylene films by a dialdehyde tether to reduce milk lactose in package [64]. Further research demonstrated that tether length and chemistry can influence retained activity of immobilized lactase [65], but more significant enhancements in retained activity can be achieved by immobilization onto nanostructures [106].

Layer-by-layer deposition allows for more enzyme to be incorporated in a biocatalytic coating compared to traditional immobilization techniques. Indeed, increasing the number of functional layers increases total protein content. Biocatalytic coatings prepared by layer-by-layer deposition of polyethylenamine, glutaraldehyde, and lactase exhibited increased protein loading with each layer [77]. However, overall activity did not increase with layers, likely a result of reduced substrate accessibility by enzyme entrapped in sub-interfacial layers. In other work, Shutava et al. layered hemoglobin, PS, and catalase in order to create a physical and chemical protective barrier [68]. Although the layer-by-layer technique has been shown to increase protein loading, diffusion of substrate to enzyme can become difficult. Increasing substrate diffusion through formed layers would improve activity retention, and therefore commercial potential, of biocatalytic coatings prepared by layer-by-layer deposition.

Photografting is often used in combination with other surface modification methods for preparation of biocatalytic coatings. Garnett et al. published an early report on immobilization of the protease trypsin to PP, PVOH, and PS by enhancing photografting with the use of metal salts [80]. Carboiimide chemistry can be used to covalently couple enzymes (via amine groups) to carboxylic acid groups introduced to a material surface via photografted polymerization [81]. Krenkova et al., functionalized poly(ethylene) glycol methacrylate to immobilize trypsin and endoproteinases for antibody analysis in enzyme reactors. In that work, 4-vinyl-2,2-dimethylazolactone was photografted to introduce a porous surface morphology with the goal of optimizing enzyme orientation to improve performance. As a result, non-specific protein binding was reduced and the protein substrate was successfully digested [82]. In other work, the hydrolase urease was immobilized to photgrafted polytetrafluoroethylene to remove urea from beverages and foods by Yamada et al., After immobilization, urease exhibited lowered activity at higher protein contents due to high protein densities as a result of grafted layers [86]. Indeed, a common challenge in enzyme immobilization is tailoring immobilization density and material chemistry to reduce adverse protein–protein and protein–surface interactions [107].

Emerging technologies in biocatalytic materials such as electrospun nanofibers and biopatterning offer new opportunities in preparing biocatalytic active packaging coatings. Electrospinning is a method by which enzymes or other active agents can be incorporated into polymer nanofibers. In one report, lactase blended into polyethylene oxide nanofibers retained up to 93% of free enzyme activity with significant retention of activity after dry storage [53]. In other work by Ge *et al.*, glucose oxidase was immobilized onto electrospun polyvinyl acetate/chitosan/tea extract fibers to reduce oxygen in packaged foods to extend shelf life [108]. The immobilized glucose oxidase in this system retained over 68% of free enzyme activity. Drug delivery with controlled release has sparked new interest in various fiber morphologies, as well as micro and nanoparticles and hydrogels. Embedding and blending is often paired with cross-linking techniques to improve retention of the support material's physical properties [109]. Biopatterning, in which specific patterns of immobilized biomolecules are defined with micron or submicron resolution, also has potential application in active packaging coatings. For example, patterning cells to the interior of microfluidic channels by photografting poly(ethylene glycol) reduced non-specific protein binding [110,111]. Creating specific patterns for enzyme immobilization can focus activity on targeted regions in a packaging matrix, thereby reducing material waste during functionalization and controlling coating costs. Because of the interdependence of the enzyme structure on biocatalytic activity, preparation of biocatalytic coatings demands a whole systems approach, with consideration given to the enzyme and the coating matrix as well as the packaged product. Consideration must be given to reducing diffusional limitations of substrate and product, as well as activity retention for increased thermostability and pH stability. The overall stability of the bound enzyme determines the success of the coating method.

4. Challenges and Perspectives

Exciting opportunities exist in the development of active packaging coatings for improving the safety, quality, and shelf life of packaged goods, which is brought about by an increasing preference and demand from consumers for additive-free foods and a continuously expanding transport chain and standards of quality driven by global commerce [112]. The global trade of active food packaging was estimated to be close to US $9,000,000,000.00 in 2011 and it is expected to be around US $12,000,000,000.00 in 2017 [113]. Nevertheless, the active food packaging market is currently widely dominated by oxygen scavenging and moisture absorption applications [112,114]. A number of challenges remain prior to the commercial adoption of the types of applications studied in this review. A key consideration in developing coatings for active packaging applications is considering toxicity and potential for regulatory approval of not just the active agent but any tether molecules or cross-linkers that may be employed [115]. Consideration must also be given to requirements of different regulatory agencies, whether under the jurisdiction of the Food and Drug Administration (in the United States), the European Food Safety Authority (in the European Union), or elsewhere. Non-migratory technologies produced by either covalent immobilization, cross-linked layer-by-layer deposition, or some photografted coatings offer a potential regulatory benefit in US regulated food packaging applications, as they would require approval as food contact materials rather than direct additives [116]. Migration testing using standardized simulants (water, 3% acetic acid, 15% ethanol, olive oil, iso-octane, and 95% ethanol) must be performed to quantify levels of migrants in packaged product systems [117]. It is also critical to consider the influence of the active coating on other performance characteristics of the packaging material. A benefit to coatings over bulk material modification is that bulk material properties should remain intact. However, the influence of the coating on processability, thermomechanical properties, barrier properties, and seal strength must be characterized [118]. Rigorous application tests must also be performed to ensure that neither material conversion steps nor end use result in delamination of active coatings. Many of the coatings technologies surveyed in this report have potential for scalability to roll-to-roll, high throughput coating operations. Finally, while incorporation of active agents and specialized packaging processes will indeed increase material cost, the opportunities for new products, enhanced safety, and reduced waste of packaged goods highlight the potential for increasing product value through smart integration of active packaging coatings.

Acknowledgments: This material is based upon work supported by the National Institute of Food and Agriculture, U.S. Department of Agriculture, under project numbers 2012-67017-30157, 2011-65210-20059, 2014-67021-21584, and 2015-67017-23119. M.J. Roman would like to acknowledge the Peter Salmon Graduate Fellowship (Department of Food Science, UMass Amherst) and Northeast Alliance for Graduate Education and the Professoriate Fellowship for their support. D.E. Wong acknowledges Agriculture and Food Research Initiative Grant No. 2015-67011-22820 for NIFA Predoctoral Fellowship support.

Author Contributions: Luis J. Bastarrachea wrote the overview of coating technologies and the antimicrobial applications sections and contributed to overall document formatting; Dana E. Wong wrote the biocatalytic applications part; Maxine J. Roman wrote the antioxidant applications section; Zhuangsheng Lin contributed the graphics and proofread the manuscript; and Julie M. Goddard wrote the challenges and perspectives section and edited the full text.

Conflicts of Interest: The authors declare no conflict of interest.

1. Appendini, P.; Hotchkiss, J.H. Review of antimicrobial food packaging. *Innov. Food Sci. Emerg. Technol.* **2002**, *3*, 113–126. [CrossRef]
2. Vermeiren, L.; Devlieghere, F.; van Beest, M.; de Kruijf, N.; Debevere, J. Developments in the active packaging of foods. *Trends Food Sci. Technol.* **1999**, *10*, 77–86. [CrossRef]
3. Suppakul, P.; Miltz, J.; Sonneveld, K.; Bigger, S.W. Active packaging technologies with an emphasis on antimicrobial packaging and its applications. *J. Food Sci.* **2003**, *68*, 408–420. [CrossRef]
4. Tian, F.; Decker, E.A.; Goddard, J.M. Controlling lipid oxidation of food by active packaging technologies. *Food Funct.* **2013**, *4*, 669–680. [CrossRef] [PubMed]
5. Ozdemir, M.; Floros, J. Active food packaging technologies. *Crit. Rev. Food Sci. Nutr.* **2004**, *44*, 185–193. [CrossRef] [PubMed]
6. Yam, K.; Takhistov, P.; Miltz, J. Intelligent packaging: Concepts and applications. *J. Food Sci.* **2005**, *70*, R1–R10. [CrossRef]
7. Kuswandi, B.; Wicaksono, Y.; Jayus; Aminah, A.; Heng, L.Y.; Ahmad, M. Smart packaging: Sensors for monitoring of food quality and safety. *Sens. Instrum. Food Qual. Saf.* **2011**, *5*, 137–146. [CrossRef]
8. Smith, J.P.; Hoshino, J.; Abe, Y. Interactive packaging involving sachet technology. In *Active Food Packaging*, 1st ed.; Rooney, M.L., Ed.; Springer: Berlin/Heidelberg, Germany, 1995; pp. 143–173.
9. Bastarrachea, L.; Dhawan, S.; Sablani, S.S. Engineering properties of polymeric-based antimicrobial films for food packaging. *Food Eng. Rev.* **2011**, *3*, 79–93. [CrossRef]
10. Han, J.H.; Floros, J.D. Casting antimicrobial packaging films and measuring their physical properties and antimicrobial activity. *J. Plast. Film Sheeting* **1997**, *13*, 287–298.
11. Vasile, C. General survey of the properties of polyolefins. In *Handbook of Polyolefins*, 2nd ed.; Vasile, C., Ed.; Marcel Dekker: New York, NY, USA, 2000; pp. 401–416.
12. Min, S.; Krochta, J.M. Edible coatings containing bioactive antimicrobial agents. In *Packaging for Nonthermal Processing of Food*; Han, J.H., Ed.; Blackwell Publishing: Hoboken, NJ, USA; IFT Press: Ames, IA, USA, 2007; pp. 29–52.
13. Han, J.H. Antimicrobial food packaging. In *Novel Food Packaging Techniques*; Ahvenainen, R., Ed.; CRC Press: Boca Raton, FL, USA; Woodhead Pub. Ltd.: Cambridge, UK, 2003; pp. 50–65.
14. Robertson, G.L. *Food Packaging: Principles and Practice*, 2nd ed.; Taylor & Francis/CRC Press: Boca Raton, FL, USA, 2006; p. 550.
15. Ratner, B.D. Surface modification of polymers: Chemical, biological and surface analytical challenges. *Biosens. Bioelectron.* **1995**, *10*, 797–804. [CrossRef]
16. Kong, F.; Hu, Y.F. Biomolecule immobilization techniques for bioactive paper fabrication. *Anal. Bioanal. Chem.* **2012**, *403*, 7–13. [CrossRef] [PubMed]
17. Chan, C.M.; Ko, T.M.; Hiraoka, H. Polymer surface modification by plasmas and photons. *Surf. Sci. Rep.* **1996**, *24*, 1–54. [CrossRef]

18. Chatelier, R.C.; Xie, X.; Gengenbach, T.R.; Griesser, H.J. Quantitative analysis of polymer surface restructuring. *Langmuir* **1995**, *11*, 2576–2584. [CrossRef]
19. Jansen, B.; Kohnen, W. Prevention of Biofilm Formation by Polymer Modification. *J. Ind. Microbiol.* **1995**, *15*, 391–396. [CrossRef] [PubMed]
20. Sato, T.; Hiruma, K.; Shirai, M.; Tominanga, K.; Haraguchi, K.; Katsuyama, T.; Shimada, T. Site-controlled growth of nanowhiskers. *Appl. Phys. Lett.* **1995**, *66*, 159–161. [CrossRef]
21. Barish, J.A.; Goddard, J.M. Topographical and chemical characterization of polymer surfaces modified by physical and chemical processes. *J. Appl. Polym. Sci.* **2011**, *120*, 2863–2871. [CrossRef]
22. Goddard, J.M.; Hotchkiss, J.H. Polymer surface modification for the attachment of bioactive compounds. *Prog. Polym. Sci.* **2007**, *32*, 698–725. [CrossRef]
23. Wagner, J.R., Jr. *Multilayer Flexible Packaging Technology and Applications for the Food, Personal Care and Over-the-Counter Pharmaceutical Industries*, 1st ed.; Elsevier Science: Oxford, NY, USA, 2010; p. 258.
24. Moy, V.T.; Florin, E.L.; Gaub, H.E. Intermolecular forces and energies between ligands and receptors. *Science* **1994**, *266*, 257–259. [CrossRef] [PubMed]
25. Hermanson, G. *Bioconjugate Techniques*, 2nd ed.; Academic Press: Boston, MA, USA, 2008; p. 1202.
26. Farkaš, P.; Bystrický, S. Chemical conjugation of biomacromolecules: A mini-review. *Chem. Pap.* **2010**, *64*, 683–695. [CrossRef]
27. Decher, G. Fuzzy nanoassemblies: Toward layered polymeric multicomposites. *Science* **1997**, *277*, 1232–1237. [CrossRef]
28. Ratner, B.D.; Hoffman, A.S. Physichochemical surface modification of materials used in medicine. In *Biomaterials Science: An Introduction to Materials in Medicine*, 3rd ed.; Ratner, B.D., Hoffman, A.S., Schoen, F.J., Lemons, J.E., Eds.; Academic Press: Oxford, UK; Waltham, MA, USA, 2013; pp. 259–276.
29. Bastarrachea, L.J.; Denis-Rohr, A.; Goddard, J.M. Antimicrobial food equipment coatings: Applications and challenges. *Annu. Rev. Food Sci. Technol.* **2015**, *6*, 97–118. [CrossRef] [PubMed]
30. Yang, Y.; Haile, M.; Park, Y.T.; Malek, F.A.; Grunlan, J.C. Super gas barrier of all-polymer multilayer thin films. *Macromolecules* **2011**, *44*, 1450–1459. [CrossRef]
31. Cerkez, I.; Kocer, H.B.; Worley, S.D.; Broughton, R.M.; Huang, T.S. N-Halamine biocidal coatings via a layer-by-layer assembly technique. *Langmuir* **2011**, *27*, 4091–4097. [CrossRef] [PubMed]
32. Bastarrachea, L.J.; McLandsborough, L.A.; Peleg, M.; Goddard, J.M. Antimicrobial N-Halamine modified polyethylene: Characterization, biocidal efficacy, regeneration, and stability. *J. Food Sci.* **2014**, *79*, E887–E897. [CrossRef] [PubMed]
33. Bastarrachea, L.J.; Peleg, M.; McLandsborough, L.A.; Goddard, J.M. Inactivation of *Listeria Monocytogenes* on a polyethylene surface modified by layer-by-layer deposition of the antimicrobial N-Halamine. *J. Food Eng.* **2013**, *117*, 52–58. [CrossRef]
34. Haynie, D.; Zhang, L.; Rudra, J.; Zhao, W.; Zhong, Y.; Palath, N. Polypeptide multilayer films. *Biomacromolecules* **2005**, *6*, 2895–2913. [CrossRef] [PubMed]
35. SPALASTM (Spray Assisted Layer-by-Layer Assembly) Coating System. Available online: http://www.agiltron.com (accessed on 20 September 2015).
36. Odian, G.G. *Principles of Polymerization*, 4th ed.; Wiley InterScience: Hoboker, NJ, USA, 2004; p. 812.
37. Carlini, C.; Angiolini, L. Polymers as free radical photoinitiators. In *Synthesis and Photosynthesis*; Springer-Verlag: Heidelberg, Germany, 1995; pp. 127–214.
38. Dunkirk, S.G.; Gregg, S.L.; Duran, L.W.; Monfils, J.D.; Haapala, J.E.; Marcy, J.A.; Clapper, D.L.; Amos, R.A.; Guire, P.E. Photochemical coatings for the prevention of bacterial colonization. *J. Biomater. Appl.* **1991**, *6*, 131–156. [CrossRef] [PubMed]
39. Matsuda, T.; Inoue, K. Novel photoreactive surface modification technology for fabricated devices. *ASAIO Trans.* **1990**, *36*, 161–164.
40. Mishra, A.; Daswal, S. Curcumin, A Novel natural photoinitiator for the copolymerization of styrene and methylmethacrylate. *J. Macromol. Sci. A* **2005**, *42*, 1667–1678. [CrossRef]
41. Zhao, J.; Lalevée, J.; Lu, H.; MacQueen, R.; Kable, S.H.; Schmidt, T.W.; Stenzel, M.H.; Xiao, P. A new role of curcumin: As a multicolor photoinitiator for polymer fabrication under household UV to red LED bulbs. *Polym. Chem.* **2015**, *6*, 5053–5061. [CrossRef]
42. Brody, A.L.; Strupinsky, E.R.; Kline, L.R. *Active Packaging for Food Applications*; Technomic Pub. Co.: Lancaster, PA, USA, 2001; p. 218.

43. Kerry, J.; Butler, P. *Smart Packaging Technologies for Fast Moving Consumer Goods*; John Wiley: Chichester, UK; Hoboken, NJ, USA, 2008; p. 340.
44. Garces, L.O.; de la Puerta, C.N. Antimicrobial packaging based on the use of natural extracts and the process to obtain this packaging. European Patent EP1657181-B1, 13 January 2010.
45. Manso, S.; Becerril, R.; Nerin, C.; Gomez-Lus, R. Influence of pH and temperature variations on vapor phase action of an antifungal food packaging against five mold strains. *Food Control* **2015**, *47*, 20–26. [CrossRef]
46. Valderrama Solano, A.C.; Rojas de Gante, C. Two different processes to obtain antimicrobial packaging containing natural oils. *Food Bioprocess Technol.* **2012**, *5*, 2522–2528. [CrossRef]
47. Minelli, M.; de Angelis, M.G.; Doghieri, F.; Rocchetti, M.; Montenero, A. Barrier properties of organic-inorganic hybrid coatings based on polyvinyl alcohol with improved water resistance. *Polym. Eng. Sci.* **2010**, *50*, 144–153. [CrossRef]
48. Lantano, C.; Alfieri, I.; Cavazza, A.; Corradini, C.; Lorenzi, A.; Zucchetto, N.; Montenero, A. Natamycin based sol-gel antimicrobial coatings on polylactic acid films for food packaging. *Food Chem.* **2014**, *165*, 342–347. [CrossRef] [PubMed]
49. Zhu, Y.; Buonocore, G.G.; Lavorgna, M. Photocatalytic activity of PLA/TiO_2 nanocomposites and TiO_2-active multilayered hybrid coatings. *Ital. J. Food Sci.* **2012**, *24*, 102–106.
50. Contini, C.; Katsikogianni, M.G.; O'Neill, F.T.; O'Sullivan, M.; Boland, F.; Dowling, D.P.; Monahan, F.J. Storage stability of an antioxidant active packaging coated with citrus extract following a plasma jet pretreatment. *Food Bioprocess Technol.* **2014**, *7*, 2228–2240. [CrossRef]
51. Contini, C.; Álvarez, R.; O'Sullivan, M.; Dowling, D.P.; Gargan, S.O.; Monahan, F.J. Effect of an active packaging with citrus extract on lipid oxidation and sensory quality of cooked turkey meat. *Meat Sci.* **2014**, *96*, 1171–1176. [CrossRef] [PubMed]
52. Contini, C.; Katsikogianni, M.G.; O'Neill, F.T.; O'Sullivan, M.; Dowling, D.P.; Monahan, F.J. PET trays coated with citrus extract exhibit antioxidant activity with cooked turkey meat. *LWT Food Sci. Technol.* **2012**, *47*, 471–477. [CrossRef]
53. Wong, D.E.; Dai, M.; Talbert, J.N.; Nugen, S.R.; Goddard, J.M. Biocatalytic polymer nanofibers for stabilization and delivery of enzymes. *J. Mol. Catal. B Enzym.* **2014**, *110*, 16–22. [CrossRef]
54. Bolumar, T.; Andersen, M.L.; Orlien, V. Antioxidant active packaging for chicken meat processed by high pressure treatment. *Food Chem.* **2011**, *129*, 1406–1412. [CrossRef]
55. Johansson, K.; Winestrand, S.; Johansson, C.; Jarnstrom, L.; Jonsson, L.J. Oxygen-scavenging coatings and films based on lignosulfonates and laccase. *J. Biotechnol.* **2012**, *161*, 14–18. [CrossRef] [PubMed]
56. Lee, C.H.; An, D.S.; Lee, S.C.; Park, H.J.; Lee, D.S. A Coating for use as an antimicrobial and antioxidative packaging material incorporating nisin and α-tocopherol. *J. Food Eng.* **2004**, *62*, 323–329. [CrossRef]
57. Winestrand, S.; Johansson, K.; Järnström, L.; Jönsson, L.J. Co-immobilization of oxalate oxidase and catalase in films for scavenging of oxygen or oxalic acid. *Biochem. Eng. J.* **2013**, *72*, 96–101. [CrossRef]
58. Makwana, S.; Choudhary, R.; Dogra, N.; Kohli, P.; Haddock, J. Nanoencapsulation and immobilization of cinnamaldehyde for developing antimicrobial food packaging material. *LWT Food Sci. Technol.* **2014**, *57*, 470–476. [CrossRef]
59. Guo, M.; Jin, T.Z.; Yang, R. Antimicrobial polylactic acid packaging films against listeria and salmonella in culture medium and on ready-to-eat meat. *Food Bioprocess Technol.* **2014**, *7*, 3293–3307. [CrossRef]
60. Theinsathid, P.; Visessanguan, W.; Kruenate, J.; Kingcha, Y.; Keeratipibul, S. Antimicrobial activity of lauric arginate-coated polylactic acid films against listeria monocytogenes and salmonella typhimurium on cooked sliced ham. *J. Food Sci.* **2012**, *77*, M142–M149. [CrossRef] [PubMed]
61. Auxier, J.A.; Schilke, K.F.; McGuire, J. Activity retention after nisin entrapment in a polyethylene oxide brush layer. *J. Food Protect.* **2014**, *77*, 1624–1629. [CrossRef] [PubMed]
62. Fernandes, S.C.M.; Sadocco, P.; Causio, J.; Silvestre, A.J.D.; Mondragon, I.; Freire, C.S.R. Antimicrobial pullulan derivative prepared by grafting with 3-aminopropyltrimethoxysilane: characterization and ability to form transparent films. *Food Hydrocoll.* **2014**, *35*, 247–252. [CrossRef]
63. Schreiber, S.B.; Bozell, J.J.; Hayes, D.G.; Zivanovic, S. Introduction of primary antioxidant activity to chitosan for application as a multifunctional food packaging material. *Food Hydrocoll.* **2013**, *33*, 207–214. [CrossRef]
64. Goddard, J.M.; Talbert, J.N.; Hotchkiss, J.H. Covalent attachment of lactase to low-density polyethylene films. *J. Food Sci.* **2007**, *72*, E36–E41. [CrossRef] [PubMed]

65. Mahoney, K.W.; Talbert, J.N.; Goddard, J.M. Effect of polyethylene glycol tether size and chemistry on the attachment of lactase to polyethylene films. *J. Appl. Polym. Sci.* **2013**, *127*, 1203–1210. [CrossRef]
66. Muriel-Galet, V.; Talbert, J.N.; Hernandez-Munoz, P.; Gavara, R.; Goddard, J.M. Covalent immobilization of lysozyme on ethylene vinyl alcohol films for nonmigrating antimicrobial packaging applications. *J. Agric. Food Chem.* **2013**, *61*, 6720–6727. [CrossRef] [PubMed]
67. Struller, C.F.; Kelly, P.J.; Copeland, N.J. Aluminum oxide barrier coatings on polymer films for food packaging applications. *Surf. Coat. Technol.* **2014**, *241*, 130–137. [CrossRef]
68. Shutava, T.G.; Kommireddy, D.S.; Lvov, Y.M. Layer-by-layer enzyme/polyelectrolyte films as a functional protective barrier in oxidizing media. *J. Am. Chem. Soc.* **2006**, *128*, 9926–9934. [CrossRef] [PubMed]
69. Anthierens, T.; Billiet, L.; Devlieghere, F.; du Prez, F. Poly(butylene adipate) functionalized with quaternary phosphonium groups as potential antimicrobial packaging material. *Innov. Food Sci. Emerg. Technol.* **2012**, *15*, 81–85. [CrossRef]
70. Soares, N.F.F.; Hotchkiss, J.H. Bitterness reduction in grapefruit juice through active packaging. *Packag. Technol. Sci.* **1998**, *11*, 9–18. [CrossRef]
71. Soares, N.; Hotchkiss, J. Naringinase immobilization in packaging films for reducing naringin concentration in grapefruit juice. *J. Food Sci.* **1998**, *63*, 61–65. [CrossRef]
72. Nunes, M.A.P.; Vila-Real, H.; Fernandes, P.C.B.; Ribeiro, M.H.L. Immobilization of naringinase in PVA-alginate matrix using an innovative technique. *Appl. Biochem. Biotechnol.* **2010**, *160*, 2129–2147. [CrossRef] [PubMed]
73. Mackiw, E.; Maka, L.; Sciezynska, H.; Pawlicka, M.; Dziadczyk, P.; Rzanek-Boroch, Z. The impact of plasma-modified films with sulfur dioxide, sodium oxide on food pathogenic microorganisms. *Package Technol. Sci.* **2015**, *28*, 285–292. [CrossRef]
74. Pinheiro, A.C.; Bourbon, A.I.; Medeiros, B.G.D.S.; da Silva, L.H.M.; da Silva, M.C.H.; Carneiro-da-Cunha, M.G.; Coimbra, M.A.; Vicente, A.A. Interactions between κ-carrageenan and chitosan in nanolayered coatings-structural and transport properties. *Carbohydr. Polym.* **2012**, *87*, 1081–1090. [CrossRef]
75. Carneiro-da-Cunha, M.G.; Cerqueira, M.A.; Souza, B.W.S.; Carvalhoc, S.; Quintas, M.A.C.; Teixeira, J.A.; Vicente, A.A. Physical and thermal properties of a chitosan/alginate nanolayered PET film. *Carbohydr. Polym.* **2010**, *82*, 153–159. [CrossRef]
76. Shutava, T.G.; Prouty, M.D.; Agabekov, V.E.; Lvov, Y.M. Antioxidant properties of layer-by-layer films on the basis of tannic acid. *Chem. Lett.* **2006**, *35*, 1144–1145. [CrossRef]
77. Wong, D.E.; Talbert, J.N.; Goddard, J.M. Layer by layer assembly of a biocatalytic packaging film: Lactase covalently bound to low-density polyethylene. *J. Food Sci.* **2013**, *78*, E853–E860. [CrossRef] [PubMed]
78. Medeiros, B.G.D.S.; Pinheiro, A.C.; Teixeira, J.A.; Vicente, A.A.; Carneiro-da-Cunha, M.G. Polysaccharide/protein nanomultilayer coatings: Construction, characterization and evaluation of their effect on "Rocha" pear (*Pyrus communis* L.) shelf-life. *Food Bioprocess Technol.* **2012**, *5*, 2435–2445. [CrossRef]
79. Arrua, D.; Strumia, M.C.; Nazareno, M.A. Immobilization of Caffeic acid on a polypropylene film: Synthesis and antioxidant properties. *J. Agric. Food Chem.* **2010**, *58*, 9228–9234. [CrossRef] [PubMed]
80. Garnett, J.L.; Jankiewicz, S.V.; Long, M.A.; Sangster, D.F. Radiation and photografting as complementary techniques for immobilizing bioactive materials. *Int. J. Radiat. Appl. Instrum. C Radiat. Phys. Chem.* **1986**, *27*, 301–309. [CrossRef]
81. Yamada, K.; Nakasone, T.; Nagano, R.; Hirata, M. Retention and reusability of trypsin activity by covalent immobilization onto grafted polyethylene plates. *J. Appl. Polym. Sci.* **2003**, *89*, 3574–3581. [CrossRef]
82. Krenkova, J.; Lacher, N.A.; Svec, F. Highly efficient enzyme reactors containing trypsin and endoproteinase LysC immobilized on porous polymer monolith coupled to MS suitable for analysis of antibodies. *Anal. Chem.* **2009**, *81*, 2004–2012. [CrossRef] [PubMed]
83. Tian, F.; Decker, E.A.; Goddard, J.M. Control of lipid oxidation by nonmigratory active packaging films prepared by photoinitiated graft polymerization. *J. Agric. Food Chem.* **2012**, *60*, 7710–7718. [CrossRef] [PubMed]
84. Roman, M.J.; Tian, F.; Decker, E.A.; Goddard, J.M. Iron chelating polypropylene films: Manipulating photoinitiated graft polymerization to tailor chelating activity. *J. Appl. Polym. Sci.* **2014**, *131*. [CrossRef]

85. Tian, F.; Decker, E.A.; McClements, D.J.; Goddard, J.M. Influence of non-migratory metal-chelating active packaging film on food quality: Impact on physical and chemical stability of emulsions. *Food Chem.* **2014**, *151*, 257–265. [CrossRef] [PubMed]
86. Yamada, K.; Iizawa, Y.; Yamada, J.; Hirata, M. Retention of activity of urease immobilized on grafted polymer films. *J. Appl. Polym. Sci.* **2006**, *102*, 4886–4896. [CrossRef]
87. Tian, F.; Decker, E.A.; Goddard, J.M. Controlling lipid oxidation via a biomimetic iron chelating active packaging material. *J. Agric. Food Chem.* **2013**, *61*, 12397–12404. [CrossRef] [PubMed]
88. Tian, F.; Roman, M.J.; Decker, E.A.; Goddard, J.M. Biomimetic design of chelating interfaces. *J. Appl. Polym. Sci.* **2015**, *132*, 1–8. [CrossRef]
89. López-Rubio, A.; Lagarón, J.M.; Ocio, M.J. *Active Polymer Packaging of Non-Meat Food Products*; John Wiley & Sons, Ltd.: West Sussex, UK, 2008; pp. 19–32.
90. Gomez-Estaca, J.; Lopez-de-Dicastillo, C.; Hernandez-Munoz, P.; Catala, R.; Gavara, R. Advances in antioxidant active food packaging. *Trends Food Sci. Technol.* **2014**, *35*, 42–51. [CrossRef]
91. Garces, O.; Nerin, C.; Beltran, J.; Roncales, P. Antioxidant active varnish. European Patent EP1477159-A1, 22 December 2003.
92. Tovar, L.; Salafranca, J.; Sánchez, C.; Nerín, C. Migration studies to assess the safety in use of a new antioxidant active packaging. *J. Agric. Food Chem.* **2005**, *53*, 5270–5275. [CrossRef] [PubMed]
93. Nerin, C.; Tovar, L.; Salafranca, J. Behaviour of a new antioxidant active film *versus* oxidizable model compounds. *J. Food Eng.* **2008**, *84*, 313–320. [CrossRef]
94. Pezo, D.; Salafranca, J.; Nerín, C. Design of a method for generation of gas-phase hydroxyl radicals, and use of HPLC with fluorescence detection to assess the antioxidant capacity of natural essential oils. *Anal. Bioanal. Chem.* **2006**, *385*, 1241–1246. [CrossRef] [PubMed]
95. Pezo, D.; Salafranca, J.; Nerín, C. Determination of the antioxidant capacity of active food packagings by in situ gas-phase hydroxyl radical generation and high-performance liquid chromatography-fluorescence detection. *J. Chromatogr. A* **2008**, *1178*, 126–133. [CrossRef] [PubMed]
96. Chatham, H. Oxygen diffusion barrier properties of transparent oxide coatings on polymeric substrates. *Surf. Coat. Technol.* **1996**, *78*, 1–9. [CrossRef]
97. Roman, M.J.; Decker, E.A.; Goddard, J.M. Performance of nonmigratory iron chelating active packaging materials in viscous model food systems. *J. Food Sci.* **2015**, *80*, 1965–1973. [CrossRef] [PubMed]
98. Fernández, A.; Cava, D.; Ocio, M.J.; Lagarón, J.M. Perspectives for biocatalysts in food packaging. *Trends Food Sci. Technol.* **2008**, *19*, 198–206. [CrossRef]
99. Brody, A.L.; Budny, J.A. Enzymes as active packaging agents. In *Active Food Packaging*; Rooney, M.L., Ed.; Blackie Academic and Professional: New York, NY, USA, 1995; pp. 174–192.
100. Gemili, S.; Yemenicioğlu, A.; Altınkaya, S.A. Development of cellulose acetate based antimicrobial food packaging materials for controlled release of lysozyme. *J. Food Eng.* **2009**, *90*, 453–462. [CrossRef]
101. Barbiroli, A.; Bonomi, F.; Capretti, G.; Iametti, S.; Manzoni, M.; Piergiovanni, L.; Rollini, M. Antimicrobial activity of lysozyme and lactoferrin incorporated in cellulose-based food packaging. *Food Control* **2012**, *26*, 387–392. [CrossRef]
102. Mendes de Souza, P.; Fernandez, A.; Lopez-Carballo, G.; Gavara, R.; Hernandez-Munoz, P. Modified sodium caseinate films as releasing carriers of lysozyme. *Food Hydrocoll.* **2010**, *24*, 300–306. [CrossRef]
103. Del Nobile, M.A.; Conte, A. *Packaging for Food Preservation*; Springer: New York, NY, USA, 2013; p. 270.
104. Andersson, M.; Andersson, T.; Adlercreutz, P.; Nielsen, T.; Hornsten, E. Toward an enzyme-based oxygen scavenging laminate. Influence of industrial lamination conditions on the performance of glucose oxidase. *Biotechnol. Bioeng.* **2002**, *79*, 37–42. [CrossRef] [PubMed]
105. Talbert, J.N.; He, F.; Seto, K.; Nugen, S.R.; Goddard, J.M. Modification of glucose oxidase for the development of biocatalytic solvent inks. *Enzyme Microb. Technol.* **2014**, *55*, 21–25. [CrossRef] [PubMed]
106. Talbert, J.N.; Goddard, J.M. Influence of nanoparticle diameter on conjugated enzyme activity. *Food Bioprod. Process.* **2013**, *91*, 693–699. [CrossRef]
107. Talbert, J.N.; Goddard, J.M. Enzymes on material surfaces. *Colloid Surf. B Biointerfaces* **2012**, *93*, 8–19. [CrossRef] [PubMed]
108. Ge, L.; Zhao, Y.; Mo, T.; Li, J.; Li, P. Immobilization of glucose oxidase in electrospun nanofibrous membranes for food preservation. *Food Control* **2012**, *26*, 188–193. [CrossRef]

109. Moskovitz, Y.; Srebnik, S. Mean-field model of immobilized enzymes embedded in a grafted polymer layer. *Biophys. J.* **2005**, *89*, 22–31. [CrossRef] [PubMed]
110. Larsen, E.K.U.; Mikkelsen, M.B.L.; Larsen, N.B. Protein and cell patterning in closed polymer channels by photoimmobilizing proteins on photografted poly(ethylene glycol) diacrylate. *Biomicrofluidics* **2014**, *8*. [CrossRef] [PubMed]
111. Larsen, E.K.U.; Mikkelsen, M.B.L.; Larsen, N.B. Facile photoimmobilization of proteins onto low-binding PEG-coated polymer surfaces. *Biomacromolecules* **2014**, *15*, 894–899. [CrossRef] [PubMed]
112. Realini, C.E.; Marcos, B. Active and intelligent packaging systems for a modern society. *Meat Sci.* **2014**, *98*, 404–419. [CrossRef] [PubMed]
113. BCC Research. The Advanced Packaging Solutions Market Value for 2017 is Projected to be nearly $44.3 Billion. 2015. Available online: http://www.bccresearch.com (accessed on 25 October 2015).
114. Day, B.P.F. Active packaging of foods. In *Smart Packaging Technologies for Fast Moving Consumer Goods*; Kerry, J.P., Butler, P., Eds.; Wiley & Sons, Ltd.: West Sussex, UK, 2008; pp. 1–18.
115. Lopez-Rubio, A.; Gavara, R.; Lagaron, J.A. Bioactive packaging: Turning foods into healthier foods through biomaterials. *Trends Food Sci. Technol.* **2006**, *17*, 567–575. [CrossRef]
116. Koontz, J.; Song, Y.; Juskelis, R.; Mehta, D. Migration database of additives and contaminants in food packaging systems for use in predictive models. In Proceedings of the 245th National Meeting of the American-Chemical-Society (ACS), New Orleans, LA, USA, 7–11 April 2013.
117. Barnes, K.A.; Sinclair, C.R.; Watson, D.H. *Chemical Migration and Food Contact Materials*, 1st ed.; CRC: Boca Raton, FL, USA, 2007; p. 464.
118. Farris, S.; Introzzi, L.; Piergiovanni, L.; Cozzolino, C.A. Effects of different sealing conditions on the seal strength of polypropylene films coated with a bio-based thin layer. *Ital. J. Food Sci.* **2011**, *23*, 111–114. [CrossRef]

© 2015 by the authors. Licensee MDPI, Basel, Switzerland. This article is an open access article distributed under the terms and conditions of the Creative Commons Attribution (CC BY) license (http://creativecommons.org/licenses/by/4.0/).

Review

Antifungal Edible Coatings for Fresh Citrus Fruit: A Review

Lluís Palou [1],*, Silvia A. Valencia-Chamorro [2] and María B. Pérez-Gago [1,3]

1 Centre de Tecnologia Postcollita (CTP), Institut Valencià d'Investigacions Agràries (IVIA), Apartat Oficial, Montcada 46113, Spain; perez_mbe@gva.es
2 Departamento de Ciencias de los Alimentos y Biotecnología, Escuela Politécnica Nacional, Quito 170517, Ecuador; silvia.valencia@epn.edu.ec
3 Fundació Agroalimed, Apartat Oficial, Montcada 46113, Spain
* Author to whom correspondence should be addressed; palou_llu@gva.es; Tel.: +34-9634-24-117; Fax: +34-9634-24-001.

Academic Editor: Stefano Farris
Received: 30 October 2015; Accepted: 1 December 2015; Published: 4 December 2015

Abstract: According to their origin, major postharvest losses of citrus fruit are caused by weight loss, fungal diseases, physiological disorders, and quarantine pests. Cold storage and postharvest treatments with conventional chemical fungicides, synthetic waxes, or combinations of them are commonly used to minimize postharvest losses. However, the repeated application of these treatments has led to important problems such as health and environmental issues associated with fungicide residues or waxes containing ammoniacal compounds, or the proliferation of resistant pathogenic fungal strains. There is, therefore, an increasing need to find non-polluting alternatives to be used as part of integrated disease management (IDM) programs for preservation of fresh citrus fruit. Among them, the development of novel natural edible films and coatings with antimicrobial properties is a technological challenge for the industry and a very active research field worldwide. Chitosan and other edible coatings formulated by adding antifungal agents to composite emulsions based on polysaccharides or proteins and lipids are reviewed in this article. The most important antifungal ingredients are selected for their ability to control major citrus postharvest diseases like green and blue molds, caused by *Penicillium digitatum* and *Penicillium italicum*, respectively, and include low-toxicity or natural chemicals such as food additives, generally recognized as safe (GRAS) compounds, plant extracts, or essential oils, and biological control agents such as some antagonistic strains of yeasts or bacteria.

Keywords: *Citrus* spp.; postharvest; disease control; fruit quality; fungicide alternatives; edible coatings; chitosan; antifungal ingredients

1. Introduction

Citrus spp. (Rutaceae) are the most widely produced fruits in the world, with a global production that exceeded 123 million tons in 2013. The list of the most important producing countries is led by China, Brazil, the United States (USA), India, Mexico, and Spain. Fruit production is mainly devoted to juice extraction, but a considerable proportion is traded as fresh entire fruit for direct consumption. Spain is the leading country for exports of fresh produce [1]. From highest to lowest production, the most cultivated citrus species are oranges (*Citrus sinensis* L.), mandarins or tangerines (*Citrus reticulata* Blanco), including clementines (*Citrus clementina* hort. ex Tanaka), satsumas (*Citrus unshiu* Marcow.), and different hybrid mandarins, lemons (*Citrus limon* (L.) Burm. f.) and limes (*Citrus aurantiifolia* (Christm.) Swingle), and grapefruits (*Citrus paradisi* Macfad.). Postharvest handling of fresh fruit in citrus packing houses is intended to commercialize fruit of maximum quality, increase their postharvest life, and reduce produce losses. In general, postharvest losses can be of physical, physiological, or pathological origin. Physical losses are typically due to rind wounds or bruises caused during harvest, transportation, or postharvest handling in the packing house. These peel injuries are not only important for causing direct losses, but also for being infection sites for economically important postharvest fungal pathogens. Other pathological losses are caused by latent pathogens that infect flowers or young fruit in the grove but develop after harvest. Likewise, some postharvest physiological losses are originated in the field and others are caused by inappropriate handling or storage conditions in the packing house or the marketplace.

Postharvest treatments with conventional synthetic waxes and/or chemical fungicides such as imazalil (IMZ), thiabendazole (TBZ), sodium ortho-phenil phenate (SOPP) or other active ingredients have been used for many years and are still currently used in citrus packing houses to preserve fresh fruit, control postharvest decay, and extend fruit shelf life. Nevertheless, the continuous application of these treatments has arisen important problems for the citrus industry such as health and environmental issues associated with chemical residues or the proliferation of pathogenic resistant strains. Updated regulations from many countries are increasingly restricting the use of agrochemicals and every day more exports markets are demanding fruit with residue levels even lower than those established by official regulations. Due to this situation, research should focus on anticipating a scenario in which conventional chemical fungicides are not available. In such a context, satisfactory postharvest decay control may be accomplished by adopting integrated disease management (IDM) programs based on comprehensive knowledge of pathogen biology and epidemiology and consideration of all preharvest, harvest, and postharvest factors with influence on disease incidence. Among the antifungal postharvest treatments alternative to synthetic chemicals that are investigated worldwide for potential inclusion in IDM programs, in this article we review the development of antifungal edible coatings as a promising novel technology intended to confront two major concerns of citrus postharvest handling, the losses due to physiological problems and the losses due to pathological problems.

2. One Solution for Two Major Citrus Postharvest Problems

2.1. Physiological Problems

Citrus fruit are non-climacteric, hence their respiration rate and ethylene production do not exhibit remarkable increase along with changes related to maturity and ripening as in climacteric fruits. However, although they have a relatively long shelf life compared to other tropical and sub-tropical fruits, they may experience important physiological postharvest losses if they are not properly handled and stored. As with other horticultural products, major postharvest losses in citrus are caused by weight loss and physiological disorders.

Water loss of citrus fruit after harvest, although not considered a physiological disorder, is responsible for loss of quality and consumer acceptability, as it results not only in direct quantitative losses (loss of salable weight), but also in losses in appearance (wilting and shriveling) and softening. In addition, most physiological disorders that affect citrus fruit tend to be related to water loss [2]. Some rind disorders that may appear under optimal, non-chilling temperatures include peel pitting, stem-end rind breakdown (SERB), and shriveling and collapse of the stem-end button. Postharvest peel pitting at non-chilling temperatures is a severe disorder that affects fruit from several citrus cultivars worldwide. Although the causes of the disorder are not fully understood, evidence indicates that altered water relations in fruit peel is a major factor contributing to this disorder. In this sense, sudden changes from low to high relative humidity (RH) after harvest induced peel pitting in "Navelina" and "Navelate" oranges [3], "Marsh" grapefruit [4] or "Fallglo" tangerines [5]. Later studies confirmed the link between alterations in the osmotic and turgor potential in the flavedo (the outer pigmented layer of the peel) and albedo (the inner white layer of the peel) with the induction of postharvest peel pitting in citrus fruit [6]. SERB involves the collapse and subsequent darkening of the epidermal tissues around the stem-end rind of the fruit. The disorder is primarily associated with low RH, particularly during the postharvest period, although preharvest conditions seem to have a critical impact on the susceptibility of fruit to SERB. In some cases, SERB has been reported to be more severe when fruit are harvested from water-stressed trees compared to non-stressed trees or those presenting nutritional imbalances in terms of nitrogen and potassium [7–9]. Shriveling and collapse of the button tissue is an age-related breakdown due to cell weakening and dehydration of mature fruit. The symptoms can vary from discolored, dried out, and extensive collapse of the fruit rind to dehydration or wilting at the stem end where the rind is thinnest [10].

Another group of physiological disorders caused by storage of citrus fruit below sub-optimal temperatures but above the freezing point is chilling injury (CI). CI is characterized by the collapse of discrete areas of the peel that form sunken lesions that tend to coalesce [11]. CI in citrus fruit may appear in various forms depending on species and cultivar and exposure conditions (temperature and duration). For instance, typical symptoms in oranges, mandarins or grapefruits can be browning of the flavedo, appearance of dark sunken areas of collapsed tissue (pitting), or appearance of soft water-soaked areas (watery breakdown), while in lemons can be browning of the albedo or peteca (a special type of rind pitting) [2,10]. In general, although CI symptoms are due to fruit storage below their optimal temperature for relatively long periods, they usually develop upon transfer to higher temperatures.

The high sensitivity of citrus fruit to the induction of physiological disorders in the peel is in many cases triggered by mechanical damage during harvest and postharvest handling. Some of these disorders include "brush burn", "zebra skin", and "oleocellosis", caused by rind abrasion, rough handling, or thorn punctures. Brush burn and zebra skin symptoms usually appear as superficial red/brown staining of the rind and, in the case of the latter, is associated with the position of the segments, whereas oleocellosis might result in dark sunken patches as cells collapse around oil glands [12].

2.2. Pathological Problems

The most important postharvest diseases of fresh citrus fruit are caused by filamentous fungi. According to the origin of the infections, they can be classified into two general groups, those that infect the fruit in the field and remain latent until their development after harvest, and those that infect the fruit through rind microwounds or injuries inflicted during fruit harvest, transportation, postharvest handling, and commercialization.

The most significant pathogens in the first group include *Lasiodiplodia theobromae* (Pat.) Griffon and Maubl. and *Phomopsis citri* H. Fawc. non Sacc. Traverso and Spessa, which cause the diseases commonly known as stem-end rots; *Alternaria citri* Ellis and N. Pierce in N. Pierce, the cause of alternaria rot or black rot; *Botrytis cinerea* Pers.:Fr., the cause of gray mold; *Colletotrichum gloeosporioides* (Penz.) Penz. and Sacc. in Penz., the cause of anthracnose; and *Phytophthora citrophthora* (R.E. Sm. and E.H. Sm.) Leonian, which cause brown rot [13,14]. In general, the incidence of these pathogens is higher in citrus production areas with abundant summer rainfall, such as Florida or Brazil, because they require rain and humid weather for inoculum production and dispersal and subsequent fruit colonization and infection.

Among wound pathogens, the most economically important are *Penicillium digitatum* (Pers.:Fr.) Sacc. and *Penicillium italicum* Wehmer, which cause the diseases known as green and blue molds, respectively. These fungi are strict wound pathogens that can only infect the fruit through rind wounds inflicted in the field before or, more commonly, during fruit harvest, or after harvest during transportation, handling in the packing house, and commercialization. Green and blue molds are very important in all citrus production areas, including those with a Mediterranean-type climate, such as Spain, California, or South Africa, because one generation of *P. digitatum* or *P. italicum* is complete in rotten fruit in 7–10 days at usual ambient conditions of 20–25 °C and their spores are easily disseminated by air currents in large amounts [15]. Hence, the source of fungal inoculum in citrus orchards and packing houses is practically continuous during the season and the fruit can become readily contaminated [16]. Furthermore, healthy citrus fruit can become unmarketable when 'soiled' with conidia of these two fungi that are loosened during handling of diseased fruit [17]. No infection occurs if the fruit rind is intact because free conidia located on the peel surface are not able to germinate. However, the spores situated in injuries that rupture oil glands or penetrate into the albedo of the peel usually bring irreversible infection within 48 h at room temperatures [18]. The germination of both *Penicillium* species inside rind wounds requires free water, nutrients, and is stimulated by volatiles emitted from the host tissue [19,20]. *Geotrichum citri-aurantii* Ferraris E.E. Butler, the cause of sour rot; *Aspergillus niger* van Tiegh, the cause of Aspergillus rot; and *Rhizopus stolonifer* (Ehrenb.:Fr.) Vuill., the cause of Rhizopus rot, are other wound pathogens that can occasionally cause important postharvest losses under specific fruit handling and environmental conditions.

3. Generalities of Antifungal Edible Coatings

Fruit coating is a common practice in citrus packing houses to replace the natural waxes that can be removed during fruit washing and handling in the packing line. Citrus commercial coatings are generically known as waxes due to the fact that composition of initial formulations was based on paraffin wax or a combination of various other waxes such as beeswax or carnauba. Typically, they are anionic microemulsions containing resins and/or waxes, such as shellac, wood rosin, candelilla wax, carnauba wax, beeswax, polyethylene, or petroleum waxes. Their main purpose is to reduce fruit weight loss, shrinkage and improve appearance, but they can also reduce the incidence of CI or other citrus rind disorders [10,21]. In some cases, however, if the coatings excessively restrict gaseous exchange during fruit storage, fruit coating can adversely affect fruit flavor due to an overproduction of volatiles associated with anaerobic conditions [22,23]. In addition, commercial citrus waxes are often amended with synthetic chemical fungicides like IMZ, TBZ, or SOPP in order to control postharvest diseases, particularly green and blue molds.

Consumer interest in health, nutrition, and food safety combined with environmental concerns has increased the interest of many research groups to develop new natural, biodegradable, edible coating formulations to replace these currently used commercial waxes, thus avoiding the use in the formulations of synthetic components such as polyethylene wax, ammonia, or morpholine. The concept of antifungal edible coatings emerges when additional ingredients with antifungal properties (low-toxicity or food-grade preservatives, biocontrol agents, *etc.*) are incorporated into these biodegradable formulations in order also to replace the use of conventional chemical fungicides for postharvest disease control.

3.1. Films vs. Coatings

Although films and coatings have the same chemical composition and sometimes are used synonymously, they refer to different concepts according to their different purpose and utilization. Films are defined as a stand-alone thin layer of materials that can be used as covers, wraps, or separation layers. Films are usually prepared by casting process, and are mainly used for determination of barrier, mechanical, solubility, and other properties provided by certain film materials. On the other hand, food coatings involve the formation of films directly on the surface of the product to which they are intended to be applied. Thus, coatings form part of the final product [24].

3.2. Components and Types of Matrixes

In general, basic components or matrix components of edible coatings are hydrocolloids (proteins or polysaccharides), lipids (waxes, acylglycerols or fatty acids), and resins. Composite coatings or blends contain a combination of polysaccharides or proteins with lipids [25–27]. Plasticizers to enhance flexibility and extensibility, and emulsifiers or surfactants to improve the stability of emulsions may also be added as matrix components. These matrixes can be directly used in foods or act as carriers of other food additives such as antioxidants, nutraceuticals, flavorings agents, *etc.* that are included to modify the functionality of the coating [25,26,28–32]. In the particular case of antifungal edible coatings, the additional ingredients are food-grade antimicrobial compounds of different nature effective in preventing or reducing fungal growth. Characteristics of main components of matrixes are the following:

- Polysaccharides can form a continuous and cohesive matrix, which is related to their chemical structure by the association through hydrogen bonding of their polymeric chains [25]. Polysaccharides contain highly polar polymers with hydroxyl groups and present a good barrier to oxygen at low RH, but low moisture barrier due to hydrophilic properties [30]. Polysaccharide materials typically used to formulate edible or biodegradable coatings include cellulose, starch, pectin, chitosan, alginate, carrageenan, pullulan, and various gums [25,30,31,33–44].
- The ability of different proteins to form edible coatings is highly dependent on their molecular characteristics: molecular weight, conformation, electrical properties, flexibility, and thermal stability [45]. Edible coatings based on proteins usually exhibit good gas barrier characteristics, but poor water barrier characteristics due to their hydrophilic character [28,46,47]. Common proteins used for edible coating formulation include, among others, corn zein, casein, wheat gluten, soy protein, whey protein, keratin, or rice bran protein [26,30,47–52]. It is important to note that some people may present allergies or protein intolerances, e.g., to wheat gluten (celiac disease) or milk protein, which could restrict the use of protein-based coatings.
- Edible films and coatings based on hydrophobic substances, such as lipids and resins, are indicated to provide a barrier to moisture and gloss to food surfaces. However, since these materials are not polymers they form films and coating with poor mechanical properties and opaque characteristics [53]. Hydrophobic substances used as components of edible coatings include a variety of animal and vegetal native oils and fats, e.g., peanut, coconut, palm, lard, tallow, *etc.*; fractionated, concentrated, and/or reconstituted oils and fats, e.g., fatty acids, mono-, di-, and triglycerides, cocoa butter substitutes, *etc.*; hydrogenated and/or transesterified oils, e.g., margarine, shortenings, *etc.*; natural vegetal and animal waxes, e.g., beeswax, candelilla, carnauba, jojoba bees, rice bran, *etc.*; non-natural waxes, e.g., paraffins, oxidized or non-oxidized polyethylene; and natural resins, e.g., asafoetida, benjoin, chicle, guarana, myrrhe, olibanum (incense), opoponax, shellac resins, wood rosin, *etc.* [41,54–57].
- Composite coatings or blends typically contain hydrocolloid components, *i.e.*, protein and/or polysaccharides, and lipids in order to combine the advantages of both types of components. Composite coatings can be produced as either bi-layer or stable emulsions. In bi-layer coatings, the lipid forms a second layer over the protein or polysaccharide layer. In emulsion coatings, the lipid is dispersed and entrapped in the supporting matrix of protein or polysaccharide [57–59]. In this type of coatings, the efficiency of lipid materials depends on the lipid structure, its chemical arrangement, hydrophobicity, physical state, and its interaction with other components of the film [53].
- Plasticizers are low molecular weight compounds of small size, high polarity, high amount of polar groups per molecule, and great distance between polar groups within a molecule. They are added to edible coating materials to decrease the intermolecular forces between polymer chains, which results in greater flexibility, elongation, toughness, and permeability [28,30,32,60]. They are, therefore, particularly indicated to form stable emulsions and improve mechanical properties when hydrocolloids and lipids are combined. Common plasticizers used for edible coatings include sucrose, glycerol, sorbitol, propylene glycol, polyethylene glycol, fatty acids, and monoglycerides. Water can also act as a plasticizer for polysaccharide and protein coatings [61].
- Emulsifiers or surfactants are agents of amphiphilic nature that interact at the water-lipid interface and reduce surface tension between the dispersed and continuous phases to improve the stability of the emulsion when hydrocolloids and lipids are combined [30]. Moreover, emulsifiers added to coating formulations promote good surface wetting, spreading, and adhesion of the coating

to the food surface. Typical emulsifiers used on edible coatings are fatty acids, ethylene glycol monostearate, glycerol monostearate, esters of fatty acids, lecithin, sucrose ester, and sorbitan monostearate, or polysorbates (tweens).

3.3. Functional Properties of Edible Coatings

The most important functional properties of edible films and coatings are edibility and biodegradability; migration, permeation, and barrier functions; and physical and mechanical protection. Such properties allow their use for food quality preservation and shelf-life extension. Furthermore, as carriers of active compounds that can be released in a controlled way, they can also provide antimicrobial activity and be applied for decay control and safety enhancement [26,29,33,35,47,62]. Edibility and biodegradability are the most beneficial characteristics of edible films and coatings. Edibility should be achieved by using food-grade ingredients for all coating components. Moreover, the whole process, facilities, and equipment should be feasible for food processing, and all components should be biodegradable and environmentally safe [24,26,63].

Edible coatings maintain food integrity and can protect coated food products against bruising, tissue damage and, in general, physical injury caused by impact, pressure, vibrations, and other mechanical factors. The ability of edible films and coatings to protect food against mechanical damage is usually assessed by determining film tensile properties such as Young's modulus (YM), which determines film stiffness as determined by ratio of pulling force/area to degree-of-film-stretch, tensile strength (TS), which indicates the pulling force per film cross-sectional area required to break the film, and elongation at break (E), which gives the degree to which the film can stretch before breaking and it is expressed as a percentage [1,10]. Other standardized mechanical examinations include compression strength, puncture strength, stiffness, tearing strength, burst strength, abrasion resistance, adhesion force, and folding endurance [24,29,35,47].

Since one of the main functions of edible coatings is to act as a protective barrier to environmental moisture, gases, flavors, aromas, or oils [26,64], other properties that are often determined on stand-alone films are water vapor permeability (WVP), oxygen permeability (OP), carbon dioxide permeability, and flavor permeability. Aroma and oil permeability are also very important for many foods but have generally received less attention. In other cases, other properties of interest are water solubility, gloss, and color.

The general protective functions of edible films and coatings may be widened with the addition to the matrixes of additional ingredients that provide new functionalities. These food additives include antioxidants, colorants, flavors, nutraceuticals, nutrients, and antimicrobial and particularly antifungal compounds [25,28–30,62,65].

3.4. Types of Antifungal Ingredients

With the exception of chitosan that presents inherent antimicrobial activity against a wide range of foodborne fungi, the antifungal effect of biopolymer-based coatings is usually achieved by the incorporation of active antimicrobial compounds to the coating formulation. Another exceptional case are gels and aqueous extracts from the leaves of the plant *Aloe vera*, which are edible and can also be used to coat fresh or minimally processed horticultural products. These coatings have been basically used for physiologic preservation of fruit after harvest, but they are also known for their antimicrobial activity [66–68]. However, in contrast to chitosan-based coatings, there is very limited information on the performance of coatings based on *A. vera* gels against fungi causing postharvest decay of citrus fruit. Early work by Saks and Barkai-Golan [69] showed significant activity of *A. vera* gels against *P. digitatum in vitro* and against green mold on grapefruits artificially inoculated with this fungus. Similar results were obtained in more recent research by [70], who found that an *A. vera*-based coating reduced green and blue molds on "Kinnow" mandarins and positively affected fruit quality during cold storage.

According to their nature, the antifungal compounds that can be added as additional ingredients to edible coatings can be classified into three categories: (1) synthetic food preservatives or GRAS (generally recognized as safe) compounds with antimicrobial activity, which include some inorganic and organic acids and their salts (carbonates, propionates, sorbates, benzoates, *etc.*), parabens (methyl and ethyl parabens) and their salts, *etc.*; (2) natural compounds such as essential oils or other natural plant extracts (cinnamon, capsicum, lemongrass, oregano, rosemary, garlic, vanilla, carvacrol, citral, cinnamaldehyde, vanillin, grape seed extracts, *etc.*) [33]; and (3) antimicrobial antagonists as biological control agents (yeasts, bacteria, and even some filamentous fungi). As they need to be allowed as food ingredients, all compounds in the first two categories should be synthetic or natural substances with known and minimal toxicological effects on mammals and the environment and they must be classified as food-grade additives or GRAS compounds by relevant regulations [71,72]. On the other hand, biocontrol agents must comply with strict and complex specific legislation, which considerably differs among different countries [73].

4. Chitosan and Chitosan-Based Citrus Coatings

Chitosan is a natural biopolymer with antimicrobial activity that has the property to form edible films and coatings [74]. It is a linear cationic polysaccharide of high molecular weight consisting of 1,4-linked 2-amino-deoxy-β-D-glucan, a partially-deacetylated derivative from chitin. Chitin is present in the exoskeletons of crustaceans (crabs, lobsters, shrimps, *etc.*) and is, after cellulose, the most abundant polysaccharide in nature [35,75]. Chitosan is produced commercially with different deacetylated grades and molecular weights, which are related to their functional properties and antimicrobial effects. Chitosan has exhibited high antimicrobial activity against a range of spoilage microorganisms, including fungi, yeasts, and bacteria. Particularly, its beneficial effect for postharvest disease reduction has been reported for a wide variety of fresh horticultural produce including citrus, apples, mango, grapes, strawberries, blueberries, lettuce, carrots, or tomatoes. Its antimicrobial activity, however, depends on several factors such as the type of chitosan, degree of acetylation, molecular weight, concentration, medium pH, target microorganism, and the presence of other ingredients in the chitosan coating [25,35,76–79]. In addition to its antimicrobial and antifungal activity, chitosan has been considered as a good candidate for postharvest treatment and long-term storage of fresh fruits and vegetables because of other important properties, such as lack of toxicity to mammals and consequent edibility, biodegradability, biocompatibility with many other compounds, and multifunctionality, greatly derived from its capacity to form coatings [62,75].

Chitosan and derivatives are currently the most assayed antifungal edible coatings for postharvest preservation of fresh citrus fruit. They have been investigated either alone or formulated with other additional antifungal ingredients. Table 1 summarizes the most important research applications, to date, of chitosan and chitosan-based coatings, including the additional antimicrobial agents, if any, the target pathogen, an assessment of the antifungal activity, and the literature reference.

Table 1. Antifungal chitosan-based edible composite coatings applied to fresh citrus fruit.

Citrus Fruit	Coating	Concentration	Antimicrobial Agent	Target Pathogen	Antimicrobial Activity [a]	Reference
"Navel" orange, Lime	Chitosan	2, 4, 6, 8 g·L^{-1}	Lemongrass oil, Citral (6, 8 mL·L^{-1})	Penicillium digitatum, Penicillium italicum	+	[80]
	Chitosan	–	Lemongrass oil, Citral	P. digitatum, P. italicum	+	
"Satsuma" mandarin	Chitosan	1.0%	–	P. digitatum	+	[81]
	Chitosan	1.0%	Clove oil (0.5 mL·L^{-1})	P. digitatum	–	
"Or" and "Mor" mandarins, "Star Ruby" grapefruit	CMC, chitosan	1.0%, 1.5%	–	–	ND	[82]
"Jincheng 447" orange	Oligochitosan	1.5%	–	Colletotrichum gloeosporioides	+	[83]
"Valencia" orange	Chitosan	0.5%	–	P. digitatum	+	[84]
"Washington Navel" orange	Chitosan	0.5%	–	P. digitatum	+	–
"Femminello" lemon	Chitosan	0.5%	–	P. digitatum	+	–
"Marsh Seedlees" grapefruit	Chitosan	0.5%	–	P. digitatum	+	–
"Navel Powell" orange	Chitosan	2%	Bergamot, thyme and tea tree oils	P. italicum	+	[85]
"Valencia" and 'Pêra Rio" oranges	Chitosan, Chitosan + TBZ	2%	–	Guignardia citricarpa	+	[34,86]
Lime	Chitosan	2, 4, 6, 8 g·L^{-1}	Citral (0, 2, 3, 4, 5 mL·L^{-1})	Geotrichum citri-aurantii	+	[87]
"Navel" orange	Chitosan	2%	–	P. digitatum, P. italicum	+	[77]

Table 1. Cont.

Citrus Fruit	Coating	Concentration	Antimicrobial Agent	Target Pathogen	Antimocrobial Activity [a]	Reference
"Murcott" tangor	Chitosan (high MW)	0.05%, 0.1%, 0.2%	-	P. digitatum, P. italicum, Botrydiplodia lecanidion, Botrytis cinerea	+	[88]
	Chitosan (low MW)	0.05%, 0.1%, 0.2%	-	P. italicum, B. lecanidion, B. cinerea	+	
"Tankan" tangor	Chitosan	0.05%–0.2%	-	P. digitatum, P. italicum	-	[78]
Lemon	Chitosan	1 mg·mL^{-1}	-	P. digitatum	-	[89]
"Fortune" mandarin, "Valencia" orange	Biorend® (chitosan comercial product)	-	-	-	+	[90]
"Eureka" lemon	Glycolchitosan	0.2%	Candida saitoana (10^8 CFU·mL^{-1})	P. digitatum	+	[91,92]
"Washington Navel" orange	Glycolchitosan	0.2%	C. saitoana (10^8 CFU·mL^{-1})	P. digitatum	+	-

4.1. Stand-Alone Chitosan Coatings

The application of chitosan to lemon fruit prior to inoculation with *P. digitatum* resulted in a near absence of fungal development in inoculated rind wounds [89]. This worker suggested that the treatment had the ability to induce the transcriptional activation of defense genes leading to the accumulation of structural and biochemical compounds at strategic sites. Chitosan against green mold caused by *P. digitatum* was tested at concentrations from 0.02% to 0.5% in both *in vitro* and *in vivo* assays. It was found in the *in vitro* tests that chitosan at concentrations higher than 0.1% totally inhibited the pathogen growth, while in the *in vivo* assays chitosan applied at the concentration of 0.5% significantly reduced green mold on "Valencia" and "Washington Navel" oranges, "Femminello" lemons and "Marsh Seedless" grapefruits. The antifungal effects of chitosan were related to direct fungitoxic activity against the pathogen [84]. In another study [77], chitosan at 2% was applied to "Navel" oranges 24 h before inoculation with *P. digitatum* or *P. italicum* and it was observed that, after 23 days at 20 °C, disease incidence (proportion of infected fruit) and severity (lesion diameter) were significantly lower on coated samples than on uncoated controls. Chitosan antifungal activity was higher against blue mold than against green mold. It was pointed out, moreover, that chitosan application enhanced the activities of several enzymes such as peroxidase (POD) and superoxide dismutase (SOD), and increased the levels of glutathione (GSH) and hydrogen peroxide (H_2O_2). The effects of low (15 kDa) and high (357 kDa) molecular weight chitosan coatings, applied at 0.05% to 0.2%, on the development of green and blue molds and the quality of "Murcott" tangerines were studied by Chien, *et al.* [88]. Low molecular weight chitosan at 0.2% showed effective antifungal activity against both molds and improved firmness, titratable acidity, ascorbic acid level, and water content of coated fruit. Moreover, the performance of low molecular weight chitosan was comparable with that of the synthetic chemical fungicide thiabendazole. In another research, the antifungal activity of chitosan (0.05%–0.2%) with two different molecular weights (92.1 kDa and 357.3 kDa) against *P. digitatum* and *P. italicum* and the effect on postharvest quality of "Tankan" fruit, were investigated [78]. The performance of chitosan depended on its type and concentration. Chitosan effectively reduced the growth of the pathogens and the percentage of decayed fruit during incubation at 24 °C. On the other hand, fruit treated with chitosan exhibited good chemical and physical properties and less decay than control fruit during long-term refrigerated storage.

Deng, *et al.* [83] studied the effect of oligochitosan (1.5%), a derivate product from the enzymatic hydrolysis of chitosan, on the development of anthracnose on oranges inoculated with *C. gloeosporioides*. The results showed lower disease incidence and severity on oligochitosan-treated than on control fruit. The authors concluded that the treatment had the ability to induce strong disease resistance in citrus fruit during storage by increasing the contents in the fruit peel of constitutive natural antifungal compounds and the activity of defense enzymes. Glycolchitosan, another chitosan derivative, was also tested, alone or in combination with other treatments, for the control of green mold on oranges and lemons [91].

Two citrus varieties, "Fortune" mandarins and "Valencia" oranges were coated with a commercial chitosan-based coating product (Biorend®, IDEBIO S.L., Salamanca, Spain), and its effects on decay, fungistatic action, and fruit quality attributes were studied. The ripening process and decay development in coated fruit were monitored by means of the magnetic resonance imaging (MRI) technique [90]. The authors concluded that this chitosan product delayed maturity and reduced decay on both citrus species, allowing the effective preservation of "Fortune" mandarins and "Valencia" oranges for 6 and 22 weeks, respectively.

Work in Brazil [86] showed the potential of chitosan coatings to prevent the development of black spot, an important quarantine disease caused by the pathogen *Phyllosticta citricarpa* (McAlpine) van de Aa, on "Valencia" oranges stored at either 25 °C for eight days or 3 °C for 21 days. The authors discussed that chitosan application stimulated disease resistance mechanisms in the fruit rind.

4.2. Chitosan Coatings Amended with other Antifungal Ingredients

The possibility of enhancing the antifungal activity of chitosan coatings by combining them with other antifungal treatments has been explored as well. Such complementary antifungal treatments can be of different nature (physical treatments [93], chemical fungicides [34], food additives [91], biocontrol antagonists [94,95], etc.) and may be applied to citrus fruit before or after the application of chitosan-based coatings. However, in this review we focus on the addition of other antifungal agents to chitosan coating matrixes that are afterwards applied to the fruit as a unique postharvest treatment. In general, chitosan-amended coatings allow a gradual release of preservatives and provide additional properties for fungal growth inhibition and fruit quality maintenance [90]. The most important of these additional antifungal ingredients are essential oils. In few cases, biocontrol agents have been tested as well.

The antifungal activity, either *in vitro* or *in vivo*, of many essential oils extracted from plants or fruits such as birch, bergamot, cumin, cinnamon, citrus, clove, lemongrass, oregano, thyme, or tea tree against major citrus pathogens, particularly *P. digitatum* and *P. italicum*, has been reported [15,96]. Among them, only a limited amount has been tested as potential ingredients of chitosan coatings. Chitosan combined with citral or lemongrass oil (at 3 mL·L^{-1} or 4 mL·L^{-1}) significantly reduced the mycelial growth and spore germination of *P. digitatum* and *P. italicum* in *in vitro* tests, while when applied to oranges and limes also provided a considerable green and blue mold reduction on fruit stored at either 20 °C or refrigeration temperatures [80]. Faten [87] reported good results with a combination of citral and chitosan for the control on lime fruit of sour rot, caused by the pathogen *G. citri-aurantii*. Reductions of disease incidence and severity as large as 89.5% and 93.5%, respectively, were obtained after the application of this coating containing chitosan (at 6 g·L^{-1} or 8.0 g·L^{-1}) and citral (at 4 mL·L^{-1} or 5 mL·L^{-1}). Both preventive (fruit coating followed by fungal inoculation) and curative (inoculation followed by coating) activity of chitosan coatings, containing or not essential oils from bergamot, thyme, or tea tree, were determined on "Powell" navel oranges against *P. italicum* [85]. It was found that, in general, the amended coatings were more effective than chitosan alone in reducing blue mold and that preventive activity was higher than curative activity. Among the tested oils, thymol showed the greatest antifungal activity when incorporated with the chitosan coating. Moreover, fruit quality attributes (acidity, pH, soluble solids, juice percentage, weight loss, firmness, color parameters, and respiration rate) of coated samples did not present any relevant changes during long-term cold storage. The inhibition by vapor contact of *A. niger* and *P. digitatum* by selected concentrations of Mexican oregano, cinnamon, or lemongrass essential oils added to chitosan, amaranth, or starch edible films was evaluated in another research [97]. The results showed that chitosan edible films incorporating Mexican oregano or cinnamon oils inhibited the two fungi at lower concentrations than those required for amaranth and starch edible films, and it was concluded that chitosan improved the release of the oil antimicrobial compounds. Sánchez-González, *et al.* [98] studied the addition of different concentrations of tea tree essential oil into chitosan films. Mechanical and optical properties and WVP of dry films, and antimicrobial activity against *P. italicum* were determined. They reported that films with a ratio of tea tree essential oil:chitosan of 1:2 provided a complete inhibition of fungal growth after five days of storage at 10 °C. Nevertheless, no every study reported synergistic activity from the addition of essential oils to chitosan coatings. Very recent work by Shao, *et al.* [81] showed that chitosan combined with clove oil was not significantly more effective than chitosan alone for the control of green mold on "Satsuma" mandarins artificially inoculated with *P. digitatum*. Furthermore, the high antifungal activity observed in *in vitro* tests was not obtained in *in vivo* tests, despite that the combination of chitosan and clove oil enhanced the activity of some fruit defense enzymes.

With reference to biocontrol agents as additional antifungal ingredients in chitosan-based coatings, El-Ghaouth, *et al.* [92] assessed the effectiveness of a 0.2% glycolchitosan coating formulated with the antagonistic yeast *Candida saitoana* against citrus green mold. They observed that this bioactive coating was superior to stand-alone glycolchitosan and *C. saitoana* treatments in controlling green mold on different orange and lemon cultivars, and the control level was equivalent to that with IMZ. The biocoating also reduced the incidence of stem-end rot caused by the pathogens L. *theobromae* or *P.citri* on naturally infected "Valencia" oranges, although in this case disease control was lower than that with IMZ. The maximum efficacy against green mold on "Eureka" lemons artificially inoculated with *P. digitatum* was obtained when citrus fruit were pretreated with the food additive sodium carbonate followed by the application of the glycolchitosan coating containing *C. saitoana* [91].

Along with the addition of essential oils or biocontrol agents, another system that has been evaluated to improve the activity of chitosan is the development of bilayer coatings comprised of chitosan and another natural polymer. This is the case of an edible composite bilayer coating formulated with carboxymethyl cellulose (CMC) and chitosan [82]. These workers applied this new coating to mandarins, oranges, and grapefruits and observed beneficial effects for fruit postharvest quality preservation. However, its effect on postharvest decay reduction was not studied.

5. Citrus Coatings Formulated with GRAS salts

Among the different GRAS compounds, organic acids and their salts are the most common synthetic antimicrobial agents tested in food systems [96]. In research conducted in our laboratory, Valencia-Chamorro, *et al.* [99] developed and optimized hydroxypropyl methylcellulose (HPMC)-lipid edible composite films formulated with a wide variety of food additives or GRAS compounds (mineral salts, organic acid salts, parabens, *etc.*) to inhibit the *in vitro* growth of the pathogens *P. digitatum* and *P. italicum*. Afterwards, the curative activity of selected coatings against green and blue molds was tested *in vivo* on oranges and mandarins artificially inoculated with the pathogens, coated, and incubated at 20 °C for a shelf life period of seven days [100]. It was found that coatings containing the GRAS salts potassium sorbate (PS), sodium benzoate (SB), sodium propionate (SP) and their mixtures were the most effective for disease reduction. Subsequent studies confirmed the antifungal activity and good effects on preservation of fruit quality of HPMC-lipid edible coatings formulated with PS, SB, SP, or their mixtures on long-term cold-stored "Valencia" oranges [101], "Ortanique" mandarins [102] and "Clemenules" mandarins [103]. In general, these coatings reduced fruit weight loss and maintained firmness without adversely affecting the overall sensory quality of coated fruit.

In research to explore the ability of GRAS salts to substitute the use of synthetic fungicides in conventional citrus commercial waxes, Youssef, *et al.* [104] reported that the salts sodium carbonate (SC) and bicarbonate (SBC), potassium carbonate (PC) and bicarbonate (PBC), ammonium bicarbonate (ABC), and PS incorporated into a commercial wax at a concentration of 6% (*w/v*) significantly reduced the incidence of postharvest diseases on naturally infected "Tarocco" and "Valencia" oranges and "Comune" clementines during cold storage followed by one week of shelf life at 20 °C. Among them, wax containing PS was the most effective to reduce decay in all tested cultivars.

6. Citrus Coatings Formulated with Essential Oils

Among plant-derived natural compounds, many essential oils have proven antimicrobial activity against different important fungal pathogens, thus gaining popularity as potential ingredients of fruit coatings. The essential oils most studied for this purpose include clove, oregano, thyme, nutmeg, basil, mustard, and cinnamon oils, and some of them have been classified as GRAS by regulators [105]. Essential oils, as stand-alone treatments for fruit preservation, can be applied as volatiles in modified atmosphere packages or prepared as aqueous solutions from plant extracts obtained with different solvents. In any case, research has frequently shown that the efficacy of these compounds in *in vivo* experiments with fruit is lower than that observed in *in vitro* tests and, in some cases, they can be phytotoxic or induce negative sensory properties to treated commodities [96,106]. Therefore, despite their important antifungal activity, commercial implementation of essential oils as stand-alone postharvest treatments is strongly restricted and their use as ingredients of edible coatings might be a potential solution for these application problems.

Good antifungal activity was achieved with an edible coating formulated with a CMC matrix amended with a stem extract of the herb *Impatiens balsamina*. The application of this coating significantly reduced total decay incidence and preserved fruit weight without adverse effects on internal quality on "Newhall" oranges stored at 5 °C for up to 100 days [107]. The mode of action of the coating was attributed to an increase of the activities of important fruit defense enzymes such as POD, SOD, chitinase (CHI), and β-1,3-glucanase (GLU). Velásquez, et al. [108] compared in laboratory trials the effect of a pectin-based edible coating amended with essential oil at different concentrations with that of commercial waxes for disease control on "Valencia" oranges inoculated with *Penicillium* sp. The pectin coating containing essential oil at 1.5% reduced decay by 83% and satisfactorily extended the shelf life of coated fruit stored at 23 °C. Nevertheless, both amended coatings and commercial waxes were not effective on fruit stored at low temperatures.

In other research works, commercial citrus waxes have been formulated with essential oils as an alternative to the use of synthetic chemical fungicides as components of these waxes. In semi-commercial and commercial trials, essential oils from the plants *Mentha spicata* and *Lippia scaberrima* incorporated into different commercial citrus waxes resulted in significant preventive activity against green mold on "Valencia" and "Tomango" oranges. Specifically, after six days of incubation at ambient temperature, essential oil from *L. scaberrima* added at 2500 mL·L^{-1} into a carnauba wax reduced by about 60% the incidence of green mold compared to control fruit coated with the carnauba wax without essential oil [109]. Moreover, the amended waxes significantly decreased orange weight loss and maintained overall fruit quality. The effectiveness of the amended coatings was related to the strong inhibitory activity against *P. digitatum* of one of the most important components of the essential oil, the terpenoid R-carvone. In laboratory tests conducted in Spain, the application of wax coatings formulated with the essential oils carvacrol or thymol satisfactorily reduced green mold incidence on lemons artificially inoculated with *P. digitatum* and, at the same time, reduced the lemon respiration rate and ethylene production [110]. Further semi-commercial trials conducted in lemon packing lines confirmed that these essential oil-amended waxes were as effective as waxes containing the conventional fungicide IMZ in reducing lemon postharvest decay and total aerobic microflora present on fruit surface. In addition, fruit weight loss and rind softening and color were equally preserved by both types of coatings [111].

In another work, citral was incorporated into a commercial carnauba wax and a concentration 10 times the minimum fungicidal concentration obtained in *in vitro* studies was required to significantly decrease the incidence of green mold on citrus fruit coated and incubated at 25 °C for 6 days [112]. Octanal, another component of citrus essentials oils, has also been recently evaluated as active ingredient of commercial waxes for postharvest coating of citrus fruit. Tao, et al. [113] determined in *in vitro* tests that the minimum fungicidal concentration (MFC) of this compound against *P. digitatum* was 1000 μL·L^{-1}. Then, they observed in *in vivo* tests that waxes amended with two times this octanal MFC effectively prevented the development of green mold on Satsuma mandarins incubated at 25 °C. They discussed that the coatings increased the activity of the enzymes SOD and catalase (CAT), but reduced the activity of phenylalanine ammonia-lyase (PAL) and POD, preventing the accumulation of H_2O_2. Commercial carnauba and/or shellac coatings amended with *Cinnamomum zeylanicum* essential oil at 0.5% (v/v) provided very high preventive activity against citrus green and blue molds, with disease reductions of about 90% with respect to uncoated control fruit treated with distilled water [114]. However, when the essential oil from *C. zeylanicum* was incorporated into polyethylene commercial waxes, disease reduction was considerably lower, indicating that the antifungal activity of the coating depended not only on the type and amount of essential oil and the volume of coating that remained on the fruit skin, but also on the formulation solubility and the particular compatibility between the essential oil and components of the wax. Among fifty-nine commercially available essential oils, workers from South Africa [115] selected lemongrass, the essential oil from the plant *Cymbopogon citrates*, as the most cost-effective compound for the control of the pathogen *G. citri-aurantii*, the cause of citrus postharvest sour rot. They developed new coating formulations based on the addition of lemongrass and *M. spicata* essential oil to a citrus commercial wax. The coating containing both essential oils, each at a concentration of 750 mL·L^{-1}, effectively controlled sour rot, but also green and blue molds, on artificially-inoculated "Valencia" oranges.

7. Citrus Coatings Formulated with Microbial Antagonists

Research in the 1990s showed that different coatings containing microbial antagonists as biocontrol agents could be of use for the control of postharvest diseases of citrus fruit. Strains of the yeasts *Candida oleophila* ([116–118], *Candida guillermondii* [119,120] and the bacterium *Pseudomonas syringae* [121] were incorporated to different coatings and assessed against green and blue molds on oranges, lemons, or grapefruits. The effectiveness of these coatings greatly depended on their ability to support populations of the biocontrol agent. Shellac and other commercial wax coatings could be toxic to the yeasts due to the addition of alcohols and basis that are used to dissolve the primary constituents. The survival of the antagonistic yeast *C. olephila* in shellac formulations significantly improved in coatings containing less than 6% alcohol and when concentrations of morpholine and ammonia were minimized. These formulations significantly improved shelf life of coated grapefruits when compared to fruit coated alone with a shellac formulation [117].

Coating formulations based on water soluble cellulose derivatives, such as HPMC or methylcellulose, amended with the biocontrol yeast *C. olephila* significantly reduced decay and prolonged the storage life of grapefruits [116]. Yeast populations were stable in the coatings, but the addition of additional food preservatives was necessary to prevent the growth of other contaminating microorganisms. Thus, the addition of 0.15% PS to these formulations controlled microbial development without reducing the population of *C. oleophila*, which continued to increase in coated fruit during a storage period of 5 weeks. Similarly, the use of commercial sucrose ester formulations as coating matrixes favored the development of *C. olephila* populations in a greater extent than shellac formulations and these coatings were more effective for decay control on grapefruits than shellac-based coatings [118]. Varying the cellulose component of the coating formulation also affected the survival of two yeast biocontrol agents, *C. guillermondii* strain US7 and *Debaryomyces* sp. strain 230. Populations of both strains during incubation at room temperatures were higher in methylcellulose matrixes. The application of a methylcellulose-based coating containing the strain US7 to naturally infected "Pineapple" and "Valencia" oranges resulted in significant decay control for the first 2 to 4 weeks of storage at 16 °C and 90% RH [120].

In very recent research, Aloui, *et al.* [122] investigated the properties and the performance of bioactive coatings formulated with matrixes of sodium alginate and locust bean gum containing cells of the killer yeast *Wickerhamomyces anomalus*. They found that the survival of the yeast in these films was very high (more than 85% of the initial population) and their barrier, mechanical, and optical properties were satisfactory. When applied to "Valencia" oranges artificially inoculated with *P. digitatum*, these coatings reduced green mold by more than 70% on fruit incubated at 25 °C. In addition, coating application effectively reduced fruit weight loss and maintained rind firmness with respect to uncoated oranges.

8. Conclusions

Increasing concerns about human health risks and environmental contamination, restricted commercial channels for conventional production, and the proliferation of resistant strains of pathogenic fungi are important problems related to the use of conventional chemical fungicides for postharvest preservation of fresh citrus fruit. These agrochemicals are often applied as ingredients of synthetic waxes, which are needed to extend fruit postharvest life by reducing transpiration and respiration and also to provide gloss and enhance fruit appearance. Cost-effective disease control in the absence of synthetic fungicides requires the implementation of global IDM strategies based, in the postharvest phase, on the adoption of alternative or complementary non-polluting antifungal treatments. Irrespective of their nature, the evaluation of these alternative treatments should generally focus on the control of green and blue molds, caused by the pathogens *P. digitatum* and *P. italicum*, respectively, since they are the most economically important citrus postharvest diseases. Among these alternatives, the development of antifungal edible coatings arises as a new safe technology with potential to overcome both citrus postharvest physiological and pathological problems.

The functionality of chitosan, with an inherent antifungal activity, and composite films and coatings based on polysaccharide- or protein-lipid matrixes can be considerably improved by the incorporation of additional antifungal ingredients such as food additives or GRAS salts, natural compounds like essential oils or other volatiles, and microbial antagonists (yeasts, bacteria) as biocontrol agents. In each particular case, these antifungal agents are typically selected according to their efficacy as stand-alone treatments against target pathogens and, in many cases, their incorporation as ingredients of coatings brings a remarkable improvement of their performance. For example, coating formulations can overcome application problems of effective agents that may induce fruit phytotoxicities or adverse sensory properties when applied alone as aqueous or gaseous treatments. Further, coating application can also increase the agent antifungal activity by regulating its temporal and spatial release or facilitating its continuous and effective contact with the target pathogen.

Due to the high economic value of the worldwide citrus trade, the development of novel antifungal edible coatings for citrus fruit is a very active research field and, as described in the present review, a considerable number of research studies reporting interesting results are available from the specialized literature and will surely increase in the next few years. Despite the substantial progress that has been accomplished in evaluating new antifungal edible coatings, their implementation is still limited, first because of the current availability of highly effective, convenient, and cheaper conventional fungicides, and second because of general limitations associated to the edible nature of food-grade coating components. Low-toxicity and/or persistence, lack of either curative or preventive activity, or excessive specificity are handicaps for commercial application. It should be taken into account, in this sense, that minimal fruit losses due to postharvest decay are required by citrus export markets and, in contrast to that of synthetic chemicals with direct fungicidal activity, the mode of action of antifungal edible coatings is usually rather fungistatic. Consequently, their effectiveness is highly dependent not only on the target pathogen but also on fruit host characteristics such as citrus species and cultivar and fruit physical and physiological condition. Of course, coating performance from the physiological point of view will also be greatly influenced by all these fruit attributes. Therefore, the development of antifungal coatings should be tailored not only to control specific diseases but also for application to specific commercially important cultivars of oranges, mandarins, lemons or grapefruits, according to particular postharvest handling procedures, fruit storage potential, and destination markets. In addition to the development of new coatings formulated with other matrixes and/or antifungal ingredients and the evaluation of new applications for existing promising coatings, future prospects include research to determine potential synergistic effects from the combined application of different types of antifungal edible coatings with other alternative disease control methods (physical, low-toxicity chemical, or biological) in a multifaceted approach within IDM strategies.

Acknowledgments: The authors thank all Spanish and international agencies that funded research on this topic. In memory of Miguel Ángel del Río, for his unconditional friendship, guidance and support.

Author Contributions: Lluís Palou drafted and organized this review article. All three authors wrote and revised the review.

Conflicts of Interest: The authors declare no conflict of interest.

1. Food and Agriculture Organization of the United Nations. FAO Statistical Yearbook 2013. Available online: http://www.fao.org/docrep/018/i3107e/i3107e03.pdf (accessed on 22 September 2015).
2. Grierson, W.; Miller, W.M. Storage of Citrus Fruit. In *Fresh Citrus Fruits*, 2nd ed.; Wardowski, W.F., Miller, W.M., Hall, D.J., Grierson, W., Eds.; Florida Science Source, Inc.: Longboak Key, FL, USA, 2006; pp. 547–581.

3. Alférez, F.; Agustí, M.; Zacarías, L. Postharvest rind staining in Navel oranges is aggravated by changes in storage relative humidity. Effect on respiration, ethylene production and water potential. *Postharvest Biol. Technol.* **2003**, *28*, 143–152. [CrossRef]
4. Alférez, F.; Burns, J.K. Postharvest peel pitting at non-chilling temperatures in grapefruit is promoted by changes from low to high relative humidity during storage. *Postharvest Biol. Technol.* **2004**, *32*, 79–87. [CrossRef]
5. Alférez, F.; Zacarías, L.; Burns, J.K. Low relative humidity at harvest and before storage at high humidity influence the severity of postharvest peel pitting in citrus. *J. Am. Soc. Hortic. Sci.* **2005**, *130*, 225–231.
6. Alférez, F.; Alquezara, B.; Burns, J.K.; Zacarías, L. Variation in water, osmotic and turgor potential in peel of 'Marsh' grapefruit during development of postharvest peel pitting. *Postharvest Biol. Technol.* **2010**, *56*, 44–49. [CrossRef]
7. Grierson, W. Physiological Disorders. In *Fresh Citrus Fruits*, 1st ed.; Wardowski, W.F., Nagy, S., Grierson, W., Eds.; AVI: New York, NY, USA, 1986; pp. 361–378.
8. Vercher, R.; Tadeo, F.R.; Almela, V.; Zaragoza, S.; Primo-Millo, E.; Agustí, M. Rind structure, epicuticular wax morphology and water permeability of 'Fortune' mandarin fruits affected by peel pitting. *Ann. Bot.* **1994**, *74*, 619–62. [CrossRef]
9. Dou, H.; Zhang, J.; Ismail, M.A.; Ritenour, M.A. Postharvest factors influencing stem-end rind breakdown (SERB) of Valencia oranges. *Proc. Fla. State. Hort. Soc.* **2001**, *114*, 164–169.
10. Porat, R.; Weiss, B.; Cohen, L.; Daus, A.; Aharoni, N. Reduction of postharvest rind disorders in citrus fruit by modified atmosphere packaging. *Postharvest Biol. Technol.* **2004**, *33*, 35–43. [CrossRef]
11. Petracek, P.D.; Kelsey, D.F.; Grierson, W. Physiological Disorders. In *Fresh Citrus Fruits*, 2nd ed.; Wardowski, W.F., Miller, W.M., Hall, D.J., Grierson, W., Eds.; Florida Science Source, Inc.: Longboak Key, FL, USA, 2006; pp. 397–419.
12. Scherrer Montero, C.R.; Schwarz, L.L.; Cunha dos Santos, L.; Pires dos Santos, R.; Bender, R.J. Oleocellosis incidence in citrus fruit in response to mechanical injuries. *Sci. Hortic.* **2012**, *134*, 227–231. [CrossRef]
13. Eckert, J.W.; Eaks, I.L. Postharvest Disorders and Diseases of Citrus Fruits. In *The Citrus Industry*; Reuter, W., Calavan, E.C., Carman, G.E., Eds.; Division of Agriculture and Natural Resources, University of California Press: Berkeley, CA, USA, 1989; pp. 179–260.
14. Snowdon, A.L. *A Color Atlas of Post-harvest Diseases and Disorders of Fruits and Vegetables, Volume 1*; General Introduction and Fruits; CRC Press: Boca Raton, FL, USA, 1990.
15. Palou, L. *Penicillium digitatum, Penicillium italicum* (Green Mold, Blue Mold). In *Postharvest Decay. Control Strategies*; Bautista-Baños, S., Ed.; Elsevier: London, UK, 2014; pp. 45–102.
16. Kanetis, L.; Förster, H.; Adaskaveg, J.E. Comparative efficacy of the new postharvest fungicides azoxystrobin, fludioxonil, and pyrimethanil for managing citrus green mold. *Plant Dis.* **2007**, *91*, 1502–1511. [CrossRef]
17. Smilanick, J.L.; Brown, G.E.; Eckert, J.W. The Biology and Control of Postharvest Diseases. In *Fresh Citrus Fruits*, 2nd ed.; Wardowski, W.F., Miller, W.M., Hall, D.J., Grierson, W., Eds.; Florida Science Source Inc.: Longboat Key, FL, USA, 2006; pp. 339–396.
18. Green, F.M. The infection of oranges by *Penicillium*. *J. Pom. Hortic. Sci.* **1932**, *10*, 184–215.
19. Lahlali, R.; Serrhini, M.N.; Friel, D.; Jijakli, M.H. *In vitro* effects of water activity, temperature and solutes on the growth rate of *P. italicum* Wehmer and *P. digitatum* Sacc. *J. Appl. Microbiol.* **2006**, *101*, 628–636. [CrossRef] [PubMed]
20. Droby, S.; Eick, A.; Macarisin, D.; Cohen, E.; Rafael, G.; Stange, R.; McColum, G.; Dudai, N.; Nasser, A.; Wisniewski, M.; Shapira, R. Role of citrus volatiles in host recognition, germination and growth of *Penicillium digitatum* and *Penicillium italicum*. *Postharvest Biol. Technol.* **2008**, *49*, 386–396. [CrossRef]
21. Bajwa, B.E.; Anjum, F.M. Improving storage performance of *Citrus reticulata* Blanco mandarins by controlling some physiological disorders. *Int. J. Food Sci. Technol.* **2007**, *42*, 459–501. [CrossRef]
22. Baldwin, E.A.; Nisperos-Carriedo, M.O.; Baker, R.A. Edible coatings for lightly processed fruits and vegetables. *HortScience* **1995**, *30*, 35–38.
23. Hagenmaier, R.; Goodner, K.; Roussef, R.; Dou, H. Storage of 'Marsh' Grapefruit and 'Valencia' oranges with different coatings. *Proc. Fla. State. Hort. Soc.* **2002**, *115*, 303–308.
24. Krochta, J.M. Proteins as raw materials for films and coatings: Definitions, Current Status, and Opportunities. In *Protein-Based Films and Coatings*; Gennadios, A., Ed.; CRC Press: Boca Raton, FL, USA, 2002; pp. 1–41.

25. Campos, C.A.; Gerschenson, L.N.; Flores, S.K. Development of edible films and coatings with antimicrobial activity. *Food Bioprocess Technol.* **2011**, *4*, 849–875. [CrossRef]
26. Han, J.H. Edible Films and Coatings: A Review. In *Innovations in Food Packaging*, 2nd ed.; Han, J.H., Ed.; Elsevier: London, UK, 2014; pp. 213–255.
27. Nisperos-Carriedo, M.O. Edible Coatings and Films Based on Polysaccharides. In *Edible Coatings and Films to Improve Food Quality*; Krochta, J.M., Baldwin, E.A., Nisperos-Carriedo, M.O., Eds.; Technomic Publishing Co. Inc.: Lancaster, PA, USA, 1994; pp. 305–335.
28. Cha, D.; Chinnan, M. Biopolymer-based antimicrobial packaging: A review. *Crit. Rev. Food Sci.* **2004**, *44*, 223–227. [CrossRef] [PubMed]
29. Dhall, R.K. Advances in edible coatings for fresh fruits and vegetables: A review. *Crit. Rev. Food Sci. Nutr.* **2013**, *53*, 435–450. [CrossRef] [PubMed]
30. Han, J.H.; Gennadios, A. Edible Films and Coatings: A review. In *Innovations in Food Packaging*, 1st ed.; Han, J.H., Ed.; Elsevier: Amsterdam, The Netherlands, 2005; pp. 239–262.
31. Pérez-Gago, M.B.; González-Aguilar, G.A.; Olivas, G.I. Edible coatings for fruits and vegetables. *Stewart Postharv. Rev.* **2010**, *6*, 1–14. [CrossRef]
32. Zaritzky, N. Edible Coatings to Improve Food Quality and Safety. In *Food Engineering Interfaces*; Aguilera, J.M., Simpson, R., Welti-Chanes, J., Bermúdez Aguirre, D., Barbosa-Cánovas, G.V., Eds.; Springer-Verlag: New York, USA, 2011; pp. 631–659.
33. Valencia-Chamorro, S.A.; Palou, L.; del Río, M.A.; Pérez-Gago, M.B. Antimicrobial edible films and coatings for fresh and minimally processed fruits and vegetables: A review. *Crit. Rev. Food Sci. Nutr.* **2011**, *51*, 872–900. [CrossRef] [PubMed]
34. Rappussi, M.C.C.; Benato, E.A.; Cia, P.; Pascholati, S.F. Chitosan and fungicides on postharvest control of guignardia citricarpa and on quality of 'Pêra Rio' oranges. *Summa Phytopathol.* **2011**, *37*, 142–144. [CrossRef]
35. Zhang, Y.; Rempel, C.; McLaren, D. Edible Coating and Film Materials: Carbohydrates. In *Innovations in Food Packaging*, 2nd ed.; Han, J.H., Ed.; Elsevier: London, UK, 2014; pp. 305–323.
36. Gol, N.B.; Patel, P.R.; Rao, T.V.R. Improvement of quality and shelf-life of strawberries with edible coatings enriched with chitosan. *Postharvest Biol. Technol.* **2013**, *85*, 185–195. [CrossRef]
37. Lin, D.; Zhao, Y. Innovations in the development and application of edible coatings for fresh and minimally processed fruits and vegetables. *Compr. Rev. Food Sci. F.* **2007**, *6*, 60–75. [CrossRef]
38. Díaz-Mula, H.M.; Serrano, M.; Valero, D. Alginate coatings preserve fruit quality and bioactive compounds during storage of sweet cherry fruit. *Food Bioprocess Technol.* **2011**, *5*, 2990–2997. [CrossRef]
39. Meng, X.; Yang, L.; Kennedy, J.F.; Tian, S. Effects of chitosan and oligochitosan on growth of two fungal pathogens and physiological properties in pear fruit. *Carbohydr. Polym.* **2010**, *81*, 70–75. [CrossRef]
40. Ferrari, C.C.; Sarantópoulos, C.I.; Carmello-Guerreiro, S.M.; Hubinger, M.D. Effect of osmotic dehydration and pectin edible coatings on quality and shelf life of fresh-cut melon. *Food Bioprocess Technol.* **2013**, *6*, 80–91. [CrossRef]
41. Pérez-Gallardo, A.; García-Almendárez, B.; Barbosa-Cánovas, G.; Pimentel-González, D.; Reyes-González, L.R.; Regalado, C. Effect of starch-beeswax coatings on quality parameters of blackberries (*Rubus* spp.). *J. Food Sci. Technol.* **2014**, *52*, 5601–5610. [CrossRef] [PubMed]
42. Pan, S.Y.; Chen, C.H.; Lai, L.S. Effect of tapioca starch/decolorized hsian-tsao leaf gum-based active coatings on the qualities of fresh-cut apples. *Food Bioprocess Technol.* **2013**, *6*, 2059–2069. [CrossRef]
43. Lima, A.M.; Cerqueira, M.A.; Souza, B.W.S.; Santos, E.C.M.; Teixeira, J.A.; Moreira, R.A.; Vicente, A.A. New edible coatings composed of galactomannans and collagen blends to improve the postharvest quality of fruits—Influence on fruits gas transfer rate. *J. Food Process. Eng.* **2010**, *97*, 101–109. [CrossRef]
44. Cerqueira, M.A.; Bourbon, A.I.; Pinheiro, A.C.; Martins, J.T.; Souza, B.W.S.; Teixeira, J.A.; Vicente, A.A. Galactomannans use in the development of edible films/coatings for food applications. *Trends Food Sci. Tech.* **2011**, *22*, 662–671. [CrossRef]
45. Vargas, M.; Pastor, C.; Chiralt, A.; McClements, D.; González-Martínez, C. Recent advance in edible coatings for fresh and minimally processed fruits. *Crit. Rev. Food Sci. Nutr.* **2008**, *48*, 496–511. [CrossRef] [PubMed]
46. Baldwin, E.A. Edible Coatings for Fresh Fruits and Vegetables: Past, Present, and Future. In *Edible Coatings and Films to Improve Food Quality*; Krochta, J.M., Baldwin, E.A., Nisperos-Carriedo, M.O., Eds.; Technomic Publishing Co. Inc.: Lancaster, PA, USA, 1994; pp. 25–64.

47. Lacroix, M.; Vu, K.D. Edible Coating and Film Materials: Proteins. In *Innovations in Food Packaging*, 2nd ed.; Han, J.H., Ed.; Elsevier: London, UK, 2014; pp. 277–304.
48. Pérez-Gago, M.B.; Serra, M.; Alonso, M.; Mateos, M.; del Río, M.A. Effect of solid content and lipid content of whey protein isolate-beeswax edible coatings on color change of fresh-cut apples. *J. Food Sci.* **2003**, *68*, 2186–2191. [CrossRef]
49. Pérez-Gago, M.B.; Serra, M.; Alonso, M.; Mateos, M.; del Río, M.A. Effect of whey protein- and hydroxypropyl methylcellulose-based edible composite coatings on color change of fresh-cut apples. *Postharvest Biol. Technol.* **2005**, *36*, 77–85. [CrossRef]
50. Baldwin, E.A.; Baker, R.A. Use of Protein in Edible Coatings for Whole and Minimally Processed Fruit and Vegetables. In *Protein-Based Films and Coatings*; Gennadios, A., Ed.; CRC Press: Boca Raton, FL, USA, 2002; pp. 501–515.
51. De S. Medeiros, B.G.; Pinheiro, A.C.; Teixeira, J.A.; Vicente, A.A.; Carneiro-da-Cunha, M.G. Polysaccharide/protein nanomultilayer coatings: Construction, characterization and evaluation of their effect on 'Rocha'pear (*Pyrus communis* L.) shelf-life. *Food Bioprocess Technol.* **2012**, *5*, 2435–2445. [CrossRef]
52. Schmid, M.; Reichert, K.; Hammann, F.; Stäbler, A. Storage time-dependent alteration of molecular interaction-property relationships of whey protein isolate-based films and coatings. *J. Mater. Sci.* **2015**, *50*, 4396–4404. [CrossRef]
53. Rhim, J.W.; Shellhammer, T.H. Lipid-Based Edible Films and Coatings. In *Innovations in Food Packaging*, 1st ed.; Han, J.H., Ed.; Elsevier: Amsterdam, The Netherlands, 2005; pp. 362–383.
54. Baldwin, E.A.; Nisperos-Carriedo, M.O.; Hagenmaier, R.D.; Baker, R.A. Use of lipids in coatings for food products. *Food Technol.* **1997**, *51*, 56–62.
55. Debeaufort, F.; Voilley, A. Lipid-Based Edible Films and Coatings. In *Edible Films and Coatings for Food Applications*; Embuscado, M., Huber, K.C., Eds.; Springer-Verlag: New York, NY, USA, 2009; pp. 135–168.
56. Galus, S.; Kadzinska, J. Food applications of emulsion-based edible films and coatings. *Trends Food Sci. Technol.* **2015**, *45*, 273–283. [CrossRef]
57. Pérez-Gago, M.B.; Rhim, J.W. Edible Coating and Film Materials: Lipid Bilayers and Lipid Emulsions. In *Innovations in Food Packaging*, 2nd ed.; Han, J.H., Ed.; Elsevier: London, UK, 2014; pp. 325–350.
58. Pérez-Gago, M.B.; Krochta, J.M. Emulsion and Bi-Layer Edible Films. In *Innovations in Food Packaging*, 1st ed.; Han, J.H., Ed.; Elsevier: Amsterdam, The Netherlands, 2005; pp. 384–402.
59. Shellhammer, T.H.; Krochta, J.M. Whey protein emulsion film performance as affected by lipid type and amount. *J. Food Sci.* **1997**, *62*, 390–394. [CrossRef]
60. Navarro-Tarazaga, M.L.; Sothornvit, R.; Pérez-Gago, M.B. Effect of plasticizer type and amount on hydroxypropyl methylcellulose-beeswax edible film properties and postharvest quality of coated plums (cv. Angeleno). *J. Agric. Food Chem.* **2008**, *56*, 9502–9509. [CrossRef] [PubMed]
61. Krochta, J. Film Edible. In *The Wiley Encyclopedia of Packaging Technology*, 2nd ed.; Brody, A.L., Marsh, K.S., Eds.; John Wiley & Sons, Inc.: New York, NY, USA, 1997; pp. 397–401.
62. Shiekh, R.A.; Malik, M.A.; Al-Thabaiti, S.A.; Shiekh, M.A. Chitosan as a novel edible coating for fresh fruits. *Food Sci. Technol. Res.* **2013**, *19*, 139–155. [CrossRef]
63. Guilbert, S.; Gontard, N. *Agro-Polymers for Edible and Biodegradable Films: Review of Agricultural Polymeric Materials, Physical and Mechanical Characteristics*, 1st ed.; Han, J.H., Ed.; Elsevier: Amsterdam, The Netherlands, 2005; pp. 263–276.
64. Krochta, J.M. Edible Protein Films and Coatings. In *Food Proteins and Their Applications*; Damodaran, S., Paraf, A., Eds.; Marcel Dekker, Inc.: New York, NY, USA, 1997; pp. 529–549.
65. Franssen, L.R.; Krochta, J.M. Edible Coating Containing Natural Antimicrobials for Processed Foods. In *Natural Antimicrobials for Minimal Processing of Foods*; Roller, S., Ed.; CRC Press: Boca Raton, FL, USA, 2000; pp. 250–262.
66. Arowora, K.A.; Williams, J.O.; Adetunji, C.O.; Fawole, O.B.; Afolayan, S.S.; Olaleye, O.O.; Adetunji, J.B.; Ogundele, B.A. Effects of *Aloe vera* coatings on quality characteristics of oranges stored under cold storage. *Greener J. Agric. Sci.* **2013**, *3*, 39–47.
67. Kumar, S.; Bhatnagar, T. Studies to enhance the shelf life of fruits using *Aloe vera* based herbal coatings: A review. *Int. J. Agric. Food Sci. Technol.* **2014**, *5*, 211–218.
68. Misir, J.; Brishti, F.H.; Hoque, M.M. *Aloe vera* gel as a novel edible coating for fresh fruits: A review. *Am. J. Food Sci. Technol.* **2014**, *2*, 93–97. [CrossRef]

69. Saks, Y.; Barkai-Golan, R. *Aloe vera* gel activity against plant pathogenic fungi. *Postharvest Biol. Technol.* **1995**, *6*, 159–165. [CrossRef]
70. Jhalegar, J.; Sharma, R.R.; Singh, D. Antifungal efficacy of botanicals against major postharvest pathogens of Kinnow mandarin and their use to maintain postharvest quality. *Fruits* **2014**, *69*, 223–237. [CrossRef]
71. European Union. Approximation of the laws of the member states concerning food additive authorized for use in foodstuffs intended for human consumption (89/107/EC). *Official J.* **1989**, *40*, 27–33.
72. United States Food and Drug Administration. Food additives permitted for direct addition to food for human consumption, subpart c. Coatings, films and related substances. *Code Fed. Reg.* **2000**, *3*, 35–41, 21 CFR 172.
73. Liu, J.; Sui, Y.; Wisniewski, M.; Droby, S.; Liu, Y. Review: Utilization of antagonistic yeasts to manage postharvest fungal diseases of fruit. *Int. J. Food Microbiol.* **2013**, *167*, 153–160. [CrossRef] [PubMed]
74. No, H.K.; Meyers, S.P.; Prinyawiwatkul, W.; Xu, Z. Applications of chitosan for improvement of quality and shelf life of foods: A review. *J. Food Sci.* **2007**, *72*, 87–100. [CrossRef] [PubMed]
75. Hafdani, F.N.; Sadeghinia, N. A review on application of chitosan as a natural antimicrobial. *World Acad. Sci. Eng. Technol.* **2011**, *50*, 252–256.
76. Aider, M. Chitosan application for active bio-based films production and potential in the food industry: Review. *LWT-Food Sci. Technol.* **2010**, *43*, 837–842. [CrossRef]
77. Zeng, K.; Deng, Y.; Ming, J.; Deng, L. Induction of disease resistance and ROS metabolism in navel oranges by chitosan. *Sci. Hortic.* **2010**, *126*, 223–228. [CrossRef]
78. Chien, P.J.; Chou, C.C. Antifungal activity of chitosan and its application to control post-harvest quality and fungal rotting of tankan citrus fruit (*Citrus tankan* Hayata). *J. Sci. Food Agric.* **2006**, *86*, 1964–1969. [CrossRef]
79. Palma-Guerrero, J.; Jansson, H.B.; Salinas, J.; López-Llorca, L.V. Effect of chitosan on hyphal growth and spore germination of plant pathogenic and biocontrol fungi. *J. Appl. Microbiol.* **2008**, *104*, 541–553. [CrossRef] [PubMed]
80. El-Mohamedy, R.S.; El-Gamal, N.G.; Bakeer, A.R.T. Application of chitosan and essential oils as alternatives fungicides to control green and blue moulds of citrus fruits. *Int. J. Curr. Microbiol. Appl. Sci.* **2015**, *4*, 629–643.
81. Shao, X.; Cao, B.; Xu, F.; Xie, S.; Yu, D.; Wang, H. Effect of postharvest application of chitosan combined with clove oil against citrus green mold. *Postharvest Biol. Technol.* **2015**, *99*, 37–43. [CrossRef]
82. Arnon, H.; Zaitsev, Y.; Porat, R.; Poverenov, E. Effects of carboxymethyl cellulose and chitosan bilayer edible coating on postharvest quality of citrus fruit. *Postharvest Biol. Technol.* **2014**, *87*, 21–26. [CrossRef]
83. Deng, L.; Zeng, K.; Zhou, Y.; Huang, Y. Effects of postharvest oligochitosan treatment on anthracnose disease in citrus (*Citrus sinensis* L. Osbeck) fruit. *Eur. Food Res. Technol.* **2014**, *240*, 795–804. [CrossRef]
84. Panebianco, S.; Vitale, A.; Platania, C.; Restuccia, C.; Polizzi, G.; Cirvilleri, G. Postharvest efficacy of resistance inducers for the control of green mold on important Sicilian citrus varieties. *J. Plant Dis. Protect.* **2014**, *121*, 177–183.
85. Cháfer, M.; Sánchez-González, L.; González-Martínez, C.; Chiralt, A. Fungal decay and shelf life of oranges coated with chitosan and bergamot, thyme, and tea tree essential oils. *J. Food Sci.* **2012**, *77*, E182–E187. [CrossRef] [PubMed]
86. Rappussi, M.C.C.; Pascholati, S.F.; Benato, E.A.; Cia, P. Chitosan reduces infection by *Guignardia citricarpa* in postharvest 'Valencia' oranges. *Braz. Arch. Biol. Technol.* **2009**, *52*, 513–521. [CrossRef]
87. Faten, M.A. Combination between citral and chitosan for controlling sour rot disease of lime fruits. *Res. J. Agri. Biol. Sci.* **2010**, *6*, 744–749.
88. Chien, P.J.; Sheu, F.; Lin, H.R. Coating citrus (Murcott tangor) fruit with low molecular weight chitosan increases postharvest quality and shelf life. *Food Chem.* **2007**, *100*, 1160–1164. [CrossRef]
89. Benhamou, N. Potential of the mycoparasite, *Verticillium lecanii*, to protect citrus fruit against *Penicillium digitatum*, the causal agent of green mold: A comparison with the effect of chitosan. *Phytopathology* **2004**, *94*, 693–705. [CrossRef] [PubMed]
90. Galed, G.; Fernández-Valle, M.E.; Martínez, A.; Heras, A. Application of MRI to monitor the process of ripening and decay in citrus treated with chitosan solutions. *J. Magn. Reson. Im.* **2004**, *22*, 127–137. [CrossRef] [PubMed]
91. El-Ghaouth, A.; Smilanick, J.L.; Wilson, C.L. Enhancement of the performance of *Candida saitoana* by the addition of glycolchitosan for the control of postharvest decay of apple and citrus fruit. *Postharvest Biol. Technol.* **2000**, *19*, 103–110. [CrossRef]

92. El-Ghaouth, A.; Smilanick, J.L.; Brown, G.E.; Ippolito, A.; Wisniewski, M.; Wilson, C.L. Application of *Candida saitona* and glycolchitosan for the control of postarvest disease of apple and citrus fruits under semi-commercial conditions. *Plant Dis.* **2000**, *84*, 243–248. [CrossRef]
93. Abbas, H.; Abassi, N.A.; Yasin, T.; Maqbool, M.; Ahmad, T. Influence of irradiated chitosan coating on postharvest quality of kinnow (*Citrus reticulata* Blanco.). *Asian J. Chem.* **2008**, *20*, 6217–6227.
94. Lu, L.; Liu, Y.; Yang, J.; Azat, R.; Yu, T.; Zheng, X. Quaternary chitosan oligomers enhance resistance and biocontrol efficacy of *Rhodosporidium paludigenum* to green mold in satsuma orange. *Carbohydr. Polym.* **2014**, *113*, 174–181. [CrossRef] [PubMed]
95. Waewthongrak, W.; Pisuchpen, S.; Leelasuphakul, W. Effect of *Bacillus subtilis* and chitosan applications on green mold (*Penicilium digitatum* Sacc.) decay in citrus fruit. *Postharvest Biol. Technol.* **2015**, *99*, 44–49. [CrossRef]
96. Palou, L.; Smilanick, J.L.; Droby, S. Alternatives to conventional fungicides for the control of citrus postharvest green and blue molds. *Stewart Postharv. Rev.* **2008**, *2*, 1–16. [CrossRef]
97. Avila-Sosa, R.; Palou, E.; Jiménez Munguía, M.T.; Nevárez-Moorillón, G.V.; Navarro Cruz, A.R.; López-Malo, A. Antifungal activity by vapor contact of essential oils added to amaranth, chitosan, or starch edible films. *Int. J. Food Microbiol.* **2012**, *153*, 66–72. [CrossRef] [PubMed]
98. Sánchez-González, L.; González-Martínez, C.; Chiralt, A.; Cháfer, M. Physical and antimicrobial properties of chitosan-tea tree essential oil composite films. *J. Food Eng.* **2010**, *98*, 443–452. [CrossRef]
99. Valencia-Chamorro, S.A.; Palou, L.; del Río, M.A.; Pérez-Gago, M.B. Inhibition of *Penicillium digitatum* and *Penicillium italicum* by hydroxypropyl methylcellulose-lipid edible composite films containing food additives with antifungal properties. *J. Agric. Food Chem.* **2008**, *56*, 11270–11278. [CrossRef] [PubMed]
100. Valencia-Chamorro, S.A.; Pérez-Gago, M.B.; del Río, M.A.; Palou, L. Curative and preventive activity of hydroxypropyl methylcellulose-lipid edible composite coatings containing antifungal food additives to control citrus postharvest green and blue molds. *J. Agric. Food Chem.* **2009**, *57*, 2770–2777. [CrossRef] [PubMed]
101. Valencia-Chamorro, S.A.; Pérez-Gago, M.B.; del Río, M.A.; Palou, L. Effect of antifungal hydroxypropyl methylcellulose (HPMC)-lipid edible composite coatings on postharvest decay development and quality attributes of cold-stored 'Valencia' oranges. *Postharvest Biol. Technol.* **2009**, *54*, 72–79. [CrossRef]
102. Valencia-Chamorro, S.A.; Pérez-Gago, M.B.; del Río, M.A.; Palou, L. Effect of antifungal hydroxypropyl methylcellulose (HPMC)-lipid edible composite coatings on penicillium decay development and postharvest quality of cold-stored 'Ortanique' mandarins. *J. Food Sci.* **2010**, *75*, 418–426. [CrossRef] [PubMed]
103. Valencia-Chamorro, S.A.; Palou, L.; del Río, M.A.; Pérez-Gago, M.B. Performance of hydroxypropyl methylcellulose (HPMC)-lipid edible composite coatings containing food additives with antifungal properties during cold storage of 'Clemenules' mandarins. *LWT-Food Sci. Technol.* **2011**, *44*, 2342–2348. [CrossRef]
104. Youssef, K.; Ligorio, A.; Nigro, F.; Ippolito, A. Activity of salts incorporated in wax in controlling postharvest diseases of citrus fruit. *Postharvest Biol. Technol.* **2012**, *65*, 39–43. [CrossRef]
105. United States Food and Drug Administration. Substances generally recognized as safe, subpart A. Essential oils, oleoresins (solvent-free), and natural extractives (including distillates). *Code Fed. Reg.* **2000**, *3*, 475–477.
106. Tripathi, P.; Dubey, N.K.; Banerji, R.; Chansouria, J.P.N. Evaluation of some essential oils as botanical fungitoxicants in management of post-harvest rotting of citrus fruits. *World J. Microbiol. Biotech.* **2004**, *20*, 317–321. [CrossRef]
107. Zeng, R.; Zhang, A.; Chen, J.; Fu, Y. Impact of carboxymethyl cellulose coating enriched with extract of *Impatiens balsamina* stems on preservation of 'Newhall' navel orange. *Sci. Hortic.* **2013**, *160*, 44–48. [CrossRef]
108. Velásquez, M.A.; Passaro, C.P.; Lara-Guzmán, O.J.; Álvarez, R.; Londono, J. Effect of an edible, fungistatic coating on the quality of the 'Valencia' orange during storage and marketing. *Acta Hortic.* **2014**, *1016*, 163–169. [CrossRef]
109. du Plooy, W.; Regnier, T.; Combrinck, S. Essential oil amended coatings as alternatives to synthetic fungicides in citrus postharvest management. *Postharvest Biol. Technol.* **2009**, *53*, 117–122. [CrossRef]
110. Pérez-Alfonso, C.O.; Martínez-Romero, D.; Zapata, P.J.; Serrano, M.; Valero, D.; Castillo, S. The effects of essential oils carvacrol and thymol on growth of *Penicillium digitatum* and *P. italicum* involved in lemon decay. *Int. J. Food Microbiol.* **2012**, *158*, 101–106. [CrossRef] [PubMed]

111. Castillo, S.; Pérez-Alfonso, C.O.; Martínez-Romero, D.; Guillén, F.; Serrano, M.; Valero, D. The essential oils thymol and carvacrol applied in the packing lines avoid lemon spoilage and maintain quality during storage. *Food Control* **2014**, *35*, 132–136. [CrossRef]
112. Fan, F.; Tao, N.; Jia, L.; He, X. Use of citral incorporated in postharvest wax of citrus fruit as a botanical fungicide against *Penicillium digitatum*. *Postharvest Biol. Technol.* **2014**, *90*, 52–55. [CrossRef]
113. Tao, N.; Fan, F.; Jia, L.; Zhang, M. Octanal incorporated in postharvest wax of Satsuma mandarin fruit as a botanical fungicide against *Penicillium digitatum*. *Food Control* **2014**, *45*, 56–61. [CrossRef]
114. Kouassi, K.H.S.; Bajji, M.; Jijakli, H. The control of postharvest blue and green molds of citrus in relation with essential oil-wax formulations, adherence and viscosity. *Postharvest Biol. Technol.* **2012**, *73*, 122–128. [CrossRef]
115. Regnier, T.; Combrinck, S.; Veldman, W.; Du Plooy, W. Application of essential oils as multi-target fungicides for the control of *Geotrichum citri-aurantii* and other postharvest pathogens of citrus. *Ind. Crop Prod.* **2014**, *61*, 151–159. [CrossRef]
116. McGuire, R.G.; Baldwin, E. Composition of cellulose coatings affect population of yeasts in the liquid formulation and on coated grapefruits. *Proc. Fla. State Hort. Soc.* **1994**, *107*, 293–297.
117. McGuire, R.G.; Hagenmaier, R. Shellac coating for grapefruits that favor biological control of *Penicillium digitatum* by *Candida oleophila*. *Biol. Control* **1996**, *7*, 100–106. [CrossRef]
118. McGuire, R.G.; Dimitroglou, D. Evaluation of shellac and sucrose ester fruit coating formulation that support biological control of post-harvest grapefruit decay. *Biocontrol Sci. Technol.* **1999**, *9*, 53–65. [CrossRef]
119. McGuire, R.G. Application of *Candida guilliermondii* in commercial citrus coatings for biocontrol of *Penicillium digitatum* on grapefruits. *Biol. Control* **1994**, *4*, 1–7. [CrossRef]
120. Potjewijd, R.; Nisperos-Carriedo, M.O.; Burns, J.K.; Parish, M.; Baldwin, E.A. Cellulose-based coatings as carriers for *Candida guilliermondii* and *Debaryomyces* sp. in reducing decay of oranges. *HortScience* **1995**, *30*, 1417–1421.
121. McGuire, R.G. Population dynamics of postharvest decay antagonist growing epiphytically and within wounds on grapefruit. *Phytopathology* **2000**, *90*, 1217–1223. [CrossRef] [PubMed]
122. Aloui, H.; Licciardello, F.; Khwaldia, K.; Hamdi, M.; Restuccia, C. Physical properties and antifungal activity of bioactive films containing *Wickerhamomyces anomalus* killer yeast and their application for preservation of oranges and control of postharvest green mold caused by *Penicillium digitatum*. *Int. J. Food Microbiol.* **2015**, *200*, 22–30. [CrossRef] [PubMed]

© 2015 by the authors. Licensee MDPI, Basel, Switzerland. This article is an open access article distributed under the terms and conditions of the Creative Commons Attribution (CC BY) license (http://creativecommons.org/licenses/by/4.0/).

Article

Chitosan-Based Coating Enriched with Hairy Fig (*Ficus hirta* Vahl.) Fruit Extract for "Newhall" Navel Orange Preservation

Chuying Chen [1], Nan Cai [1], Jinyin Chen [1,2,*], Xuan Peng [2] and Chunpeng Wan [1,*]

[1] Jiangxi Key Laboratory for Postharvest Technology and Nondestructive Testing of Fruits & Vegetables, Collaborative Innovation Center of Postharvest Key Technology and Quality Safety of Fruits and Vegetables, Jiangxi Agricultural University, Nanchang 330045, China; cy.chen@jxau.edu.cn (C.C.); wq1252733770@163.com (N.C.)
[2] Pingxiang University, Pingxiang 337055, China; pengx1104@163.com
* Correspondence: jinyinchen@126.com (J.C.); chunpengwan@jxau.edu.cn (C.W.); Tel.: +86-791-8381-3158 (J.C. & C.W.)

Received: 25 October 2018; Accepted: 4 December 2018; Published: 4 December 2018

Abstract: A novel coating based on 1.5% chitosan (CH), enriched with or without hairy fig (*Ficus hirta* Vahl.) fruit extract (HFE), was applied to "Newhall" navel orange for improving the preservation effect. Changes in physicochemical indexes were analyzed over 120 days of cold storage. Uncoated fruit were used as the control. The CH-HFE coating, based on 1.5% CH enriched with HFE, exhibited the best preservation effect and showed the lowest decay rate (5.2%) and weight loss (5.16%). The CH-HFE coating could postpone the ripening and senescence of navel oranges, and maintain higher fruit quality by inhibiting respiration, decreasing the accumulation of malondialdehyde (MDA), and enhancing the activities of protective enzyme, including superoxide dismutase (SOD), peroxidase (POD), chitinase (CHI), and β-1,3-glucanase (GLU), which suggests that CH-HFE coating has the potential to improve the postharvest quality of "Newhall" navel orange and prolong the storage life.

Keywords: chitosan; edible coating; hairy fig fruits; navel oranges; physicochemical responses

---

1. Introduction

"Newhall" navel orange (*Citrus sinensis* L., Osbeck) fruits enjoy great popularity in China and around the world, for their good taste, abundant vitamin, C and widespread availability [1]. However, there are serious problems restricting the sound development of citrus production. Postharvest losses, frequent decay caused by a variety of plant pathogenic fungi, and a lack of scientific management, are issues of concern that need to be solved; of these, postharvest decay is the biggest hurdle that needs urgent resolution.

Pathogen infection is an important factor that affects citrus fruit postharvest physiology, disease resistance, and metabolism. Blue and green mold of citrus, caused by *Penicillium italicum* and *Penicillium digitatum*, respectively, are the most economically important postharvest diseases of citrus, and cause heavy losses during storage, transportation, and marketing, thus debasing the commodity value of harvested fruits [2]. The two *Penicillium* molds may cause 60%–80% decay losses under ambient conditions [3], which leads to severe economic losses for exporting countries. At present, the primary means for controlling of the two *Penicillium* molds still relies mainly on the use of chemical fungicides, especially imazalil (IMZ), prochloraz, thiabendazole (TBZ), calcium polysulfide pyrimethanil, or different mixtures of these compounds [4–7]. Increasing public concern of chemical residues on human health and environmental pollution, due to excessive use of chemical fungicides, have prompted investigation of alternative strategies for reducing postharvest decay and maintaining citrus fruit postharvest quality, without any human, environmental, or plant toxicity [8–10]. A variety of plant-derived compounds have been recognized and generally regarded as safe (GRAS) substances for their antifungal activities, and are being used for controlling postharvest fungal rotting of fresh horticultural products [11–13]. In recent years, numerous researchers have documented the antifungal effects of plant extracts or essential oils for reducing postharvest disease development caused by pathogenic fungus, leading to heavy losses and serious deterioration of citrus fruits [14–17].

The hairy fig is a deciduous plant widely distributed in southern China as a traditional plant resource used as medicinal and edible food by Hakka people. It is a clearly recognizable plant for its five-fingered leaf shape and mature fruit that resembles wild peach (Figure 1). Moreover, the fruits of hairy fig are a famous herb used by Hakka people in Chinese folk medicine for inhibiting tumor growth, promoting lactation, as anticoagulant, and for improving fatigue resistance [18,19]. Currently, our previous research demonstrated that HFE has strong antifungal activity against *P. italicum* and *P. digitatum* in vitro conditions [20,21]. Not surprisingly, chitosan coating enriched with HFE provided an enhanced antifungal activity, and it is likely that research will broaden the practicability of the botanical fungicide. Development of antimicrobial/antioxidant coatings from polysaccharides, such as chitosan, have been studied extensively, whereas few works have been conducted on developing films using the combination of chitosan and natural plant extract-based antifungal components. The purpose of this research was to evaluate the effect of chitosan coating enriched with or without HFE for cold-stored navel orange preservation.

Figure 1. The (**A**) leaves and (**B**) fruits of hairy fig (*Ficus hirta* Vahl.).

2. Materials and Methods

2.1. Materials

Fruits of navel oranges (*Citrus Sinensis* L. Osbeck cv. Newhall), used throughout this study, were harvested at the peak of the harvesting season from an orchard situated in the southeast of Ganzhou city (Jiangxi, China). The fruits were picked on the basis of consistent size (240–280 g) and uniform color (citrus color index, 3.5–4.8). Fruits with any mechanical injury, blemish, or diseases were discarded.

2.2. Extraction of HFE

The fruits of hairy fig (origin: Guangdong, China) were purchased from Huafeng herbs store in Zhangshu (Jiangxi, China) and powdered in a grinder (less than 20 mesh) after drying below 40 °C. The HFE was obtained using an ultrasonic-assisted method, as described below [21]. The air-dried mixture of 100 g powder sample was suspended in 3.1 L 90% ethanol (v/v) at 51 °C with ultrasonic-assisted extraction (40 kHz) for 65 min. The HFE was filtered and concentrated by vacuum distillation at 45 °C, using a Buchi rotary evaporator (R210, Buchi, Labortechnik AG, Flawil, Switzerland). The remaining solution was dissolved in sterile water and made up to 100 mL at a concentration of 1 g mL^{-1} (raw herb/solvent, w/v), then stored at 4 °C for further use.

2.3. Preparation of CH-HFE and Chitosan Coatings

Chitosan solution (1.5%, w/v) was prepared by dissolving 15.0 g of chitosan (the degree of deacetylation of 90%, Sinopharm Chemical Reagent Co., Ltd., Shanghai, China) in 800 mL of acetic acid solution (0.5%, v/v). Crude HFE extract (35 mL) was put into the coating, with agitation for 1 h. The pH of the solution was adjusted with 1.0 M NaOH to pH 5.4, and the total volume of the solution was made up to 1000 mL. Chitosan coating of the same concentration (1.5%, w/v) was prepared in the same way.

2.4. Navel Orange Treatment and Storage

The selected fruits were washed with tap water and air-dried at room temperature (25 ± 1 °C), then coated by dipping in chitosan coating and/or HFE for 1 min (CH-HFE and 1.5% chitosan coating, respectively), while the control group was dipped in 0.5% acetic acid solution (pH 5.4). After drying, the coated, as well as control, fruits were individually film (18 cm × 15 cm, Lingqu fresh packaging products Co. Ltd., Guilin, China)-packaged and pre-cooled (10–12 °C, 12 h). Finally, all fruits were stored at 5 ± 0.5 °C, and 85%–90% relative humidity (RH) for 120 days. The procedures for the coating and control group were performed three times of total 560 navel oranges per treatment. At each sampling point (15, 30, 45, 60, 75, 90, 105, and 120 days), each replicates of 10 navel oranges were randomly picked out from the coating and control group for analyzing total soluble solid (TSS) content, titratable acid (TA) content, vitamin C (VC) content, total sugar content, respiration rate, MDA content, as well as protective enzyme activities.

2.5. Measurement of Physicochemical Indexes

Fruit decay rate was visually evaluated using the same 80 navel oranges per treatment per replicate, during the storage period of semi-monthly, and expressed as the percentage of rotted fruits. Navel oranges with apparent disease spots were considered to be decayed. Weight loss was measured by weighing the same 20 fruits during storage every 15 days, and the data are means of 20 samples $\pm$ SE.

TSS, TA, VC, and total sugar were analyzed after completely mixing the orange juice from 10 fruits in the coated and control groups. TSS was measured using a V RA-250 WE digital brix-meter (KYOTO, Tokyo, Japan) and expressed as a percentage. TA and VC were determined by titration with 0.1 M NaOH, and 2,6-dichlorophenol indophenol, respectively. Total sugar content was measured using the anthrone colorimetric method [22].

2.6. Assay of Respiration Rate and MDA Content

The respiration rate was determined based on a method described by our previous study [23]. Six fruits, per treatment per replicate, were weighed before being sealed in an airtight plastic container (internal diameter of 27.5 cm, 30.0 cm high) at 25 °C. The increased CO_2 concentration in the container was monitored by using a GHX-3051H infrared CO_2 fruit and vegetable breathing apparatus (Jingmi Scientific LLC., Shanghai, China). Respiration rate, measured by CO_2 production, was expressed as mg kg^{-1} h^{-1}.

The MDA content in coating and control groups was measured according to the method of Hodges et al. [24]. Pericarp tissues from 10 fruits were ground in a MM 400 frozen grinder (Retsch GmbH., Arzberg, Germany), and 2.0 g of powder were homogenized in 25 mL of ice-cold 50 mM phosphate buffer (pH 7.8) containing 1 mM EDTA and 2% (w/v) PVP, and centrifuged at 12,000 g (5804R, Eppendorf, Hamburg, Germany) for 20 min at 4 °C. Afterwards, 2 mL of the collected supernatant was mixed with 2 mL of 0.5% (w/v) thiobarbituric acid (TBA), and further incubated in boiling water for 30 min. After being cooled and centrifuged at 6000 g (5804R, Eppendorf) for 10 min, the absorbance of supernatant was measured at three different wavelengths (450, 532, and 600 nm) using a M5 Multiscan Spectrum microplate reader (Molecular Devices Corporation, Sunnyvale, CA, USA). The MDA content was calculated according to the formula $(6.452 \times (A_{532} - A_{600}) - 0.559 \times A_{450})$, and expressed as mmol g^{-1} frozen weight (FW).

2.7. Determination of Protective Enzymes Activities

Aliquots of powder (2.0 g) were homogenized with various ice-cold extraction buffers to prepare extracts for assay of the following protective enzymes: 10 mL of 50 mM ice-cold phosphate buffer (pH 7.8) containing 1 mM EDTA, 5 mM DTT, and 2% (w/v) PVP for superoxide dismutase (SOD, EC 1.15.1.1); 8 mL of 100 mM ice-cold phosphate buffer (pH 7.5) containing 1 mM polyethylene glycol (PEG), 4% (w/v) PVP, and 1% (w/v) Triton X-100 for peroxidase (POD, EC 1.11.1.7); 10 mL of 100 mM ice-cold sodium acetate buffer (pH 5.2) containing 5 mM β-mercaptoethanol, 1 mM PEG, 1 mM EDTA, 4% (w/v) PVP, and 0.5% (w/v) Triton X-100 for chitinase (CHI, EC 3.2.1.14) and β-1,3-glucanase (GLU, EC 3.2.1.73). All homogenates were centrifuged at 12,000 g (5804R, Eppendorf) for 30 min at 4 °C. The supernatants were then collected and used for the enzyme activity assays.

SOD activity was assayed by measuring its ability to inhibit the photoreduction of nitroblue tetrazolium (NBT) according to the method of Sala and Lafuente, with slight modifications [25]. The reaction mixture consisted of 1.5 mL PBS (50 mM), 0.3 mL Met (130 mM), 0.3 mL NBT (0.75 mM), 0.3 mL EDTA-Na$_2$ (0.1 mM), 0.3 mL riboflavin (20 μM), 0.1 mL enzyme extract, and 0.5 mL distilled water in a total volume of 3.3 mL. The mixtures were illuminated by light (4000 Lx) for 20 min at 28 °C, and the absorbance was then determined at 560 nm (UV-1800, Shimadzu, Tokyo, Japan). One unit of SOD activity was defined as the amount of enzyme that would inhibit 50% of NBT photoreduction, and expressed as U min^{-1} g^{-1} FW.

POD activity was based on the determination of guaiacol oxidation at 470 nm in the presence of H$_2$O$_2$. Collected supernatant (100 μL) was mixed with 3.0 mL of 25 mM guaiacol and 200 μL of 50 mM H$_2$O$_2$. Oxidation of guaiacol was determined at 470 nm for 3 min at 25 °C. One unit of POD activity was defined as an increment of 0.01 in absorbance per minute at 470 nm, and expressed as U min^{-1} g^{-1} FW.

CHI and GLU activities were assayed by the method described by Abeles et al. [26] using chitinase and laminarin as substrate. One unit of CHI activity was defined as the amount of enzyme that catalyzed the produce of 1 nmol of N-acetyl-D-glucosamine per hour at 585 nm, and expressed as U h^{-1} g^{-1} FW. One unit of GLU activity was defined as the amount of enzyme that produced a reducing sugar equivalent to 1 nmol glucose equivalents per hour at 540 nm, and expressed as U h^{-1} g^{-1} FW.

2.8. Statistical Analysis

Three biological replicates per treatments were done, and the effect of CH-HFE coating on "Newhall" navel orange preservation was analyzed using variance analysis ($p < 0.05$). Duncan's multiple range test was used to determine the mean differences. The data are represented as the mean with standard error (SE); these were calculated from physical and chemical experiments, which were performed in three replications.

3. Results and Discussion

3.1. Decay Rate and Weight Loss

Decay rate is a distinctly important indicator that affects the preservation effect of horticultural produces. As shown in Figure 2A, CH-HFE and 1.5% chitosan coatings delayed the appearance of fungal infection in comparison to uncoated group, which started to decay from 60 days of storage. Many of the uncoated orange fruits (10.4%) were rotten at the termination of storage, while fruits treated by CH-HFE and 1.5% chitosan coating exhibited significantly lower decay rate than the control group at the level of $p < 0.05$ (5.2% and 6.8%, respectively). As the data in Figure 2B show, for the whole time of experiment, the percentage of weight loss in control group was higher than in both coating treatments, with high weight loss (9.12%) of uncoated orange fruits at the end of storage. However, orange fruits treated with CH-HFE coating displayed the lowest weight loss at 5.16%, which was much lower than that of 1.5% chitosan and control fruits throughout the 120-day storage period.

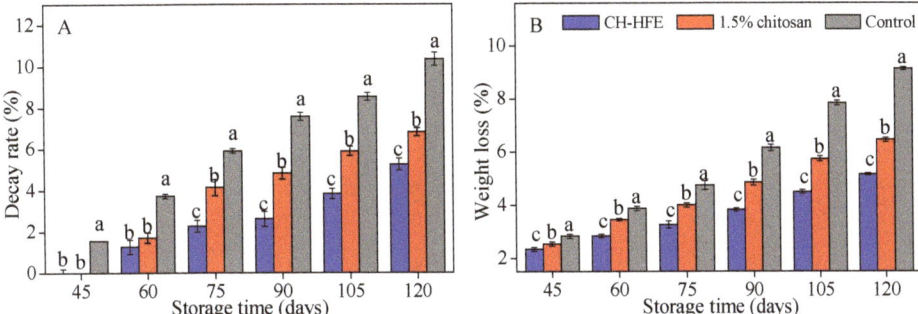

Figure 2. Changes in decay rate (**A**) and weight loss (**B**) of navel orange fruits stored at 5 °C for 120 days. Bars indicate standard error of three replicates.

3.2. Fruit Quality

TSS, TA, total sugar, and VC were the key determinants of citrus fruit quality. Figure 3 showed that the content of TSS, TA, total sugar, and VC varied within coated and uncoated groups as storage progressed. As illustrated in Figure 3A,C, the content of TSS and total sugar increased continuously during the early stage of storage, and decreased slightly in the subsequent storage period. The content of TSS and total sugar in the control group reached the highest level at 30 days, while the content of TSS and total sugar in coating groups reached their peaks at 60 days, and were delayed for 30 days more than the control group. During the late 60 days of storage stage, the content of TSS and total sugar in coated fruits was significantly higher ($p < 0.05$) than the control group. The TA content of all samples fell greatly after 120 days of storage (Figure 3B), and the value was significantly higher ($p < 0.05$) in CH-HFE-coated treated fruit, compared to 1.5% chitosan coating and the control group. During the storage, the control group showed a greater TA loss. The low level of TA in the control group, relative to coating treatments, suggested that the coating delayed ripening by providing a semi-permeable film around the fruits. The CH-HFE and 1.5% chitosan-coated fruits showed a slight increase in VC content during the first 30 days, reached their peak concentration (56.87 mg/100 g and 54.04 mg/100 g), followed by a decline in the subsequent storage period, while the control group presented a gradual decrease in VC content during the whole storage (Figure 3D). Under the storage conditions, the VC content in CH-HFE coated fruits was significantly higher ($p < 0.05$) than 1.5% chitosan-coated and control group fruits.

3.3. Respiration Rate and MDA Content

As illustrated in Figure 4A, the respiration rate decreased gradually, and reached its valley value at 60 days after storage, and recovered in the subsequent storage period. The valley value in coating-treated fruits was significantly lower ($p < 0.05$) than that in control group. During storage, the respiration rate of CH-HFE-coated fruits was significantly lower ($p < 0.05$) than those with 1.5% chitosan coating and the control group.

MDA is the final product of lipid peroxidation, and its content has been used as one of the direct indexes of cell oxidative damage [27]. As shown in Figure 4B, MDA content increased during storage, and a significant difference was shown between the coating treatments and the control group. At the end of the storage (120 days), the MDA content of CH-HFE, 1.5% chitosan, and control groups reached their maximum (2.47, 2.85, and 3.25 mmol g^{-1}, respectively). The MDA content of control fruits was significantly higher than that of CH-HFE and 1.5% chitosan coatings, 23.7% and 12.3%, respectively.

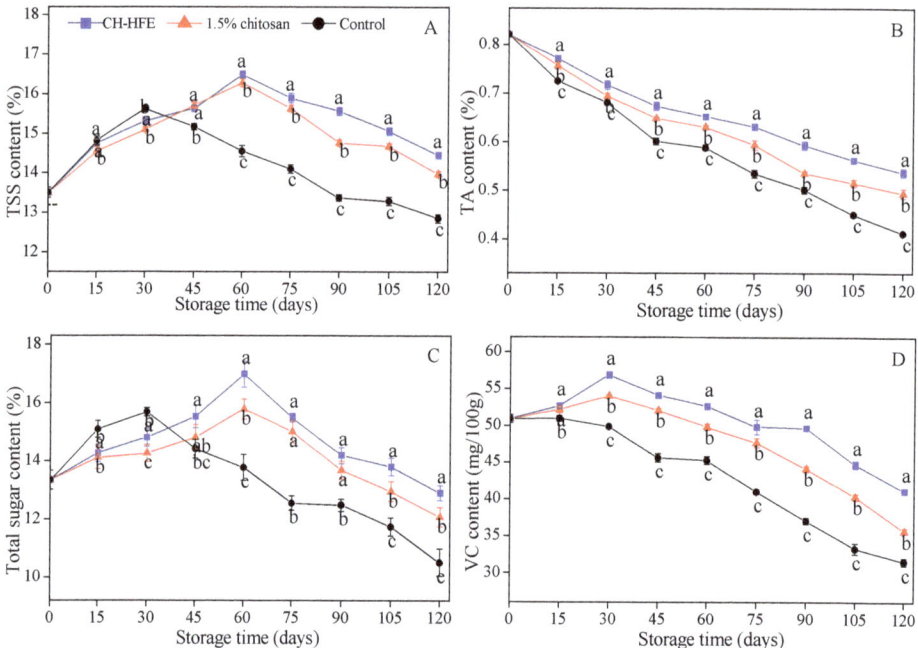

Figure 3. Changes in total soluble solid (TSS) (**A**), titratable acid (TA) (**B**), total sugar (**C**), and vitamin C (VC) content (**D**) of navel orange fruits stored at 5 °C for 120 days. Each value represents the mean ± SE of three replicates, each consisting of 30 randomly selected fruits.

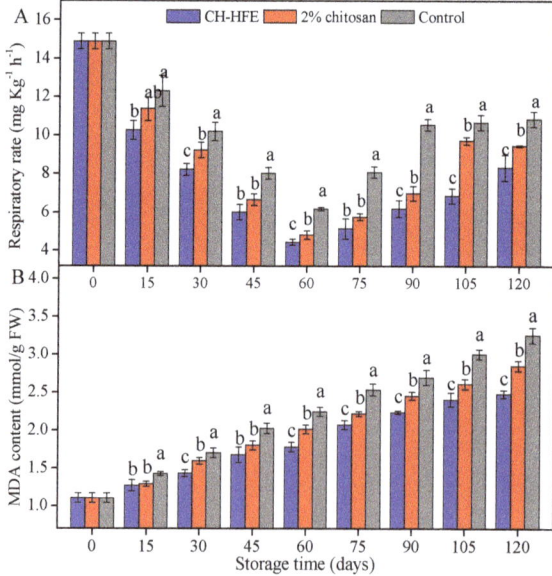

Figure 4. Changes in respiration rate (**A**) and malondialdehyde (MDA) content (**B**) of navel orange fruits stored at 5 °C for 120 days. Bars indicate standard error of three replicates, each consisting of 30 randomly selected fruits.

3.4. Activities of Protective Enzymes

SOD, POD, CHI, and GLU were the main protective enzymes to alleviate lipid peroxidation and delay fruit senescence. Figure 5 showed the activities of SOD, POD, CHI, and GLU varied within coated and uncoated groups, as storage progressed. As illustrated in Figure 5A,B, the activities of SOD and POD gradually increased to reach their peak, and then dropped quickly for the following day. The activities of SOD and POD in the control group reached peak values at 30 days and 60 days, while peak values of SOD and POD activities in coating groups were significantly higher ($p < 0.05$) than the control group, and delayed for 30 days. Moreover, SOD and POD activities in the CH-HFE coated fruits was significantly higher ($p < 0.05$) than those with 1.5% chitosan coating and the control group, during the later stage of storage. The CHI activity decreased slightly during the first 30 days, and then increased notably at the end of storage period (Figure 5C). The coated fruits had a relatively high CHI activity, and the average activity of CHI of the CH-HFE and 1.5% chitosan-coated treatment increased to approximately 1.30 and 1.19 times higher than that in the control group. The GLU activity of all samples fell greatly after 120 days of storage (Figure 5D), and the value was significantly higher ($p < 0.05$) in CH-HFE-coated treated fruit compared to 1.5% chitosan coating and the control group.

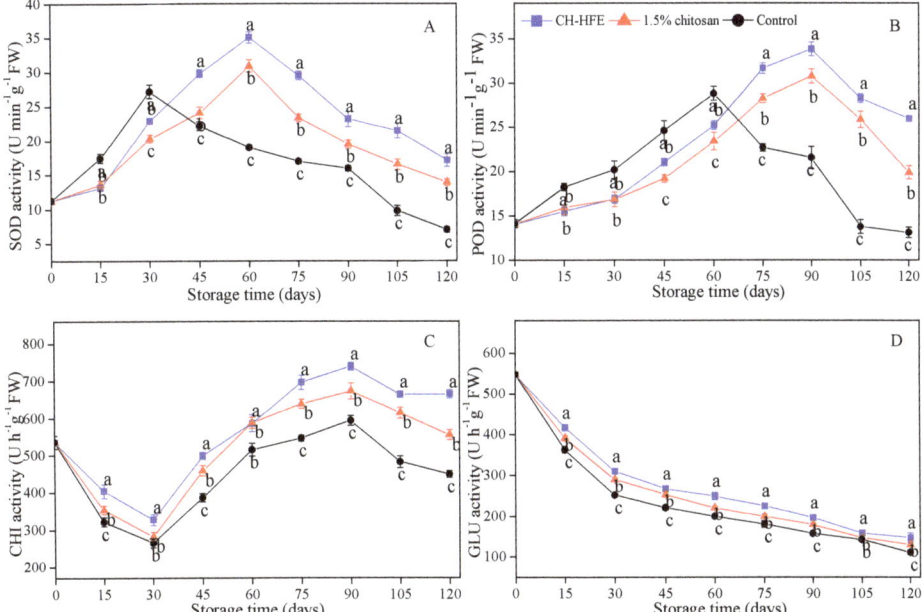

Figure 5. Changes in the activities of SOD (**A**), POD (**B**), CHI (**C**), and GLU (**D**) of navel orange fruits stored at 5 °C for 120 days. Each value represents the mean ± SE of three replicates selected randomly from 30 fruits each treatment.

4. Discussion

As a medicinal plant and food, hairy fig (*Ficus hirta* Vahl.) fruits have strong antifungal, antioxidant, and antiseptic effects [28,29]. Our previous studies have been demonstrated that HFE was able to significantly inhibit both *P. italicum* and *P. digitatum* in vitro [21,30]. In the present work, we found that HFE had an abundance of antioxidants and more effective DPPH, OH, ABTS radical scavenging activities, and Fe^{3+} reducing power than ascorbic acid in different solvent extracts (data not shown). A similar result has been reported where the roots of *Ficus hirta* Vahl. showed prominent anti-inflammatory activity [31]. A recent in vitro test has shown that HFE had a broad antifungal spectrum and high efficacy in controlling fungal growth of fungal pathogens, such as *P. italicum*, *P. digitatum*, and *Geotrichum citri-aurantii* and *Alternaria citri* in citrus, *Botryosphaeria dothidea*, *Phomopsis* sp., and *Botrytis cinerea* in kiwifruits, *Alternaria alternata* in pears, and *Phomopsis vexans* in eggplants [21]. HFE has been shown to be a natural and safe fungistat when applying to control blue mold in citrus, and incorporated into alginate-based edible coatings to enhance Nanfeng mandarin preservation [28]. According to these results, there is no doubt that HFE can be safely used as a natural preservative for horticultural products.

Decay rate and weight loss are the main distinctly important indicators that affect the preservation effect of fresh fruit and vegetables, mainly due to postharvest diseases by pathogen infection, and the loss of water caused by transpiration [32,33]. Edible coatings provide a semi-permeable film around fruit, reducing disease incidence and depressing the rate of dehydration and respiration [28,34]. Coating "Newhall" navel oranges with CH-HFE is clearly effective in conferring a beneficial barrier to pathogen infection and water loss; thus, the decreased rate of fruit decay and weight loss in CH-HFE-coated fruit was evaluated during cold storage in this study. Our results are supported by Shah et al. [35], where fruit rotting intensity and weight loss of Kinnow mandarin can be reduced by coating with carboxymethyl cellulose (CMC) or guar gum-based coatings enriched with silver nanoparticles. Apart from citrus fruit, chitosan-based coatings containing natural antifungal agents, such as plant extracts or essential oils, have been effective at controlling the rate of fruit decay and weight loss from other horticultural products, including strawberry [36–39], guava [40,41], grape [42–44], tomato, and cucumber [45–47].

In general, TSS and TA, known as a respiration substrate, are continually consumed, and affect firsthand the taste flavor of navel oranges. Meanwhile, VC is a nutritional component of citrus fruits, and is also an important antioxidant which scavenges active oxygen. Due to the post-maturing action of citrus fruit, the fruit quality of harvested "Newhall" navel oranges, after a certain period of storage time, was improved, and the content of nutritional components were also increased. As the film acts as a gas barrier, the respiration process slows down the consumption of nutritional components, thus, coating treatment has a positive effect on the increase or maintenance of VC and other components [33]. In the present study, we found that the changes of CH-HFE and 1.5% chitosan coatings on nutritional components, such as TSS, TA, and total sugar of "Newhall" navel oranges are probably due to the slowing down of respiration, hence effectively delaying fruit ripening and senescence. It is shown by us, that CH-HFE and 1.5% chitosan coatings result in a beneficial semi-permeable film surrounding the fruit, modifying the internal atmosphere by elevating CO_2 and/or reducing O_2, and delaying the degradation rate of nutritional components in navel orange fruits during the mid–later storage period. Our results are in accordance with those of Chen et al. [23] and Shah et al. [35], where a slow decline in TSS, TA, and VC was recorded in citrus fruits treated with CMC composite coatings. However, other studies have demonstrated that the levels of total soluble solids and acidity, and flavor preferences in "Navel" oranges and "Star Ruby" grapefruits treated with CMC/chitosan bilayer coating, were the same as in the uncoated fruits [48].

Coated "Newhall" oranges generally had lower respiration rate than the uncoated fruits (Figure 5A), likely due to the modification of internal atmosphere by elevating CO_2 and/or reducing O_2 [28]. Similar observations have been reported by Arnon et al. [48], Chien et al. [49], Contreras-Oliva et al. [50], and Xu et al. [51] in citrus fruits coated with chitosan-based edible coatings. In our research, the incorporation of HFE into 1.5% chitosan coating formulation resulted in a significantly lower respiration rate than 1.5% chitosan coating alone, and uncoated fruits, during the whole storage. Similarly, the respiration rate of citrus fruits coated with CMC edible coatings containing clove oil, *Impatiens balsamina* L. extract, and silver nanoparticles, has been shown to decrease, compared to uncoated fruits [23,35,52].

In this present study, the activities of protective enzymes, such as SOD, POD, CHI, and GLU, are enhanced by CH-HFE coating, which suggests that CH-HFE coating induces natural resistance in "Newhall" orange fruit by activating the fruit's antioxidant and defense-related system to control the level of reactive oxygen species (ROS) and delay fruit senescence (Figure 5A–D). Chen et al. [21,28] and Zeng et al. [18] have also found that *Ficus hirta* Vahl. extracts have a powerful antifungal activity against *Penicillium* spp. and abundant antioxidants, as well as inhibitory efficacy for HeLa cells. Chen and co-authors also maintain that an alginate-based edible coating containing HFE can significantly increase the activities of antioxidant and defense-related enzymes and enhance the preservation effect of Nanfeng mandarin [28]. An advantage of the activities of protective enzymes has been reported, previously, for navel oranges treated with a 1.5% CMC coating enriched with *Impatiens balsamina* L. extract, strawberry treated with a 1.0% chitosan coating containing lemon essential oil, guava treated with 2.0% sodium alginate and 1.0% chitosan coatings enriched with pomegranate peel extract, and cherry tomato treated with a 1.0% chitosan coating incorporating grapefruit seed extract [37,40,46,52]. Thus, our results may imply that these different protective enzymes in the antioxidant and defense-related system may be collectively induced by CH-HFE coating in navel orange fruits. The induced activities of protective enzymes may be a crucial part of the mechanism of CH-HFE coating in enhancing disease resistance and prolonging storage-life for navel orange preservation.

5. Conclusions

Edible coatings containing natural antifungal agents, such as plant extracts and essential oils, are being used widely to enhance the preservation effect of horticultural products by prolonging their storage life. In this study, our work indicates that the application of plant extracts, such as hairy fig fruit extract (HFE) with chitosan, could enhance the postharvest quality of navel orange. A combination of HFE into chitosan-based coatings not only resulted in a lower percentage of rotting fruits and weight loss, but also in maintenance of good fruit quality of navel orange, by depressing the respiration rate and, thereby, delaying fruit senescence. Moreover, protective enzymes, including SOD, POD, CHI, and GLU in CH-HFE coating treatment, had obviously higher than those activities of the control group or 1.5% chitosan coating alone. According to the results from this study, CH-HFE coating treatment showed the best preservation effect for harvested "Newhall" navel orange. The probable preservation mechanism of CH-HFE coating of navel oranges is mainly through reducing the percentage of fruit decay, weight loss, and MDA content, depressing the respiration rate, and maintaining a high level of fruit quality and protective enzymes. In short, our work indicates that CH-HFE coating has become a natural and safe alternative treatment for enhancing postharvest quality and prolonging the storage life of navel orange.

Author Contributions: Conceptualization, J.C. and W.C.; Methodology, C.C. and X.P.; Data Analysis, C.C. and Z.N.; Writing—Original Draft Preparation, C.C.; Writing—Review and Editing, W.C.; Project Administration, J.C.; Funding Acquisition, J.C., C.C. and W.C.

Funding: This research was funded by National Natural Science Foundation of China (No. 31760598), and Natural Science Foundation and Advantage Innovation Team Project of Jiangxi Province (No. 20171BAB214031 and No. 20181BCB24005).

Conflicts of Interest: The authors declare no conflict of interest.

References

1. Palma, A.; D'Aquino, S.; Vanadia, S.; Angioni, A.; Schirra, M. Cold quarantine responses of 'Tarocco' oranges to short hot water and thiabendazole postharvest dip treatments. *Postharvest Biol. Technol.* **2013**, *78*, 24–33. [CrossRef]
2. Zeng, K.; Deng, Y.; Ming, J.; Deng, L. Induction of disease resistance and ROS metabolism in navel oranges by chitosan. *Sci. Hortic.* **2010**, *126*, 223–228. [CrossRef]
3. Moscoso-Ramírez, P.A.; Montesinos-Herrero, C.; Palou, L. Control of citrus postharvest *Penicillium* molds with sodium ethylparaben. *Crop Prot.* **2013**, *46*, 44–51. [CrossRef]

4. Montesinos-Herrero, C.; Smilanick, J.L.; Tebbets, J.S.; Walse, S.; Palou, L. Control of citrus postharvest decay by ammonia gas fumigation and its influence on the efficacy of the fungicide imazalil. *Postharvest Biol. Technol.* **2011**, *59*, 85–93. [CrossRef]
5. Smilanick, J.L.; Sorenson, D. Control of postharvest decay of citrus fruit with calcium polysulfide. *Postharvest Biol. Technol.* **2001**, *21*, 157–168. [CrossRef]
6. Hao, W.; Zhong, G.; Hu, M.; Luo, J.; Weng, Q.; Rizwan-ul-Haq, M. Control of citrus postharvest green and blue mold and sour rot by tea saponin combined with imazalil and prochloraz. *Postharvest Biol. Technol.* **2010**, *56*, 39–43. [CrossRef]
7. D'Aquino, S.; Fadda, A.; Barberis, A.; Palma, A.; Angioni, A.; Schirra, M. Combined effects of potassium sorbate, hot water and thiabendazole against green mould of citrus fruit and residue levels. *Food Chem.* **2013**, *141*, 858–864. [CrossRef] [PubMed]
8. Talibi, I.; Boubaker, H.; Boudyach, E.; Ait Ben Aoumar, A. Alternative methods for the control of postharvest citrus diseases. *J. Appl. Microbiol.* **2014**, *117*, 1–17. [CrossRef]
9. Da Cruz Cabral, L.; Fernández Pinto, V.; Patriarca, A. Application of plant derived compounds to control fungal spoilage and mycotoxin production in foods. *Int. J. Food Microbiol.* **2013**, *166*, 1–14. [CrossRef]
10. Karim, H.; Boubaker, H.; Askarne, L.; Cherifi, K.; Lakhtar, H.; Msanda, F.; Boudyach, E.H.; Ait Ben Aoumar, A. Use of Cistus aqueous extracts as botanical fungicides in the control of Citrus sour rot. *Microb. Pathog.* **2017**, *104*, 263–267. [CrossRef]
11. Palou, L.; Ali, A.; Fallik, E.; Romanazzi, G. GRAS, plant- and animal-derived compounds as alternatives to conventional fungicides for the control of postharvest diseases of fresh horticultural produce. *Postharvest Biol. Technol.* **2016**, *122*, 41–52. [CrossRef]
12. Tripathi, P.; Dubey, N.K. Exploitation of natural products as an alternative strategy to control postharvest fungal rotting of fruit and vegetables. *Postharvest Biol. Technol.* **2004**, *32*, 235–245. [CrossRef]
13. Camele, I.; Altieri, L.; De Martino, L.; De Feo, V.; Mancini, E.; Rana, G.L. In vitro control of post-harvest fruit rot fungi by some plant essential oil components. *Int. J. Mol. Sci.* **2012**, *13*, 2290–2300. [CrossRef] [PubMed]
14. Li Destri Nicosia, M.G.; Pangallo, S.; Raphael, G.; Romeo, F.V.; Strano, M.C.; Rapisarda, P.; Droby, S.; Schena, L. Control of postharvest fungal rots on citrus fruit and sweet cherries using a pomegranate peel extract. *Postharvest Biol. Technol.* **2016**, *114*, 54–61. [CrossRef]
15. Vitoratos, A.; Bilalis, D.; Karkanis, A.; Efthimiadou, A. Antifungal Activity of Plant Essential Oils Against *Botrytis cinerea*, *Penicillium italicum* and *Penicillium digitatum*. *Not. Bot. Horti Agrobot.* **2013**, *41*, 86–92. [CrossRef]
16. Jhalegar, M.J.; Sharma, R.R.; Singh, D. In vitro and in vivo activity of essential oils against major postharvest pathogens of Kinnow (*Citrus nobilis* × *C. deliciosa*) mandarin. *J. Food Sci. Technol.* **2015**, *52*, 2229–2237. [CrossRef]
17. Askarne, L.; Talibi, I.; Boubaker, H.; Boudyach, E.; Msanda, F.; Saadi, B.; Serghini, M.; Ait Ben Aoumar, A. In vitro and in vivo antifungal activity of several Moroccan plants against *Penicillium italicum*, the causal agent of citrus blue mold. *Crop Prot.* **2012**, *40*, 53–58. [CrossRef]
18. Zeng, Y.W.; Liu, X.Z.; Lv, Z.C.; Peng, Y.H. Effects of *Ficus hirta* Vahl. (Wuzhimaotao) extracts on growth inhibition of HeLa cells. *Exp. Toxicol. Pathol.* **2012**, *64*, 743–749. [CrossRef]
19. Ya, J.; Zhang, X.Q.; Wang, Y.; Zhang, Q.W.; Chen, J.X.; Ye, W.C. Two new phenolic compounds from the roots of *Ficus hirta*. *Nat. Prod. Res.* **2010**, *24*, 621–625. [CrossRef]
20. Wan, C.P.; Chen, C.Y.; Li, M.X.; Yang, Y.X.; Chen, M.; Chen, J.Y. Chemical constituents and antifungal activity of *Ficus hirta* Vahl. fruits. *Plants* **2017**, *6*, 44. [CrossRef]
21. Chen, C.Y.; Wan, C.P.; Peng, X.; Chen, Y.H.; Chen, M.; Chen, J.Y. Optimization of antifungal extracts from *Ficus hirta* fruits using response surface methodology and antifungal activity tests. *Molecules* **2015**, *20*, 19647–19659. [CrossRef] [PubMed]
22. Helrich, K. *Official Methods of Analysis of the Association of Official Analytical Chemists*; Association of Official Analytical Chemists: Arlington, VA, USA, 1990.
23. Chen, C.Y.; Zheng, J.P.; Wan, C.P.; Chen, M.; Chen, J.Y. Effect of carboxymethyl cellulose coating enriched with clove oil on postharvest quality of 'Xinyu' mandarin oranges. *Fruits* **2016**, *71*, 319–327. [CrossRef]
24. Hodges, D.M.; DeLong, J.M.; Forney, C.F.; Prange, R.K. Improving the thiobarbituric acid-reactive-substances assay for estimating lipid peroxidation in plant tissues containing anthocyanin and other interfering compounds. *Planta* **1999**, *207*, 604–611. [CrossRef]

25. Sala, J.M.; Lafuente, M.A.T. Antioxidant enzymes activities and rindstaining in 'Navelina' oranges as affected by storage relative humidity and ethylene conditioning. *Postharvest Biol. Technol.* **2004**, *31*, 277–285. [CrossRef]
26. Abeles, F.B.; Bosshart, R.P.; Forrence, L.E.; Habig, W.H. Preparation and purification of glucanase and chitinase from bean leaves. *Plant Physiol.* **1971**, *47*, 129–134. [CrossRef] [PubMed]
27. Xu, W.T.; Peng, X.L.; Luo, Y.B.; Wang, J.A.; Guo, X.; Huang, K.L. Physiological and biochemical responses of grapefruit seed extract dip on 'Redglobe' grape. *LWT Food Sci. Technol.* **2009**, *42*, 471–476. [CrossRef]
28. Chen, C.Y.; Peng, X.; Zeng, R.; Chen, M.; Wan, C.P.; Chen, J.Y. Ficus hirta fruits extract incorporated into an alginate-based edible coating for Nanfeng mandarin preservation. *Sci. Hortic.* **2016**, *202*, 41–48. [CrossRef]
29. Yi, T.; Chen, Q.L.; He, X.C.; So, S.W.; Lo, Y.L.; Fan, L.L.; Xu, J.; Tang, Y.; Zhang, J.Y.; Zhao, Z.Z. Chemical quantification and antioxidant assay of four active components in Ficus hirta root using UPLC-PAD-MS fingerprinting combined with cluster analysis. *Chem. Cent. J.* **2013**, *7*, 1752–1760. [CrossRef] [PubMed]
30. Wan, C.; Han, J.; Chen, C.; Yao, L.; Chen, J.; Yuan, T. Monosubstituted benzene derivatives from fruits of *Ficus hirta* and their antifungal activity against phytopathogen *Penicillium italicum*. *J. Agric. Food Chem.* **2016**, *64*, 5621–5624. [CrossRef] [PubMed]
31. Cheng, J.; Yi, X.; Wang, Y.; Huang, X.; He, X. Phenolics from the roots of hairy fig (*Ficus hirta* Vahl.) exert prominent anti-inflammatory activity. *J. Funct. Foods* **2017**, *31*, 79–88. [CrossRef]
32. Grande-Tovar, C.D.; Chaves-Lopez, C.; Serio, A.; Rossi, C.; Paparella, A. Chitosan coatings enriched with essential oils: Effects on fungi involve in fruit decay and mechanisms of action. *Trends Food Sci. Technol.* **2018**, *78*, 61–71. [CrossRef]
33. Xing, Y.; Xu, Q.; Yang, S.; Chen, C.; Tang, Y.; Sun, S.; Zhang, L.; Che, Z.; Li, X. Preservation mechanism of chitosan-based coating with cinnamon oil for fruits storage based on sensor data. *Sensors* **2016**, *16*, 1111. [CrossRef] [PubMed]
34. Sapper, M.; Chiralt, A. Starch-based coatings for preservation of fruits and vegetables. *Coatings* **2018**, *8*, 152. [CrossRef]
35. Shah, S.W.A.; Jahangir, M.; Qaisar, M.; Khan, S.A.; Mahmood, T.; Saeed, M.; Farid, A.; Liaquat, M. Storage stability of Kinnow fruit (*Citrus reticulata*) as affected by CMC and guar gum-based silver nanoparticle coatings. *Molecules* **2015**, *20*, 22645–22661. [CrossRef] [PubMed]
36. Velickova, E.; Winkelhausen, E.; Kuzmanova, S.; Alves, V.D.; Moldão-Martins, M. Impact of chitosan-beeswax edible coatings on the quality of fresh strawberries (*Fragaria ananassa* cv Camarosa) under commercial storage conditions. *LWT Food Sci. Technol.* **2013**, *52*, 80–92. [CrossRef]
37. Perdones, A.; Sánchez-González, L.; Chiralt, A.; Vargas, M. Effect of chitosan–lemon essential oil coatings on storage-keeping quality of strawberry. *Postharvest Biol. Technol.* **2012**, *70*, 32–41. [CrossRef]
38. Shahbazi, Y. Application of carboxymethyl cellulose and chitosan coatings containing *Mentha spicata* essential oil in fresh strawberries. *Int. J. Biol. Macromol.* **2018**, *112*, 264–272. [CrossRef] [PubMed]
39. Ventura-Aguilar, R.I.; Bautista-Baños, S.; Flores-García, G.; Zavaleta-Avejar, L. Impact of chitosan based edible coatings functionalized with natural compounds on *Colletotrichum fragariae* development and the quality of strawberries. *Food Chem.* **2018**, *262*, 142–149. [CrossRef] [PubMed]
40. Nair, M.S.; Saxena, A.; Kaur, C. Effect of chitosan and alginate based coatings enriched with pomegranate peel extract to extend the postharvest quality of guava (*Psidium guajava* L.). *Food Chem.* **2018**, *240*, 245–252. [CrossRef] [PubMed]
41. Lima Oliveira, P.D.; de Oliveira, K.Á.R.; Vieira, W.A.d.S.; Câmara, M.P.S.; de Souza, E.L. Control of anthracnose caused by *Colletotrichum* species in guava, mango and papaya using synergistic combinations of chitosan and *Cymbopogon citratus* (D.C. ex Nees) Stapf. essential oil. *Int. J. Food Microbiol.* **2018**, *266*, 87–94. [CrossRef]
42. Sánchez-González, L.; Pastor, C.; Vargas, M.; Chiralt, A.; González-Martínez, C.; Cháfer, M. Effect of hydroxypropylmethylcellulose and chitosan coatings with and without bergamot essential oil on quality and safety of cold-stored grapes. *Postharvest Biol. Technol.* **2011**, *60*, 57–63. [CrossRef]
43. Guerra, I.C.D.; de Oliveira, P.D.L.; Santos, M.M.F.; Lúcio, A.S.S.C.; Tavares, J.F.; Barbosa-Filho, J.M.; Madruga, M.S.; de Souza, E.L. The effects of composite coatings containing chitosan and Mentha (*Piperita* L. or x *villosa* Huds) essential oil on postharvest mold occurrence and quality of table grape cv. Isabella. *Innov. Food Sci. Emerg. Technol.* **2016**, *34*, 112–121. [CrossRef]

44. Dos Santos, N.S.T.; Athayde Aguiar, A.J.A.; de Oliveira, C.E.V.; Veríssimo de Sales, C.; de Melo e Silva, S.; Sousa da Silva, R.; Stamford, T.C.M.; de Souza, E.L. Efficacy of the application of a coating composed of chitosan and *Origanum vulgare* L. essential oil to control *Rhizopus stolonifer* and *Aspergillus niger* in grapes (*Vitis labrusca* L.). *Food Microbiol.* **2012**, *32*, 345–353. [CrossRef] [PubMed]
45. Athayde, A.J.A.A.; Oliveira, P.D.L.D.; Guerra, I.C.D.; Conceição, M.L.D.; Lima, M.A.B.D.; Arcanjo, N.M.O.; Madruga, M.S.; Berger, L.R.R.; Souza, E.L.D. A coating composed of chitosan and *Cymbopogon citratus* (Dc. Ex Nees) essential oil to control *Rhizopus* soft rot and quality in tomato fruit stored at room temperature. *J. Hortic. Sci. Biotechnol.* **2016**, *91*, 582–591. [CrossRef]
46. Won, J.S.; Lee, S.J.; Park, H.H.; Song, K.B.; Min, S.C. Edible coating using a chitosan-based colloid incorporating grapefruit seed extract for cherry tomato safety and preservation. *J. Food Sci.* **2018**, *83*, 138–146. [CrossRef] [PubMed]
47. Mohammadi, A.; Hashemi, M.; Hosseini, S.M. Postharvest treatment of nanochitosan-based coating loaded with *Zataria multiflora* essential oil improves antioxidant activity and extends shelf-life of cucumber. *Innov. Food Sci. Emerg.* **2016**, *33*, 580–588. [CrossRef]
48. Arnon, H.; Zaitsev, Y.; Porat, R.; Poverenov, E. Effects of carboxymethyl cellulose and chitosan bilayer edible coating on postharvest quality of citrus fruit. *Postharvest Biol. Technol.* **2014**, *87*, 21–26. [CrossRef]
49. Chien, P.-J.; Sheu, F.; Lin, H.-R. Coating citrus (Murcott tangor) fruit with low molecular weight chitosan increases postharvest quality and shelf life. *Food Chem.* **2007**, *100*, 1160–1164. [CrossRef]
50. Contreras-Oliva, A.; Perez-Gago, M.B.; Rojas-Argudo, C. Effects of chitosan coatings on physicochemical and nutritional quality of clementine mandarins cv. 'Oronules'. *Food Sci. Technol. Int.* **2012**, *18*, 303–315. [CrossRef]
51. Xu, D.; Qin, H.; Ren, D. Prolonged preservation of tangerine fruits using chitosan/montmorillonite composite coating. *Postharvest Biol. Technol.* **2018**, *143*, 50–57. [CrossRef]
52. Zeng, R.; Zhang, A.S.; Chen, J.Y.; Fu, Y.Q. Impact of carboxymethyl cellulose coating enriched with extract of *Impatiens balsamina* stems on preservation of 'Newhall' navel orange. *Sci. Hortic.* **2013**, *160*, 44–48. [CrossRef]

© 2018 by the authors. Licensee MDPI, Basel, Switzerland. This article is an open access article distributed under the terms and conditions of the Creative Commons Attribution (CC BY) license (http://creativecommons.org/licenses/by/4.0/).

Article

Fabrication and Testing of PVA/Chitosan Bilayer Films for Strawberry Packaging

Yaowen Liu [1,2,*,†], Shuyao Wang [1,†], Wenting Lan [1] and Wen Qin [1]

1   College of Food Science, Sichuan Agricultural University, Yaan 625014, China; shuyaow@126.com (S.W.); 18227593253@163.com (W.L.); qinwen1967@yahoo.com.cn (W.Q.)
2   School of Materials Science and Engineering, Southwest Jiaotong University, Chengdu 610031, China
*   Correspondence: lyw@my.swjtu.edu.cn; Tel.: +86-835-8763-4068
†   These authors contributed equally to the work.

Received: 15 June 2017; Accepted: 13 July 2017; Published: 25 July 2017

Abstract: Strawberry packaging based on four different chitosan–poly(vinylalcohol) blend films with chitosan contents of 0 wt %, 20 wt %, 25 wt %, and 30 wt % was tested. The samples were stored at 18 ± 2 °C and 60% ± 5% relative humidity for six days. Strawberry quality was evaluated during and after storage. Strawberries packaged using these films showed significant differences in weight loss and firmness, decay percentage, titratable acidity, total soluble solids, and ascorbic acid content when compared to non-packaged strawberries. The 25 wt % bilayer film showed the best performance in terms of delaying changes in strawberries. The findings suggest that these 25 wt % chitosan films can used to extend strawberry shelf lives while maintaining quality levels.

Keywords: bilayer films; strawberry; packaging

1. Introduction

The strawberry is among the most popular fruits worldwide, and is thus referred to as the "fruit queen". Strawberries possess high levels of antioxidant activity and vitamin E, vitamin C, β-carotene, and phenolic compounds such as anthocyanins, which benefit consumer health [1]. However, strawberries are perishable and have very short postharvest lives (around five days at 4 °C) because of their susceptibility to mechanical damage, physiological deterioration, and lack protective rinds, which can exhibit symptoms of pathogens [2]. Previous research has reported that Botrytis cinerea determines strawberry shelf lives, so the inhibition of microbial growth can prolong sellable periods [3]. Some researchers have used fungicides to prevent postharvest rot by spraying them on the strawberries several times. However, the presence of residues limits the broad use of fungicides [4]. Controlling CO_2 and O_2 levels can also reduce the incidence of strawberry decay. Modified atmospheres with low O_2 and high CO_2 concentrations are very effective at inhibiting microbial growth and reducing the decay of fresh produce, including strawberries [5]. However, some experiments have shown that prolonged exposure of strawberries to high CO_2 concentrations can cause the development of unwanted flavors [6]. To slow metabolic processes and reduce deterioration prior to transport, low temperatures are widely used to reduce spoilage and extend the lives of fresh fruits and vegetables. However, it is inconvenient and expensive to use transportable freezers or ice to control strawberry temperatures.

Biologically active packaging has become an effective method of controlling fungal decay, and several studies have investigated the potential of natural polymers for food protection. Most of these types of macromolecules can be processed into films or coated onto fruit surfaces. Owing to their high (CO_2/O_2) permselectivities and ability to act as partial moisture barriers, these films can reduce fruit respiration and transpiration rates [7]. Biologically active packages include polysaccharides such as chitosan (CH). CH is a linear polysaccharide that consists of a low acetyl substituted form of chitin and a natural carbohydrate copolymer. CH offers several advantages such as biodegradability, broad availability, and non-toxicity [8]. CH also has good anti-fungal activity against several postharvest pathogens, particularly the grey mold that is one of the main causes of strawberry deterioration and postharvest decay [9]. Moreover, studies have demonstrated that CH can be mixed with other polymers to increase the shelf lives of fresh strawberries [10]. The use of liquid CH with strawberries would be effective, but would require high proportions, which would result in bad flavors, oily textures, and the potential for allergic reactions after consuming large amounts [11].

Unfortunately, pure CH films have high water vapor permeabilities and poor mechanical properties, and so cannot act as inert barriers between the product and the environment. In order to fulfill all necessary fruit packaging requirements, several methods of mixing CH with other polymers such as polylactic acid, poly(lactic-co-glycolic acid), polyethylene, etc. in composites have been developed. Poly(vinylalcohol) (PVA) is a hydroxyl-rich, semi-crystalline polymer. It can be processed using aqueous methods, and has several interesting physical properties that arise from the presence of hydroxyl groups and hydrogen bond formation. It is non-toxic, biocompatible and biodegradable, and offers good mechanical properties and chemical stability. Moreover, PVA/CH blends have good physical and antibacterial properties because of specific intermolecular interactions between PVA and CH. PVA/CH blend films represent a better choice for strawberry packaging than coating CH on the fruit surface or refrigeration. These blends are especially appropriate for transportation due to their good mechanical and oxygen barrier properties [12].

In order to prolong the shelf life of strawberries and reduce the hazard of ingesting chemical reagents, edible coating materials have been widely studied. PVA/CH blend films with various CH contents on fresh strawberries were determined in this study. The films were characterized based on scanning electron microscopy (SEM), mechanical tests, oxygen permeability (OP) and water vapor permeability (WVP). Non-packaged strawberries and strawberries packaged with four different films were compared and analyzed in terms of color, weight loss, decay percentage (% decay), firmness, soluble solid content (SSC), titratable acidity (TA), and sensory characteristics.

2. Results and Discussion

2.1. SEM Analysis

The weight ratios of PVA:CH of 80:20, 75:25, and 70:30 were denoted as PVA/CH-2, PVA/CH-2.5, and PVA/CH-3, respectively [13]. The morphologies of the PVA film, PVA/CH-2, PVA/CH-2.5, and PVA/CH-3 are seen clearly in the SEM images (Figure 1). PVA film presents good structural integrity, smoothness, flatness, and crack-free states. CH shows even distributions in PVA/CH-2, and PVA/CH-2.5 films, demonstrating the high compatibility of the two polymers and a compact structure lacking phase separation [14]. No air bubbles, pores, cracks, or droplets are observed, further confirming the high compatibility of the two polymers, similar to that observed by Tripathi et al. [15]. However, when the CH content is increased, it appears visibly as rough areas. The rough areas of the PVA/CH-3 show scaly structures. The roughness is because the CH molecules disrupt the compact structure of the PVA matrix. The incorporation of CH minimizes the free volume of the matrix, condensing the microstructure of the film [16]. These structural properties of the dried CH matrices may offer a larger surface area, and therefore, better matrix–solvent interactions, allowing for faster solvent uptake. This leads to dissociation, with the most obvious result of scaly structures [15].

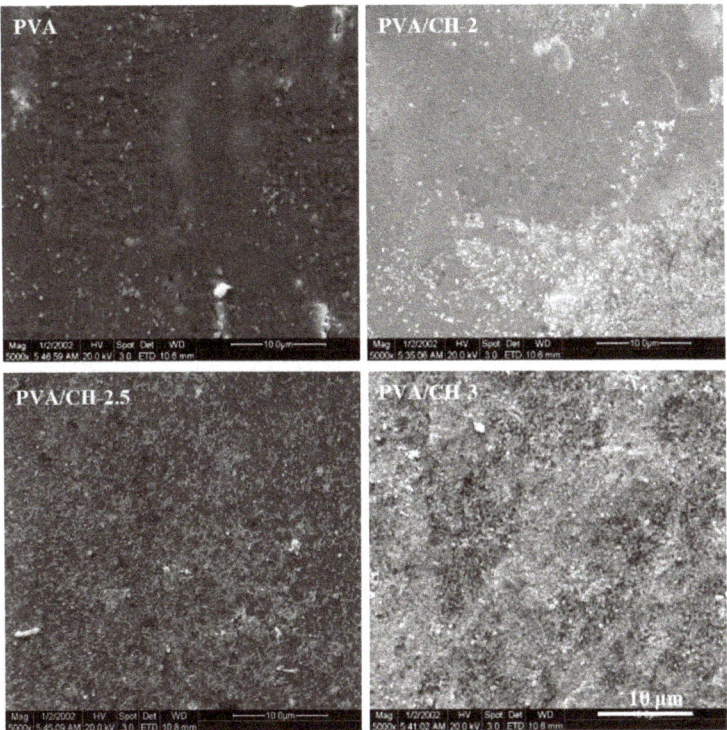

Figure 1. Scanning electron microscopy (SEM) morphologies of pure Poly(vinylalcohol) (PVA) film and different PVA/CH films.

2.2. FT-IR Analysis

Figure 2 presents the characteristic attenuated total reflectance–Fourier transform infrared (ATR–FTIR) spectra corresponding to CH, PVA, and PVA/CH-2.5. For PVA film, the bands at approximately 3455 cm^{-1} and 1630 cm^{-1} are assigned to the stretching and bending vibrations of the hydroxyl group, respectively [17]. The band corresponding to asymmetric stretching vibrations of the methylene group occurs at approximately 2933 cm^{-1}. The band at approximately 1096 cm^{-1} corresponds to C–O stretching in the acetyl groups present on the PVA backbone [18]. Notably, the bands of CH are assigned to the saccharide structure at 1166 cm^{-1}, 1077 cm^{-1}, and 1018 cm^{-1}. The strong amino characteristic bands at 3430 cm^{-1}, 1660 cm^{-1}, and 1290 cm^{-1} are assigned to hydroxyl stretching, amide I, and amide II, respectively [19]. The spectra of PVA/CH-2.5 indicates clear increases in the intensity of the band at 3380 cm^{-1}, attributed to hydroxyl group stretching vibrations of PVA, with a secondary amide group of CH. The band at approximately 1077 cm^{-1} indicates the presence of a hydroxyl group with polymeric association and a secondary amide. The band at 1450 cm^{-1}, appearing for weight fractions of 25% CH, is assigned to C=N pyridine ring vibrations. This confirmed the complexation between the PVA and CH [19]. For all PVA/CH films, the characteristic bands are similar to those of PVA; strong amorphous carbonyl stretching vibrations of PVA remain constant in all films. The peak intensity increases with increased CH contents. The characteristic shape of the CH spectrum is changed, and the peak shifts to a lower frequency range because of hydrogen bonding between the hydroxyl groups of PVA and hydroxyl, or amine groups of CH in the blended films [20]. Kim et al. also reported that the crystallization-sensitive band of PVA at 1140 cm^{-1} was observed with a similar intensity without significant changes in frequency for PVA and blended films [21].

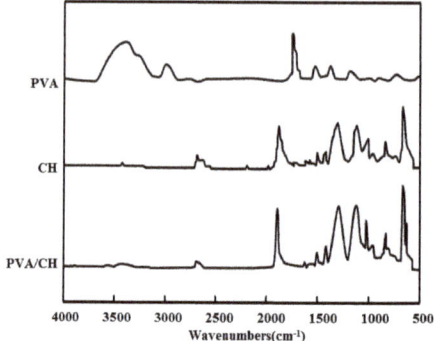

Figure 2. Attenuated total reflectance–Fourier transform infrared (ATR–FTIR) spectra of PVA, CH, and PVA/CH-2.5 films.

2.3. Mechanical Properties

Table 1 shows the mechanical properties of pure PVA film and PVA with different CH weight ratios. The ultimate stress of PVA film is 25.08 ± 4.32 MPa, and the specific deformation is 72.97% ± 7.26%. When the concentration of CH in the PVA/CH films is increased, the ultimate stress of the films is also increased, while the elongation at break of the films is decreased. Compare to the pure PVA film, the stress of PVA/CH film increases from 29.76 ± 4.81 MPa to 35.39 ± 5.35 MPa, while the strain decreases from 77.13% ± 7.94% to 72.91% ± 8.17%. Bispo et al. proposed that films formed from polymer blends present intermediate values of maximum stress compared with those of films comprising the pure components [22]. Therefore, the mechanical properties of the PVA/CH films may suggest that the PVA polymer has greater influence on the tensile stress and maximum specific deformation. The increases in CH content cause change in stress and strain, due to the thermodynamic immiscibility and inherent incompatibility between thermoplastic polymers and CH [23]; the CH also disrupts the intramolecular hydrogen bonding of the PVA molecules [23]. Moreover, as the CH addition is increased, this changed the intermolecular hydrogen bonds formed between the amine and hydroxyl groups of CH [24]. However, PVA/CH-2.5 shows the best mechanical properties (Young's modulus 63.57 ± 6.98 MPa; those results proved that blend miscibility of the two polymers as a function of molecular interactions between PVA and CH [25].

Table 1. Mechanical properties of pure PVA film and PVA/CH with different CH weight ratios.

Samples	Thickness (μm)	Stress (MPa)	Strain (%)	Young's Modulus (MPa)
PVA	373.1 ± 12.8 [a]	25.08 ± 4.32 [a]	82.97 ± 7.26 [a]	56.11 ± 5.84 [a]
PVA/CH-2	396.4 ± 13.4 [a]	29.76 ± 4.81 [a]	77.13 ± 7.94 [b]	59.53 ± 6.42 [a]
PVA/CH-2.5	418.6 ± 14.7 [b]	35.39 ± 5.35 [b]	75.33 ± 8.44 [b]	63.57 ± 6.98 [b]
PVA/CH-3	433.9 ± 15.1 [b]	31.44 ± 5.12 [a]	72.91 ± 8.17 [b]	58.45 ± 6.35 [a]

Notes: Values with the same letter are not statistically different, according to Duncan's Multiple Range Test at $p < 0.05$; [a, b], means with the same letter in the same column are not significant different ($p > 0.05$).

2.4. Oxygen Permeability and Water Vapor Permeability

The OPs of packaging materials are of considerable importance in food preservation. Low OPs indicate excellent oxygen barrier properties. Table 2 shows the OP and WVP values for all films. The OP of the pure PVA film is 0.12 ± 0.04 cm^2 m^{-2} atm^{-1} day^{-1} MPa. The addition of CH reduces the abilities of the blend films to act as oxygen barriers. The OPs of PVA/CH-2 and PVA/CH-2.5 increase slightly, from 0.15 ± 0.07 to 0.16 ± 0.08 cm^2 m^{-2} atm^{-1} day^{-1} MPa. This can be attributed to the lower crystallinities of the blended films with low CH concentrations [26]. The OPs increase rapidly with the CH content when the latter is over 30 wt %, resulting in blend films with poorer oxygen barrier properties.

Table 2. Oxygen permeability and water vapor permeation of different films.

Samples	Oxygen Permeability ($cm^2\ m^2\ day^{-1}\ MPa^{-1}$)	Water Vapor Permeation ($g\ cm^{-1}\ s^{-1}\ Pa^{-1}$)
PVA	0.12 ± 0.04 [b]	10.11 ± 2.14 [b]
PVA/CS-2	0.15 ± 0.07 [b]	12.43 ± 3.72 [b]
PVA/CS-2.5	0.16 ± 0.08 [b]	14.93 ± 4.09 [b]
PVA/CS-3	0.39 ± 0.11 [a]	22.99 ± 5.57 [a]

Notes: Values with the same letter are not statistically different, according to Duncan's Multiple Range Test at $p < 0.05$; [a, b], means with the same letter in the same column are not significant different ($p > 0.05$).

The WVPs of PVA/CH-2 and PVA/CH-2.5 films are 12.43 ± 3.72, and 14.93 ± 4.09 $g\ cm^{-1}\ s^{-1}\ Pa^{-1}$, respectively. This demonstrates that the WVPs of the composite films are higher than that of the pure PVA film (10.11 ± 2.14 $g\ cm^{-1}\ s^{-1}\ Pa^{-1}$); however, the differences are not significant. PVA/CH-3 film exhibits significantly higher WVPs ($p < 0.05$) than PVA. Those results indicated that the interaction between PVA and CH can increase the water vapor barrier properties. Furthermore, increasing the CH content clearly increases the WVPs of the films. The hydrophilic nature of CH favors the transport of water molecules through the film [27], and composite films with higher CH contents exhibit lower crystallinities [28].

2.5. pH Values

Table 3 compared the pH values of packaged and unpackaged strawberries. The pH values of unpackaged strawberries increased significantly with the increase of storage time. The pH values of unpackaged strawberries reached to 3.99 ± 0.29 after six days. Han et al. demonstrated that the pH values of fruit are related to the fruit senescence [29]. A possible explanation for this might be the utilization of organic acids of fruit during respiration [30]. The pH difference between day 2 and day 4 was not statistically significant for all packaged groups. Martinez–Ferrer et al. considered that might be the low conversion rate of organic acids during respiration [30]. The pH values of strawberries packaged with PVA/CH-2, PVA/CH-2.5 and PVA/CH-3 were significantly different than the pH values of strawberries packaged with the PVA sample at day 6. The findings of the current study are consistent with those of Cong et al., who found that CH coating containing natamycin slowed the changes in pH values of Hami melons [31]. At six days, the pH values of fruits packaged with PVA/CH-2.5 was 3.65 ± 0.16, and the change was minimum (3.41 ± 0.08 for one day). A possible explanation for this might be that relatively small changes of O_2 and CO_2 levels resulted in slight changes of the pH values for strawberries [32].

Table 3. pH analysis of strawberries treated with different films during storage.

Treatments	pH						
	0 Day	1 Day	2 Days	3 Days	4 Days	5 Days	6 Days
Unpackaged	3.41 ± 0.08 [b]	3.53 ± 0.12 [b]	3.64 ± 0.16 [b]	3.70 ± 0.18 [b]	3.76 ± 0.20 [b]	3.88 ± 0.24 [a]	3.99 ± 0.29 [a]
PVA	3.41 ± 0.08 [b]	3.49 ± 0.11 [b]	3.57 ± 0.14 [b]	3.64 ± 0.16 [b]	3.71 ± 0.18 [b]	3.77 ± 0.20 [a]	3.83 ± 0.23 [a]
PVA/CH-2	3.41 ± 0.08 [b]	3.47 ± 0.10 [b]	3.53 ± 0.13 [b]	3.57 ± 0.14 [b]	3.62 ± 0.16 [b]	3.69 ± 0.18 [b]	3.76 ± 0.20 [b]
PVA/CH-2.5	3.41 ± 0.08 [b]	3.46 ± 0.09 [b]	3.51 ± 0.12 [b]	3.55 ± 0.14 [b]	3.59 ± 0.05 [b]	3.61 ± 0.16 [b]	3.65 ± 0.16 [b]
PVA/CH-3	3.41 ± 0.08 [b]	3.46 ± 0.09 [b]	3.51 ± 0.12 [b]	3.57 ± 0.15 [b]	3.61 ± 0.16 [b]	3.68 ± 0.18 [b]	3.74 ± 0.20 [b]

Notes: Values with the same letter are not statistically different, according to Duncan's Multiple Range Test at $p < 0.05$; [a, b], means with the same letter in the same column are not significant different ($p > 0.05$).

2.6. Weight Loss

Owing to respiration and transpiration, strawberries are highly susceptible to the rapid loss of water. In our studies, weight loss was tracked over the storage period of six days. Figure 3a shows that all strawberries experience progressive weight loss during the storage period. The packaged samples experience significantly less weight loss than non-packaged samples after two days ($p < 0.05$). This is because the major cause of weight loss among non-packaged samples is the migration of water from the fruit to the environment [33], and the packaging materials serve as semipermeable barriers that block oxygen, carbon dioxide, and moisture. This reduces respiration, water loss, and oxidation [34]. Since PVA has the lowest water vapor permeability, strawberries packaged with it exhibit the lowest weight loss. PVA/CH-2.5 exhibits the lowest weight loss among the PVA/CH bilayer films. This is because PVA/CH-2.5 has higher oxygen permeability, and the hydrophilic CH interacts with water molecules to increase the transport of water vapor. Hence, the packaging of strawberries in bilayer films is clearly effective in providing a physical barrier to moisture loss, and therefore in retarding respiration and fruit shriveling [35].

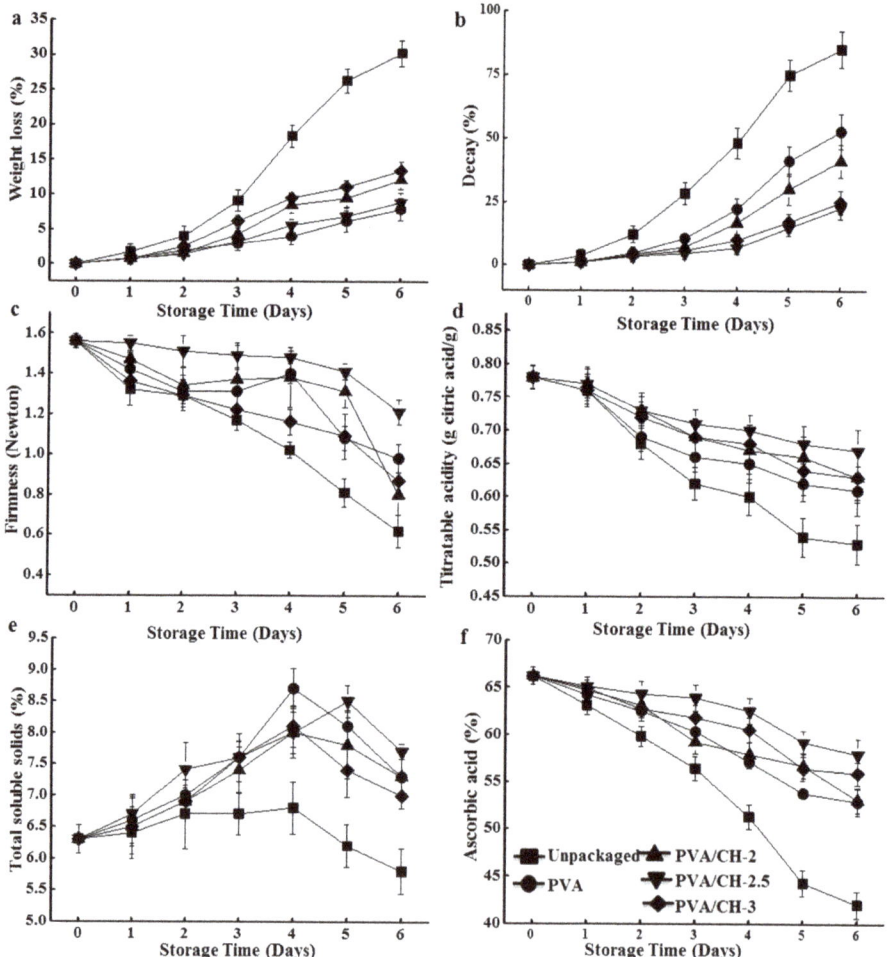

Figure 3. Effect of unpackaged PVA and different PVA/CH films on the (a) weight loss, (b) decay, (c) firmness, (d) titratable acid, (e) total soluble solids and (f) ascorbic acid quality parameters of strawberries during storage times.

2.7. Decay Percentage

Strawberries are highly perishable fruits with high postharvest physiological activities that limit their shelf lives. Generally, the infected areas increase gradually with storage time, as shown in Figure 3b. The decay percentages of the packaged strawberries range from 22.7% ± 3.2% to 52.9% ± 5.7%, but the non-packaged strawberries experience a decay of 85.1% ± 7.4% after 6 days. The packaging significantly reduces strawberry decay during the six days test ($p < 0.05$). PVA/CH films and pure PVA film can decrease the rate at which the fruit rots as well. Some studies have indicated that CH appears to have multiple functions, as it interferes directly with fungal growth and activates several biological processes in plant tissues [36]. PVA/CH-2.5 reduces the degree of decay better than other packaging materials throughout the storage period. This may be because of its WVP and OP, which may affect the growth environment for bacteria.

2.8. Firmness

Loss of texture is one of the main factors that limits quality and the postharvest shelf lives of fruit and vegetables. Therefore, texture is an important strawberry quality parameter. During ripening, strawberries soften considerably due to degradation of the middle lamella of the cell wall. Cortical parenchyma cells, cell wall strengths, cell-to-cell contact [37], and cellular turgor can also influence firmness [38]. The changes in strawberry firmness are shown in Figure 3c. The firmness of all strawberries was significantly lower among the non-packaged samples than among the packaged fruit ($p < 0.05$). This leads to increased water loss and fungal infections among strawberries that lack packaging protection, which leads to more pronounced tissue senescence and broken cell walls [39]. After four days, the firmness curve exhibits a turning point. The firmness of all packaged strawberries begins to decrease significantly due to senescence, which softens the fruit via pectin hydrolysis, and depolymerization degradation of the cell wall [40]. After six days, the loss of firmness is approximately 48% among non-packaged fruit. The losses are 31.5% ± 9.7%, 16.2% ± 4.7%, 10.4% ± 2.4%, and 20.6% ± 5.1% in PVA, PVA/CH-2, PVA/CH-2.5 and PVA/CH-3 packaged fruit samples, respectively. The PVA/CH-2.5 samples do not experience significant surface softening when compared to the fresh strawberries, and the film is effective with regard to firmness retention. It also has been reported that chitosan coatings and other biopolymers are selective O_2 and CO_2 barriers, and thus can modify the internal atmosphere and slow the respiration rates of fresh fruits and vegetables [41].

2.9. Titrable Acidity (TA)

Organic acids are among the most important components of the flavor of a strawberry. Changes in acidity are significantly affected by the rate of metabolism, especially respiration. Respiration consumes organic acid, and therefore acidity declines during storage. This is also the main cause of fruit senescence [10]. The effect of packaging on the TA of strawberries is shown in Figure 3d. At the end of the storage period, the TA of non-packaged strawberries decreases significantly faster ($p < 0.05$) than packaged samples. This is because the packaging can modify the internal atmosphere around the strawberries, and may therefore delay the utilization of organic acids. The decrease in the TA of the PVA-packaged fruit is less (23.4%) than that of the unpackaged fruit (36.5%), but more than that of fruit protected by PVA/CH films (from 17.6% to 19.8%). This suggests that bilayer films are better than pure PVA materials at delaying decreases in TA during storage. The PVA/CH-2.5 bilayer film is the most effective at maintaining higher TA levels after six days.

2.10. Total Soluble Solids (TSS)

Total soluble solids (TSS) is an important parameter that affects fruit quality and consumer acceptability. Changes in the TSSs of strawberry samples with storage time are shown in Figure 3e. The TSS increases significantly during four days of storage. Starting on the fifth day, the TSS begins to decrease rapidly in PVA-packaged samples (from 8.7% to 7.3%) due to hydrolysis. Experiments also show that adding CH to the coating formulation helps to maintain higher TSS accumulation at the end of the storage period. After six days, the PVA/CH bilayer film slows the conversion that reduces sugar levels in the strawberries. The PVA/CH-2.5 sample retains the highest TSS in the current work, which can be explained by the considerable loss of water experienced by strawberries during storage at room temperature. The PVA/CH-2.5 bilayer film offers the most suitable environment for strawberry preservation, as previously noted with regard to weight loss.

2.11. Ascorbic Acid

Obviously, the ascorbic acid content gradually declines with postharvest elongation in both packaged and non-packaged fruits (Figure 3f). All packaging materials inhibited ascorbic acid loss in packaged fruit. With the PVA/CH-2.5 film, the ascorbic acid content decreased from 66.2 ± 4.8 mg/100 g (one day) to 57.9 ± 3.6 mg/100 g (six days), while the unpackaged samples showed significantly ($p < 0.05$) lower amounts of ascorbic acid (42.4 ± 2.9 mg/100 g, six days). Some studies have also shown that the incorporation of CH can slow the deteriorative oxidation of ascorbic acid in fruit [10]. During storage, strawberry decay and the low CH content increase ascorbic acid degradation.

2.12. Sensory

Sensory acceptance scores, including those for appearance, color, odor, flavor, texture, and overall acceptability were measured after six days and are shown in Figure 4. The appearance acceptability scores of the control strawberries are significantly different from those of the packaged strawberries after two days. The control and PVA-packaged strawberries have unacceptable scores after four days of storage, while the PVA/CH films maintain acceptable scores (greater than five) until five days. The color acceptability scores of the packaged strawberries do not change significantly until after three days of storage. After three days, the color acceptability of samples other than PVA/CH-2.5 decrease significantly. The odor acceptability scores of the control strawberries fail consumer acceptance after four days. The odor acceptability scores of the PVA/CH-2 and PVA/CH-3 samples are almost equal, with the former having the highest scores. The control has the lowest odor acceptability score after six days. The flavor acceptability scores of all strawberries decrease significantly during the six days. The scores of PVA-packaged strawberries are greater than those of the control strawberries, and are the highest among all of the samples after two days. Both control and PVA-packaged strawberries are unacceptable after five days. Although the flavor scores of PVA/CH film-packaged strawberries decrease during the 6 days of testing, they remain acceptable. The overall acceptance scores of the control strawberries decrease during storage. The PVA and PVA/CH-2-packaged strawberries are not acceptable after five days. However, the PVA/CH-2.5 and PVA/CH-3 samples are still acceptable at the end of the storage period. This finding agrees with those of Sangsuwan et al., who reported that CH beads loaded with lavender essential oil can extend the mold-free storage lives of strawberries stored at 7 °C from two days (control) to eight days with acceptable overall sensory scores [43].

The result of sensory evaluation was similar to the appearance changes of strawberries. As shown in Figure 5, both packed and unpacked strawberries have begun to decay after four days except the group packed with PVA-CH-2.5 film. The decay of the strawberries could have been prevented by PVA-CH-2.5 film until the 6th day. The mold and yeast counts during the 6 days storage also proved the similar conclusion.

As shown in Figure 6, the strawberries packaged with PVA/CH films presented a significantly lower amount of mold and yeast growth than the uncoated strawberries ($p < 0.05$). Moreover, the mold and yeast reduction was more evident in the strawberries that were packaged with high concentration CH due to the antimicrobial capacity of CH, especially against the fungi and yeast spoilage of strawberries [44]. CH potentially causes severe cellular damage in mold and yeast by altering the synthesis of fungal enzymes [45], inducing morphological changes, and causing structural alterations and molecular disorganization in fungal cells [46]. Similar results were observed by

Valenzuela et al. [44]. The results achieved here also indicated that PVA/CH-2.5 film can not only maintain the nutritive and organoleptic properties of strawberries, it could also reduce the amount of mold and yeast during storage times. The high effectiveness of PVA/CH-2.5 is due to the ionic and hydrophilic interaction between PVA and CH, which increased the availability of the amino groups of the CH to the antimicrobial properties [44].

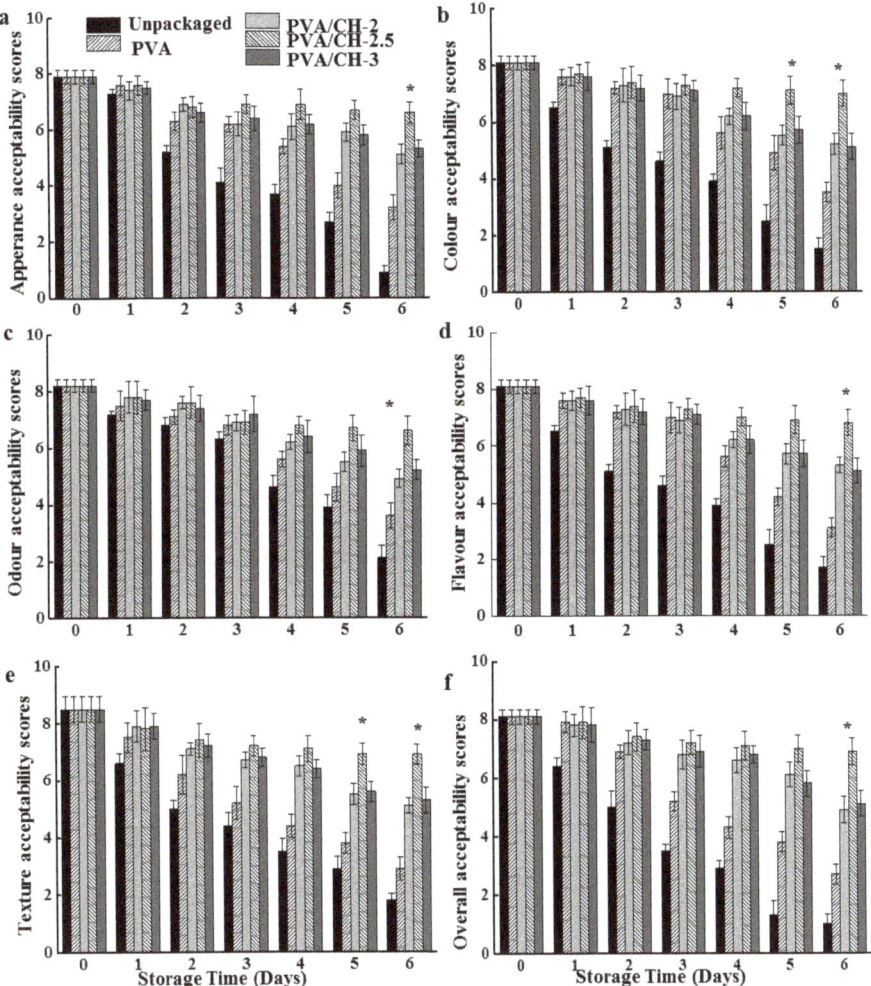

Figure 4. The sensory scores of (**a**) appearance acceptability; (**b**) color acceptability; (**c**) odor acceptability; (**d**) flavor acceptability; (**e**) texture acceptability and (**f**) overall acceptability of strawberries during storage times.

Figure 5. Digital images of the appearance and inner changes of strawberries during storage times.

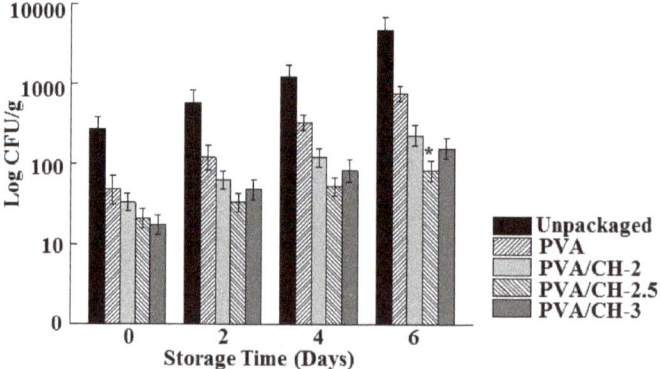

Figure 6. Evolution of mold and yeast counts of unpackaged and packaged strawberries with different PVA/CH films during storage times.

3. Materials and Methods

3.1. Materials

CH (food grade, 90% degree of deacetylation, molecular mass of 165,000 Da, low viscosity) was obtained from Shangdong Aokang Biological Technology (Jinan, Shangdong, China). PVA with molecular weight of 124,000–186,000 was purchased from Sigma Aldrich (98%, St. Louis, MO, USA). Acetic acid (glacial 100%, water solution) was obtained from Guoling Instrument Inc. (Dongguan, China) All other chemical reagents used were of analytical grade and obtained from Chengdu Kelong Reagent Co. (Chengdu, China), unless otherwise indicated. Strawberries were harvested from an orchard in Shuangliu, Sichuan. The selected fruit was of uniform size and color and free of physical damage and fungal infection. They were washed with 1% sodium hypochlorite for 1 min, then rinsed with distilled water and allowed to dry. Packaging experiments were carried out on the same day.

3.2. PVA/CH Film Preparation

PVA was dissolved in boiling water at 100 °C and stirred for 8 h to produce a clear, 10 wt % solution. It was degassed in a vacuum desiccator and poured into a Teflon pan (20 cm × 20 cm). The solution was dried in an oven at 50 °C for four days to ensure the removal of residual solvents, and the film was peeled from the pan and kept in a vacuum oven until use. To produce a completely homogenized CH solution, various CH mass ratios were dissolved in 0.8% acetic acid and stirred overnight. Briefly, a CH solution in acetic acid/water/ethanol was loaded into a syringe equipped with a metal capillary. A rotating metal plate collector was placed approximately 5 cm from the capillary tip. The solution was extruded from the needle tip at a constant flow rate of 10 µL/min using a precision pump (Zhejiang University Medical Instrument Company, Hangzhou, China). The applied voltage was set to approximately 15 kV using a high-voltage statitron (Tianjing High Voltage Power Supply Company, Tianjing, China). Low-humidity conditions were created by using a dehumidifier during the electrospray process. After CH deposited on both surfaces, the films were dried, first in a vacuum at 50 °C for 24 h and then at 18 ± 2 °C.

3.3. Film Characterization

The pure PVA, PVA/CH-2, PVA/CH-2.5, and PVA/CH-3 bilayer films' thickness were measured with a ZUS-4 micrometer (Yue Ming Small Machine Co., Changchun, China). The thickness measurements were taken at 10 random positions on each film, and the mean was calculated. The mean value for the thickness was used in calculating the film oxygen permeability (OP) and mechanical properties. The morphologies of the films were examined using SEM (FEI Quanta 200, Eindhoven, The Netherlands) equipped with a field-emission gun and Robinson detector, after the samples were vacuum-coated with thin layers of gold to minimize the charging effect. Attenuated total reflectance–Fourier transform infrared (ATR-FTIR) spectrometry (ATR-FTIR, Nicolet 5700, Thermo Nicolet Instrument Corp., Madison, WI, USA) was used to identify the chemical structures of the PVA/CH composite films and the possible interactions between their components. A small section cut from each composite film was used. The samples were analyzed with a resolution of 4 cm^{-1}, an aperture setting of 6 mm, a scanner velocity of 2.2 kHz, a background scan time of 32 s, a sample scan time of 32 s, and a total of 100 scans per sample, in the range of 400 to 4000 cm^{-1}. The composite films were cut into rectangles of 50 mm in length and 5 mm in width for PVA/CH fibers. Tensile testing was performed using a universal testing machine (UTM, Instron 5583, Norwood, MA, USA) with the crosshead speed of 5 mm/min and a 30 mm gauge length. The average value from five measured samples is reported for subsequent analysis.

3.4. Strawberries Preparation and Packaging

The strawberries were randomly divided into four groups and conducted with three replicates, with 10 strawberries in each treatment. The weight ratio of PVA:CH of 80:20, 75:25, and 70:30 were denoted as PVA/CH-2, PVA/CH-2.5, and PVA/CH-3, respectively. Ten strawberries (250 ± 10 g) were packed in sealed PVA or PVA/CH bags (18 cm × 18 cm). These samples were placed in conditions of 18 ± 2 °C and 60 ± 5% RH. The quality of both packaged and control strawberries were evaluated after one day, two days, three days, four days, five days, and six days of storage.

3.5. Oxygen and Water Vapor Permeability

Permeability tests were performed using a PERME TM OX2/231 Permeability Tester from Labthink Instruments Co., Ltd. (Jinan, China), using oxygen as the test gas (RH 50%) at a temperature of 18 ± 2 °C. Nitrogen was used as the oxygen carrier. The oxygen flow rate was fixed at 20 mL/min, while that of nitrogen was 10 mL/min. The WVPs of the films were determined by a Perme VAC-V1 (Labthink, Jinan, China) at 18 ± 2 °C and 60% RH.

3.6. Strawberry Quality during Storage

3.6.1. Total Soluble Solids (TSS) and pH Measurement

Each group of strawberries was collected. Juice was obtained by homogenizing strawberries in a blender, and then filtering through a cheese cloth. The pH values of samples were determined with a pH meter (Sartorius PP-50, Goettingen, Germany) [47]. The TSS was determined using a digital refractometer (Atago, Tokyo, Japan).

3.6.2. The Weight Loss

Strawberry weight loss was measured using a balance (±0.0001 g) and expressed according to the following formula:

$$\text{Weight loss rate (\%)} = (W_0 - W_T)/W_0 \times 100\% \tag{1}$$

where W_0: the weight of the fresh strawberries, W_T: the weight of the strawberries after storage for specific time intervals.

3.6.3. Microbiological Analyses and Decay Percentage

The microbiological analyses were performed by the official standard method [44]. Mold and yeast counts were expressed as logarithm colony-forming units per gram of strawberries (log CFU/g). Decay of the strawberries was recorded during each day of the experiment. Visible gray mold infections and wounded areas were photographed using a camera (Canon PowerShot S2IS, Canon Inc., Tokyo, Japan), and the images were processed using Adobe Photoshop CS 5.1 (Adobe Systems Inc., San Jose, CA, USA). The infected/wounded areas of each individual fruit were expressed as a decay percentage, and the mean wounded area content was calculated at each CH concentration level. The experiments were performed with five replications.

3.6.4. Firmness

The firmness of non-packaged and film-packaged strawberries was measured at 18 ± 2 °C using a texture analyzer with a 5 kg load cell. A single, whole strawberry was placed on a floating platform and a 2 mm diameter flat head stainless steel cylindrical probe was set to penetrate 6 mm into the fruit at 1 mm/s [43]. After the firmness measurements, the strawberries were cut into small pieces and homogenized for 2 min at high speed using a hand-held blender. The homogenized strawberry puree was divided into three groups of 5 g, 5 g, and 10 g.

3.6.5. Titratable Acidity (TA)

For the TA analyses, 5 g of the homogenized strawberry puree was put into 50 mL volumetric flask with the distilled water constant volume for 30 min and then filtered. Taking 20 mL filtrate titrated using 0.01 mol/L NaOH. The total titratable acidity of the diluted puree was calculated using the formula:

$$TA = (V(NaOH) \times 0.1 \times 0.064 \times V_0)/(m \times V_1) \qquad (2)$$

$V(NaOH)$ is the mL of NaOH spent for titration, 0.1 is the molarity of NaOH solution, 0.064 is conversion factor for citric acid; V_0 is the volumetric of the volumetric flask; m is filtrate volume. The results were expressed as percent citric acid (g citric acid/100 g dry weight). For determining soluble solids content (SSC), 5 g of homogenized strawberry puree was suspended in 10 mL distilled water and then filtered by filter paper.

3.6.6. Ascorbic Acid Determination

10 g of the homogenized strawberry puree was suspended in a 50 mL volumetric flask with 10 mL of 1% HCl and distilled water. The sample was then extracted at 1200 rad/min and centrifuged for 10 min. The clean supernatant was collected and 1 mL put in each of two 50 mL volumetric flasks: (A) containing 2 mL of 10% HCl and (B) containing 4 mL of 1 mol/L NaOH, both with distilled water. Absorption photometry of the sample was conducted against a control at 243.3 nm using an absorption spectrophotometer (MAPADA, Instruments. Co., Ltd., Shanghai, China). The levels of ascorbic acid were estimated from a standard curve prepared from pure ascorbic acid readings.

3.6.7. Preliminary Sensory Evaluation

Strawberries from each set of treatment conditions were served to panelists in a random order. The evaluations were performed under ambient conditions at about $18 \pm 2\,^\circ\text{C}$ and $60\% \pm 5\%$ RH in a sensory evaluation room. The quality attributes (appearance, color, odor, flavor, texture, and overall acceptability) of the strawberries were evaluated by 14 trained assessors, using a nine-point hedonic scale where 9 indicates excellence and freshness; 7 is very good; 5 is good but indicates the limit of marketability; 3 is fair, indicating the limit of usability; and 1 means poor or unusable. Panelists used water to cleanse their mouths between samples [44].

3.7. Statistical Analysis

Multiple samples were tested, and the results were reported as the mean ± the standard deviation. The values were submitted to analysis of variance and the means were separated by Duncan's Multiple Range Test (SuperANOVA, Abacus Concepts, Inc., Berkeley, CA, USA). A p value of < 0.05 was considered significant.

4. Conclusions

PVA/CH blends exhibit better morphological and mechanical properties than neat PVA. They slow metabolic processes and delay strawberry ripening, as indicated by reductions in weight loss as well as retention of firmness, decay percentage, TA, TSS, and ascorbic acid content. PVA/CH bilayer films showed a beneficial effect on the maintenance of strawberry quality and freshness throughout the storage period. The PVA/CH-2.5 bilayer film balances the benefits of PVA and CH and may be the best choice for improving the quality and shelf lives of fresh strawberries for up to six days at $18 \pm 2\,^\circ\text{C}$ and a relative humidity of $60\% \pm 5\%$, with acceptable overall sensory scores.

Acknowledgments: We acknowledge the Natural Science Fund of Education Department of Sichuan province (16ZB0044 and 035Z1373).

Author Contributions: Yaowen Liu and Shuyao Wang developed the original idea and the protocol, abstracted and analyzed data, wrote the manuscript, and is guarantor. Wenting Lan and Wen Qin contributed to the development of the protocol, abstracted data, and prepared the manuscript.

Conflicts of Interest: The authors declare no conflict of interest.

References

1. Van, D.V.F.; Tarola, A.M.; üemes, D.G.; Pirovani, M.E. Bioactive compounds and antioxidant capacity of Camarosa and Selva Strawberries (*Fragaria* × *ananassa* Duch). *Food* **2013**, *2*, 120–131.
2. Atress, A.S.H.; El-Mogy, M.M.; Aboul-Anean, H.E.; Alsanius, B.W. Improving strawberries fruit storability by edible coating as carrier of thymol or calcium chloride. *J. Hortic. Sci. Ornam. Plant.* **2010**, *2*, 88–97.
3. Juliana, A.; Maria, R.; Rafael, A.; Jerri, G.; Annemiek, S.; Adam, L.G. Evaluation of chlorine dioxide as an antimicrobial against Botrytis cinerea in California strawberries. *Food Packag. Shelf Life* **2016**, *9*, 45–54.
4. Feliziani, E.; Landi, L.; Romanazzi, G. Preharvest treatments with chitosan and other alternatives to conventional fungicides to control postharvest decay of strawberry. *Carbohydr. Polym.* **2015**, *132*, 111–117. [CrossRef] [PubMed]
5. Harder, F.R.; Elgar, H.J.; Watkins, C.B.; Jackson, P.J.; Hallett, I.C. Physical and mechanical changes in strawberry fruit after high carbon dioxide treatments. *Postharvest Biol. Technol.* **2000**, *19*, 139–146.
6. Li, C.; Kader, A.A. Residual effects of controlled atmospheres on postharvest physiology and quality of strawberries. *J. Am. Soc.* **1989**, *114*, 629–634.
7. Krochta, J.M.; Elizabeth, A.B.; Myrna, O. *Edible Coatings and Films to Improve Food Quality*; Technomic Publishing Co.: Lancaster, PA, USA; pp. 201–277.
8. Lozano-Navarro, J.I.; Díaz-Zavala, N.P.; Velasco-Santos, C.; Martínez-Hernández, A.L.; Tijerina-Ramos, B.I.; García-Hernández, M.; Rivera-Armenta, J.L.; Páramo-García, U.; Reyes-de la Torre, A.I. Antimicrobial, optical and mechanical properties of chitosan-starch films with natural extracts. *Int. J. Mol. Sci.* **2017**, *18*, 997. [CrossRef] [PubMed]
9. Romanazzi, G.; Nigro, F.; Ippolito, A. Effectiveness of pre and postharvest chitosan treatments on storage decay of strawberries. *Riv. Di Fruttic. e di Ortofloric.* **2000**, *62*, 71–75.
10. Khalifa, I.; Barakat, H.; El-Mansy, H.A.; Soliman, S.A. Improving the shelf-life stability of apple and strawberry fruits applying chitosan-incorporated olive oil processing residues coating. *Food Packag. Shelf Life* **2016**, *9*, 10–19. [CrossRef]
11. Velickova, E.; Winkelhausen, A.; Kuzmanova, S.; Alves, V.; Moldão-Martins, M. Impact of chitosan-beeswax edible coatings on the quality of fresh strawberries (*Fragaria ananassa* cv Camarosa) under commercial Storage Conditions. *LWT-Food Sci. Technol.* **2013**, *62*, 80–92. [CrossRef]
12. Hu, D.; Wang, L. Fabrication of antibouterial blend film from poly(vinyl alcohol)and quaternized chitosan for packaging. *Mater. Res. Bull.* **2016**, *78*, 46–52. [CrossRef]
13. Hyder, M.N.; Chen, P. Pervaporation dehydration of ethylene glycol with chitosan–poly(vinyl alcohol) blend membranes: Effect of CS–PVA blending ratios. *J. Membr. Sci.* **2009**, *340*, 171–180. [CrossRef]
14. Abdelrazek, E.M.; Elashmawi, I.S.; Labeeb, S. Chitosan filler effects on the experimental characterization, spectroscopic investigation and thermal studies of PVA/PVP blend films. *Phys. B Condens. Matter* **2010**, *405*, 2021–2027. [CrossRef]
15. Tripathi, S.; Mehrotra, G.K.; Dutta, P.K. Physicochemical and bioactivity of cross-linked chitosan-PVA film for food packaging applications. *Int. J. Biol. Macromol.* **2009**, *45*, 372–376. [CrossRef] [PubMed]
16. Batista, K.A.; Lopes, F.M.; Yamashita, F.; Fernandes, K.F. Lipase entrapment in PVA/Chitosan biodegradable film for reactor coatings. *Mater. Sci. Eng. C Mater.* **2013**, *33*, 1696–1701. [CrossRef] [PubMed]
17. Li, X.; Goh, S.H.; Lai, Y.H.; Wee, A.T.S. Miscibility of carboxyl-containing polysiloxane/poly(vinylpyridine) blends. *Polymer* **2000**, *41*, 6563–6571. [CrossRef]
18. Abdelaziz, M.; Abdelrazek, E.M. Effect of dopant mixture on structural, optical and electron spin resonance properties of polyvinyl alcohol. *Phys. B Condensed Matter* **2007**, *390*, 1–9. [CrossRef]
19. Cao, S.; Shi, Y.; Chen, G. Blend of chitosan acetate salt with poly(n-vinyl-2-pyrrolidone): Interaction between chain-chain. *Mater. Res. Bull.* **1998**, *41*, 553–559. [CrossRef]
20. Farris, S.; Introzzi, L.; Biagioni, P.; Holz, T.; Schiraldi, A.; Piergiovanni, L. Wetting of biopolymer coatings: Contact angle kinetics and image analysis investigation. *Langmuir* **2011**, *27*, 7563–7574. [CrossRef] [PubMed]
21. Kim, J.H.; Kim, J.Y.; Lee, Y.M.; Kim, K.Y. Properties and swelling characteristics of cross-linked poly(vinyl alcohol) chitosan blend membrane. *J. Appl. Polym. Sci.* **1992**, *45*, 1711–1717. [CrossRef]
22. Mispo, V.M.; Mansur, A.A.; Barbosa-Stancioli, E.F.; Mansur, H.S. Biocompatibility of nanostructured chitosan/poly(vinyl alcohol) blends chemically crosslinked with genipin for biomedical applications. *J. Biomed. Nanotechnol.* **2010**, *6*, 166–175.

23. Quirozcastillo, J.M.; Rodríguezfélix, D.E.; Grijalvamonteverde, H.; Del, C.T.; Plascenciajatomea, M.; Rodríguezfélix, F.; Pedro, H.F. Preparation of extruded polyethylene/chitosan blends compatibilized with polyethylene-graft-maleic anhydride. *Carbohydr. Polym.* **2014**, *101*, 1094–1100. [CrossRef] [PubMed]
24. Pelissari, F.M.; Yamashita, F.; Mve, G. Extrusion parameters related to starch/chitosan active films properties. *Int. J. Food Sci. Technol.* **2011**, *46*, 702–710. [CrossRef]
25. Hang, A.T.; Tae, B.; Park, J.S. Non-woven mats of poly(vinyl alcohol)/chitosan blends containing silver nanoparticles: Fabrication and characterization. *Carbohydr. Polym.* **2010**, *82*, 472–479. [CrossRef]
26. Laxmeshwar, S.S.; Viveka, S.; Madhu Kumar, D.J.; Dinesha; Bhajanthri, R.F.; Nagaraja, G.K. Preparation and characterization of modified cellulose fiber-reinforced polyvinyl alcohol/polypyrrolidone hybrid film composites. *Macromol. Sci. A* **2012**, *49*, 639–647. [CrossRef]
27. Fernandezsaiz, P.; Soler, C.; Lagaron, J.M.; Ocio, M.J. Effects of chitosan films on the growth of Listeria monocytogenes, Staphylococcus aureus and Salmonella spp. in laboratory media and in fish soup. *Int. J. Food Microbiol.* **2010**, *137*, 287–294. [CrossRef] [PubMed]
28. Kanatt, S.R.; Rao, M.S.; Chawla, S.P.; Sharma, A. Active chitosan–polyvinyl alcohol films with natural extracts. *Food Hydrocoll.* **2012**, *29*, 290–297. [CrossRef]
29. Han, C.; Zhao, Y.; Leonard, S.W.; Traber, M.G. Edible coatings to improve storability and enhance nutritional value of fresh and frozen strawberries (*Fragaria* × *ananassa*) and raspberries (*Rubus ideaus*). *Postharvest Biol. Technol.* **2004**, *33*, 67–78. [CrossRef]
30. Martinez-Ferrer, M.; Harper, C.; Perez-Muroz, F.; Chaparro, M. Modified atmosphere packaging of minimally processed mango and pineapple fruits. *J. Food Sci.* **2002**, *67*, 3365–3371. [CrossRef]
31. Cong, F.S.; Zhang, Y.G.; Dong, W.Y. Use of surface coatings with natamycin to improve the storability of Hami melon at ambient temperature. *Postharvest Biol. Technol.* **2007**, *46*, 71–75. [CrossRef]
32. Almenar, E.; Del-Valle, V.; Hernández-Muñoz, P.; Lagarón, J.M.; Catalá, R.; Gavara, R. Equilibrium modified atmosphere packaging of wild strawberries. *J. Sci. Food Agric.* **2007**, *87*, 1931–1939. [CrossRef]
33. Duan, J.; Wu, R.; Strik, B.C.; Zhao, Y. Effect of edible coatings on the quality of fresh blueberries (Duke and Elliott) under commercial storage conditions. *Postharvest Biol. Technol.* **2011**, *59*, 71–79. [CrossRef]
34. Maqbool, M.; Ali, A.; Alderson, P.G.; Zahid, N.; Siddiqui, Y. Effect of a novel edible composite coating based on gum arabic and chitosan on biochemical and physiological responses of banana fruits during cold storage. *J. Agric. Food Chem.* **2011**, *59*, 5474–5482. [CrossRef] [PubMed]
35. Abugoch, L.E.; Tapia, C.; Villamán, M.C.; Yazdanipedram, M.; Díazdosque, M. Characterization of quinoa protein-chitosan blend edible films. *Food Hydrocoll.* **2011**, *25*, 879–886. [CrossRef]
36. Bai, R.K.; Huang, M.Y.; Jiang, Y.Y. Selective permeabilities of chitosan-acetic acid complex membrane and chitosan-polymer complex membranes for oxygen and carbon dioxide. *Polym. Bull.* **1998**, *20*, 83–88. [CrossRef]
37. Perkins-Veazie, P. Growth and ripening of strawberry fruit. *Hortic. Rev.* **2010**, *17*, 267–297.
38. Harker, F.R.; Redgwell, R.J.; Hallett, I.C.; Murray, S.H.; Carter, G. Texture of fresh fruit. *Hortic. Rev.* **1997**, *8*, 13–18.
39. Velickova, E.; Winkelhausen, E.; Kuzmanova, S.; Moldão-Martins, M.; Alves, V.D. Characterization of multilayered and composite edible films from chitosan and beeswax. Characterization of multilayered and composite edible films from chitosan and beeswax. *LWT Food Sci. Technol.* **2015**, *21*, 83–93.
40. Villarreal, N.M.; Rosli, H.G.; Martínez, G.A.; Civello, P.M. Polygalacturonase activity and expression of related genes during ripening of strawberry cultivars with contrasting fruit firmness. *Postharvest Biol. Technol.* **2008**, *47*, 141–150. [CrossRef]
41. Hernández-Muñoz, P.; Almenar, E.; Ocio, M.J.; Gavara, R. Effect of calcium dips and chitosan coatings on postharvest life of strawberries (*fragaria* × *ananassa*). *Postharvest Biol. Technol.* **2006**, *39*, 247–253. [CrossRef]
42. Gol, N.B.; Patel, P.R.; Rao, T.V.R. Improvement of quality and shelf-life of strawberries with edible coatings enriched with chitosan. *Postharvest Biol. Technol.* **2013**, *85*, 185–195. [CrossRef]
43. Sangsuwan, J.; Pongsapakworawat, T.; Bangmo, P.; Sutthasupa, S. Effect of chitosan beads incorporated with lavender or red thyme essential oils in inhibiting botrytis cinerea and their application in strawberry packaging system. *LWT-Food Sci. Technol.* **2016**, *74*, 14–20. [CrossRef]
44. Valenzuela, C.; Tapia, C.; López, L.; Escalona, V.; Abugoch, L. Effect of edible quinoa protein-chitosan based films on refrigerated strawberry (*Fragaria* × *ananassa*) quality. *Electron. J. Biotechnol.* **2015**, *6*, 406–411. [CrossRef]

45. El Ghaouth, A.; Arul, J.; Asselin, A.; Benhamou, N. Antifungal activity of chitosan on post-harvest pathogens: Induction of morphological and cytological alterations in Rhizopus stolonifer. *Mycol. Res.* **1992**, *96*, 769–779. [CrossRef]
46. Bautista-Baños, S.; An, H.L.; Velázquez Del Valle, M.G.; Hernández-López, M.; Ait Barka, E.; Bosquez-Molina, E.; Bosquez-Molina, E.; Wilson, C.L. Chitosan as a potential natural compound to control pre and postharvest diseases of horticultural commodities. *Crop. Prot.* **2006**, *25*, 108–118. [CrossRef]
47. Aday, M.S.; Caner, C. The applications of active packaging and chlorine dioxide for extended shelf life of fresh strawberries. *Packag. Technol. Sci.* **2011**, *24*, 123–136. [CrossRef]

© 2017 by the authors. Licensee MDPI, Basel, Switzerland. This article is an open access article distributed under the terms and conditions of the Creative Commons Attribution (CC BY) license (http://creativecommons.org/licenses/by/4.0/).

Article

Effective Postharvest Preservation of Kiwifruit and Romaine Lettuce with a Chitosan Hydrochloride Coating

Elena Fortunati [1,*], Geremia Giovanale [2], Francesca Luzi [1], Angelo Mazzaglia [2], Josè Maria Kenny [1], Luigi Torre [1] and Giorgio Mariano Balestra [2,*]

1 Department of Civil and Environmental Engineering, University of Perugia, UdR INSTM, Strada di Pentima 4, 05100 Terni, Italy; francesca.luzi85@gmail.com (F.L.); jose.kenny@unipg.it (J.M.K.); luigi.torre@unipg.it (L.T.)
2 Department of Agriculture and Forestry Science (DAFNE), University of Tuscia, Via S. Camillo De Lellis snc, 01100 Viterbo, Italy; g.giovanale@unitus.it (G.G.); angmazza@unitus.it (A.M.)
* Correspondence: elena.fortunati@unipg.it (E.F.); balestra@unitus.it (G.M.B.); Tel.: +39-0744-492921 (E.F.); +39-0761-357474 (G.M.B.); Fax: +39-0744-492950 (E.F.); +39-0761-357434 (G.M.B.)

Academic Editors: Stefano Farris and Lluís Palou
Received: 28 August 2017; Accepted: 9 November 2017; Published: 11 November 2017

Abstract: Kiwifruits and romaine lettuce, among the most horticulturally-consumed fresh products, were selected to investigate how to reduce damage and losses before commercialization. The film-forming properties, physico-chemical, and morphological characteristics, as well as the antimicrobial response against *Botrytis cinerea* and *Pectobacterium carotovorum* subsp. *carotovorum* of chitosan hydrochloride (CH)-based coatings were investigated. The results underlined the film-forming capability of this CH that maintained its physico-chemical characteristics also after dissolution in water. Morphological investigations by FESEM (Field Emission Scanning Electron Microscopy) underlined a well-distributed and homogeneous thin coating (less than 3–5 µm) on the lettuce leaves that do not negatively affect the food product functionality, guaranteeing the normal breathing of the food. FESEM images also highlighted the good distribution of CH coating on kiwifruit peels. The in vitro antimicrobial assays showed that both the mycelial growth of *Botrytis cinerea* and the bacterial growth of *Pectobacterium carotovorum* subsp. *carotovorum* were totally inhibited by the presence of CH, whereas in vivo antimicrobial properties were proved for 5–7 days on lettuce and until to 20–25 days on kiwifruits, demonstrating that the proposed coating is able to contrast gray mold frequently caused by the two selected plant pathogens during postharvest phases of fruit or vegetable products.

Keywords: chitosan hydrochloride; coating; edible film; food safety; postharvest; antimicrobial properties; *Botrytis cinerea*; *Pectobacterium carotovorum* subsp. *carotovorum*; rotting

1. Introduction

Harvested fruits and vegetables, when infected by degrading microbes, inevitably undergo a reduction of economic value, a significant reduction of the food shelf-life, and often become insecure for human safety [1]. In 2011, a FAO (Food and Agriculture Organization of the United Nations) report quantified postharvest losses for about one-third of the fresh consumables produced worldwide [2], which explains the great efforts of researchers and industries to counteract these costs in recent decades.

The main strategy to control postharvest rot provides for the use of fungicides [3]. However, their use can be both non-economical, when the cost of treatment exceeds the loss from rot, and dangerous, because of the risk to select resistant strains by reiterated applications [4]. Moreover, the risk for human health and environment pushes urgently for new, effective, and safer control strategies against postharvest diseases. Biological control with antagonistic microorganisms is one of the most promising alternatives, within the development of technologies able to preserve fresh horticultural products from dangerous microorganisms (bacteria and fungi) during postharvest [5–7]. The use of natural compounds or less aggressive additives as chemical extracts was also recently investigated [8–11]. Alternative postharvest solutions are becoming essential to preserve the freshness and quality of food products, and the application of edible coatings lend themselves to this aim, with promising results already reported in preserving the quality of products [12–14].

Recently, the use of nontoxic materials to develop edible coatings with the idea to preserve human health was proposed [15–17]. Specifically, edible coatings are thin layers of non-toxic materials, often extracted from animal or vegetal source, and applied directly on to the food surface. The use of edible coatings to preserve food product quality is a relatively low cost and environmentally friendly strategy with several advantages, including biodegradability, as well as the possibility to obtain a semi-permeable barrier against gases and water vapor, reducing microbial attack [18]. Furthermore, edible coatings can be combined with natural or synthetic active principles to prevent microbial decay in a more effective manner [19]. Natural additives, such as essential oils and fruit seed extracts have demonstrated good antimicrobial and antifungal activity with biopolymeric compatibility, thus lending themselves to be added into edible coating formulations [20]. Plant extracts offer advantages in terms of low production costs, low toxicity, and good biodegradability [21]. Additionally, most of the extracts are rich in polyphenols which can improve the antioxidant properties of the edible coatings [22].

In this research activity, chitosan hydrochloride (CH) film-forming solution was considered as an active coating for kiwifruit and lettuce and its effect as an antimicrobial agent was investigated. Chitosan hydrochloride is used as basic substance in plant protection; it is purposed in conformity with the legal provisions of Regulation (EC) No 1107/2009 [23]. Recently, chitosan and its byproducts, like oligochitosan, have been investigated to control postharvest diseases [17]. Chitosan is a natural, nontoxic in a range of toxicity tests, biocompatible, safe, and biodegradable natural alkaline polysaccharide derived from the de-acetylation of chitin [24], widely applied in biotechnology, medicine, water treatment, food science, and agriculture [25]. In the agricultural sector, chitosan has been used in leaf, seed, vegetable, and fruit coatings, and in plant protection [26]. Chitosan is obtained by deacetylating chitin comprising copolymers of β(1,4)-glucosamine and N-acetyl-d-glucosamine, and it can be extracted from wastes of the food-processing industry (shrimps, shells of crabs, krill, and lobsters) by using a concentrated basic solution combined with high temperatures [27]. Chitosan was classified as safe (GRAS) by the US Food and Drug Administration (FDA) in 2001 and it presents bacteriostatic and fungistatic properties [28] that perfectly address the active edible coating concept to preserve the freshness and to avoid the microbial growth on the surface of vegetable and fruit products [11]. The scavenging ability of chitosan is associated with the presence of active hydroxyl and amino groups in the polymer chains. The hydroxyl groups in the polysaccharide units can react with free radicals, and according to free radical theory, the amino groups of chitosan can react with free radicals to form additional stable macroradicals [29].

The aim of this work is to analyze the effect of chitosan hydrochloride-based edible coating for the preservation of fresh vegetables and fruits, specifically of lettuce and kiwifruits, from soft rots and gray mold caused, respectively, by *Pectobacterium carotovorum* subsp. *carotovorum* and *Botrytis cinerea* during postharvest storage. The effectiveness of this innovative treatment was evaluated both in vitro and in vivo on kiwifruits of *Actinidia deliciosa* cv. Hayward and on heads of Romaine lettuce (*Lactuca sativa* var *longifolia*).

2. Materials and Methods

2.1. Materials

Romaine lettuce was collected in Central Italy, Lazio region, Romaine lettuce, as a dark leafy green, is rich source of vitamin K, vitamin A, antioxidants, and a moderate source of folate and iron. It is particularly requested from consumers and, so, its cultivation and production in the USA recently increased up to 40% [30].

Kiwifruits were collected in Central Italy, Lazio region (Cisterna di Latina, LT), the most important Italian kiwifruit area, where around 8000 hectares of *Actinidia* spp are cultivated. The *Actinidia* cultivation and the kiwifruit production show a trend of an exponential increase that has reached nearly 100,000 hectares, and Italy is the world largest producer of kiwifruit with commercial value, excluding China.

Chitosan hydrochloride (CH—M_w = 60,000 Da, degree of deacetylation = 80%–90%) was supplied by Sigma-Aldrich® (St. Louis, MO, USA).

2.2. Preparation and Characterization of Film Coating-Forming Solution

Chitosan hydrochloride solution (1 g/L) was prepared by dissolving the polymer in deionized water under magnetic stirring for 2 h at room temperature (RT). The chitosan solution was cast onto a Petri dish cover by Teflon® (The Chemours Company, Wilmington, DE, USA) and dried at RT for 24 h. The preparation of the CH-based film before the coating application on food product surface was useful to evaluate the morphological, chemical, and thermal characteristics.

The microstructure of CH pellet and produced film (surface and fractured surfaces) were investigated by FESEM (Supra 25-Zeiss, Carl Zeiss Microscopy GmbH, Jena, Germany) after gold sputtering and by using an accelerating voltage of 5 kV. Fourier infrared (FTIR) spectra of chitosan powder and chitosan film were analyzed by using a Jasco FTIR 615 spectrometer (Jasco Inc, Easton, MD, USA) in the 400–4000 cm^{-1} range, in transmission mode. The chitosan hydrochloride pellet was analyzed using KBr discs while a few drops of chitosan solution were casted on a silicon wafer and investigated. Thermogravimetric measurements (TGA) were performed by using a Seiko Exstar 6300 (RT Instruments Inc., Woodland, CA, USA). Heating scans from 30 to 900 °C at 10 °C·min^{-1} under nitrogen atmosphere were performed for the chitosan pellet and film.

2.3. In Vitro Antimicrobial Assays

The in vitro antimicrobial activity of chitosan hydrochloride was assayed by preliminary incorporation of the active ingredient in respective media for each of the two considered pathogens (both bacterium and fungus). Chitosan hydrochloride was dissolved at 1% (w/v) concentration in sterile distilled water under magnetic stirring for 2 h at RT. Then, 100 mL of the solution was added to 1 L of nutrient agar (NA) medium immediately before it was poured into the Petri dishes at a temperature of 40–45 °C, to obtain a final concentration of 1 g/L in the medium. Parallel controls were maintained with corresponding amounts of sterile distilled water mixed with NA medium.

Antimicrobial assays were developed by using aggressive micro-organisms on cultivated plants and able to cause severe damage and economic losses, especially during postharvest phases. Specifically, a high virulent strain of the fungal pathogen *Botrytis cinerea*, able to colonize host tissue by a relevant mycelium development and causing extended gray mold, and the collapse of kiwifruits during their storage/commercialization [31]. Additionally, a highly virulent bacterial strain of *Pectobacterium carotovorum* subsp. *carotovorum* able, due to its enzymatic properties, to destroy external and internal tissues of several vegetables, such as lettuce, was selected and used [32].

Concerning *B. cinerea*, the known strain CBS 120091 (obtained from Westerdijk Fungal Biodiversity Institute, Utrecht, The Netherlands) was stored in the microbial collection of the Department of Agriculture and Forestry Science (DAFNE) at the University of Tuscia, Viterbo, Italy. After the revitalization on potato dextrose agar (PDA) medium, plugs 5 mm in diameter were excised from a freshly-grown culture and positioned at the center of the Petri dishes filled with the PDA medium incorporated with CH, and then incubated at 24 ± 1 °C for seven days. The treatment, as for control without CH, was tested in triplicate. The radial growth of mycelium was measured along perpendicular axes from inoculum point at the center of the plate (four measures per each plate) after one, two, three, four, and seven days after inoculation.

With respect to *P. c.* subsp. *carotovorum*, the known bacterial strain used (DSM 30184 from Leibniz Institute DSMZ-German Collection of Microorganisms and Cell Cultures, Braunschweig, Germany) in the experiment was stored in the same microbial collection of DAFNE. After revitalization on NA supplemented by 5% of sucrose (NAS), bacterial colonies were collected and suspended in sterile distilled water (SDW) at a concentration of 1×10^6 CFU/mL. Then, 100 µL aliquots of the bacterial suspension were homogeneously distributed on the surface of Petri dishes filled with the NA medium incorporated with CH, as described, and then incubated at 27 ± 1 °C for 48 h. The treatment, as for control without CH, was tested in triplicate. The growth of bacterial colonies was assessed 48 h after inoculation.

2.4. Coating Application and in Vivo Tests on Kiwifruits and Lettuce

The in vivo tests, both those on kiwifruit and on lettuce, have been set to reproduce, at best, what is provided in common postharvest treatments. The application of CH coating, indeed, was simply obtained by full immersion in chitosan hydrochloride solution (1 g/L) after a previous washing in tap water. Then kiwifruits and vegetables were dried at room temperature.

More specifically, for kiwifruit, the wounds, normally caused by detachment from the plant during harvest, were simulated by a cut with a sterile scalpel at the petiole. Then, fruits were immediately immersed for 30 s in a bath containing the chitosan hydrochloride solution, or the other active ingredients chosen for the comparison, or pure water as a control. According to the market analysis, Fenhexamid, as chemical active ingredient present in the most used commercial formulation (Teldor® Plus, New Boston, NH, USA) was selected and utilized, dissolved in water at the commercially-suggested dose (1.2 g/L) for comparative treatment. Then, 100 µL aliquots of a suspension containing 1×10^6 conidia/mL of *B. cinerea* in sterile distilled water, as assessed by a Thoma cell counting chamber, were distributed by pipette on the surface of the cut petiole. Inoculated fruits were then transferred into a storage box at 4 °C. Six homogenous fruits were used per each treatment: chitosan solution, one commercial formulation, and SDW for both positive (inoculated fruits) and negative (not inoculated fruits) controls, for a total of 30 fruits. The entire experiment was repeated twice. The effectiveness of treatments was evaluated by means of a quantitative scale for gray mold severity, in which 0 is for totally healthy fruits, 1 is for fruits with 1%–20% surface damage, 2 is for fruits with 21%–40% surface damage, 3 is for fruits with 41%–60% surface damage, 4 is for fruits with 61%–80% surface damage, and 5 is for fruits with >81% surface damage. Fruits were examined and data collected at 1, 7, 14, 17, 20, 23, and 25 days after inoculation. Data were statistically analyzed by Tukey's HSD multiple comparison test with $p = 0.01$.

On Romaine lettuce (*Lactuca sativa* L. var. *longifolia*), five heads were first cut off by sterile scalpel at the taproot, to imitate what happens during harvest, washed in a water bath, and treated with chitosan hydrochloride solution. As a reference for the effectiveness of the treatment (positive control), five heads were alternatively treated with a sodium hypochlorite solution (1 mL/L), which is a common postharvest treatment for lettuce, or not further treated, as negative control (five heads). After the treatments, the lettuce heads were dried at room temperature and then inoculated on the fresh cut by spraying about 500 µL of a bacterial suspension (1×10^6 CFU/mL). A set of five plants were not subjected to inoculum as an additional control. After the inoculation, lettuce heads were kept at 90% RH and 26 °C to ensure environmental conditions ideal for bacterial proliferation. The entire experiment was repeated twice. The disease progression on lettuce heads was checked at one, three, and five days after inoculation, by means of a scale of damaging on the taproot cut surface, in which 0 is for totally healthy cut, 1 is for 1%–20% surface damage, 2 is for 21%–40% surface damage, 3 is for 41%–60% surface damage, 4 is for 61%–80% surface damage, and 5 is for >81% surface damage. Data were statistically analyzed by Tukey's HSD multiple comparison test with p = 0.01 by DSAASTAT software (Version 1.022).

Finally, the presence and microstructure of chitosan coatings on lettuce leaves and cores, and on kiwifruit peel, were investigated by FESEM (Supra 25-Zeiss, Carl Zeiss Microscopy GmbH, Jena, Germany) after gold sputtering and by using an accelerating voltage of 5 kV.

3. Results and Discussion

3.1. Morphological, Chemical, and Thermal Properties of Polymer Pellet and Chitosan-Based Film Coating

The bioactivity of chitosan is a function of its physico-chemical properties which also have an effect on its film-forming characteristics [33]. For this reason, morphological, chemical, and thermal properties of a polymer pellet and chitosan-based coating film were investigated to study the applicability of the selected commercial polymer grade as a coating with the final goal to preserve the food quality and safety during the storage, transportation, and market period.

The microstructure of the CH pellet and produced film (surface and cross-section) were investigated by FESEM in order to prove the film-forming capability of the selected grade of polymer in water and investigate the morphological and physico-chemical properties of the obtained film-coating before its application on the fresh food. Chitosan hydrochloride powder is a white or off-white odorless, semitransparent, and amorphous material. FESEM micrographs of the CH pellet at different magnifications are shown in Figure 1. The CH powder appeared agglomerated into flakes with irregular shape and having dimensions ranging between 10 and 100–200 µm (Figure 1a,b) [34]. Chitosan hydrochloride solution was then prepared by dissolving the polymer, at a specific concentration (1 g/L), in deionized water, by magnetic stirrer. As demonstrated by the inset in Figure 1b, this grade of polymer presented a perfect ability to dissolve in water, forming a clear and stable solution characterized by a pH value of 6.22 [17]. For this ability to easily dissolve in a sustainable solvent, chitosan hydrochloride is widely used in the food industry as a preservative and can keep fruits and vegetables fresh. After the dissolution, the obtained CH solution was cast onto a Petri dish and dried in order to test the film-forming capability of the selected commercial polymer. Figure 1c,d show the surface topography of the obtained film at different magnifications. The film surface was characterized by a well-defined porous structure, with interconnected pores that are also present on the cross-sections of the obtained film, as shown in Figure 1e,f. FESEM investigation of the film fractured surfaces, in fact (Figure 1e,f), underlined a tri-dimensional porous structure on the film thickness, whereas thin films (4.8 ± 0.3 µm thick), typical of medium-low molecular weight polymer, were obtained. This morphology will guarantee the breathing of fruit-vegetable products of fundamental importance during the storage, transportation, and market period for their correct conservation.

Furthermore, before the application of the CH coating on fruit and vegetable products, chemical characterization by FTIR and the investigation of the thermal stability by TGA, before and after the dissolution in water of CH powder, were performed and compared. Figure 2a shows the FTIR spectra for the chitosan pellet and film, whereas Table 1 shows the assignment of the main peaks at each wavenumber. The results underlined that not particular alterations on the chemical properties of CH powder were induced by the dissolution in water since all the main characteristic bands are present in both the CH pellet and film spectrum. The main characteristic bands were assigned to saccharide structures at 900, 1020, and 1155 cm^{-1} and strong amino characteristic bands at around 1635 and 1570 cm^{-1} assigned to amide I and amide II bands, respectively. Furthermore, the peak at about 1250 cm^{-1} corresponds to the amino group, whereas 2884 cm^{-1} corresponds to the C–H stretching [35,36]. All these cited peaks, characterizing both the CH pellet and film spectra, underlined that the film-forming procedure does not affect the chemical properties of the selected polymer; however, a shift of the OH stretching, centered at 3430 cm^{-1} for the chitosan pellet and at 3370 cm^{-1}

for the film was registered (Table 1), and a more intense associated band (Figure 2a) was highlighted for the CH film, induced by the solvent casting procedure in water. This result was also confirmed by TGA analysis.

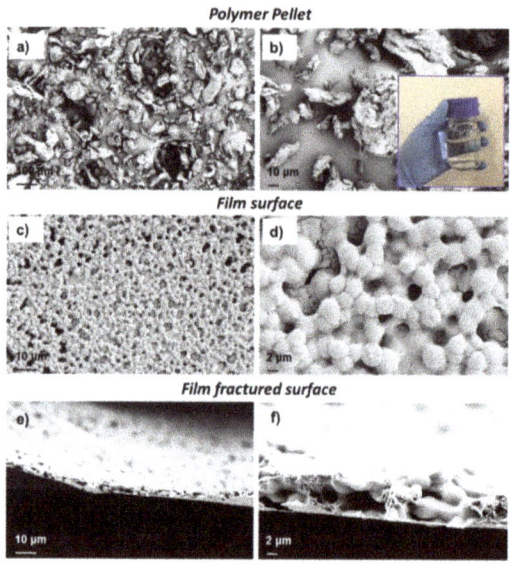

Figure 1. FESEM investigation of (**a**,**b**) chitosan pellet, (**c**,**d**) surface, and (**e**,**f**) fractured surface of the chitosan hydrochloride-based coating. Visual image of chitosan hydrochloride solution (inset in **b**).

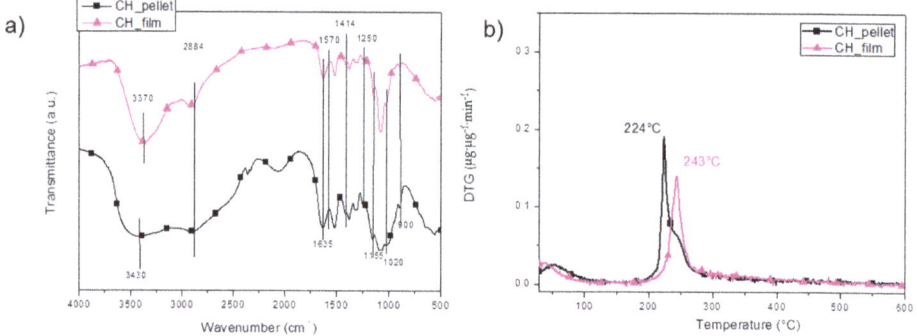

Figure 2. (**a**) FTIR spectra and (**b**) DTG thermograms of the chitosan hydrochloride pellet and film.

The derivate (DTG) curves of weight loss for the TGA tests of the chitosan pellet and film are reported in Figure 2b. Two significant weight loss stages were observed in the DTG curve for CH powder. The small weight loss at 50–120 °C (about 10%) is due to the loss of adsorbed and bound water, while the second main weight loss step between 200 and 300 °C and centered at 224 °C (showing a shoulder at 240–250 °C) was attributed to the thermal degradation of CH, as previously reported by several authors [36,37]. The CH film thermogram showed a similar behavior with the first small signal (at around 100 °C) and the second main peak always in the same region (200–300 °C). However, a shift to a higher temperature of the main degradation peak (centered at 243 °C for chitosan film) was registered and due to a re-arrangement of the polymer chain during the film-forming procedure that induced and increased in the final thermal stability of the obtained structure.

Table 1. FTIR bands of chitosan hydrochloride.

Wavenumber (cm^{-1})	Assignment
3430 (pellet) 3370 (film)	–OH stretching
2884	C–H stretching
1635	Amide I
1570	Amide II
1414	–CH$_2$ bending
1250	amino group
900, 1020, and 1155	Saccharide structures

After the characterization, the CH solution at the specific selected concentration of 1 g/L was applied on lettuce and kiwifruits as described. The ability of chitosan hydrochloride to form a homogeneous film, its appearance and distribution on both vegetables (here, lettuce as an example) and fruit (kiwi) were investigated by FESEM at different magnifications and compared with the same food products treated with water as a control. Furthermore, in the case of lettuce, evidence of both leaves and taproot were reported since the taproot surface represents a preferential method for pathogen attack.

Figure 3A reports the evidence of lettuce leaves treated with water characterized by the typical stomatal aperture (Figure 3A(a,b)).

The arrows in Figure 3A(c,d) indicated the stomatal aperture partially covered by the CH coating. The FESEM images underlined a well distributed and homogeneous coating on the lettuce leaves, as well as the lettuce taproot (Figure 3B(c,d)) with the typical tri-dimensional porous structure as reported and discussed in Figure 1c–f (see inset in Figure 3B(d)). The FESEM images also underlined that, although the CH coating was well distributed on the vegetable's surface, it should not negatively affect the food product functionality since the stomatal aperture was quite evident after the application of the coating, guaranteeing the normal breathing of the food due to the porous structure of the applied CH coating.

Comparable results were also obtained for kiwifruit peel treated with the CH coating and always compared with water-treated kiwifruit products, used as a control (Figure 4). FESEM images underlined the well-obtained distribution of CH coating that perfectly adhered and covered the kiwifruit peels (see arrows in Figure 4c,d) maintained, also in this case, the porous structure (see inset in Figure 4d). Furthermore, although in all cases a thin film (less than 3–5 µm) was obtained due the low molecular weight of the selected commercial-grade polymer, the coating was able to thoroughly seal lettuce and kiwifruit peels, potentially guaranteeing good conservation of the food products.

In any case, the main goal of the present work was to prove the film-forming characteristics of the selected chitosan hydrochloride-grade polymer and to prove its potential efficiency, in vitro and in vivo, against specific pathogens causing fresh food deterioration, as following discussed. No direct evidence of the shelf-life or food qualities and vegetable/fruit organoleptic characteristics during storage were addressed here or studied, which could represent the main goals for future work.

3.2. Antimicrobial In Vitro Activity of Chitosan Hydrochloride

The results of in vitro radial growth of B. cinerea on PDA medium (control) or on PDA amended with chitosan hydrochloride (CH) are shown in Figure 5. The differences between CH treatment and control in terms of radial growth of the fungus (B. cinerea) at different times are reported, showing an inhibition of about 50%.

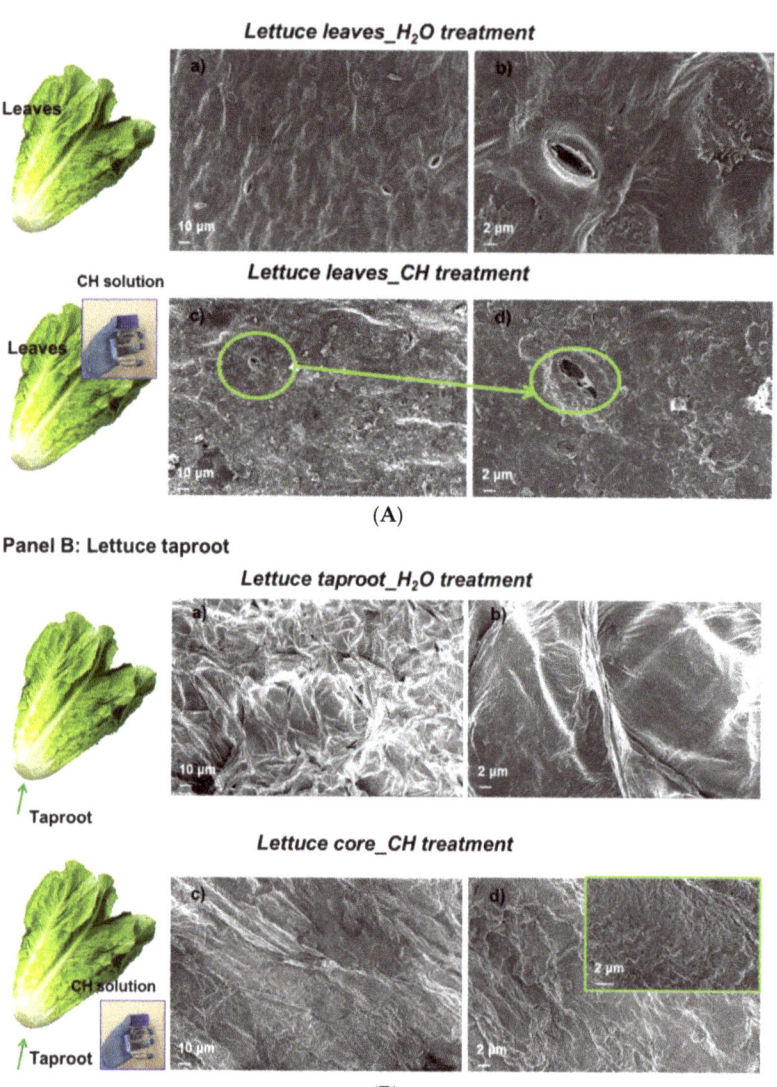

Figure 3. Evidence of CH coating on (**A**) lettuce leaves and (**B**) lettuce taproot treated with (**a,b**) water as control and with (**c,d**) CH solution.

Kiwi fruit

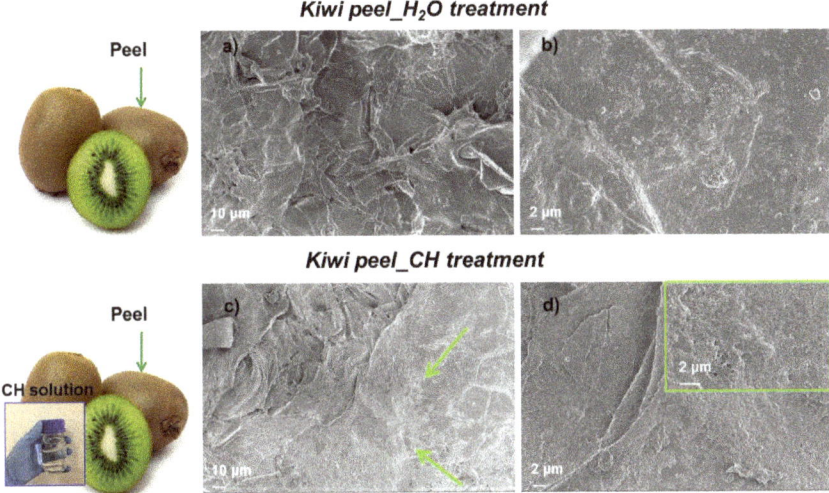

Figure 4. Evidence of CH coating on kiwifruit peel treated with (**a**,**b**) water as control and with (**c**,**d**) CH solution.

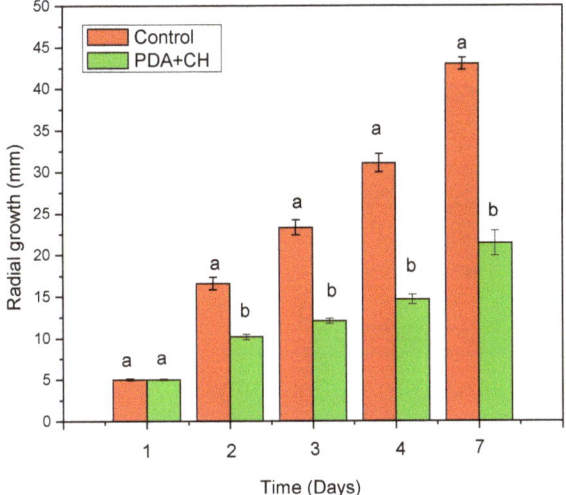

Figure 5. In vitro radial growth of *Botrytis cinerea* on PDA medium (control) or on PDA amended with chitosan hydrochloride (CH). For each evaluation date, columns with different letters are significantly different according to Tukey's HSD test ($p = 0.01$). Significant differences between two bars are marked with different letters; if there is no significant difference between two bars they get the same letter.

Figure 6 shows, visually, the antimicrobial effect of chitosan hydrochloride against *B. cinerea*. The image (Figure 6b) well underlined as the mycelial growth of *B. cinerea* was almost totally inhibited by the incorporation of chitosan hydrochloride solution in the medium until the end of the test.

The mycelial mat was rarefied and weak due to the CH presence (Figure 6b) whereas it appears completely homogeneous in the control (Figure 5a).

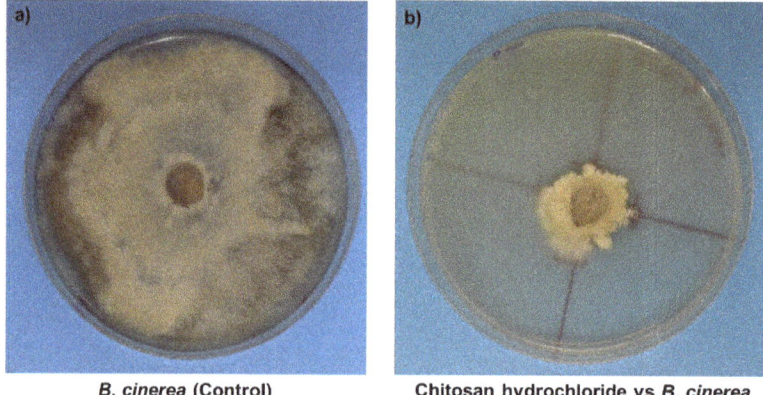

B. cinerea (Control) Chitosan hydrochloride vs B. cinerea

Figure 6. In vitro results after 7 days, with an evident inhibition of radial growth of *B. cinerea* by chitosan hydrochloride. *Botrytis cinerea* (**a**) and Chitosan hydrochloride vs *Botrytis cinerea* (**b**).

Figure 7 shows the antimicrobial effect of chitosan hydrochloride in vitro against *P. c.* subsp. *carotovorum* bacterium. As previously discussed for the fungal pathogen, the bacterial growth of *P. c.* subsp. *carotovorum* was also totally depleted by CH (data not shown), showing a very important activity to inhibit the growth of the colonies of this dangerous bacterial plant pathogen (Figure 7b).

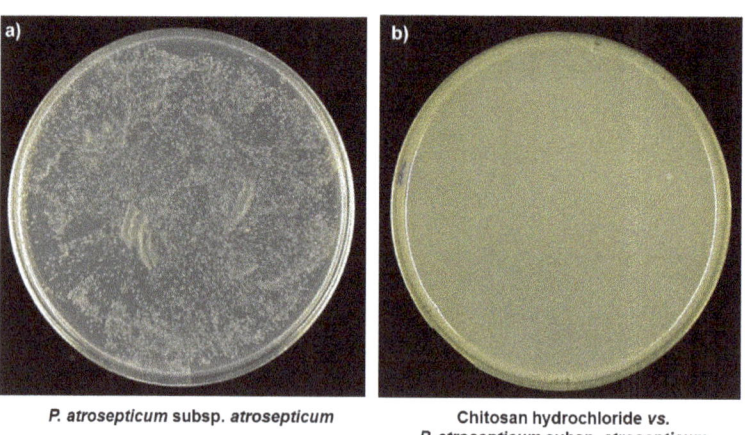

P. atrosepticum subsp. atrosepticum   Chitosan hydrochloride vs.
                                        P. atrosepticum subsp. atrosepticum

Figure 7. In vitro growth of colonies of *Pectobacterium carotovorum* subsp. *carotovorum* (**a**) in NA medium (control) or (**b**) in NA amended with CH ($p = 0.01$).

3.3. Antimicrobial Activity on Stored Fruits and Vegetables: In Vivo Assays

The results of in vivo antimicrobial tests against the *B. cinerea* by using CH on kiwifruits, are summarized in Figure 8. Until the 25th day after artificial inoculation, kiwifruits treated with chitosan hydrochloride showed a lower level of disease severity with respect to the control kiwifruits and to those treated with chemical compounds, both at external and internal levels (Figure 8).

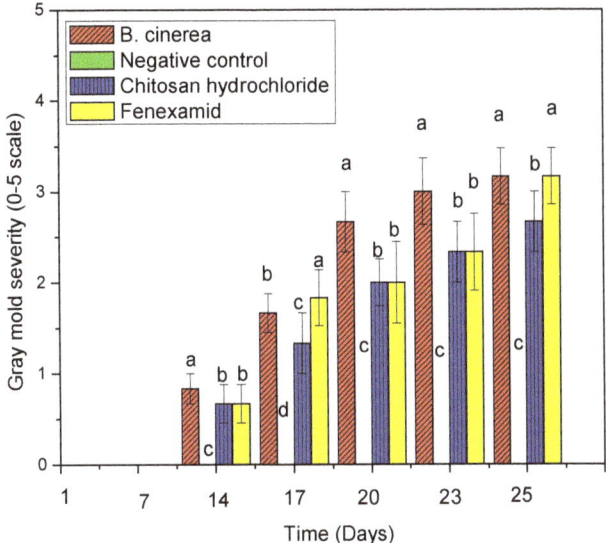

Figure 8. Gray mold severity on kiwifruits artificially inoculated with *B. cinerea* after treatment with the fungicide fenexamid or CH. Negative controls were not inoculated. For each evaluation date, columns with different letters are significantly different according to Tukey's HSD test ($p = 0.01$). Green histograms related to negative controls (only sterile distilled water, SDW) are absent because non-microbial growth was recorded in this thesis. Significant differences among the bars are marked with different letters; if there is no significant difference among the bars they get the same letter.

The kiwifruits, inoculated with this fungal pathogen, but not protected by any chemical product, started showing the typical gray mold at the petiole end, where the inoculum was deposited, starting from day 10. After that, the tissue rotting continued to expand reaching, after 25 days, an average of 3.2 in the symptom scale, which corresponds to about 70% of the entire surface of the fruit. The initial symptoms on kiwifruits treated with chitosan was slightly delayed, appearing around the 15th day but, interestingly, the rotting process was limited and was markedly less than those observed on kiwifruit controls and to those on kiwifruits submitted to chemical treatment.

The in vivo antimicrobial assays of the CH coating applied on lettuce surfaces with respect to *P. c.* subsp. *carotovotum* bacterial plant pathogen were investigated. The results, after seven days (data not shown), demonstrated an important reduction of damage (value 2–3), induced by chitosan hydrochloride-based coating. The damage, in the presence of the active coating, appeared (see Figure 9), in fact, much less severe that those recorded for the controls (only inoculated by *P. c.* subsp. *carotovotum*, 10^6 CFU/mL) even if the resulted in being much less effective with respect to the damage level developed on/in the samples treated with chemical compounds (value 1–2).

All the results obtained by in vitro and in vivo antimicrobial tests were statistically significant. On the basis of our evidence, the selected grade of chitosan hydrochloride, able to form a stable solution in a green solvent as water and to form a homogeneous coating on food, could be preventively applied to prolong the postharvest shelf-life of important fresh products. Considering the high concentrations used for fungal and bacterial artificial inoculations (1×10^6 conidia/mL and 1×10^6 CFU/mL, respectively) it is reasonable to consider the effectiveness of the chitosan hydrochloride-based coating against these plant pathogens up to five days on Romaine lettuce and up to 20 days on kiwifruits. Mechanisms by which chitosan hydrochloride coating reduced the decay of lettuce and kiwifruits with respect to *P. c.* subsp. *carotovorum* and to *B. cinerea*, not even studied here in depth, seems to be related to its bacteriostatic/fungistatic properties. Until now, antimicrobial activity of chitosan has been widely studied against clinically-important microorganisms; this can be considered a new contribution with respect to dangerous foodborne (bacteria and fungi) plant pathogens as other active principles of natural origin recently resulted in being able to control plant pathogens in greenhouses and in open fields [38–40]. Finally, it is well known that gray mold of kiwifruits is mainly caused by latent or wound infections that are produced in the orchard [41]; thus, effective postharvest control means need to provide curative activity (the treatment should be applied after the pathogen inoculation). Here the potentiality of the chitosan hydrochloride treatment, as a preventive strategy, in contrast to *B. cinerea*, was considered and studied to reduce further treatments. Future research activities will be also addressed for curative activity.

Figure 9. In vivo symptoms (external and internal damages) developed after five days due to artificial inoculation by *Pectobacterium carotovorum* subsp. *carotovorum* (DSM 30184 [Leibniz Institute German Collection of Microorganisms and Cell Cultures, Germany]) bacterial plant pathogen on Romaine lettuce after preventive treatments by chitosan hypochloride and chemical compound (Sodium hypochlorite) respect to the control thesis (treated only by *P. c.* subsp. *carotovorum* and, as control negative, only by SDW). (**a**) Control positive: *P. c.* subsp. *carotovorum* bacterial plant pathogen at 1×10^6 CFU/mL; (**b**) Control negative, SDW; (**c**) Chitosan hydrochloride (1 g/L) solution; and (**d**) Sodium hypochlorite (1 mL/L) solution ($p = 0.01$). Chitosan hydrochloride and Sodium hypochlorite (Sigma-Aldrich®, St. Louis, MO, USA).

4. Conclusions

The present work revealed interesting results of chitosan hydrochloride-based coatings on the safety of horticultural fresh products, such as lettuce and kiwifruits. Commercial grade chitosan hydrochloride was selected here as a polymer for food coating and its film-forming properties, physico-chemical and morphological characteristics, as well as antimicrobial response against *B. cinerea* and *P. c.* subsp. *carotovorum* were investigated. The results underlined the film-forming capability of this grade of chitosan, which maintained its physico-chemical characteristics after dissolution in water and which formed a thin and well-distributed coating on both kiwifruit and lettuce. The in vitro antimicrobial assays showed that both the mycelial growth of *B. cinerea* and the bacterial growth of *P. c.* subsp. *carotovorum* were totally inhibited by the presence of CH, whereas in vivo antimicrobial properties were proved for 5–7 days on lettuce and until 20–25 days on kiwifruits demonstrating that chitosan-based coating is able to contrast gray mold frequently caused by the two selected plant pathogens during the postharvest phases of both fruit or vegetable products. On *B. cinerea*, chitosan has also shown to inhibit the spore germination and this is an additional positive function that this organic material can express with respect to this dangerous pathogen [42]. Chitosan applications, in combination with essential oils, were also recently tested with respect to different bacterial plant pathogens and so our study assumes a particular relevance to improve sustainable strategies able to reduce the negative impact of different plant pathogens during postharvest phases [43]. The obtained results contribute to the idea and need of novel greener strategies and approaches for food quality and safety preservation. By using natural compounds, like chitosan hydrochloride, interesting opportunities emerge to limit the damage caused by dangerous plant pathogens on vegetable and fruit production after harvesting, reducing or avoid remarkable economic losses and preserving final products.

Author Contributions: Elena Fortunati, Giorgio Mariano Balestra, Josè Maria Kenny, and Luigi Torre conceived and designed the experiments; Elena Fortunati, Geremia Giovanale, and Francesca Luzi performed the experiments; Elena Fortunati, Angelo Mazzaglia, and Giorgio Mariano Balestra analyzed the data; Elena Fortunati and Giorgio Mariano Balestra wrote the paper.

Conflicts of Interest: The authors declare no conflict of interest.

References

1. Lugtenberg, B.; Rozen, D.E.; Kamilova, F. Wars between microbes on roots and fruits. *F1000Research* **2017**, *6*, 343. [CrossRef] [PubMed]
2. Gastavsson, J.; Cederberg, C.; Sonesson, U. *Global Food Losses and Food Waste*; Food and Agriculture Organization (FAO) of the United Nations: Rome, Italy, 2011.
3. Eckert, J.W.; Ogawa, J.M. The chemical control of postharvest diseases: Deciduous fruit, berries, vegetables and root/tuber crops. *Annu. Rev. Phytopathol.* **1988**, *26*, 433–469. [CrossRef]
4. Spotts, R.A.; Cervantes, L.A. Populations, pathogenicity, and benomyl resistance of *Botrytis* spp., *Penicillium* spp., and Mucor piriformis in packinghouses. *Plant Dis.* **1986**, *70*, 106–108. [CrossRef]
5. Wilson, C.L.; Pusey, P.L. Potential for biological control of postharvest plant diseases. *Plant Dis.* **1985**, *69*, 375–378. [CrossRef]
6. Wilson, C.L.; Wisniewski, M.E. Biological control of postharvest diseases of fruits and vegetables: An emerging technolygy. *Annu. Rev. Phytopathol.* **1989**, *27*, 425–441. [CrossRef]
7. Sharma, R.R.; Singh, D.; Singh, R. Biological control of postharvest diseases of fruits and vegetables by microbial antagonists: A review. *Biol. Control* **2009**, *50*, 205–221. [CrossRef]
8. Yang, W.; Fortunati, E.; Dominici, F.; Giovanale, G.; Mazzaglia, A.; Balestra, G.M.; Kenny, J.M.; Puglia, D. Synergic effect of cellulose and lignin nanostructures in PLA based systems for food antibacterial packaging. *Eur. Polym. J.* **2016**, *79*, 1–12. [CrossRef]
9. Mazzaglia, A.; Fortunati, E.; Kenny, J.K.; Torre, L.; Balestra, G.M. Nanomaterials in Plant Protection. In *Nanotechnology in Agriculture and Food Science*; Axelos, M.A.V., Van de Voorde, M., Eds.; Wiley-VCH Verlag GmbH & Co. KGaA: Weinheim, Germany, 2017; pp. 115–134.
10. Fortunati, E.; Benincasa, P.; Balestra, G.M.; Luzi, F.; Mazzaglia, A.; Del Buono, D.; Puglia, D.; Torre, L. Revalorization of barley straw and husk as precursors for cellulose nanocrystals extraction and their effect on PVA_CH nanocomposites. *Ind. Crops Prod.* **2016**, *92*, 201–217. [CrossRef]
11. Luzi, F.; Fortunati, E.; Giovanale, G.; Mazzaglia, A.; Torre, L.; Balestra, G.M. Cellulose nanocrystals from *Actinidia deliciosa* pruning residues combined with carvacrol in PVA_CH films with antioxidant/antimicrobial properties for packaging applications. *Int. J. Biol. Macromol.* **2017**, *104*, 43–55. [CrossRef] [PubMed]
12. Badawy, M.E.I.; Rabea, E.I. Potential of the biopolymer chitosan with different molecular weights to control postharvest gray mold of tomato fruit. *Postharvest Biol. Technol.* **2009**, *51*, 110–117. [CrossRef]
13. Davila-Avina, J.E.J.; Villa-Rodríguez, J.; Cruz-Valenzuela, R.; Rodríguez-Armenta, M.; Espino-Díaz, M.; Ayala-Zavala, J.F.; Gonzalez-Aguilar, G. Effect of edible coatings, storage time and maturity stage on overall quality of tomato fruits. *Am. J. Agric. Biol. Sci.* **2011**, *6*, 162–171. [CrossRef]

14. Perdones, A.; Escriche, I.; Chiralt, A.; Vargas, M. Effect of chitosan–lemon essential oil coatings on volatile profile of strawberries during storage. *Food Chem.* **2016**, *197*, 979–986. [CrossRef] [PubMed]
15. Lee, S.Y.; Lee, S.J.; Choi, D.S.; Hur, S.J. Current topics in active and intelligent food packaging for preservation of fresh foods. *J. Sci. Food Agric.* **2015**, *95*, 2799–2810. [CrossRef] [PubMed]
16. Romanazzi, G.; Feliziani, E.; Santini, M.; Landi, L. Effectiveness of postharvest treatment with chitosan and other resistance inducers in the control of storage decay of strawberry. *Postharvest Biol. Technol.* **2013**, *75*, 24–27. [CrossRef]
17. Jongsri, P.; Wangsomboondee, T.; Rojsitthisak, P.; Seraypheap, K. Effect of molecular weights of chitosan coating on postharvest quality and physicochemical characteristics of mango fruit. *LWT Food Sci. Technol.* **2016**, *73*, 28–36. [CrossRef]
18. Dhall, R.K. Advances in edible coatings for fresh fruits and vegetables: A review. *Crit. Rev. Food Sci. Nutr.* **2013**, *53*, 435–450. [CrossRef] [PubMed]
19. Falguera, V.; Quintero, J.P.; Jimenez, A.; Munoz, J.A.; Ibarza, A. Edible films and coatings: Structures, active functions and trends in their use. *Trends Food Sci. Technol.* **2011**, *22*, 292–303. [CrossRef]
20. Kanmani, P.; Rhim, J.W. Antimicrobial and physical-mechanical properties of agar-based films incorporated with grapefruit seed extract. *Carbohydr. Polym.* **2014**, *102*, 708–716. [CrossRef] [PubMed]
21. Maswada, H.F.; Abdallah, S.A. In vitro activity of three geophytic plant extracts against three post-harvest pathogenic fungi. *Pak. J. Biol. Sci.* **2013**, *23*, 1698–1705. [CrossRef]
22. Silva-Weiss, A.; Bifani, V.; Ihl, M.; Sobral, P.J.A.; Gomez-Guillen, M.C. Polyphenol-rich extract from murta leaves on rheological properties of film-forming solutions based on different hydrocolloid blends. *J. Food Eng.* **2014**, *140*, 28–38. [CrossRef]
23. *Review Report for the Basic Substance Chitosan Hydrochloride Finalised in the Standing Committee on the Food Chain and Animal Health at its meeting on 20 March 2014—In View of the Approval of Chitosan Hydrochloride as Basic substance in Accordance with Regulation (EC) No 1107/2009*; European Commission: Brussels, Belgium, 2014.
24. Carlson, R.P.; Taffs, R.; Davison, W.M.; Stewart, P.S. Anti-biofilm properties of chitosan coated surfaces. *J. Biomater. Sci. Polym. Ed.* **2008**, *19*, 1035–1046. [CrossRef] [PubMed]
25. Kumar, R.M.N.V. A review of chitin and chitosan applications. *React. Funct. Polym.* **2000**, *46*, 1–27. [CrossRef]
26. Devlieghere, F.; Vermeulen, A.; Debevere, J. Chitosan: Antimicrobial activity, interactions with food components and applicability as a coating on fruit and vegetables. *Food Microbiol.* **2004**, *21*, 703–714. [CrossRef]
27. Tharanathan, R.N.; Kittur, F.S. Chitin—The undisputed biomolecule of great potential. *Crit. Rev. Food Sci. Nutr.* **2003**, *43*, 61–87. [CrossRef] [PubMed]
28. Benhabiles, M.S.; Salah, R.; Lounici, H.; Drouiche, N.; Goosen, M.F.A.; Mameri, N. Antibacterial activity of chitin, chitosan and its oligomers prepared fromshrimp shell waste. *Food Hydrocoll.* **2012**, *29*, 48–56. [CrossRef]
29. Yen, M.T.; Tseng, Y.H.; Li, R.C.; Mau, J.L. Antioxidant properties of fungal chitosan from shiitake stipes. *LWT Food Sci. Technol.* **2007**, *40*, 255–261. [CrossRef]
30. Davidson, A. *The Oxford Companion to Food*; Oxford University Press: Oxford, UK, 1999.
31. Elad, Y.; Williamson, B.; Tudzinski, P.; Delen, N. *Botrytis: Biology, Pathology and Control*; Springer: Berlin, Germany, 2007; p. 403.
32. Ma, B.; Hibbing, M.E.; Kim, H.S.; Reedy, R.M.; Yedidia, I.; Breuer, J.; Breuer, J.; Glasner, J.D.; Perna, N.T.; Kelman, A.; et al. Host range and molecular phylogenies of the soft rot enterobacterial genera *pectobacterium* and *dickeya*. *Phytopathology* **2007**, *97*, 1150–1163. [CrossRef] [PubMed]
33. No, H.K.; Meyers, S.P.; Prinyawiwatkul, W.; Xu, Z. Applications of chitosan for improvement of quality and shelf life of food: A Review. *J. Food Sci. Technol.* **2007**, *72*, 87–100. [CrossRef] [PubMed]
34. Bonilla, J.; Fortunati, E.; Vargas, M.; Chiralt, A.; Kenny, J.M. Effects of chitosan on the physicochemical and antimicrobial properties of PLA films. *J. Food Eng.* **2013**, *119*, 236–243. [CrossRef]
35. Lawrie, G.; Keen, I.; Drew, B.; Chandler-Temple, A.; Rintoul, L.; Fredericks, P.; Grøndahl, L. Interactions between Alginate and Chitosan Biopolymers Characterized Using FTIR and XPS. *Biomacromolecules* **2007**, *8*, 2533–2541. [CrossRef] [PubMed]
36. Bonilla, J.; Fortunati, E.; Atarés, L.; Chiralt, A.; Kenny, J.M. Physical, structural and antimicrobial properties of poly vinyl alcohol–chitosan biodegradable films. *Food Hydrocoll.* **2014**, *35*, 463–470. [CrossRef]

37. Chen, C.H.; Wang, F.Y.; Mao, C.F.; Liao, W.T.; Hsieh, C.D. Studies of chitosan: II. Preparation and characterization of chitosan/poly(vinyl alcohol)/gelatin ternary blend films. *Int. J. Biol. Macromol.* **2008**, *43*, 37–42. [CrossRef] [PubMed]
38. Quattrucci, A.; Ovidi, E.; Tiezzi, A.; Vinciguerra, V.; Balestra, G.M. Biological control of tomato bacterial speck using Punica granatum fruit peel extract. *Crop Prot.* **2013**, *46*, 18–22. [CrossRef]
39. Fortunati, E.; Rescignano, N.; Botticella, E.; La Fiandra, D.; Renzi, M.; Mazzaglia, A.; Balestra, G.M. Effect of poly(DL-lactide-co-glycolide) nanoparticles or cellulose nanocrystals-based formulations on *Pseudomonas syringae* pv. *tomato* (Pst) and tomato plant development. *J. Plant Dis. Prot.* **2016**, *123*, 301–310. [CrossRef]
40. Rossetti, A.; Mazzaglia, A.; Muganu, M.; Paolocci, M.; Sguizzato, M.; Esposito, E.; Cortesi, R.; Balestra, G.M. Microparticles containing gallic and ellagic acids for the biological control of bacterial diseases of kiwifruit plants. *J. Plant Dis. Prot.* **2017**, *124*, 269–278. [CrossRef]
41. Michailides, T.J.; Morgan, D.P. New technique predicts gray mold in stored kiwifruit. *Calif. Agric.* **1996**, *50*, 34–40. [CrossRef]
42. El Gaouth, A.; Arul, J.; Grenier, J.; Asselin, A. Antifungal activity of chitosan on two postharvest pathogens of strawberry fruits. *Phytopathology* **1992**, *82*, 398–402. [CrossRef]
43. Badawy, M.E.I.; Rabea, E.I.; Taktak, N.E.M.; El-Nouby, M.A.M. The antibacterial activity of chitosan products blended with monoterpenes and their biofilms against plant pathogenic bacteria. *Hindawi Publ. Corp. Sci.* **2016**, *2016*, 1796256. [CrossRef] [PubMed]

© 2017 by the authors. Licensee MDPI, Basel, Switzerland. This article is an open access article distributed under the terms and conditions of the Creative Commons Attribution (CC BY) license (http://creativecommons.org/licenses/by/4.0/).

Article

Effect of an Edible Coating Based on Chitosan and Oxidized Starch on Shelf Life of *Carica papaya* L., and Its Physicochemical and Antimicrobial Properties

Monserrat Escamilla-García [1], María J. Rodríguez-Hernández [1], Hilda M. Hernández-Hernández [2], Luis F. Delgado-Sánchez [1], Blanca E. García-Almendárez [1], Aldo Amaro-Reyes [1] and Carlos Regalado-González [1,*]

[1] Department of Food Research and Postgraduate Studies, C.U., Autonomous University of Querétaro, Cerro de las Campanas S/N, Las Campanas, Santiago de Querétaro 76010, Mexico; moneg14@hotmail.com (M.E.-G.); majose0880@gmail.com (M.J.R.-H.); felipedelgado.ibt@gmail.com (L.F.D.-S.); blancag31@gmail.com (B.E.G.-A.); aldoamaro@gmail.com (A.A.-R.)
[2] CONACyT-Center of Research and Assistance in Technology and Design of Jalisco State (CIATEJ), Av. Normalistas 800, Colonias de la Normal, Guadalajara, Jalisco 44270, Mexico; hhernandez@ciatej.mx
* Correspondence: carlosr@uaq.mx or regcarlos@gmail.com; Tel.: +52-442-123-8332

Received: 9 August 2018; Accepted: 6 September 2018; Published: 7 September 2018

Abstract: Papaya production plays an important economic role in Mexico's economy. After harvest, it continues to ripen, leading to softening, skin color changes, development of strong aroma, and microbial spoilage. The objective of this work was to apply an active coating of chitosan–starch to increase papaya shelf life and to evaluate physicochemical and antimicrobial properties of the coating. Papaya surfaces were coated with a chitosan-oxidized starch (1:3 w/w) solution and stored at room temperature (25 ± 1 °C) for 15 days. Variables measured were color, titratable acidity, vitamin C, pH, soluble solids, volatile compounds by gas chromatography, texture, homogeneity by image analysis, and coating antimicrobial activity. At the end of the storage time, there were no significant differences ($p > 0.05$) between coated and uncoated papayas for pH (4.3 ± 0.2), titratable acidity (0.12% ± 0.01% citric acid), and soluble solids (12 ± 0.2 °Bx). Papaya firmness decreased to 10 N for coated and 0.5 N for uncoated papayas. Volatile compounds identified in uncoated papaya (acetic acid, butyric acid, ethyl acetate, ethyl butanoate) are related to fermentation. Total microbial population of coated papaya decreased after 15 days, whereas population of uncoated papaya increased. This active coating permitted longer shelf life of papaya than that of the uncoated fruit.

Keywords: *Carica papaya* L.; edible coatings; chitosan; starch; image analysis

1. Introduction

Papaya (*Carica papaya* L.) is a perennial plant with a rapid growth, and it provides fruits for more than twenty years [1]. Papaya is considered one of the most important fruits worldwide because of its high contents of ascorbic acid, provitamin A, calcium, and carotenoids [2]. Mexico is the third largest producer worldwide of this fruit with 951,922 t in 2016 [3]. It is a climacteric fruit that shows short life after harvest, and it is susceptible to fungi contamination mainly by *Colletotrichum gloeosporioides* responsible for anthracnose disease, which reduces its shelf life, severely depreciating the market value of the fruit [4]. Papaya maturation after harvest involves various metabolic processes; first, papaya is mature, hard and inedible, and after a few days of being harvested, the edible fruit becomes sweet, soft, and aromatic. However, an uncontrolled fruit maturation process can cause pulp softening, changes in skin and pulp color, and strong aroma development; all this is related to increased ethylene production causing postharvest losses [5].

Edible coatings have emerged as a new technology for safe maintenance and the improvement of quality of fresh fruits by their application immediately after harvest to reduce water loss and mechanical and microbial damage, prevent favorable volatiles losses, and delay senescence [6]. Coatings used for fruits preservation must have precisely balanced gas permeability properties for a normal exchange of CO_2/O_2, limited permeability to water vapor to inhibit the escape of moisture, antimicrobial activity, and good adhesion to product surface [7]. According to their composition, edible coatings may be classified into three groups: polysaccharides, proteins, and lipids as their major compounds. Coatings based on polysaccharides are characterized by providing a minimum moisture barrier, whereas gas barrier properties induce desirable atmosphere modification and increased shelf life without creation of severe anaerobic conditions. Polysaccharides commonly used in edible coatings are starch, dextrin, pectin, cellulose and its derivatives, chitosan, alginate, carrageenan, gellan, etc. [8].

Chitosan is a biopolymer obtained from chitin deacetylation in an alkaline medium, consisting of β-(1–4)-2-acetamido-D-glucose and β-(1–4)-2-amino-D-glucose units [9]. It is biodegradable, biocompatible, and nontoxic, and shows antimicrobial activity and good film-forming properties. Applied as a coating on fruit and vegetable surfaces, it reduces respiration rate by regulating gases permeability [10]. Starch is frequently used in edible coatings preparation due to its chemical, physical, and functional characteristics [11]. It comprises two structures, one linear (amylose) and the other branched (amylopectin), that are assembled naturally in granular form (size of ~1 μm to 100 μm) [12]. Amylose tends to be orientated in parallel due to its linearity, resulting in hydrogen bonds between hydroxyl groups reducing polymer affinity for water and allowing films and gels formation [11]. Other components may be added to edible coatings to extend their applications, such as antimicrobial agents, antioxidants, texture modifiers, dyes, flavorings, nutrients, spices, surfactants, emulsifiers, and plasticizers, among others [13].

Ethyl lauroyl arginate (LAE) is a generally recognized as safe (GRAS) cationic surfactant that is water soluble and provides antimicrobial activity against fungi and Gram-positive and Gram-negative bacteria, due to its cationic surfactant nature. It specifically affects the negatively charged microbial proteins of cell membranes or enzymatic systems, causing their denaturation and leading to microbial inhibition or death [14]. Nisin is a bacteriocin from *Lactococcus lactis* subsp. *lactis* considered as GRAS that is used in the food industry as a preservative and consists of unusual and distinctive amino acids post-translationally modified: lanthionine bound to a thioether, 3-methylanthionine, 2,3-didehydroalanine, and 2,3-didehydrobutirin. Gram-positive bacteria such as *Mycobacterium*, *Staphylococcus* spp., *Clostridium* sp., *Listeria* sp., and *Bacillus* sp. are very sensitive to this antibiotic [15]. The objective of this work was to evaluate the effect of an edible coating based on chitosan and oxidized starch on the shelf life of *Carica papaya* L., and the physicochemical and antimicrobial properties of the active coating.

2. Materials and Methods

2.1. Materials

Medium molecular weight chitosan from shrimp shells (375 kDa, >75% deacetylation), glycerol (≥99.5%), and Tween 80 were purchased from Sigma-Aldrich (St. Louis, MO, USA); lactic acid (85%) was obtained from Fermont (León, Mexico). Oxidized starch was supplied by Ingredion (San Juan del Río, Mexico), LAE was a gift from Lamirsa (Barcelona, Spain). Violet red bile agar, papa dextrose agar, plate count agar, and casein peptone were acquired from BD Bioxon (Distrito Federal, México). Papayas (*Carica papaya* L.) were harvested from Tecoman, Colima, Mexico.

2.2. Methods

2.2.1. Papaya Harvest

One hundred papayas (*C. papaya* L. var. Maradol) were harvested from Tecoman, Colima, Mexico at maturation stage 1, defined as a yellow stripe near fruit apex (70%–80% of yellow surface) [16], and all fruits were similar in size and without any apparent physical damage.

2.2.2. Coating Preparation

Papaya edible coating was produced according to Escamilla-García et al. [17] with some modifications. One chitosan solution and one starch paste slurry were made in parallel. Chitosan (1% w/v) in lactic acid (0.5% v/v) was produced with constant agitation at 80 °C during 60 min; the starch paste slurry was prepared with 3.5% (w/v) oxidized starch in distilled water heated at 90 °C and stirred for 30 min. Starch:chitosan preparations were mixed in 3:1 (w/w) ratio and stirred for 5 min at room temperature; glycerol was added as plasticizer at chitosan:glycerol ratio of 1:1 (w/w), and stirred for 10 min. Previous experiments showed that a combination of 3.750 µg/mL of nisin and 0.0625 mg/mL of LAE were enough to inhibit coliforms, mesophilic bacteria, and yeasts, and thus were used as antimicrobial agents [18].

2.2.3. Coating Application

Papayas were immersed in distilled water for 1 min to eliminate solid residues and impurities, followed by 1 min immersion in neutral electrolyzed water (NEW) developed at the Center for Research and Technological Development in Electrochemistry of Querétaro (CIDETEQ), at a concentration of 200 ppm total available chlorine. NEW is obtained by electrolysis of diluted NaCl solutions (0.1%–1% w/v) without the use of a membrane separating anode and cathode. Several studies have shown that highly oxidizing chlorine species are present in NEW, mainly hypochlorous acid, hypochlorite ions, chlorine dioxide, and ozone, which provide high bactericidal activity [19–21]. Oxidizing species are rapidly reduced during treatments, and there is no need for rinsing. Water was removed by drying at room temperature. Coating was applied on papaya peel using a sterile sponge (Figure 1), and properties of coated papayas were evaluated at 0, 5, 10, and 15 days of storage at room temperature. Uncoated papayas were used as control and were analyzed at the same storage times.

Figure 1. Coating application.

2.2.4. Papayas Characterization

For papaya characterization, analyses of whole papayas (without peeling) were carried out to determine volatile compounds, wt % loss (dry basis, db), image analysis, color, and antimicrobial activity. Physicochemical parameters, such as % acidity (db), texture, vitamin C (db), pH, and soluble solids, were determined in papaya pulp.

Texture and Weight Loss

Texture was evaluated using a texturometer (Stable Micro Systems, TAXT2i, Surrey, UK) on papaya pulp pieces (2 cm × 2 cm × 3 cm) with and without coating. The force required to penetrate 5 mm of pulp was determined at 0, 5, 10, and 15 days of storage. Determinations were conducted in 6 different sections of the fruit, using a 5 mm probe and test speed of 4 mm/s with automatic return. Fruit weight losses, with and without coating, were recorded during storage time (0, 5, 10, 15 days) using a scale (Torrey, PCR-20, Monterrey, Mexico).

pH, Soluble Solids, Titratable Acidity, and Vitamin C

Ten g of papaya pulp was weighed and homogenized in 100 mL of distilled water, and from each sample, 3 pH measurements were taken using a calibrated potentiometer (Hanna, Mod. 209, Woonsocket, RI, USA) at 0, 5, 10, and 15 days of fruit storage, with and without coating [22]. Soluble solids of papaya samples with and without coating were determined by an Abbe refractometer (Atago, DR-A1, Tokyo, Japan) at the different storage times. Titratable acidity was evaluated following AOAC method 942.15A, in which 5 mL of diluted papaya juice in 95 mL distilled water was titrated, with 0.1 M NaOH, and expressed as citric acid percentage [23].

Color

Color changes of papaya peel surface were evaluated at the different storage times (0, 5, 10, and 15 days) with and without coating. For this test, sections in which color was determined were marked; thus, measurements were always conducted in the same section. Parameters measured were L^* (Luminosity), a^* (green to red), and b^* (blue to yellow) using a Minolta CR400 colorimeter (Konica Minolta Sensing, Osaka, Japan), a light source D65, and 10° angle [24].

Volatile Compounds

Volatile compounds determination was carried out according to García-Aguilar et al. [25]. Fruits were placed for 2 h before analysis in hermetic containers. The solid phase microextraction method was applied in this analysis using a pre-equilibrium time of 15 min at 50 °C. Solid-phase microextraction was performed with a 75 mm divinylbezene/carboxene/polydimethylsiloxane fiber (Supelco, Bellefone, VA, USA). The fiber was exposed to the interior of the hermetic containers for 20 min at room temperature (~25 °C), and then placed in the injection port of a gas chromatograph (Mod. 7890A, Agilent Technologies, Santa Clara, CA, USA) coupled to a quadrupole mass spectrometer (Agilent, 5975C) for 45 min at 230 °C for analytes desorption. Compounds identification was carried out using NIST/EPA/NIH Mass Spectra Library version 1.7 (Gaithersburg, MD, USA), considering a similarity >80%. Peak area was obtained by Equation (1):

$$A = W_{1/2}h \tag{1}$$

where A is peak area, $W_{1/2}$ is width of half peak height, and h is peak height.

2.2.5. Image Analysis

Papaya ripening changes were evaluated by image analysis (IA). Samples were illuminated using four TL-D deluxe fluorescent lamps, daylight 18 W (Philips, TL-D, Eindhoven, The Netherlands), and color temperature of 6500 K (D65). Lamps (60 cm long) were arranged in a square shape 35 cm above the sample at an angle of 45°, and pictures were taken from apical and equatorial areas of the papaya. A digital camera Nikon D3200 (Tokyo, Japan) of 24.2 Mpixels resolution was used to take images without zoom or flash, and were processed following Arzate-Vázquez et al. [26]. Using Image J 1.51j8 program [27], images were changed to 8-bit (250 × 250 pixels, TIFF format) and contrast, correlation, entropy, angular momentum, and fractal dimension parameters were obtained. The gray level co-occurrence matrix complement was used to evaluate papaya texture. Fractal dimensions (FD) were estimated by counting changes of differential boxes from gray level images.

2.2.6. Microbiological Analysis

Total coliforms, fungi and yeasts, and mesophilic aerobic bacteria were quantified. Papayas with and without coating at 0, 5, 10, and 15 days of storage were analyzed in triplicate. Samples were prepared according to NOM-110-SSA1-1994 [28] with some modifications. Papaya weight was recorded, then the whole fruit was placed in a sterile plastic bag containing 100 mL of sterile peptone water (1 g casein peptone, 8.5 g NaCl/L, pH 7.2 ± 0.1), and stirred for 3 min to recover the microorganisms, which were appropriately diluted to carry out population counts (CFU/mL) in triplicate experiments.

Total Coliforms

Total coliforms determination was performed by pouring 1 mL of each sample on violet red bile agar, using decimal dilutions where necessary, and incubated at 35 °C for 48 h. Red or violet colonies (0.5 mm) surrounded by a precipitated bile area were quantified and expressed as CFU/g [29].

Molds and Yeasts

Molds and yeasts analysis was carried out according to Pérez-Grijalva et al. [30] with some modifications. Samples of 1 mL were poured into potato dextrose agar (pH = 3.5 ± 0.1) and incubated at 25 ± 1 °C; colonies (CFU/g) were evaluated after 3 days of incubation.

Mesophilic Aerobic Bacteria

Mesophilic aerobic bacteria count was performed as described by Alsharjabi et al. [29], using plate count agar, pouring 1 mL of sample and incubating for 48 h at 37 °C. Results were reported as Log_{10} CFU/g.

2.2.7. Statistical Analysis

Samples were analyzed in triplicate. Data were evaluated by one-way analysis of variance (ANOVA) and significant differences were analyzed by the Tukey test ($p < 0.05$).

3. Results and Discussion

3.1. Edible Coating Characterization

3.1.1. Texture and Weight Loss

Firmness and weight loss changes of papaya with and without coating at different storage times are shown in Figure 2. Coated fruits required a larger force to penetrate 5 mm of pulp than the uncoated ones, which after 10 days of storage showed firmness loss of 92.02%, whereas coated papaya showed firmness reduction of 47.36%. This behavior is similar to that reported by Cortez-Vega et al. [31], who applied a fish-protein-based coating on minimally processed papaya. According to firmness values, uncoated papaya achieved its final stage of ripening after 5 days of storage (close to 10 N), while coated fruit reached this stage after 15 days [4]. Figure 2b shows that uncoated papaya exhibited an increasing trend in weight loss with storage time, up to 33.3% after 15 days of storage, whereas the coated fruit did not display any significant change in weight during 15 days.

Similar results were obtained by Adetunji et al. [32], who applied a layer of chitosan on papaya surface, but did find weight losses even with this coating. Hazarika et al. [33] applied coatings of carboxymethyl cellulose, chitosan, and *Aloe vera* on papaya, which again decreased but not suppressed weight losses.

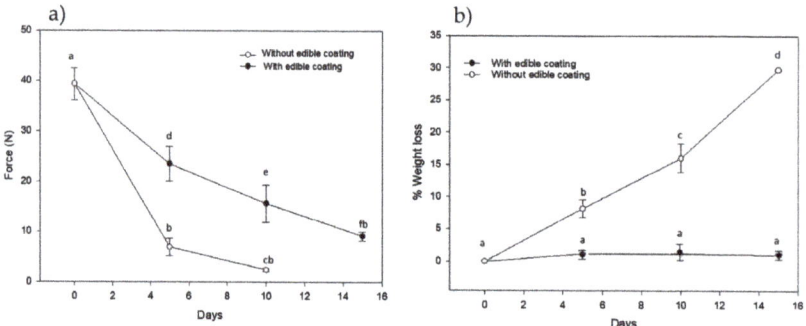

Figure 2. Physical changes during the ripening of papaya with and without coating during storage at room temperature: (**a**) papaya pulp firmness; (**b**) weight loss. a–f: used on top of reported values, indicate that if the same letter appears at different times, the values compared are not significantly different ($p > 0.05$).

3.1.2. pH, Soluble Solids, Titratable Acidity, and Vitamin C

Along the storage period of papayas, the acidity showed a decreasing trend, but after 5 and 10 days, coated papaya exhibited significantly ($p < 0.05$) lower acidity than the uncoated fruits (Figure 3a). However, at the end of storage time, there was no significant difference in this parameter. A similar behavior was reported by Ali et al. [34], who applied different chitosan concentrations as coatings to papaya, and found that at least 2% chitosan (w/v) was required to create an internal atmosphere to reduce acidity changes. Thus, in the present study, the applied coating did not show this feature due to similar acidity after 15 days of storage for coated and uncoated papayas.

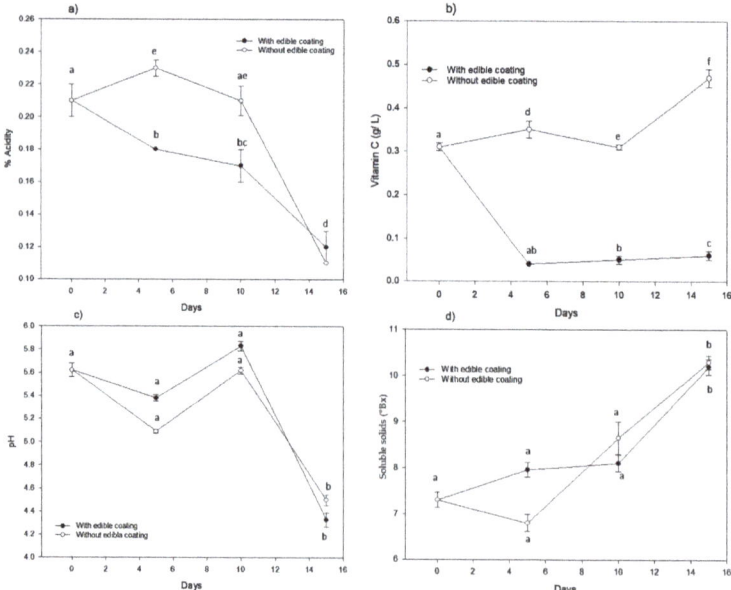

Figure 3. Chemical changes during papaya ripening, coated and uncoated, during storage time at room temperature: (**a**) titratable acidity (% cítric acid); (**b**) vitamin C (g/kg); (**c**) pH; (**d**) soluble solids. a–f: used on top of reported values, indicate that if the same letter appears at different times, the values compared are not significantly different ($p > 0.05$).

Vitamin C content is a characteristic factor in the papaya ripening process that is related to oxidative degradation during maturation stages [35]. Figure 3b reveals a decrease in vitamin C of coated papaya after 5 days of storage, whereas the opposite was shown by the uncoated fruit, which agrees with Yurena et al. [36], who reported that vitamin C increased with maturation stage. Similar behavior was described by Wall [37] for papaya from different cultivars. Vitamin C increase may be related to increased lipid peroxidation, since this oxidative phenomenon induces intensification of antioxidant compounds such as ascorbic acid [38].

Up to 10 days of storage for both coated and uncoated papayas showed a similar trend in pH values, and after 15 days both treatments showed a sharp pH reduction significantly different from the initial value, but similar between them (Figure 3c). Reduction of pH during storage time was attributed to organic acids production, which is related to the papaya ripening process [31]. Soluble solids of coated and uncoated papayas did not show any significant difference ($p < 0.05$) during the storage time, increasing significantly in relation to the initial value after 15 days (Figure 3d). Jayanthunge et al. [39] obtained similar results by storing papaya in microperforated polyvinyl chloride containers, and suggested that soluble solids increased as a result of the different metabolic activities in tissues leading to pectin decomposition and carbohydrates hydrolysis into simple sugars during storage.

3.1.3. Color

Papaya samples color changed from mostly green to yellow-red over time (Figure 4), associated with chlorophyll degradation [5]. Uncoated papaya reached yellow-red coloration and did not reveal significant changes during the fifteen days of storage (Supplementary Figure S1), whereas coated papaya showed a gradual change from green to yellow-red as storage time increased (Figure S2). Figure 4b shows chromatic change in a^* parameter, where uncoated papaya reached a red color from day 5, without any more significant changes in the remaining storage time, while coated papaya gradually acquired this coloration.

Figure 4c shows significant changes for coated papaya in the b^* chromatic value from 5–10 days, while uncoated papaya does not show significant changes from 5 to 15 days. According to Barragán-Iglesias et al. [40], there are five papaya ripening levels based on color and days after anthesis (DAA). Physiological maturity is reached at 135 DAA and the surface is green; ripeness ready for consumption occurs at 156 DAA, and 100% of the surface is orange-red.

They defined three intermediate stages depending on the extent of orange-red color of the fruit surface: 143 DAA (25%), 149 DAA (50%), and 153 DAA (75%). From the evaluated color of papaya samples, it is concluded that the application of coating delayed papaya ripening reaching maturity for consumption after 15 days, whereas uncoated papaya reached this stage after 5 days.

3.1.4. Volatile Compounds

The volatile compounds identified are listed in Table 1, and are in agreement with those obtained in other studies [31,41–43]. These results indicate that coating application affects the papaya ripening process. Fuggate et al. [41] and Almora et al. [44] reported papaya chemical changes during the intermediate stages of ready-for-consumption ripeness, where the amount of ethyl butanoate, ethyl hexanoate, and alcohols increased. These increments were also noted for coated and uncoated papayas, but the uncoated fruit produced ethyl butanoate after 5 days, whereas coated papaya produced it after 10 days, and its concentration after 15 days was higher in uncoated papaya. In relation to ethyl hexanoate, it was detected after 15 days of storage, with higher concentration in the uncoated papayas.

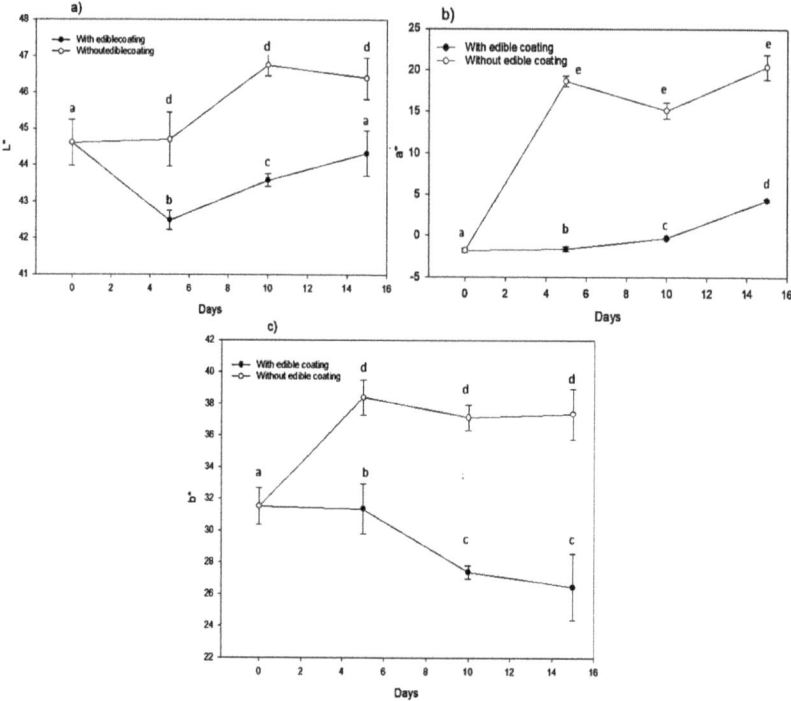

Figure 4. Papaya color changes in storage with and without edible coating: (**a**) L^*; (**b**) a^*; (**c**) b^*. a–d: used on top of reported values, indicate that if the same letter appears at different times, the values compared are not significantly different ($p > 0.05$).

Table 1. Volatile compounds generated during papaya ripening with and without edible coating.

		0 Day	5 Days		10 Days		15 Days	
No.	Compound	Control	With Edible Coating	Without Edible Coating	With Edible Coating	Without Edible Coating	With Edible Coating	Without Edible Coating
		\multicolumn{7}{c}{Peak area $\times 10^{-7}$}							
1	4-Methyloctane	9.36	11.70	6.98	12.05	6.54	12.29	6.06
2	1-Heptene, 2-methyl-	0.58	0.59	0.57	0.35	0.33	0.17	0.15
3	1-Hexene, 2,5-dimethyl-	0.33	0.34	033	0.33	0.29	0.21	0.13
4	1-Butanol	–	2.38	–	2.30	–	2.27	1.30
5	2-Butenoic acid, ethyl ester	0.37	0.35	0.38	0.35	0.38	0.35	0.38
6	2-Ethylhexanol	–	–	–	–	–	1.89	1.87
7	2-Ethylpentane	–	–	–	6.15	6.08	2.28	1.07
8	α-Methylfuran	–	–	1.24	–	1.33	–	1.68
9	Acetoin	–	–	–	–	–	5.79	6.11
10	Acetone	–	–	1.64	1.88	1.56	1.97	1.90
11	Acetic acid	–	–	–	–	–	1.40	5.19
12	Benzene	0.49	0.58	0.39	0.43	0.27	0.38	0.17
13	Benzothiazole	0.36	0.31	0.41	0.35	–	0.20	0.39
14	Benzyl Alcohol	–	0.64	0.63	0.77	0.75	0.72	0.67
15	Benzyl cyanide	0.15	0.15	0.14	0.15	0.14	0.15	0.14
16	Benzyl nitrile	–	–	–	0.18	–	0.24	–
17	Butanoic acid	–	–	–	1.07	–	1.25	–
18	Butyric acid	12.20	21.90	222.00	22.80	254.00	30.10	271.00
19	Caproic acid	–	–	3.18	2.27	2.40	1.87	1.23
20	Cetyl chloride	0.80	0.96	0.65	1.02	0.72	1.11	0.82
21	cis-Linalool oxide	–	–	4.95	–	5.21	–	5.84
22	Cyclohexane, methyl-	–	–	–	1.11	–	1.48	–
23	Limonene	–	1.13	–	1.69	–	1.85	–
24	Decane	–	1.12	0.98	4.16	2.57	6.13	5.71
25	Dibutyl formamide	–	–	115.00	0.01	200.00	0.009	243.00
26	Diethyl Phthalate	0.62	0.63	0.04	0.84	0.24	1.34	0.40
27	Dimethyl sulfide	–	0.30	0.25	0.25	0.24	0.17	0.13
28	Dodecane	–	–	0.69	–	0.76	–	0.83
29	Dodecane, 4,6-dimethyl-	–	–	–	0.69	–	0.81	–

Table 1. Cont.

No.	Compound	0 Day Control	5 Days With Edible Coating	5 Days Without Edible Coating	10 Days With Edible Coating	10 Days Without Edible Coating	15 Days With Edible Coating	15 Days Without Edible Coating
				Peak area $\times 10^{-7}$				
30	Ethanol	18.00	34.10	18.69	11.80	22.40	8.31	31.50
31	Ethyl Acetate	–	29.10	–	18.70	–	12.90	6.29
32	Ethyl butanoate	–	–	23.00	41.40	41.50	34.70	161.00
33	Ethyl hexanoate	–	–	–	–	–	1.72	5.07
34	Ethyl propanoate	–	–	–	–	–	0.37	0.56
35	Formic acid, pentyl ester	–	0.39	–	0.40	–	0.42	–
36	Heptane	14.00	17.60	10.50	4.35	5.32	6.35	2.52
37	Hexadecane	–	–	–	0.11	–	0.34	–
38	Isoheptane	–	–	0.17	0.06	0.05	0.09	0.07
39	Isobutenylcarbinol	–	–	–	0.06	2.97	0.09	0.32
40	Isopentyl alcohol, acetate	0.83	0.24	1.42	0.11	1.84	0.08	1.95
41	Isopreno	–	–	0.15	–	0.21	–	0.71
42	Isothymol methyl ether	–	–	–	–	–	1.83	1.53
43	Linalool	1.79	0.33	3.24	0.08	3.41	0.08	3.63
44	Methyl acetate	–	–	4.96	–	5.02	–	5.21
45	Methyl benzoate	0.78	0.70	0.85	0.42	1.23	0.22	1.74
46	Methylene Chloride	–	–	–	1.76	–	1.94	–
47	Methyl crotonate	–	–	4.97	–	5.03	–	5.71
48	Methyl nerolate	–	–	0.26	–	0.31	–	0.37
49	Methyl propanoate	0.74	1.14	0.96	1.44	1.04	1.66	1.12
50	Methyl salicylate	–	–	–	–	–	0.13	0.17
51	Morpholine, 4-octadecyl-	–	–	0.46	–	0.52	–	0.63
52	N,N-Dibutylacetamide	–	–	0.14	–	0.22	–	0.27
53	Nonane	–	–	0.48	–	0.53	–	0.61
54	Nonane, 2,5-dimethyl-	–	–	4.38	–	4.51	–	4.63
55	Nonane, 2,2,4,4,6,8,8-heptamethyl-	0.24	2.41	2.27	3.31	3.12	4.12	4.26
56	Nonane, 3-methyl-5-propyl-	–	0.22	–	0.28	–	0.33	–
57	Octane	1.05	0.25	0.66	0.30	0.72	0.39	0.89
58	Oxime-, methoxy-phenyl	2.34	3.54	2.48	4.50	3.71	0.63	0.64
59	Pentadecane	–	1.67	–	1.74	–	1.95	–
60	Penta ethylene glycol monododecyl ether	–	–	0.42	–	0.55	–	0.76
61	Phenol	0.46	–	–	0.25	–	0.42	–
62	Propanoic acid, ethyl ester	–	1.32	–	1.62	–	1.74	–
63	Propanoic acid, 2-methyl-, ethyl ester	0.25	0.26	0.24	0.76	0.39	1.11	0.44
64	Toluene	–	0.62	0.57	0.63	1.14	0.44	0.61

Among alcohols related to papaya ripening [44] are 2-ethylhexanol and benzyl alcohol, which increased in similar quantities with storage time in coated and uncoated papayas. Ethanol was detected in both coated and uncoated papayas after 5 days, but concentration was higher in uncoated fruits at all storage times. According to Lee et al. [45], compounds such as 2-ethyl hexanol, acetic acid, butyric acid, ethyl acetate, ethyl butanoate, ethyl hexanoate, and methyl acetate are related to the papaya fermentation process. Uncoated papayas generated ethyl butanoate after 5 days, whereas coated fruits generated it after 10 days. Acetic acid appeared after 15 days in both treatments, whereas butyric acid generation was about 10 times higher in uncoated than in coated papayas (Table 1). In addition, higher amounts of these compounds were detected in uncoated papayas at all storage times. It must be pointed out that only uncoated papayas generated methyl acetate from day 5 until the end of storage time. Thus, these results indicate that coating application delays papaya fermentation.

3.2. Microbiological Analysis

Microbial contamination is an important cause of papaya postharvest losses. Average initial values in papaya surface were 35.3 ± 5.5 colony-forming units (CFU)/g of total coliforms, 548.5 ± 97.3 CFU/g of mesophilic aerobic bacteria, 239.75 ± 28.5 CFU/g of yeasts, and 0 molds. Figure 5a shows total coliforms population, where a decrease is observed for the first 10 days in coated papaya, whereas population increased in uncoated papaya. A similar behavior is observed for mesophilic aerobic bacteria, where coating application allowed the reduction of 2 Log_{10} CFU g^{-1} after 15 days of storage, while uncoated papaya increased their population by 2 Log_{10} CFU g^{-1} (Figure 5b). Coated papaya showed a fungal population <10 UFC g^{-1}, whereas the uncoated fruits at the end of storage exhibited a population of 206 CFU g^{-1} (Figure 5c). Antimicrobial activity was most noticeable in yeasts (Figure 5d); coated papaya revealed a population decrease of 3 Log_{10} CFU g^{-1} versus population increase to about 4.7 Log_{10} CFU g^{-1} for uncoated papayas, showing a completely spoilt fruit (Figure S1d). Similar results have been obtained using starch-based edible coatings added with nisin and LAE, achieving inhibition against bacteria and fungi [17,18].

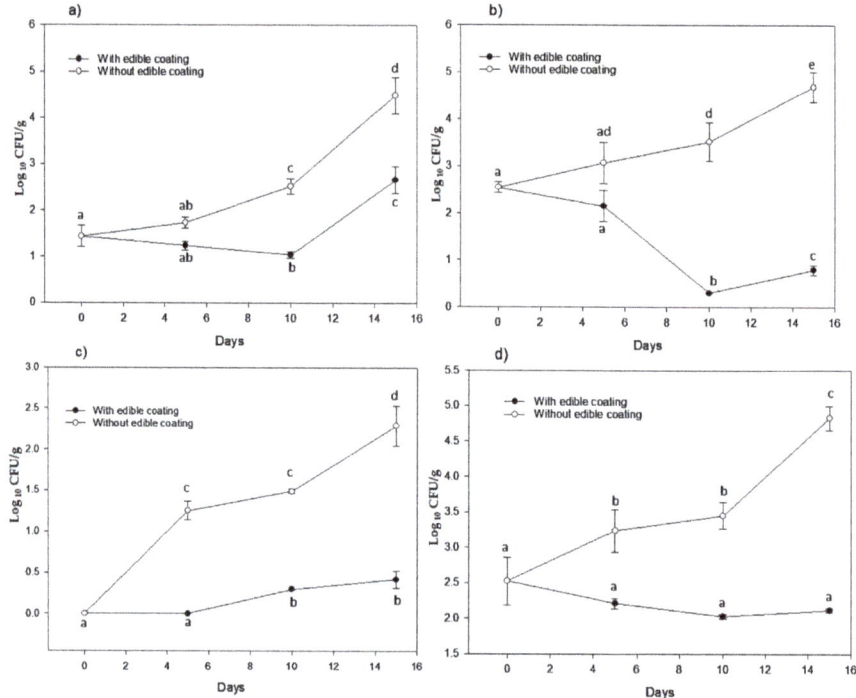

Figure 5. Antimicrobial activity of papaya with and without edible coating: (**a**) total coliforms; (**b**) mesophilic aerobic bacterial; (**c**) fungi; (**d**) yeasts. a–e: used on top of reported values, indicate that if the same letter appears at different times, the values compared are not significantly different ($p > 0.05$).

3.3. Image Analysis

Images showing textural changes in coated and uncoated papayas are shown in Figure 6A. Entropy is an indicator of the complexity of the texture image, and measures the disorder or randomness [46]. Entropy values decreased gradually for coated and uncoated papayas during the first days, being significantly different ($p < 0.05$) until 10 days of storage. After this time, coating application permitted lower entropy values than those of uncoated papaya, indicating a surface with less damage. Fruit texture is an important factor that can be used to determine the stage of ripening, because during this process, a detachment of parenchyma cell wall takes place [47]. These results indicate that coated papaya displays a cell membrane system more ordered and with less mobility.

The angular second moment (ASM) is a parameter that measures the homogeneity of an image [48], and ASM showed a time-dependent increment (Figure 6B). Samples with edible coating showed higher values than those without it. Images with edible coating presented higher homogeneity values due to the stage of ripening, and in uncoated papayas due to fungi present on the papayas' surface.

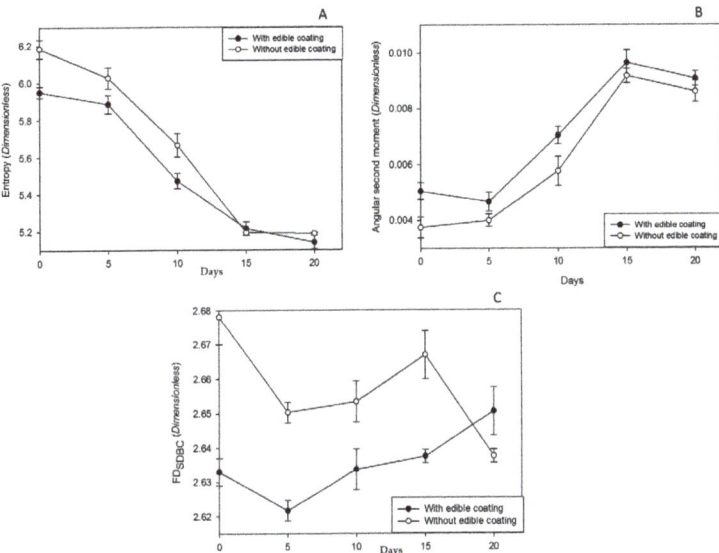

Figure 6. Image analysis of papaya with and without edible coating: (**a**) entropy; (**b**) angular second moment; (**c**) fractal dimension.

The fractal dimension (FD) or fractal texture is a measure of the degree of roughness of the images; the higher the values of fractal dimension, the rougher or more complex the gray level images (Figure 7), while lower FD values can be associated with simpler or smoother images. After 20 days, the FD of uncoated papaya is smaller than that of the coated fruit, and roughness was attributed to fungal growth. Samples without edible coating showed more complex images than coated papayas (Figure 6C); this could be due to changes derived from the papayas' maturation.

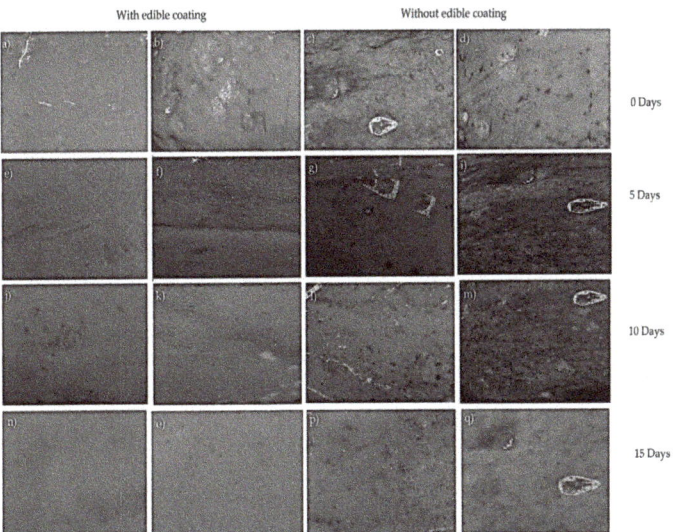

Figure 7. Grayscale papaya crops (8 bits) after different days of storage. 0 days (initial time): with coating (**a,b**); without coating (**c,d**). 5 days: with coating (**e,f**); without coating (**g,i**). 10 days: with coating (**j,k**); without coating (**l,m**). 15 days: with coating (**n,o**); without coating (**p,q**).

4. Conclusions

Edible coatings exhibited a positive effect on papaya shelf life kept at room temperature, preserving its properties during a longer storage time than uncoated fruits. Coating helped to provide larger papaya pulp firmness, indicating that uncoated papaya reached a final stage of ripening after 5 days, whereas the coated fruit reached this stage after 15 days at room temperature. Volatile compounds characteristic of papaya fermentation, such as ethyl butanoate, appeared after 5 days, whereas coated fruits generated it after 10 d. In addition, butyric acid generation was about 10 times higher in uncoated than in coated papayas throughout the 15 days of storage. Images of papaya surfaces with edible coating showed higher homogeneity values than the uncoated fruit at all storage times. Microbial population of papaya surfaces decreased during storage in the coated fruits, whereas the opposite occurred in uncoated papayas. Therefore, this coating can successfully increase papaya shelf life.

Supplementary Materials: The following are available online at http://www.mdpi.com/2079-6412/8/9/318/s1, Figure S1: Color development during storage of uncoated papaya, Figure S2: Color development during storage of coated papaya.

Author Contributions: Conceptualization, M.E.-G.; Methodology, M.J.R.-H. and L.F.D.-S.; Visualization, H.M.H.-H.; Validation, B.E.G.-A. and A.A.-R.; Writing-Review & Editing, C.R.-G.

Funding: This research was financed through the project CB-166751 from CONACyT, and AMEXCID, SRE, File B/ITA/17/001, Mexico. The publication fee was funded by Programa para el Desarrollo Profesional Docente de la Secretaría de Educación Pública of Mexico.

Acknowledgments: The authors also wish to thank the Centro de Investigación en Ciencia Aplicada y Tecnología Avanzada (CICATA) for providing the facilities for this research work and for technical assistance.

Conflicts of Interest: The authors declare no conflict of interest.

References

1. Ojo, O.A.; Ojo, A.B.; Awoyinka, O.; Ajiboye, B.O.; Oyinloye, B.E.; Osukoya, O.A.; Olayide, I.I.; Ibitayo, A. Aqueous extract of *Carica papaya* Linn. roots potentially attenuates arsenic induced biochemical and genotoxic effects in Wistar rats. *J. Tradit. Complement. Med.* **2018**, *8*, 324–334. [CrossRef] [PubMed]
2. Ferraz, T.M.; Rodrigues, W.P.; Netto, A.T.; de Oliveira Reis, F.; Peçanha, A.L.; de Assis Figueiredo, F.A.M.M.; de Sousa, E.F.; Glenn, D.M.; Campostrini, E. Comparison between single-leaf and whole-canopy gas exchange measurements in papaya (*Carica papaya* L.) plants. *Sci. Hortic.* **2016**, *209*, 73–78. [CrossRef]
3. FAOSTAT. Available online: http://fenix.fao.org/faostat/internal/en/#data/QC (accessed on 3 August 2018).
4. Zillo, R.R.; da Silva, P.P.M.; de Oliveira, J.; da Glória, E.M.; Spoto, M.H.F. Carboxymethylcellulose coating associated with essential oil can increase papaya shelf life. *Sci. Hortic.* **2018**, *239*, 70–77. [CrossRef]
5. Hamzah, H.M.; Osman, A.; Tan, C.P.; Ghazali, F.M. Carrageenan as an alternative coating for papaya (*Carica papaya* L. cv. Eksotika). *Postharvest. Biol. Technol.* **2013**, *75*, 142–146. [CrossRef]
6. Ncama, K.; Magwaza, L.S.; Mditshwa, A.; Tesfay, S.Z. Plant-based edible coatings for managing postharvest quality of fresh horticultural produce: A review. *Food Packag. Shelf Life* **2018**, *16*, 157–167. [CrossRef]
7. Arnon-Rips, H.; Poverenov, E. Improving food products' quality and storability by using layer by layer edible coatings. *Trends Food Sci. Technol.* **2018**, *75*, 81–92. [CrossRef]
8. Yousuf, B.; Qadri, O.S.; Srivastava, A.K. Recent developments in shelf-life extension of fresh-cut fruits and vegetables by application of different edible coatings: A review. *LWT-Food Sci. Technol.* **2018**, *89*, 198–209. [CrossRef]
9. Lima Oliveira, P.D.; de Oliveira, K.Á.R.; dos Santos Vieira, W.A.; Câmara, M.P.S.; de Souza, E.L. Control of anthracnose caused by *Colletotrichum* species in guava, mango and papaya using synergistic combinations of chitosan and *Cymbopogon citratus* (D.C. ex Nees) Stapf. essential oil. *Int. J. Food Microbiol.* **2018**, *266*, 87–94. [CrossRef] [PubMed]
10. Hajji, S.; Younes, I.; Affes, S.; Boufi, S.; Nasri, M. Optimization of the formulation of chitosan edible coatings supplemented with carotenoproteins and their use for extending strawberries postharvest life. *Food Hydrocoll.* **2018**, *83*, 375–392. [CrossRef]

11. Guimarães, I.C.; dos Reis, K.C.; Menezes, E.G.T.; Rodrigues, A.C.; da Silva, T.F.; de Oliveira, I.R.N.; de Barros Vilas Boas, E.V. Cellulose microfibrillated suspension of carrots obtained by mechanical defibrillation and their application in edible starch films. *Ind. Crops Prod.* **2016**, *89*, 285–294. [CrossRef]
12. Zhu, F. Modifications of starch by electric field based techniques. *Trends Food Sci. Technol.* **2018**, *75*, 158–169. [CrossRef]
13. Tavassoli-Kafrani, E.; Shekarchizadeh, H.; Masoudpour-Behabadi, M. Development of edible films and coatings from alginates and carrageenans. *Carbohydr. Polym.* **2016**, *137*, 360–374. [CrossRef] [PubMed]
14. Pezo, D.; Navascués, B.; Salafranca, J.; Nerín, C. Analytical procedure for the determination of Ethyl Lauroyl Arginate (LAE) to assess the kinetics and specific migration from a new antimicrobial active food packaging. *Anal. Chim. Acta* **2012**, *745*, 92–98. [CrossRef] [PubMed]
15. Khan, I.; Oh, D.-H. Integration of nisin into nanoparticles for application in foods. *Innov. Food Sci. Emerg. Technol.* **2016**, *34*, 376–384. [CrossRef]
16. Albertini, S.; Reyes, A.E.L.; Trigo, J.M.; Sarriés, G.A.; Spoto, M.H.F. Effects of chemical treatments on fresh-cut papaya. *Food Chem.* **2016**, *190*, 1182–1189. [CrossRef] [PubMed]
17. Escamilla-García, M.; Reyes-Basurto, A.; García-Almendárez, B.E.; Hernández-Hernández, E.; Calderón-Domínguez, G.; Rossi-Márquez, G.; Regalado-González, C. Modified starch-chitosan edible films: Physicochemical and mechanical characterization. *Coatings* **2017**, *7*, 224. [CrossRef]
18. Sánchez-Ortega, I.; García-Almendárez, B.E.; Santos-López, E.M.; Reyes-González, L.R.; Regalado, C. Characterization and antimicrobial effect of starch-based edible coating suspensions. *Food Hydrocoll.* **2016**, *52*, 906–913. [CrossRef]
19. Arevalos-Sánchez, M.; Regalado, C.; Martin, S.E.; Domínguez-Domínguez, J.; García-Almendárez, B.E. Effect of neutral electrolyzed water and nisin on Listeria monocytogenes biofilms, and on listeriolysin O activity. *Food Control* **2012**, *24*, 116–122. [CrossRef]
20. Arevalos-Sánchez, M.; Regalado, C.; Martin, S.E.; Meas-Vong, Y.; Cadena-Moreno, E.; García-Almendárez, B.E. Effect of neutral electrolyzed water on lux-tagged Listeria monocytogenes EGDe biofilms adhered to stainless steel and visualization with destructive and non-destructive microscopy techniques. *Food Control* **2013**, *34*, 472–477. [CrossRef]
21. Rahman, S.M.E.; Khan, I.; Oh, D.H. Electrolyzed water as a novel sanitizer in the food industry: Current trends and future perspectives. *Compr. Rev. Food Sci. Food Saf.* **2016**, *15*, 471–490. [CrossRef]
22. Waghmare, R.B.; Annapure, U.S. Combined effect of chemical treatment and/or modified atmosphere packaging (MAP) on quality of fresh-cut papaya. *Postharvest. Biol. Technol.* **2013**, *85*, 147–153. [CrossRef]
23. Dhital, R.; Mora, N.B.; Watson, D.; Kohli, P.; Choudhary, R. Efficacy of limonene nano coatings on post-harvest shelf life of strawberries. *LWT-Food Sci. Technol.* **2018**, *97*, 124–134. [CrossRef]
24. Mishra, B.B.; Gautam, S.; Chander, R.; Sharma, A. Characterization of nutritional, organoleptic and functional properties of intermediate moisture shelf stable ready-to-eat *Carica papaya* cubes. *Food Biosci.* **2015**, *10*, 69–79. [CrossRef]
25. García-Aguilar, L.; Rojas-Molina, A.; Ibarra-Alvarado, C.; Rojas-Molina, J.I.; Vázquez-Landaverde, P.A.; Luna-Vázquez, F.J.; Zavala-Sánchez, M.A. Nutritional value and volatile compounds of black cherry (*Prunus serotina*) seeds. *Molecules* **2015**, *20*, 3479–3495. [CrossRef] [PubMed]
26. Arzate-Vázquez, I.; Chanona-Pérez, J.J.; de Jesus Perea-Flores, M.; Calderón-Domínguez, G.; Moreno-Armendáriz, M.A.; Calvo, H.; Godoy-Calderón, S.; Quevedo, R.; Gutiérrez-López, G. Image processing applied to classification of avocado variety Hass (*Persea americana* Mill.) during the ripening process. *Food Bioprocess Technol.* **2011**, *4*, 1307–1313. [CrossRef]
27. Schneider, C.A.; Rasband, W.S.; Eliceiri, K.W. NIH Image to ImageJ: 25 years of image analysis. *Nat. Methods* **2012**, *9*, 671–675. [CrossRef] [PubMed]
28. NORMA Official Mexicana NOM-110-SSA1-1994, Bienes y Servicios. Available online: http://www.salud.gob.mx/unidades/cdi/nom/110ssa14.html (accessed on 28 June 2018).
29. Alsharjabi, F.A.; Al-Qadasi, A.M.; Al-Shorgani, N.K. Bacteriological evaluation of weaning dried foods consumed in Taiz City, Republic of Yemen. *J. Saudi Soc. Agric. Sci.* **2017**, in press. [CrossRef]
30. Pérez-Grijalva, B.; Herrera-Sotero, M.; Mora-Escobedo, R.; Zebadúa-García, J.C.; Silva-Hernández, E.; Oliart-Ros, R.; Pérez-Cruz, C.; Guzmán-Gerónimo, R. Effect of microwaves and ultrasound on bioactive compounds and microbiological quality of blackberry juice. *LWT-Food Sci. Technol.* **2018**, *87*, 47–53. [CrossRef]

31. Cortez-Vega, W.R.; Pizato, S.; de Souza, J.T.A.; Prentice, C. Using edible coatings from Whitemouth croaker (*Micropogonias furnieri*) protein isolate and organo-clay nanocomposite for improve the conservation properties of fresh-cut 'Formosa' papaya. *Innov. Food Sci. Emerg. Technol.* **2014**, *22*, 197–202. [CrossRef]
32. Adetunji, C.O.; Ogundare, M.O.; Ogunkunle, A.T.J.; Kolawole, O.M.; Adetunji, J.B. Effects of edible coatings from xanthum gum produced from xanthomonas campestris pammel on the shelf life of *Carica papaya* Linn fruits. *Asian J. Agric. Biol.* **2014**, *2*, 8–13.
33. Hazarika, T.; Lalthanpuii, M.D. Influence of edible coatings on physico-chemical characteristics and shelf-life of papaya (*Carica papaya*) fruits during ambient storage. *Indian J. Agric. Sci.* **2017**, *87*, 1077–1083.
34. Ali, A.; Muhammad, M.T.M.; Sijam, K.; Siddiqui, Y. Effect of chitosan coatings on the physicochemical characteristics of Eksotika II papaya (*Carica papaya* L.) fruit during cold storage. *Food Chem.* **2011**, *124*, 620–626. [CrossRef]
35. Siriamornpun, S.; Kaewseejan, N. Quality, bioactive compounds and antioxidant capacity of selected climacteric fruits with relation to their maturity. *Sci. Hortic.* **2017**, *221*, 33–42. [CrossRef]
36. Hernández, Y.; Lobo, M.G.; González, M. Determination of vitamin C in tropical fruits: A comparative evaluation of methods. *Food Chem.* **2006**, *96*, 654–664. [CrossRef]
37. Wall, M.M. Ascorbic acid, vitamin A, and mineral composition of banana (*Musa* sp.) and papaya (*Carica papaya*) cultivars grown in Hawaii. *J. Food Compost. Anal.* **2006**, *19*, 434–445. [CrossRef]
38. Jimenez, A.; Creissen, G.; Kular, B.; Firmin, J.; Robinson, S.; Verhoeyen, M.; Mullineaux, P. Changes in oxidative processes and components of the antioxidant system during tomato fruit ripening. *Planta* **2002**, *214*, 751–758. [CrossRef] [PubMed]
39. Jayathunge, K.G.L.R.; Gunawardhana, D.K.S.N.; Illeperuma, D.C.K.; Chandrajith, U.G.; Thilakarathne, B.M.K.S.; Fernando, M.D.; Palipane, K.B. Physico-chemical and sensory quality of fresh cut papaya (*Carica papaya*) packaged in micro-perforated polyvinyl chloride containers. *J. Sci. Technol.* **2014**, *51*, 3918–3925. [CrossRef] [PubMed]
40. Barragán-Iglesias, J.; Méndez-Lagunas, L.L.; Rodríguez-Ramírez, J. Ripeness indexes and physicochemical changes of papaya (*Carica papaya* L. cv. Maradol) during ripening on-tree. *Sci. Hortic.* **2018**, *236*, 272–278. [CrossRef]
41. Fuggate, P.; Wongs-Aree, C.; Noichinda, S.; Kanlayanarat, S. Quality and volatile attributes of attached and detached 'Pluk Mai Lie' papaya during fruit ripening. *Sci. Hortic.* **2010**, *126*, 120–129. [CrossRef]
42. Lieb, V.M.; Esquivel, P.; Castillo, E.C.; Carle, R.; Steingass, C.B. GC–MS profiling, descriptive sensory analysis, and consumer acceptance of Costa Rican papaya (*Carica papaya* L.) fruit purees. *Food Chem.* **2018**, *248*, 238–246. [CrossRef] [PubMed]
43. Barreto, G.P.M.; Fabi, J.P.; De Rosso, V.V.; Cordenunsi, B.R.; Lajolo, F.M.; do Nascimento, J.R.O.; Mercadante, A.Z. Influence of ethylene on carotenoid biosynthesis during papaya postharvesting ripening. *J. Food Compost. Anal.* **2011**, *24*, 620–624. [CrossRef]
44. Almora, K.; Pino, J.A.; Hernández, M.; Duarte, C.; González, J.; Roncal, E. Evaluation of volatiles from ripening papaya (*Carica papaya* L., var. Maradol roja). *Food Chem.* **2004**, *86*, 127–130. [CrossRef]
45. Lee, P.-R.; Ong, Y.-L.; Yu, B.; Curran, P.; Liu, S.-Q. Profile of volatile compounds during papaya juice fermentation by a mixed culture of Saccharomyces cerevisiae and Williopsis saturnus. *Food Microbiol.* **2010**, *27*, 853–861. [CrossRef] [PubMed]
46. Chang, L.; He, S.; Liu, Q.; Xiang, J.; Huang, D. Quantifying muskmelon fruit attributes with A-TEP-based model and machine vision measurement. *J. Integr. Agric.* **2018**, *17*, 1369–1379. [CrossRef]
47. Chen, A.; Yang, Z.; Zhang, N.; Zhao, S.; Chen, M. Quantitative evaluation and prediction for preservation quality of cold shocked cucumber based on entropy. *Innov. Food Sci. Emerg. Technol.* **2016**, *35*, 58–66. [CrossRef]
48. Calvo, H.; Moreno-Armendáriz, M.A.; Godoy-Calderón, S. A practical framework for automatic food products classification using computer vision and inductive characterization. *Neurocomputing* **2016**, *175*, 911–923. [CrossRef]

© 2018 by the authors. Licensee MDPI, Basel, Switzerland. This article is an open access article distributed under the terms and conditions of the Creative Commons Attribution (CC BY) license (http://creativecommons.org/licenses/by/4.0/).

Article

Antimicrobial Performance of Two Different Packaging Materials on the Microbiological Quality of Fresh Salmon

Manuela Rollini [1,*], Tim Nielsen [2], Alida Musatti [1], Sara Limbo [1], Luciano Piergiovanni [1], Pilar Hernandez Munoz [3] and Rafael Gavara [3]

1. DeFENS, Department of Food, Environmental and Nutritional Sciences, Università degli Studi di Milano, Via G. Celoria 2, Milano 20133, Italy; alida.musatti@unimi.it (A.M.); sara.limbo@unimi.it (S.L.); luciano.piergiovanni@unimi.it (L.P.)
2. SP Food and Bioscience, Ideon, Lund 22370, Sweden; Tim.Nielsen@sp.se
3. Packaging Lab., Instituto de Agroquímica y Tecnología de Alimentos, IATA-CSIC, Paterna 46980, Spain; phernan@iata.csic.es (P.H.M.); rgavara@iata.csic.es (R.G.)
* Corresponding author: manuela.rollini@unimi.it; Tel.: +39-02-50319150

Academic Editor: Sheryl Barringer
Received: 11 December 2015; Accepted: 20 January 2016; Published: 27 January 2016

Abstract: In the present research the antimicrobial activity of two active packaging materials on the spoilage microbiota of fresh salmon fillets was tested. A PET-coated film (PET: Polyethylene Terephthalate) containing lysozyme and lactoferrin was tested in parallel with a carvacrol-coextruded multilayer film. Salmon fillet samples were stored up to four days at 0 and 5 °C, comparatively. The carvacrol multilayer film was found effective in preventing mesophiles and psychrotrophs at shorter storage time and at lower temperature (4.0 compared to 5.0 log CFU/g in the control sample—CFU: Colony Forming Units). Lysozyme/lactoferrin-coated PET was instead efficient in decreasing H_2S-producing bacteria at longer storage time and higher temperature (2.7 instead of 4.7 log CFU/g in the control sample). Even if is not intended as a way to "clean" a contaminated food product, an active package solution can indeed contribute to reducing the microbial population in food items, thus lowering the risk of food-related diseases.

Keywords: active packaging; carvacrol; coextrusion; lysozyme; lactoferrin; coating; salmon

1. Introduction

Salmon is an important product of aquaculture: 1,400,000 ton were produced in 2010 with a value of more than seven billion US dollars. In 2009, the main producers of Atlantic salmon were Norway, Chile, the EU and Canada [1]. In the EU the main farmed species is Atlantic salmon, accounting for 93% of the total aquaculture production. The EU is very dependent on the rest of the world for salmon since it imports 80% of its supply from other countries, and 80% of that from Norway.

Fresh seafood is characterized by a relatively short shelf-life and is typically spoiled by aerobic Gram-negative bacteria. In fish, the spoilage process is well documented and consists of autolytic degradation by fish enzymes and the production of unpleasant odors and flavors as a result of microbial action [2]. Typically, in the chilled seafood supply chain, microbial-mediated changes dominate the spoilage process [3]. The bacteria responsible for spoilage in marine fish vary according to the harvest environment, the degree of cross-contamination and the preservation methods applied post-harvest. The primary spoilage bacteria in aerobically packed fish are Gram-negatives from the genera *Pseudomonas* and *Shewanella* while in modified atmospheres, they are *Photobacterium* as well as lactic acid bacteria (LAB), such as *Lactobacillus* and *Carnobacterium* [3,4]. *Shewanella putrefaciens* and

Pseudomonas spp. become the main producers of the volatile compounds associated with spoilage, such as trimethylamine (TMA), ammonia and sulphides. TMA is particularly responsible for the unpleasant odor of spoiled fish, and is a common index of seafood quality. However, the changes in sensory attributes often occur before products are hygienically spoiled [5].

During storage, spoilage bacteria are selected primarily as a result of the physical and chemical condition in the products; however, seafood spoilage obviously involves growth of the microorganisms to a high amount (>10^6–10^7 CFU/g) and the interaction between groups of microorganisms may influence their growth and metabolism [6]. In particular the high iron-binding capacity of the *Pseudomonas* siderophores may cause this bacterial group to be positively selected, as well as LAB inhibit the growth of other bacteria due to the formation of lactic acid and/or bacteriocins or by competition for nutrients [7].

A significant support to the fight against microbial spoilage may derive from food packaging, which not only acts as a barrier against moisture, water vapor, gases and solutes, but may also serve as a carrier of active substances, such as antimicrobials, in active packaging. Active packaging is defined as an integrated system in which the package, product and environment interact to prolong shelf-life or to enhance safety and/or quality of food products [8].

The antimicrobial features of food packaging materials can be achieved by different strategies: among others, the incorporation in the bulky polymer of migrating compounds, grafting of antimicrobial moieties, and immobilization of antimicrobial agents on the surface of the material in direct contact with the food are the most widely adopted routes. However, direct incorporation in the plastic polymer matrix is not a feasible approach when dealing with antimicrobial agents that are highly sensitive to package production conditions; indeed, high processing pressure and high temperature, or incompatibility with the packaging material, can inactivate the active agents [9]. Alternative production methods have recently been considered. In particular, coating technology has gained increasing attention due to its promising potential as a valid route to generate antimicrobial packaging materials [10].

The aim of the present research was to evaluate the antimicrobial effectiveness of two active packaging materials on the spoilage microbiota of fresh salmon fillets. In particular, a PET-coated film containing lysozyme and lactoferrin was tested in parallel with a carvacrol-coextruded multilayer film. These natural compounds have been selected for their interesting antimicrobial performance evidenced in the frame of the European funded project NAFISPACK—Natural Antimicrobials for Innovative and Safe Packaging (EU212544). Salmon samples after packaging were stored up to four days at 0 and 5 °C, comparatively, to show any significant effect of the developed materials on spoilage microbiota. These two temperatures were chosen as the first represents the one applied by the company to deliver salmon samples to the market, while the second is commonly used by consumers for home storage.

2. Experimental Section

2.1. Antimicrobial Films

Lysozyme-Lactoferrin (LZ-LF) water solutions were coated together onto PET: the formulation included gelatin as the main biopolymer network, glycerol as plasticizer, and a lipid phase (a monoglyceride acetylate) as slipping agent, in order to avoid blocking of the film coils during the unwinding operations on the reels. The total dry matter was 19.1 wt %, where LZ and LF accounted for 3 wt % (on the total weight). The procedure has been up-scaled in a pilot plant coating machine at a speed of 10 m/min. The removal of the solvent (water) was achieved by the combined effect of IR lamps and a flux of mild air. Right before the coating deposition, the plastic web underwent an in-line corona treatment to improve surface wettability.

Carvacrol (4.8%) was instead incorporated in a coextruded multilayer film 73 μm thick, made up of two external layers of polypropylene, two tie layers, and an internal layer of ethylene–vinyl alcohol copolymer with a 29% ethylene molar content (EVOH-29), prepared as reported elsewhere [11]. Briefly, films were obtained by extrusion processes using a flat sheet die 500 mm wide and three single screw extruders (Dr. Collin GmbH, Edersberg, Germany) with 30 mm of screw diameter and L/D ratio of 30. Die, feeding block, and transfer lines were maintained at 250 °C. Temperature of chill roll was set at 45 °C and an additional airknife was used. Line speed was 9.5 m/min.

2.2. Sample Preparation

Fresh salmon fillets cut to cubes (*ca.* 120 g) were packaged in laminated material made up of polyethylene/polyamide/polyethylene in Bremerhaven (Germany) and then transported on ice to the University of Milan, where they arrived the next day (Figure 1). Fillets were repackaged into pouches (20 cm × 12 cm) made from a high barrier multilayer film after being wrapped in sheets of the antimicrobial films. The transfer of the samples from the original package to the experimental packages was made in sterile conditions under a safety cabinet. Samples were all packaged under vacuum. Reference samples were also prepared in an identical way, without wrapping in antimicrobial films.

Figure 1. Sample of salmon fillets, cut into cubes, employed in shelf-life trials.

After packaging, samples were all stored at 0 °C and analyses were performed at days 0, 1 and 4. At day 1, some samples were moved to 5 °C, where they were left until the microbiological analysis.

2.3. Microbiological Analysis

Microbial analyses were performed at days 0, 1 and 4 after packaging. Samples were taken from the surface of the samples (2–3 mm) and transferred aseptically and weighed in a sterile Stomacher bag, diluted with peptone water (PW) (Scharlab, Barcelona, Spain) and blended in Stomacher (IUL S.L., Barcelona, Spain) for 6 min. Ten-fold dilution series in PW of the obtained suspensions were made and plated on selective solid media: TSA (Merck, Germany) for mesophiles and psychrotrophs, MEA (Sigma Aldrich, St. Louis, MO, USA) for yeasts and fungi, *Pseudomonas* agar base (Himedia, India) for *Pseudomonas* spp., VRBLA (Violet Red Bile Agar, Merck, Germany) for coliforms, MRS agar (Scharlab, Barcelona, Spain) for lactic acid bacteria, and Lyngby Iron agar (Oxoid, UK) for H_2S-producing bacteria. Colonies were counted after incubation at 30 °C for 24 h for mesophiles, 10 °C for 10 days for psychrotrophs, 30 °C for five days for yeasts and fungi, 25 °C for 24 h for *Pseudomonas*, 37 °C for 24 h for coliforms, 25 °C for five days for lactic acid bacteria and 20 °C for three days for H_2S-producing bacteria. Counts were performed in triplicate and reported as logarithms of the number of colony forming units (log CFU/g salmon), and means and standard deviations were calculated.

3. Results and Discussion

Aerobic mesophiles and psychrotrophs as well as H_2S-producing bacteria were the prevalent population in the salmon at time of packaging (3.3–3.6 log CFU/g), followed by LAB (2.2 log CFU/g and *Enterobacteriaceae* (1.5 log CFU/g) (Table 1). According to the International Commission on Microbiological Specifications for Foods, most aquatic animals at the time of harvest have microbial counts in the range of 2 to 5 log CFU/g [12]. In the present study the initial value of microbial load was in the same range, as also reported by other authors [13]. After one day of storage at 0 °C, the psychrotroph population increased up to 5.5 log CFU/g in control samples. Salmon packed in carvacrol active films was characterized by a reduced psychrotroph population (3.9 log CFU/g), while LZ-LF–coated films were ineffective (5.8 log CFU/g) (Table 2). Gram-negative psychrotrophic bacteria are the major group of microorganisms responsible for the spoilage of fresh fish at chilled temperatures [2]. In this study, mesophiles in the first 24 h remained in the range 3.9–4.1 log CFU/g, without significant difference between the control sample and those stored in the active packages.

Table 1. Microbial population (log CFU/g) present in salmon at time of packaging.

Microorganisms	Microbial Count (log CFU/g)
Aerobic mesophiles	3.6 ± 0.2
Aerobic psychrotrophs	3.5 ± 0.1
Enterobacteriacae	1.5 ± 0.4
Lactic acid bacteria	2.2 ± 0.2
Pseudomonas	<2
H_2S-producing bacteria	3.3 ± 0.2

Table 2. Microbial population (log CFU/g) in salmon packed in two different antimicrobial materials and in one reference material (control) after storage for one day at 0 °C.

Microorganisms	Control	Carvacrol	LZ LF
Aerobic mesophiles	3.9 ± 0.7	4.1 ± 0.6	4.0 ± 0.5
Aerobic psychrotrophs	5.5 ± 0.4	3.8 ± 0.5	5.8 ± 0.2
Enterobacteriaceae	1.6 ± 0.2	1.6 ± 0.5	1.6 ± 0.2
Lactic acid bacteria	2.2 ± 0.3	2.2 ± 0.4	3.1 ± 0.1
Pseudomonas	<2	<2	<2
H_2S-prod. Bacteria	2.7 ± 0.6	3.6 ± 0.4	2.1 ± 0.2

At day 1 some samples were moved to 5 °C while, in parallel, other samples were kept at 0 °C. All samples were then stored for up to four days, one day more than the normal shelf-life suggested by the company.

After four days at 0 °C, salmon packed in carvacrol films maintained psychrotroph and mesophile populations at low levels, *i.e.*, around 4.0 log CFU/g, compared to 5.0 log CFU/g in control samples (Table 3). Also, LZ-LF–coated films showed an interesting performance, reducing the populations up to 3.0 and 3.7 log CFU/g, respectively. Note also, at this storage temperature, the efficacy of the two active films in reducing the *Pseudomonas* population of approximately 1 log cycle compared to control, from 3.2 to 2.2–2.3 log CFU/g.

Salmon samples stored for four days at 5 °C without active packaging were characterized by mesophile and psychrotroph population of 5.3 log CFU/g. Carvacrol active films were effective in maintaining only the mesophile population at a low level of 4.6 log CFU/g, while LZ-LF–coated PET reduced both the populations down to 4.5 and 3.8 log CFU/g, respectively. It must be noted, also, that in this last case there was a reduction of H_2S-producing bacteria to 2.7 instead of 4.7 log CFU/g as seen in the control samples. This positive observation that LZ/LF–coated PET films could prevent growth of H_2S-producing bacteria is of actual relevance since this type of microorganism has been identified as the most potent in causing rejection of fresh salmon fillets, due to the production of off-odors during growth [4].

Table 3. Microbial population (log CFU/g) in salmon packed in two different antimicrobial materials and in one reference material (control) after storage for four days at 0 °C and 5 °C, comparatively.

Temperature	Microorganisms	Control	Carvacrol	LZ LF
0 °C	Aerobic mesophiles	5.1 ± 0.2	3.9 ± 0.2	3.7 ± 0.3
	Aerobic psychrotrophs	5.0 ± 0.3	4.0 ± 0.1	3.0 ± 0.1
	Enterobacteriaceae	2.0 ± 0.0	1.9 ± 0.3	2.2 ± 0.2
	Lactic acid bacteria	2.9 ± 0.2	2.6 ± 0.1	2.0 ± 0.1
	Pseudomonas	3.2 ± 0.5	2.2 ± 0.2	2.3 ± 0.6
	H$_2$S-prod. bacteria	3.3 ± 0.2	3.4 ± 0.4	3.4 ± 0.4
5 °C	Aerobic mesophiles	5.3 ± 0.6	4.6 ± 0.6	4.5 ± 0.3
	Aerobic psychrotrophs	5.3 ± 0.6	6.9 ± 0.4	3.8 ± 0.2
	Enterobacteriaceae	1.9 ± 0.0	2.5 ± 0.5	2.2 ± 0.2
	Lactic acid bacteria	2.8 ± 0.5	3.2 ± 0.4	3.1 ± 0.2
	Pseudomonas	2.5 ± 0.1	2.8 ± 0.3	2.2 ± 0.2
	H$_2$S-prod. bacteria	4.7 ± 0.6	4.0 ± 0.7	2.7 ± 0.5

Enterobacteriaceae were also found to be members of the microbial association implicated in the spoilage of fresh sliced salmon during refrigerated storage. This finding is in agreement with results reported for different fish species, including fresh Atlantic salmon [14] as well as rainbow trout [15], in which *Enterobacteriaceae* were determined as a part of the microbial population at the end of the product shelf-life under refrigerated storage. In this study, *Enterobacteriaceae* were always found in low levels, less than 3 log CFU/g; although *Enterobacteriaceae* can grow at low temperatures, their proliferation was slow during refrigerated storage, possibly because their growth rate is lower than that of other Gram-negative psychrotrophic spoilers [16].

The LAB population was always found lower than 3.2 log CFU/g, and was not influenced by the active packaging employed. The low LAB count in this study was expected since they tend to grow slowly at refrigeration temperatures [13].

The obtained data satisfied the recommended microbiological limits for fresh and frozen fish reported in [12], which are defined only for aerobic plate counts performed at 20–25 °C (5.5 to 7 log CFU/g) and for *E. coli* (1 to 2.7 log CFU/g). The first data reflect handling practices in the fish industry, from shipboard to market delivery, while the second parameter is considered an indicator of contamination and, when present in high numbers, suggests temperature abuse in product handling.

The present research was aimed at investigating the efficacy of two antimicrobial food packaging materials on microbial spoilage of fresh salmon fillets. Incorporation of natural antimicrobials into food packaging materials to control the growth of spoilage and pathogenic organisms has been researched for the last decades. As regards fish products and shelf-life, the vast majority of data relates to the possibility of applying antimicrobial edible films in contact with the fish surface. Jasour *et al.* [17] evaluated the effect of an edible coating based on chitosan coated with the lactoperoxidase system (LPS) on the quality and shelf-life extension of rainbow trout during refrigerated storage at 4 °C. Results indicated that antimicrobial coating was found efficient in reducing *Shewanella putrefaciens*, *Pseudomonas fluorescens* as well as psychrotrophic and mesophilic bacterial populations compared to the control sample.

Few authors have investigated the performance of an antimicrobial package prepared by applying a coating procedure. Gomez Estaca *et al.* [18] produced a complex gelatin-chitosan film incorporating clove essential oil which was applied to fish during chilled storage: the growth of microorganisms was drastically reduced in Gram-negative bacteria, especially *Enterobacteriaceae*, while LAB remained constant for much of the storage period. Neetoo and Mahomoodally [19] compared the antimicrobial efficacy against *Listeria monocytogenes* in smoked salmon fillets of films or direct coatings incorporating nisin (Nis) and sodiumlactate (SL), sodiumdiacetate (SD), potassiumsorbate (PS), and/or sodium benzoate (SB) in binary or ternary combinations on cold smoked salmon. Surface treatments incorporating Nis (25,000 IU/mL) in combination with PS (0.3%) and SB (0.1%) had the highest inhibitory activity, reducing the population of L. *monocytogenes* by a maximum of 3.3 log CFU/cm^2 (films) and 2.9 log CFU/cm^2 (coatings) relative to control samples after 10 days of storage at 21 °C. During refrigerated storage, coatings were more effective in inhibiting growth of L. *monocytogenes* than their film counterparts. Cellulose-based coatings incorporating Nis, PS, and SB reduced the population of L. *monocytogenes*, and anaerobic and aerobic spoilage microbiota by a maximum of 4.2, 4.8, and 4.9 log CFU/cm^2, respectively, after four weeks of refrigerated storage.

In the present study the carvacrol-incorporated multilayer film was found effective in preventing mesophiles and psychrotrophs at shorter storage time and at lower temperature, while lysozyme-lactoferrin–coated PET was mostly efficient in decreasing H$_2$S-producing bacteria at longer storage time and higher temperature. Even if it is not intended as a way to "clean" a contaminated food product, an antimicrobial package solution can indeed contribute to reducing the microbial population in food items, thus lowering the decay of organoleptic features and increasing shelf life.

4. Conclusions

This work provides examples of active food packages, in which the antimicrobial compounds, *i.e.*, the volatile carvacrol and the soluble association lysozyme-lactoferrin, were applied, the first incorporated in a coextruded multilayer film while the second association was coated onto PET. Application of these materials on actual salmon samples under conditions similar to those foreseeable for a future practical use gave positive results, in particular for the reduction of H$_2$S-producing bacteria in coated films. Future work will indicate whether the antimicrobial-loaded films used here may find other practical uses (e.g., liners or wraps) and whether they will be effective to improve the safety and to extend the shelf-life of other food products.

Acknowledgments: Work supported by funds of the EU project FP7-NAFISPACK 212544.

Author Contributions: Tim Nielsen and Luciano Piergiovanni conceived and designed the experiments; Rafael Gavara and Pilar Hernandez-Munoz prepared the films; Tim Nielsen organized all the salmon dispatches from Bremerhaven to Milano; Alida Musatti, Manuela Rollini and Sara Limbo performed the shelf life experiments; Manuela Rollini wrote the paper.

Conflicts of Interest: The authors declare no conflict of interest.

References

1. Food and Agriculture Organization (FAO), Fisheries and Aquaculture Department. *The State of World Fisheries and Aquaculture 2012*; Food and Agriculture Organization of the United Nations: Rome, Italy, 2012.
2. Gram, L.; Huss, H.H. Microbiological spoilage of fish and fish products. *Int. J. Food Microbiol.* **1996**, *33*, 121–137. [CrossRef]
3. Emborg, J.; Laursen, B.G.; Rathjen, T.; Dalgaard, P. Microbial spoilage and formation of biogenic amines in fresh and thawed modified atmosphere-packed salmon (*Salmo salar*) at 2 °C. *J. Appl. Microbiol.* **2002**, *92*, 790–799. [CrossRef] [PubMed]
4. Dalgaard, P.; Gram, L.; Huss, H.H. Spoilage and shelf-life of cod fillets packed in vacuum or modified atmospheres. *Int. J. Food Microbiol.* **1993**, *19*, 283–294. [CrossRef]
5. Franzetti, L.; Martinoli, S.; Piergiovanni, L.; Galli, A. Influence of active packaging on the shelf life of minimally processed fish products in a modified atmosphere. *Packag. Technol. Sci.* **2002**, *14*, 267–274. [CrossRef]
6. Gram, L.; Dalgaard, P. Fish spoilage bacteria: Problems and solutions. *Curr. Opin. Microbiol.* **2002**, *13*, 262–266. [CrossRef]
7. Gram, L.; Ravn, L.; Rasch, M.; Bruhn, J.B.; Christensen, A.B.; Givskov, M. Food spoilage-interactions between food spoilage bacteria. *Int. J. Food Microbiol.* **2002**, *78*, 79–97. [CrossRef]
8. Suppakul, P.; Miltz, J.; Sonneveld, K.; Bigger, S.W. Active packaging technologies with an emphasis on antimicrobial packaging and its application. *J. Food Sci.* **2003**, *68*, 408–420. [CrossRef]

9. Perez-Perez, C.; Regalado-González, C.; Rodríguez-Rodríguez, C.; Barbosa-Rodríguez, J.R.; Villaseñor-Ortega, F. Incorporation of antimicrobial agents in food packaging films and coatings. In *Advances in Agricultural and Food Biotechnology*; Guevara-Gonzalez, R.G., Torres-Pacheco, I., Eds.; Research Signpost: Trivandrum, Kerala, India, 2006; pp. 193–216.
10. Farris, S.; Piergiovanni, L. Emerging coating technologies for food and beverage packaging materials. In *Emerging Food Packaging Technologies: Principles and Practice*; Yam, K., Lee, D.S., Eds.; Woodhead Publishing Ltd.: Oxford, UK, 2012; pp. 274–302.
11. Cerisuelo, J.P.; Bermudez, J.M.; Aucejo, S.; Català, R.; Gavara, R.; Hernandez-Munoz, P. Describing and modeling the release of an antimicrobial agent from an active PP/EVOH/PP package for salmon. *J. Food Eng.* **2013**, *116*, 352–361. [CrossRef]
12. International Commission of Microbiological Specifications for Foods (ICMSF). Sampling plans for fish and shellfish. In *Microorganisms in Foods 2. Sampling for Microbiological Analysis: Principles and Specific Application*; ICMSF: Toronto, ON, Canada, 1986; pp. 181–196.
13. Sallam, K.I. Antimicrobial and antioxidant effects of sodium acetate, sodium lactate and sodium citrate in refrigerated sliced salmon. *Food Cont.* **2007**, *18*, 566–575. [CrossRef] [PubMed]
14. Amanatidou, A.; Schluter, O.; Lemkau, K.; Gorris, L.G.M.; Smid, E.J.; Knorr, D. Effect of combined application of high pressure treatment and modified atmospheres on the shelf life of fresh Atlantic salmon. *Innov. Food Sci. Emerg. Technol.* **2000**, *1*, 87–98. [CrossRef]
15. Chytiri, S.; Chouliara, I.; Savvaidis, I.N.; Kontominas, M.G. Microbiological, chemical, and sensory assessment of iced whole and filleted aquacultured rainbow trout. *Food Microbiol.* **2004**, *21*, 157–165. [CrossRef]
16. Papadopoulos, V.; Chouliara, I.; Badeka, A.; Savvaidis, I.N.; Kontominas, M.G. Effect of gutting on microbiological, chemical, and sensory properties of aquacultured sea bass (*Dicentrarchus. labrax*) stored in ice. *Food Microbiol.* **2003**, *20*, 411–420. [CrossRef]
17. Jasour, M.S.; Ehsani, A.; Mehryar, L.; Naghibi, S.S. Chitosan coating incorporated with the lactoperoxidase system: An active edible coating for fish preservation. *J. Sci. Food Agric.* **2014**, *95*, 1373–1378. [CrossRef] [PubMed]
18. Gómez-Estaca, J.; López de Lacey, A.; López-Caballero, M.E.; Gómez-Guillén, M.C.; Montero, P. Biodegradable gelatin-chitosan films incorporated with essential oils as antimicrobial agents for fish preservation. *Food Microbiol.* **2010**, *27*, 889–896. [CrossRef] [PubMed]
19. Neeto, H.; Mahomoodally, F. Use of antimicrobial films and edible coatings incorporating chemical and biological preservatives to control growth of *Listeria monocytogenes* on cold smoked salmon. *BioMed Res. Int.* **2014**, *2014*. [CrossRef] [PubMed]

© 2016 by the authors. Licensee MDPI, Basel, Switzerland. This article is an open access article distributed under the terms and conditions of the Creative Commons Attribution (CC BY) license (http://creativecommons.org/licenses/by/4.0/).

Article

Development of Antibacterial Composite Films Based on Isotactic Polypropylene and Coated ZnO Particles for Active Food Packaging

Clara Silvestre [1,*], Donatella Duraccio [2], Antonella Marra [1], Valentina Strongone [1] and Sossio Cimmino [1]

1. Institute of Polymers, Composites and Biomaterials, Consiglio Nazionale delle Ricerche (IPCB/CNR) Via Campi Flegrei 34, Naples 80078, Italy; antonella.marra@ipcb.cnr.it (A.M.); valentina.strongone@outlook.it (V.S.); sossio.cimmino@cnr.it (S.C.)
2. Istituto per le Macchine Agricole e Movimento Terra, Consiglio Nazionale delle Ricerche (IMAMOTER-CNR) Strada delle Cacce 73, Torino 10135, Italy; donatella.duraccio@cnr.it
* Correspondence: clara.silvestre@cnr.it; Tel.: +39-081-867-5067; Fax: +39-081-867-5230

Academic Editor: Stefano Farris
Received: 11 December 2015; Accepted: 15 January 2016; Published: 22 January 2016

Abstract: This study was aimed at developing new films based on isotactic polypropylene (iPP) for food packaging applications using zinc oxide (ZnO) with submicron dimension particles obtained by spray pyrolysis. To improve compatibility with iPP, the ZnO particles were coated with stearic acid (ZnOc). Composites based on iPP with 2 wt % and 5 wt % of ZnOc were prepared in a twin-screw extruder and then filmed by a calender. The effect of ZnOc on the properties of iPP were assessed and compared with those obtained in previous study on iPP/ZnO and iPP/iPPgMA/ZnO. For all composites, a homogeneous distribution and dispersion of ZnOc was obtained indicating that the coating with stearic acid of the ZnO particles reduces the surface polarity mismatch between iPP and ZnO. The iPP/ZnOc composite films have relevant antibacterial properties with respect to *E. coli*, higher thermal stability and improved mechanical and impact properties than the pure polymer and the composites iPP/ZnO and iPP/iPP-g-MA/ZnO. This study demonstrated that iPP/ZnOc films are suitable materials for potential application in the active packaging field.

Keywords: isotactic polypropylene; zinc oxide; properties; active packaging; composites

1. Introduction

It is extensively reported that the nano/microparticles of zinc oxide (ZnO) exhibit antibacterial activity against gram-positive and gram-negative bacteria. This activity does not require the presence of UV light (unlike TiO_2), being stimulated by visible light, and it is inversely dependent on particle size [1–5]. The mechanisms responsible for the antibacterial activity are not fully understood. Distinctive mechanisms that have been put forward in the literature are listed as the following: direct contact of ZnO with cell walls, resulting in destructing bacterial cell integrity [6–8], liberation of antimicrobial ions, mainly Zn^{2+} ions [8], and reactive oxygen species formation [9–11].

There has been a great deal of interest in the antimicrobial property of ZnO for food packaging applications, as a viable solution for stopping infectious diseases [12–14]. Due also to low cost, ZnO particles (with sub-micro and nano-dimensions) are therefore ideal fillers for polymers to be applied in the field of active food packaging. Thus, ZnO particles have been incorporated into a number of different polymers used in food packaging [15,16], such as low-density polyethylene (LDPE) [17,18], isotactic polypropylene (iPP) [19–24], polyamide (PA) [18,25], and polylactic acid (PLA) [26].

Recently, the influence of ZnO particles, obtained by spray pyrolysis with submicron dimensions [22,27,28], on the structure, morphology, thermal stability, photo stability, and mechanical and antibacterial properties of (iPP)/ZnO composites was investigated [22,24]. The addition of ZnO particles imparts improvements on the photodegradation resistance of iPP to ultraviolet irradiation and the composites exhibit significant antibacterial activity against *Escherichia coli*. This activity is dependent on exposure time and composition.

On the other hand, it was noticed that due to the surface polarity mismatch between iPP and ZnO, agglomeration phenomena of the ZnO particles occur and that these phenomena cause a decrease in the mechanical and other functional properties of iPP/ZnO composites with respect to plain iPP [22–24].

The main problem to be solved in adding ZnO nano/microparticles to an iPP matrix seems therefore related to the formation of agglomerated domains that occur because of the strong intermolecular interactions among the ZnO particles in combination with their high surface area. This prevents transfer of their superior properties to the composite. Good dispersion has been reported for some polar polymers [18,25], but ZnO dispersion in non-polar polymers, such as iPP, during melt processing remains a challenge.

A largely proposed strategy, to improve dispersion consists in adding a compatibiliser, containing groups suitable for interaction with the two components [29–33]. Following this strategy in a previous paper, polypropylene grafted with maleic anhydride (PPgMA) [24] was selected as the most promising candidate as a compatibiliser between iPP and ZnO. In particular, the influence of three PPgMA (with different MW and MA% content) added to iPP/ZnO 98/2 wt % on the structure, morphology, mechanical, thermal, barrier properties and antibacterial activity against *E. coli* was investigated with the aim to verify if the compatibiliser PPgMA could be beneficial in order to increase the dispersion of ZnO in an iPP matrix in order to have films with improved properties. It was found that the presence of this compatilizer improves the dispersion of the particles in the matrix, but, at the same time, does not cause any enhancement in the barrier and mechanical properties and indeed reduces the antibacterial activity with respect to iPP/ZnO. An important aspect found in this study is that the more the ZnO are well embedded in polymer material, the more the antibacterial activity decreases, probably because the surface of the particle available for contact with the solution decreases.

An alternative methodology to improve the dispersion consists of modifying the particles' surfaces with groups suitable for interaction with the matrix.

The main purpose of this paper is to develop new films based on iPP for applications in the food industry as active packaging using coated ZnO particles to improve compatibility between the organic phase and inorganic one. In particular, the surface of the ZnO particles, obtained by spray pyrolysis, was coated with stearic acid (ZnOc). Objective of the paper is also to assess the influence of the coating process on the structure, morphology, and thermal stability of the zinc oxide particles.

2. Experimental

2.1. Materials and Sample Preparation

The materials used in this work are: (1) isotactic polypropylene (iPP, Moplen X30S), in pellets, kindly supplied by Basell (Ferrara, Italy), with melt flow index = 9 dg·min^{-1} (2.16 kg, 230 °C), $M_w = 3.5 \times 10^5$ and $M_n = 4.7 \times 10^4$; (2) zinc oxide coated with stearic acid (ZnOc) (white powder). The ZnO particles were synthesized using a preindustrial spray scale pyrolysis platform at the Pylote in Toulouse-France and then coated with stearic acid. This technique [22,27,28] provides many advantages compared to other techniques of preparation: the simplicity of the process, high purity of the powders obtained, more uniform chemical composition, narrow size distribution, better regularity in shape and the ability to synthesize multicomponent materials. The coating of the ZnO particles was performed by preparing a solution of stearic acid and ZnO (1:10) in isopropanol under stirring for 12 h. The powder was recovered by centrifuge and dried in an oven at a temperature of 70 °C.

2.2. Composite Preparation

The powders of ZnOc were mixed with iPP at two different compositions: 2% and 5% (see Table 1), in a twin screw extruder Collin ZK 25 ($D = 25$ mm and $L/D = 24$).

Table 1. Isotactic polypropylene/Zinc oxide coated with stearic acid (iPP/ZnOc) mixtures.

Sample	iPP (wt %)	ZnOc (wt %)
iPP	100	—
iPP/2%ZnOc	98	2
iPP/5%ZnOc	95	5

The temperature setting of the extruder from the hopper to the die was: 180/195/195/190/180 °C. The screw speed of the dispenser was 20 rpm while the speed of the extruder screws was 25 rpm.

Films of iPP and iPP/ZnOc were obtained by compression molding in a press at 210 °C and 100 bars. The films had a thickness of about 110–120 µm.

2.3. Characterization

The following technologies were used to determine the properties of the films:

1) FT-IR Spectroscopy

Infrared spectra of the compression molded films were recorded with a PerkinElmer FT-IR spectrometer, model Paragon 500 equipment (PerkinElmer, Boston, MA, USA). The IR spectra were recorded in the range 4000–800 cm^{-1} with 4 cm^{-1} resolution and 20 scans.

2) Wide-Angle X-ray Diffraction

Wide-angle X-ray diffraction (WAXD) measurements were conducted using a Philips XPW diffractometer (Philips, Almelo, The Netherlands) with CuKa radiation (1.542 Å) filtered by nickel. The scanning rate was 0.02°/s and the scanning angle was from 5° to 45°. The ratio of the area under the crystalline peaks and the total area multiplied by 100 was taken as the crystalline percentage degree.

3) Thermogravimetric Analysis

The thermal stability of the blends was examined by thermogravimetric analysis (TGA), using a PerkinElmer-Pyris Diamond apparatus (PerkinElmer, Boston, MA, USA) with a heating rate of 10 °C/min in air. Two measurements were performed for each sample.

4) Scanning Electron Microscopy

The surface analysis was performed using SEM, Fei Quanta 200 FEG (Fei, Hillsboro, OR, USA), on particles of the powders and on cryogenically fractured surfaces of composites. Before the observation, samples were coated with an Au/Pd alloy using an E5 150 SEM coating unit.

5) Tensile Tests

Dumbbell-shaped specimens were cut from the compression molded films and used for the tensile measurements. Stress–strain curves were obtained using an Instron machine, Model 4505 (Instron, Torino, Italy) at room temperature (25 °C) at a crosshead speed of 5 mm/min. Ten tests were performed for each composition.

6) Charpy Impact Test

The impact test allows for determining the degree of toughness of a polymer. Charpy impact tests were performed by using a pendulum CEAST (CEAST, Torino, Italy) with appropriate software for processing the data. The tests were carried out at room temperature on slabs obtained by compression moulding. Rectangular samples, with width of 3 mm, thickness of about 4 mm and length of 5 cm were used. The samples were cut from slabs obtained by using the same conditions adopted for the preparation of films.

7) Analysis UV-Visible Spectrometric

The UV-Visible spectrometry is useful for evaluating the ability of a material to minimize radiation potentially dangerous to the packaged food. The instrument used is a spectrometer Shimadzu UV-2101PC (Shimadzu, Columbia, MD, USA). UV-Vis spectra were recorded in transmission in the range 200–850 nm.

8) Antibacterial Test

The antimicrobial activity of the iPP/ZnOc composites was evaluated using *E. coli* DSM 498T (DSMZ, Braunschweig, Germany) as test microorganisms. The evaluation was performed using the ASTM Standard Test Method E 2149-10 [34]. The preparation of the bacterial inoculum required to grow a fresh 18-h shake culture of *E. coli* DSM 498T in a sterile nutrient broth (LB composition for 1 L: 10 g of triptone, 5 g of yeast extract and 10 g of sodium chloride) The colonies were maintained according to good microbiological practice and examined for purity by creating a streak plate. The bacterial inoculum was diluted using a sterile buffer solution (composition for one litre: 0.150 g of potassium chloride, 2.25 g of sodium chloride, 0.05 g of sodium bicarbonate, 0.12 g of $CaCl_2 \cdot 6H_2O$ and pH = 7) until the solution reached an absorbance of 0.3 ± 0.01 at 600 nm, as measured spectrophotometrically. This solution, which had a concentration of 1.5×10^8–3.0×10^8 colony forming units/mL (CFUs/mL), was diluted with the buffer solution to obtain a final concentration of 1.5×10^6–3×10^6 CFUs/mL, that was the working bacterial dilution.

The experiments were performed in 50 mL sterilized flasks. One gram of the film was maintained in contact with 10 mL of the working bacterial dilution. After 2 min, 100 mL of the working bacterial dilution was transferred to a test tube, which was followed by serial dilution and plating out on Petri dishes (10 mm × 90 mm) in which the culture media was previously poured. The Petri dishes were incubated at 35 °C for 24 h. These dishes represented the T_0 contact time. The flasks were then placed on a wrist-action shaker for 1 h, 24 h, 48 h, 5 days and 10 days. The bacterial concentration in the solutions at these time points was evaluated by again performing serial dilutions and standard plate counting techniques. Three experiments were performed for each composition. The number of colonies in the Petri dish after incubation was converted into the number of colonies that form a unit per millilitre (CFUs/m) of buffer solution in the flask. The percentage reduction (R%) was calculated using the following formula:

$$R\% \text{ (CFU/mL)} = [(B - A)/B] \times 100, \qquad (1)$$

where A = CFUs/mL for the flask containing the sample after the specific contact time and B = CFUs/mL at T_0.

3. Results and Discussion

3.1. Analysis of ZnO and ZnOc Particles

Before blending the ZnOc particles with iPP, an investigation of properties of the ZnO and ZnOc particles has been performed, to assess the amount of stearic acid present on the ZnO particles and the influence of the coating process on the structure, morphology, and thermal stability of the zinc oxide particles. The content of stearic acid present on the surface of the coated particles was evaluated through thermogravimetric analysis. Figure 1 reports the thermogravimetric curves of ZnO and ZnOc particles recorded during the heating rate of 20 °C/min in air from room temperature to 700 °C.

In the entire T range, for ZnO particles, no weight loss was observed. In the case of ZnOc, after the thermal treatment, a weight reduction of about 9% is found. Considering that ZnO does not undergo degradation, it can be concluded that the weight reduction percentage observed for ZnOc corresponds probably to the percentage of stearic acid present on the surface of the particles.

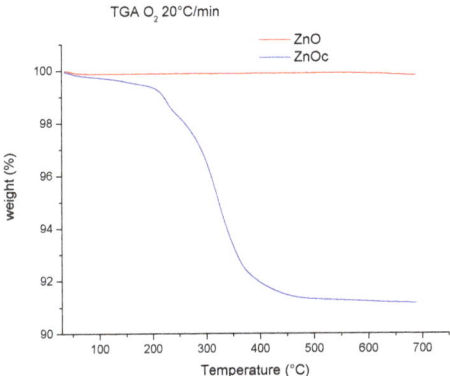

Figure 1. Thermostability curve of ZnO and ZnOc particles.

To study the influence of the stearic acid on the crystalline structure of the ZnO particles, the spectra of X-ray diffraction at high angles were recorded. As shown in Figure 2, the spectrum of the particles of ZnOc presents the same peaks as those of ZnO, suggesting that the coating does not alter the crystalline structure of the particles (zincite) [35].

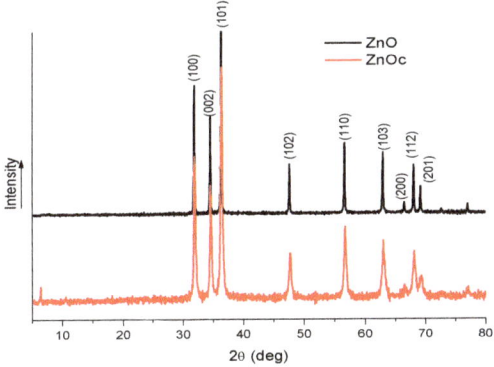

Figure 2. X-ray diffraction pattern of ZnO and ZnOc particles.

In order to recognize the functional groups present, the interaction between the stearic acid and the ZnO particles and to obtain information on the shape of the particles, FTIR analysis was been performed.

Figure 3 reports the FTIR spectrum of ZnOc and ZnO. For the coated sample, different bands can be observed, in particular according to literature [35–38]:

- at 2916 and 2848 cm^{-1}: these vibration bands can be assigned to the "stretching" of the symmetric and asymmetric aliphatic group CH_2;
- at 1460 cm^{-1}: this band is assigned to the vibration of "bending" of aliphatic groups CH_2 and CH_3 of stearic acid;
- at 1540 and 1384 cm^{-1}: these bands are assigned to the asymmetric and symmetric vibrations of the carboxylate group of the stearic acid;

- at around 454 cm^{-1}: These bands give information about the shape of the particles. It is interesting to go deeper into the bands at points 3 and 4.

In particular, the bands at 1540 and 1384 cm^{-1} (region of absorption of the carbonyl C=O) can be related to the coordination behavior of the carboxylate group when it forms complexes with metals. These bands can give important information on the nature of the link between ZnO and the carboxylate group (COO–) of stearic acid.

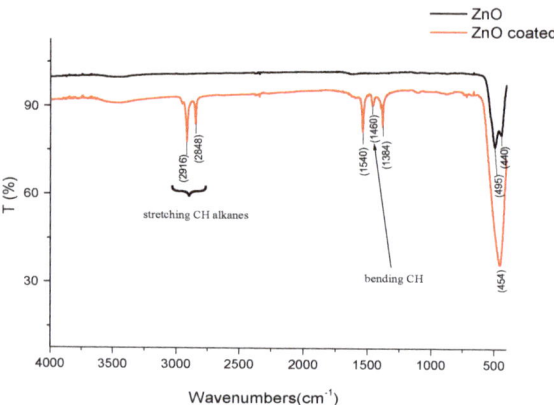

Figure 3. FT-IR spectrum of ZnO and ZnOc powder.

According to literature, the carboxylate group has versatile coordination behavior, when it forms coordination complexes with metals. It can be ionic, monodentate, bidentate chelating or bridging. Measuring the frequency of asymmetric, (ν_{as}(COO–)) and symmetric bands (ν_{as}(COO–) and ν_s(COO–)) and the magnitude of their separation, Δ, ($\Delta = \nu_{as}$(COO–) $- \nu_s$(COO–)), the mode of the carboxylate binding with ZnO can be determined [36–38]. Generally, depending on the value of Δ, the following order is proposed for the coordination of carboxylates of divalent metals: Δ(chelating) < Δ(bridge) < Δ(ionic) < Δ(monodentate).

Finally, the assignment of the type of link is done comparing the $\Delta_{(experimental)}$ with that of the corresponding sodium salt ($\Delta_{(sodium\ salt)}$) with the following rules:

- if $\Delta_{(experimental)} \ll \Delta_{(sodium\ salt)}$ is bidentate chelating coordination;
- if $\Delta_{(experimental)} \leqslant \Delta_{(sodium\ salt)}$ is bidentate coordination to bridge;
- if $\Delta_{(experimental)} > \Delta_{(sodium\ salt)}$ coordination is monodentate.

From the figure, $\Delta_{(experimental)}$ = (1540 − 1384) cm^{-1} = 156 cm^{-1}. According to the criterion above, and taking into account that from literature data, the sodium stearate as Δ equal to 138 cm^{-1} [35], it can be concluded that the coordination is monodentate.

The band at 454 cm^{-1} can give information about the shape of the particles. Using the theory of dielectric media [34,36], the single band in ZnOc indicates particles with a spherical shape. It is interesting to make a comparison with the spectrum of the uncoated particles where the two bands indicate the presence of a structure with a mainly prismatic shape.

The morphology of the ZnO and ZnOc particles was studied using scanning electron microscopy (SEM). Figure 4 shows SEM micrographs of ZnO and ZnOc respectively. From the micrographs, it is clear that ZnO particles are characterized by a hexagonal crystal structure, as already emerged from FTIR analysis, with a smooth surface. Moreover, the ZnO particles seem to have a strong tendency to form agglomerates. Contrary ZnOc particles, more homogeneously dispersed in the matrix, have a spherical shape. Comparing the dimensions of the kinds of particles, the size of the ZnO particles ranges between 250 to 500 nm while that of the particles of ZnOc varies between 1 to 1.2 µm.

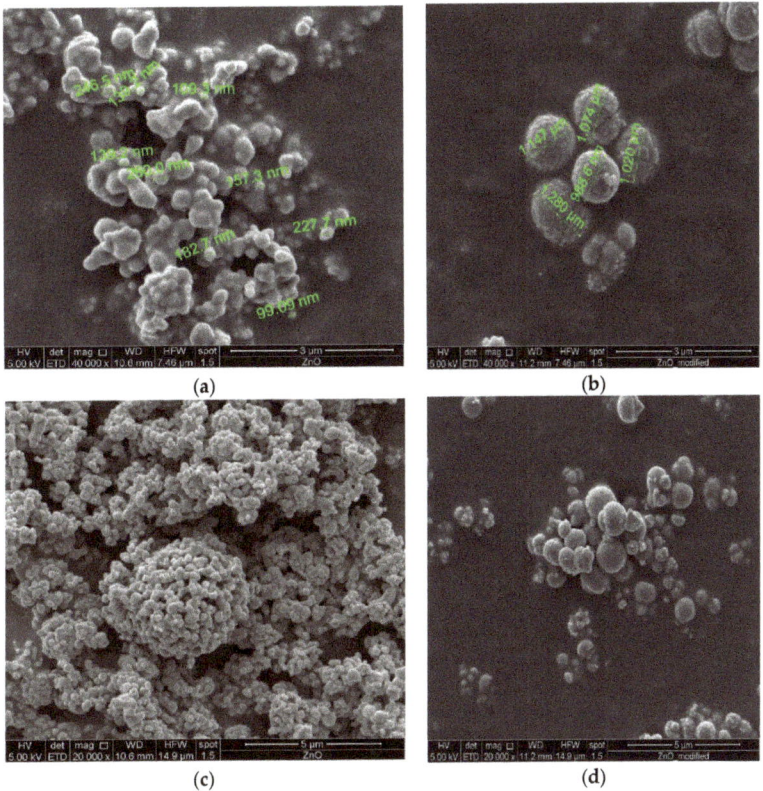

Figure 4. SEM micrographs of (**a,c**) ZnO and (**b,d**) ZnOc powders. (a,b) 40000×; (c,d) 20000×.

3.2. Analysis of the iPP/ZnOc Composites

3.2.1. Structure and Morphology

The WAXD patterns of iPP and iPP/ZnOc composites are reported in Figure 5. All samples show the presence of the peak at $2\theta = 18°$–$19°$ characteristic of α form of iPP [39,40]. The sample iPP/5%ZnOc is characterized also by a small percentage of the form β, highlighted by the presence of the peak at $2\theta = 16°$.

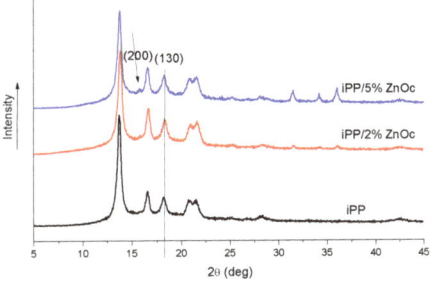

Figure 5. WAXD spectrum of iPP and iPP/ZnOc composites.

UV-Vis spectra are reported in Figure 6. For all samples, an absorption band at 280 nm is observed, probably due to the presence of stabilizer added to commercial iPP. For the samples containing ZnOc, an absorption band is also observed in the region around 385 nm, as indicated by arrows. This band is due to the inherent capacity of the ZnO particles to absorb the UV light [29,30,35,36].

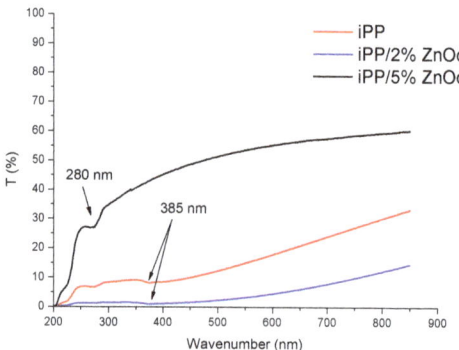

Figure 6. UV-Vis spectrum of iPP and iPP/ZnOc samples.

Figure 7 shows SEM micrographs of the fractured surface of iPP, iPP/2%ZnOc, iPP/5%ZnOc. It is possible to note that the particles are fairly distributed within the polymer matrix. Only a few aggregates with a size of about 5 μm can be observed.

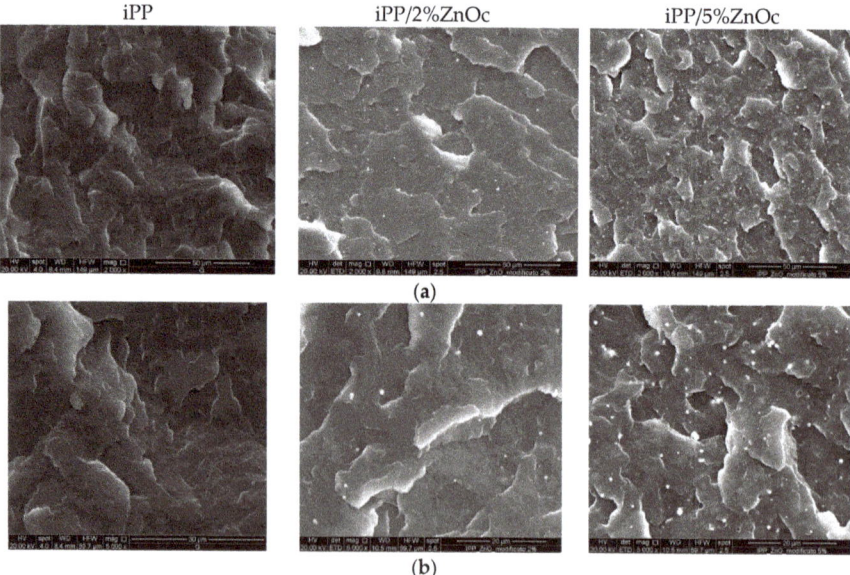

Figure 7. Micrographs of iPP and iPP/ZnOc pellets fractured in liquid nitrogen at magnification (**a**) 2000× and (**b**) 5000×.

Comparing the results with those obtained for the system iPP/ZnO and iPP/PPgMA/ZnO [24] at a given composition, it seems that the coating of the ZnO with stearic acid favors a better dispersion and distribution of the particles in the iPP matrix and prevents the formation of agglomerates.

3.2.2. Thermostability

Figure 8 shows the % weight loss of the samples as function of temperature for iPP and iPP/ZnOc samples, whereas Table 2 reports the values of the temperature at the inflection point of the curve of Figure 8 detected at the maximum of the peak of the first derivative and which corresponds to the maximum rate of the degradation of the sample.

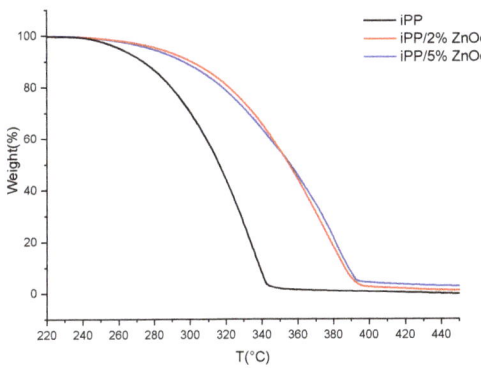

Figure 8. Thermostability curve of iPP and iPP/ZnOc composites.

Table 2. Thermal degradation temperature at which the degradation rate is maximum (T_{max}).

Sample	T_{max} (°C)
iPP	335
iPP/2%ZnOc	375
iPP/5%ZnOc	381

For the two composites, a delay in the temperature of starting degradation, compared to iPP, and a consistent increase of the Tmax are observed. Taking into account that the presence of uncoated ZnO at the same composition did not have consistent influence on the thermostability of iPP, as reported in the paper at reference [22], the increase in thermal stability should be attributed to the stearic acid that coats the ZnO particles. As it was reported in a previous section (see Figure 1), the degradation of the stearic acid starts before the degradation of iPP. The degradation products of the stearic acid probably act as a barrier for the degradation of the matrix also slowing the diffusion of the degradation products of iPP in the sample causing an increase of the thermal stability of the iPP/ZnOc composites.

3.2.3. Mechanical and Impact Properties

Figure 9 shows the stress-strain curves of iPP and iPP/ZnOc composites, whereas Table 3 reports the values of mechanical parameters, (Young modulus (E), stress and strain at the yield point (σ_y, ε_y), and at break (σ_b, ε_b)).

It can be seen that all the samples has the typical behavior of a semi-crystalline polyolefin, with the phenomenon of yield strength, cold drawing, fiber elongation and final break of the fibers.

From the values shown in the Table 3, it can be observed that: (1) the two composite films have similar values of Young modulus and strain at the yield point but higher than those of plain iPP; (2) the elongation at yield and the stress at break point can be considered similar for the three samples (the differences are inside the experimental error), whereas the elongation at break decreases with the addition of ZnOc. Comparing these results with those reported in reference [22], where ZnO not coated was used, it can be observed that the composites with ZnOc present improved mechanical properties.

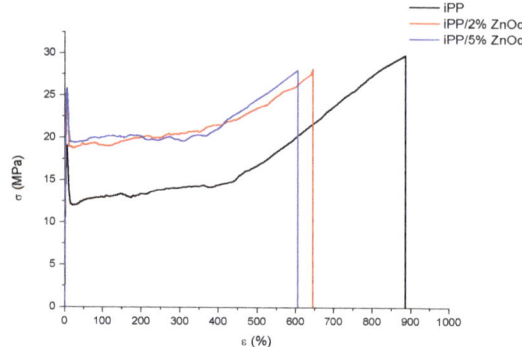

Figure 9. Stress–strain curves of iPP and iPP/ZnOc films.

Table 3. Stress–strain parameters of iPP and iPP/ZnOc composites.

Sample	E (MPa)	σ_y (MPa)	ε_y (%)	σ_b (MPa)	ε_b (%)
iPP	1350 ± 100	19 ± 3	7 ± 2	30 ± 3	890 ± 65
iPP/2%ZnOc	1537 ± 44	25 ± 2	7 ± 1	28 ± 4	645 ± 54
iPP/5%ZnOc	1515 ± 79	26 ± 1	7 ± 1	28 ± 3	605 ± 76

Table 4 shows the values of the impact test, in particular the values of the force (F) that the pendulum lost on impact with the sample, the energy (U) absorbed by the samples at the break and the toughness (T). The results demonstrate that the presence of ZnOc increases the toughness; in fact, the toughness of iPP/5%ZnOc sample is 26% higher than that of iPP.

Table 4. Impact tests values for iPP and iPP/ZnOc composites.

Sample	F (N)	E (J)	T (kJ/m^2)
iPP	73 ± 7	0.032 ± 0.008	1.91 ± 0.35
iPP/2%ZnOc	84 ± 3	0.033 ± 0.005	2.07 ± 0.22
iPP/5%ZnOc	89 ± 5	0.042 ± 0.004	2.41 ± 0.14

3.2.4. Antibacterial Properties

In Figure 10, the antimicrobial effect against *E. coli* is presented as a function of time for the different composites. Without ZnO particles, the reference concentration of the micro-organism is measured to be ~2 × 10^6. After 1 h, no change in the concentration was observed for all samples. By increasing the time, a decrease in the *E. coli* concentration is observed for the composites. The effect is more evident for the iPP/5%ZnOc composite. Significant variations in concentration are observed increasing the contact time and ZnOc content. After 24 h, the concentration of *E. coli* decreases to 8.8 × 10^5 for iPP/2%ZnOc and 1.7 × 10^5 for iPP/5%ZnOc. After 48 h, the bacterial concentration was significantly decreased for the iPP/5%ZnOc sample (2 × 10^3 CFU/mL). The sample iPP/2%ZnOc reaches similar values after five days.

The values of percentage reduction (%R) of *E. coli* for all samples at different contact times are reported in Table 5. Neat iPP exhibits no bactericidal activity, and R was observed to be zero up to day 10. The iPP/5%ZnOc composite exhibited maximum reduction, 99.9%, after 48 h.

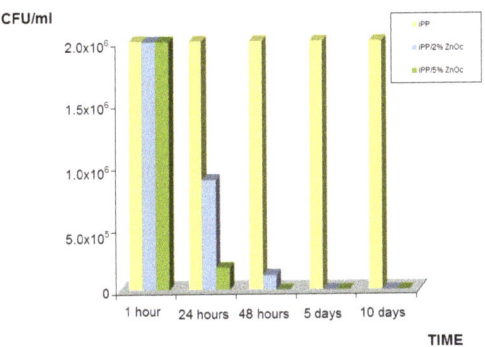

Figure 10. Effect of time and filler content on the antibacterial activity of iPP and iPP/2%–5% ZnOc composites.

Table 5. Percent reduction of E.coli at different times of contact with films of iPP and iPP/2%–5% ZnOc.

Sample	%R (t = 1 h)	%R (t = 24 h)	%R (t = 48 h)	%R (t = 5 days)	%R (t = 10 days)
iPP	0	0	0	0	0
iPP/2%ZnOc	0	55.61 ± 0.01	94.00 ± 0.01	99.99 ± 0.01	99.99 ± 0.01
iPP/5%ZnOc	0	91.12 ± 0.01	99.99 ± 0.01	99.99 ± 0.01	99.99 ± 0.01

In Table 6 a comparison of the bacterial activity of the three systems, iPP/ZnO, iPP/ZnOc and iPP/PPgMA/ZnO, at the same ZnO content (2%) is reported. From this table, it is clearly confirmed that:

- in the iPP/ZnOc system the ZnOc particles maintain their antibacterial properties against *E. coli*, with respect to the uncoated particles;
- in the system iPP/PPgMA/ZnO, the ZnO particles that are linked to the maleic anhydride groups of PPgMA [24], do not display similar antibacterial activity at least up to 48 h. Probably, the PP chains of the PPgMA, due to the link between MA and ZnO, cover the ZnO particles and hinder the antibacterial activity.

Table 6. Percent reduction of *E. coli* at 48 h and 5 days of contact for iPP and different films at 2% ZnO, (adapted from Table 5 of this paper and references [22,24]).

Sample	%R (t = 48 h)	%R (t = 5 days)
iPP	0	0
iPP/2%ZnO	90	99.99
iPP/2%ZnOc	94	99.99
iPP/PP$_{(9k)}$gMA$_{(4.8)}$ */2%ZnO	65	NA
iPP/PP$_{(65k)}$gMA$_{(1.4)}$ **/2%ZnO	60	NA
iPP/PP$_{(95k)}$gMA$_{(0.5)}$ ***/2%ZnO	31 ± 5	NA

* PP$_{(9k)}$gMA$_{(4.8)}$ Mw = 9100 MA (wt %) = 4.8; ** PP$_{(65k)}$gMA$_{(1.4)}$ Mw = 65,000 MA (wt %) = 1.4; ***/PP$_{(95k)}$gMA$_{(0.5)}$ Mw = 95,000 MA (wt %) = 0.5.

The results obtained from the analysis of antibacterial properties allow us to conclude that the particles of ZnO with stearic acid have relevant antibacterial property against *E. coli*, similar to that of ZnO and that the coating of the particles does not have a negative influence as the coating of the particles with PPgMA [24].

4. Conclusions

This work had as its final objective the preparation of the film based on the isotactic polypropylene matrix intended for packaging food, with improved properties by the addition of ZnO particles coated with stearic acid (ZnOc). The latter has the function of compatibilization between the inorganic metal oxide particles phase and the organic matrix of the isotactic polypropylene. The samples were prepared in a twin-screw extruder and then filmed by a compression molding.

It was observed that the stearic acid coating on the ZnO particles reduces the surface polarity mismatch between iPP and ZnO and allows the formation of a composite with fair distribution of particles.

The principal achievement of the novel composites is the strong antibacterial activity against *E. coli*: the bacterial concentration decreases with increasing concentration of ZnOc and the contact time between the film and the bacterial solution. After 48 h, the bacterial reduction was significantly decreased for the sample containing 5% of ZnOc (R = 99.99%); for the sample iPP/2% ZnOc, it reaches these values after five days.

Moreover, the iPP/ZnOc composites present improvement of the thermal stability, tensile parameters (Young modulus and stress at the yield point) and impact properties with respect to neat iPP, iPP/ZnO, and iPP/PPgMA/ZnO at least for samples containing 2% ZnO [22,24]. It also has to be underlined that the films containing ZnOc show an absorption in the region around 385 nm, confirming that the ZnOc particles also have a shielding effect to UV radiation as those of ZnO as reported in the literature.

On the base of the results obtained, it can be stated that the methodology proposed (using of novel ZnO particles obtained by spray pyrolysis; coating the ZnO with stearic acid and optimization of the processing conditions) is innovative, because no literature is available (at our knowledge) on the properties of such an iPP/ZnOc system prepared directly by melt mixing. Moreover, it is also very efficient in preventing formation of agglomerated domains and in providing a system with improved properties.

In conclusion, the composites iPP/5% ZnOc films have relevant antibacterial property against *E. coli*, higher thermal stability and improved mechanical and impact properties than the pure iPP film so that they are suitable for application in the food industry as active packaging films.

Acknowledgments: The research described herein was partially supported by the European Community's Seventh Framework Programme (ERA-Net Susfood-CEREAL "Improved and resource efficiency throughout the post-harvest chain of fresh-cut fruits and vegetable"), Italian Health Ministry (Progetti Ricerca Corrente/2013 "Impiego di nanopackaging innovativo nella filiera della carne: valutazione dell'efficacia antibatterica e della sicurezza d'uso"), and Italian Ministry of Foreign Affair (Bilateral Project Italy/Quebec 2014–2016 "Sviluppo di nanomateriali ecosostenibili per l'Imballaggio alimentare adatti alla sterilizzazione per radiazione"). Jannette Dexpert-Ghys and Marc Verelst from The Centre d'Élaboration de Matériaux et d'Etudes Structurales (CEMES-CNRS), Toulouse-France, are kindly thanked for supplying ZnO particles obtained by spray-pyrolysis. Ida Romano of the Istituto di Chimica Biomolecolare (CNR), Pozzuoli (NA) Italy is kindly thanked for performing the antimicrobial tests.

Author Contributions: Clara Silvestre and Sossio Cimmino supervised the research program. Clara Silvestre, Sossio Cimmino and Donatella Duraccio designed the setup. Antonella Marra, Valentina Strongone and Donatella Duraccio performed the experiments. All authors contributed to the analysis of the presented experiments and correlation of the different means of investigations. Antonella Marra and Clara Silvestre wrote the initial draft. Clara Silvestre and Sossio Cimmino coordinated the revisions of the draft in the final form.

Conflicts of Interest: The authors declare no conflict of interest.

References

1. Stoimenov, P.K.; Klinger, R.L.; Marchin, G.L.; Klabunde, K.J. Metal oxide nanoparticles as bactericidal agents. *Languimuir* **2002**, *18*, 6679–6686. [CrossRef]
2. Jones, N.; Ray, B.; Ranjit, K.T.; Manna, A.C. Antibacterial activity of ZnO nanoparticle suspensions on a broad spectrum of microorganisms. *FEMS Microbiol. Lett.* **2008**, *279*, 71–76. [CrossRef] [PubMed]

3. Sirelkhatim, A.; Mahmud, S.; Seeni, A.; Kaus, N.H.M.; Ann, L.C.; Bakhori, S.K.M.; Habsah, H.; Dasmawati, M. Review on Zinc oxide nanoparticles: Antibacterial activity and toxicity mechanism. *Nano-Micro Lett.* **2015**, *7*, 219–242. [CrossRef]
4. Padmavathy, N.; Vijayaraghavan, R. Enhanced bioactivity of ZnO nanoparticles—An antimicrobial study. *Sci. Technol. Adv. Mater.* **2008**, *9*. [CrossRef]
5. Yamamoto, O. Influence of particle size on the antibacterial activity of zinc oxide. *Int. J. Inorg. Mater.* **2001**, *3*, 643–646. [CrossRef]
6. Brayner, R.; Ferrari-Iliou, R.; Brivois, N.; Djediat, S.; Benedetti, M.F.; Fiévet, F. Toxicological impact studies based on *Escherichia coli* bacteria in ultrafine ZnO nanoparticles colloidal medium. *Nano Lett.* **2006**, *6*, 866–870. [CrossRef] [PubMed]
7. Zhang, L.; Jiang, Y.; Ding, Y.; Daskalakis, N.; Jeuken, L.; Povey, M.; O'Neill, A.J.; York, D.W. Mechanistic investigation into antibacterial behaviour of suspensions of ZnO nanoparticles against *E. coli*. *J. Nanopart. Res.* **2010**, *12*, 1625–1636. [CrossRef]
8. Li, M.; Zhu, L.; Lin, D. Toxicity of ZnO nanoparticles to *Escherichia coli*: Mechanism and the influence of medium components. *Environ. Sci. Technol.* **2011**, *45*, 1977–1983. [CrossRef] [PubMed]
9. Sawai, J.; Shoji, S.; Igarashi, H.; Hashimoto, A.; Kokugan, T.; Shimizu, M.; Kojima, H. Hydrogen peroxide as an antibacterial factor in zinc oxide powder slurry. *J. Ferment. Bioeng.* **1998**, *86*, 521–522. [CrossRef]
10. Lipovsky, A.; Nitzan, Y.; Gedanken, A.; Lubart, R. Antifungal activity of ZnO nanoparticles—The role of ROS mediated cell injury. *Nanotechnology* **2011**, *22*. [CrossRef] [PubMed]
11. Yamamoto, O.; Sawai, J.; Sasamoto, T. Change in antibacterial characteristics with doping amount of ZnO in MgO-ZnO solid solution. *Int. J. Inorg. Mater.* **2000**, *2*, 451–454. [CrossRef]
12. Silvestre, C.; Duraccio, D.; Cimmino, S. Food packaging based on polymer nanomaterials. *Prog. Polym. Sci.* **2011**, *36*, 1766–1782. [CrossRef]
13. Silvestre, C.; Cimmino, S. *Ecosustainable Polymer Nanomaterials for Food Packaging: Innovative Solutions, Characterization Needs, Safety and Environmental Issues*; Silvestre, C., Cimmino, S., Eds.; CRC Press, Taylor & Francis Group: Boca Raton, FL, USA, 2013.
14. Lagaron, M.; Ocio, M.J.; Lopez-Rubio, A.J. *Antimicrobial Polymers*; Yam, K.L., Lee, D.S., Eds.; John Wiley & Son: Hoboken, NJ, USA, 2012.
15. Matei, A.; Cernica, I.; Cadar, O.; Roman, C.; Schiopu, V. Synthesis and characterization of ZnO—Polymer nanocomposites. *Int. J. Mater. Form.* **2008**, *1*, 767–770. [CrossRef]
16. Huang, C.; Chen, S.; Wei, W.C.J. Processing and property improvement of polymeric composites with added ZnO nanoparticles through microinjection molding. *J. Appl. Polym. Sci.* **2006**, *102*, 6009–6016. [CrossRef]
17. Emamifar, A.; Kadivar, M.; Shahedi, M.; Solimanian-Zad, S. Effect of nanocomposite packaging containing Ag and ZnO on reducing pasteurization temperature of orange juice. *J. Food Process. Preserv.* **2012**, *36*, 104–112. [CrossRef]
18. Droval, G.; Aranberri, I.; Bilbao, A.; German, L.; Verelst, M.; Dexpert-Ghys, J. Antimicrobial activity of nanocomposites: Poly(amide) 6 and low density poly(ethylene) filled with zinc oxide. *E-Polymers* **2008**, *128*, 1–13. [CrossRef]
19. Lepot, N.; van Bael, M.K.; van den Rul, H.; D'Haen, J.; Peeters, R.; Franco, D.; Mullens, J. Influence of incorporation of ZnO nanoparticles and biaxial orientation on mechanical and oxygen barrier properties of polypropylene films for food packaging. *J. Appl. Polym. Sci.* **2015**, *120*, 1616–1623. [CrossRef]
20. Chandramouleeswaran, S.; Mhaske, S.T.; Kathe, A.A.; Varadarajan, P.V.; Prasad, V.; Vigneshwaran, N. Functional behaviour of polypropylene/ZnO-soluble starch nanocomposites. *Nanotechnolohy* **2007**, *18*. [CrossRef]
21. Tang, J.; Wang, Y.; Liu, H.; Belfiore, A. Effects of organic nucleating agents and zinc oxide nanoparticles on isotactic polypropylene crystallization. *Polymer* **2004**, *45*, 2081–2091. [CrossRef]
22. Silvestre, C.; Cimmino, S.; Pezzuto, M.; Marra, A.; Ambrogi, V.; Dexpert-Ghys, J.; Verelst, M.; Augier, S.; Romano, I.; Duraccio, D. Preparation and characterization of isotactic polypropylene/zinc oxide microcomposites with antibacterial activity. *Polym. J.* **2013**, *45*, 938–945. [CrossRef]
23. Duraccio, D.; Silvestre, C.; Pezzuto, M.; Cimmino, S.; Marra, A. Polypropylene and polyethylene-based nanocomposites for food packaging applications. In *Ecosustainable Polymer Nanomaterials for Food Packaging*; Silvestre, C., Cimmino, S., Eds.; CRC Press, Taylor & Francis Group: Boca Raton, FL, USA, 2013; pp. 143–167.

24. Cimmino, S.; Duraccio, D.; Marra, A.; Pezzuto, M.; Romano, I.; Silvestre, C. Effect of compatibilisers on mechanical, barrier and antimicrobial properties of iPP/ZnO nano/microcomposites for food packaging application. *J. Appl. Packag. Res.* **2015**, *7*, 108–127.
25. Erem, A.D.; Ozcan, G.; Skrifvars, M. Antibacterial activity of PA6/ZnO nanocomposite fibers. *Text. Res. J.* **2011**, *81*, 1638–1646. [CrossRef]
26. Murariu, M.; Paint, Y.; Murariu, O.; Raquez, J.M.; Bonnaud, L.; Dubois, P. Current progress in the production of PLA–ZnO nanocomposites: Beneficial effects of chain extender addition on key properties. *J. Appl. Polym. Sci.* **2015**, *132*. [CrossRef]
27. Krunks, M.; Mellikov, E. Zinc oxide thin films by the spray pirolysis method. *Thin Solid Films* **1995**, *270*, 33–36. [CrossRef]
28. Alavi, S.; Caussat, B.; Couderc, J.P.; Dexpert-Ghys, J.; Joffin, N.; Neumeyer, D.; Verelst, M. Spray pyrolysis synthesis of submicronic particles. Possibilities and limits. *Adv. Sci. Technol.* **2003**, *30*, 417–424.
29. Cimmino, S.; Silvestre, C.; Duraccio, D.; Pezzuto, M. Effect of hydrocarbon resin on the morphology and mechanical properties of isotactic polypropylene/clay composites. *J. Appl. Polym. Sci.* **2011**, *119*, 1135–1143. [CrossRef]
30. Kaci, M.; Benhamida, A.; Cimmino, S.; Silvestre, C.; Carfagna, C. Waste and virgin LDPE/PET blends compatibilized with an Ethylene-Butyl Acrylate-Glycidyl Methacrylate (EBAGMA) Terpolymer, 1. *Macromol. Mater. Eng.* **2005**, *290*, 987–995. [CrossRef]
31. Utracki, L.A. Compatibilization of polymer blends. *Can. J. Chem. Eng.* **2002**, *80*, 1008–1016. [CrossRef]
32. Akbar, B.; Bagheri, R. Influence of compatibilizer and processing conditions on morphology, mechanical properties, and deformation mechanism of PP/Clay nanocomposite. *J. Nanomater.* **2012**, *8*. [CrossRef]
33. Bastarrachea, L.J.; Wong, D.E.; Roman, M.J.; Lin, Z.; Goddard, J.M. Active packaging coatings. *Coatings* **2015**, *5*, 771–791. [CrossRef]
34. ASTM E 2149-10: *Standard Test Method for Determining the Antimicrobial Activity of Immobilized Antimicrobial Agents under Dynamic Contact Conditions*; ASTM: West Conshohocken, PA, USA, 2001.
35. Wang, Z.L. Zinc oxide nanostructures: Growth, properties and applications. *J. Phys. Condens. Matter* **2004**, *16*, 829–858. [CrossRef]
36. Zeleňák, V.; Vargová, Z.; Györyová, K. Correletion of infrared spectra of zinc(II) carboxtlates with their structures. *Spectrochim. Acta Part A* **2007**, *66*, 262–272. [CrossRef] [PubMed]
37. Capelle, H.A.; Britcher, L.G.; Morris, G.E. Sodium stearate absorbtion onto titania pigment. *J. Colloid Interface Sci.* **2003**, *268*, 293–300. [CrossRef] [PubMed]
38. Andrés Vergés, M.; Mifsud, A.; Serna, C.J. Formation of rod-like zinc oxide microcrystals in homogeneous solution. *J. Chem. Soc. Farday Trans.* **1990**, *86*, 959–963. [CrossRef]
39. Silvestre, C.; Cimmino, S.; di Pace, E. Morphology of polyolefins. In *Handbook of Polyolefins*, 2nd ed.; Vasile, C., Ed.; Marcel Dekker: New York, NY, USA, 2000; pp. 175–206.
40. Silvestre, C.; Cimmino, S.; Triolo, R. Structure, morphology and crystallization of a random ethylene-propylene copolymer. *J. Polym. Sci. Part B Polym. Phys.* **2003**, *41*, 493–500. [CrossRef]

© 2016 by the authors. Licensee MDPI, Basel, Switzerland. This article is an open access article distributed under the terms and conditions of the Creative Commons Attribution (CC BY) license (http://creativecommons.org/licenses/by/4.0/).

Article

Photoactivated Self-Sanitizing Chlorophyllin-Containing Coatings to Prevent Microbial Contamination in Packaged Food

Gracia López-Carballo, Pilar Hernández-Muñoz and Rafael Gavara *

Packaging Group, Instituto de Agroquímica y Tecnología de Alimentos, IATA-CSIC, Av. Agustin Escardino 7, 46980 Paterna, Spain; glopez@iata.csic.es (G.L.-C.); phernan@iata.csic.es (P.H.-M.)
* Correspondence: rgavara@iata.csic.es; Tel.: +34-9-6390-0022

Received: 27 August 2018; Accepted: 19 September 2018; Published: 19 September 2018

Abstract: Chlorophyllins are semi-synthetic porphyrins obtained from chlorophyll that—when exposed to visible light—generate radical oxygen substances with antimicrobial activity. In this work, chlorophyllins incorporated with polyethylene (PE), polyvinyl alcohol (PVOH), (hydroxypropyl)methyl cellulose (HPMC), and gelatin (G) were formulated for application as coatings in packages providing antimicrobial activity after photoactivation. First, the antimicrobial properties of two porphyrins (sodium magnesium chlorophyllin, E-140, and sodium copper chlorophyllin, E-141) were analyzed against *L. monocytogenes* and *Escherichia coli*. The results indicated that E-140 was more active than E-141 and that chlorophyllins were more effective against Gram-positive bacteria. In addition, both chlorophyllins were more efficient when irradiated with halogen lamps than with LEDs, and they were inactive in dark conditions. Then, coatings on polyethylene terephthalate (PET) film were prepared, and their effect against the test bacteria was similar to that shown previously with pure chlorophyllins, i.e., greater activity in films containing E-140. Among the coating matrices, those based on PE presented the least effect (1 log reduction), whereas PVOH, HPMC, and G were lethal (7 log reduction). The self-sanitizing effect of these coatings was also analyzed by contaminating the surface of the coatings and irradiating them through the PET surface, which showed high efficiency, although the activity of the coatings was limited to *L. monocytogenes*. Finally, coated films were applied as separators of bologna slices. After irradiation, all the films showed count reductions of *L. monocytogenes* and the usual microbial load; the gelatin coating was the most effective, with an average of 3 log reduction.

Keywords: porphyrin; chlorophyllin; active coating; antimicrobial; photoactivation; self-sanitizing; bologna

---

1. Introduction

Recent reports from the World Health Organization (WHO) have pointed out the high incidence of foodborne diseases as a public health problem [1]. The globalization of food supply has presented new challenges for food safety and has contributed to the internationalization of the public health problem of foodborne diseases [1]. Foodborne pathogens are usually eliminated by thermal treatment during food processing [2]. These methods can inactivate pathogenic microorganisms but could also reduce nutrients or modify organoleptic and sensory properties. To overcome these disadvantages, nonthermal treatments, such as high pressure, ultraviolet or pulsed light, or irradiation, have been developed. However, they are not as effective as thermal ones, have a high application cost [3], and are not always accepted by consumers [4].

In the last few years, many research groups have been paying attention to antimicrobial food packaging, a nonthermal treatment that involves the development of active antimicrobial films and coatings to act as protective barriers against pathogens present in food products. Polymer films and coatings can be used as carriers for a wide range of food additives, including various antimicrobials that could reduce the risk of pathogen growth on food surfaces and thus prolong product shelf life. In this context, there is increasing demand for natural, healthier additives. Chlorophyllins (water-soluble sodium magnesium chlorophyllin, E-140, and water-soluble sodium copper chlorophyllin, E-141) are semisynthetic porphyrins obtained from chlorophyll and used as colorants in dietary supplements and in cosmetics [2].

When these molecules are exposed to visible light in air, they generate singlet oxygen and radical oxygen substances, which have antimicrobial activity [2,5]. The cationic 5,10,15,20-tetrakis(1-methylpyridinium-4-yl) porphyrin tetra-iodide as photosensitizer has been reported to provide effective photoinactivation of *Pseudomonas syringae* pv. *actinidiae* in kiwifruit leaves under sunlight irradiation without damaging the plant [5]. Other authors have shown that chlorophyllin-based photosensitization reduced mesophilic bacteria and inoculated strains of food pathogens on cherry tomatoes without producing harmful effects on the nutritional quality of the tomatoes or their antioxidant activity [6].

These molecules are not toxic, and they do not present any effect in dark conditions. There are few studies about the use of porphyrins in antimicrobial applications. Porphyrin films with nylon or cellulose have been developed for industrial, household, and medical applications [7,8]. Antimicrobial activity of phthalocyanine-dyed paper has also been tested against *E. coli* and *A. baylyi* by illuminating for 1 h under lights of various intensities.

However, the application of polymer films or coatings carrying these antimicrobial compounds to avoid microbial contamination and improve the shelf life of food products has scarcely been reported. Previous studies with chlorophyllin-based films and coatings have shown their effectiveness in preventing microbial contamination of cooked frankfurters inoculated with *S. aureus* and *L. monocytogenes* [2]. The antimicrobial activity of chlorophyllins has been found to be dependent on the choice of light source: quartz lamps, near-infrared lamps, halogen lamps, slide projector lamps, special photodynamic lamps, or special light-emitting diode lights [9].

The aim of this work was the development of antimicrobial photosensitizer coatings based on synthetic and biological polymer matrices with various degrees of hydrophilicity, such as gelatin (G), polyvinyl alcohol (PVOH), polyethylene (PE) and (hydroxypropyl) methyl cellulose (HPMC), incorporated with two chlorophyllins and exposed to the radiation of halogen lamps and LED lamps as activation sources. These films were applied on bologna slices to test their antimicrobial effectiveness.

2. Materials and Methods

2.1. Reactive Agents

Water-soluble sodium magnesium chlorophyllin, E-140, and water-soluble sodium copper chlorophyllin, E-141, were obtained from Natracol (ROHA Europe S.L.U., Torrente, Spain). Gelatin from porcine skin (G) and (hydroxypropyl) methyl cellulose (HPMC) were supplied by Sigma (Barcelona, Spain). Low-density polyethylene (LDPE) emulsion (50% w/w) Aquaseal® 2200 (PE) was kindly provided by Paramelt B.V. (Heerhugowaard, The Netherlands), and Gohsenol AH17 polyvinyl alcohol was kindly provided by the Nippon Synthetic Chemical Company (Osaka, Japan).

2.2. Bacterial Strains

Bacterial strains were obtained from the Spanish Type Culture Collection (Valencia, Spain): *L. monocytogenes* CECT 911 (ATCC 19112) was used as a model of a Gram-positive bacterium and *Escherichia coli* CECT 434 (ATCC 25922) as a model of a Gram-negative bacterium. The strains were stored in Tryptone soy broth (TSB) with glycerol at $-80\ °C$. For experimental use, the stock cultures were maintained by regular subculture on agar medium slants at $4\ °C$ and transferred every month. A loopful of each strain was transferred to 10 mL of TSB and incubated at $37\ °C$ overnight to obtain early stationary phase cells.

Bacteria were spread on selective media, PALCAM agar and Mueller Hinton agar, for *L. monocytogenes* and *E. coli*, respectively.

2.3. Methods

2.3.1. Minimum Inhibitory Concentration and Minimum Bactericidal Concentration

Minimum inhibitory concentration (MIC) and minimum bactericidal concentration (MBC) of porphyrin were determined for *E. coli* and *L. monocytogenes*. The MIC is defined as the amount or concentration of active compound in which growth inhibition is observed. The MBC is the concentration in which there is no growth of microorganisms [10]. The procedure, which we have called the "drops method", consisted of adding 100 µL of microorganism (10^7 CFU/mL) and 10 µL of chlorophyllin prepared at different concentrations ranging between 0.0016 mg/mL and 15 mg/mL. Chlorophyllins were activated by two lighting systems: three 500 W Haloline 64702 halogen lamps (Osram) or five 50 W 81.575/Dia LED lights (Electro DH) for 15 min. Their light spectra differ greatly: LED has its maximum radiation in a thin band close to 450 nm and a medium-intensity wide band with a maximum at 550 nm and an unimportant contribution in the infrared region, while halogen light increases in intensity with wavelength, having its maximum in the infrared region [11]. As control samples, identical experiments were also carried out with (a) samples with chlorophyllins stored in dark conditions and (b) an illuminated sample without chlorophyllins. After photosensitization, the suspension of microorganisms was recovered, plated in agar medium, and incubated overnight at 37 °C. Cytotoxicity was calculated as the difference between colony forming units (CFU) counted with (photosensitizer and illumination) and without photodynamic treatment. Experiments were carried out in triplicate.

2.3.2. Antimicrobial Coating Preparation

Various matrices were employed as chlorophyllin carriers: gelatin (G), polyethylene (PE), polyvinyl alcohol (PVOH), and (hydroxypropyl) methyl cellulose (HPMC).

G coating-forming solution was prepared by dissolving 10 g of gelatin in 100 mL of 50% (v/v) aqueous ethanol for 2 h at 75 °C and adding 25% of glycerol (with respect to polymer dry mass) as plasticizer.

PE coating emulsion was prepared from a commercial emulsion, previously homogenized for 10 min in an ultrasonic bath. Then, 5% of 2-propanol was added to improve wettability, and the mixture was stirred for 10 min using a magnetic stirrer at room temperature.

PVOH coating-forming solution was prepared by dissolving 4 g in 100 mL of deionized water with 10% of glycerol as plasticizer and stirring for 2 h at 75 °C.

HPMC coating-forming solution was prepared by dissolving 2.5 g of HPMC in 100 mL of 50% (v/v) aqueous ethanol with 20% of glycerol as plasticizer and stirring for 2 h at 75 °C.

G, PE, PVOH, and HPMC coatings were prepared on a 30 µm poly (ethylene terephthalate) (PET) film. The PET film was set on a glass surface and corona-treated (Model BD-20V corona treater, Electro-Technic Products, Chicago, IL, USA) to improve coating adherence. Ten grams of the film-forming solutions was spread using a 50 µm extension bar (Linlab, Logroño, Spain), and the samples were placed in a homemade drying tunnel equipped with a 2500 W heat source and a 30 W fan for 3 min until they were completely dry. Coating thickness was calculated as the difference between the film thicknesses measured with a micrometer before and after the coating process. Finally, the coatings were stored in glass desiccators containing silica gel at 22 ± 2 °C prior to use.

2.3.3. Color Measurements

The color of the coated films was measured using a Konica Minolta CM-3500d spectrophotometer (Konica Minolta Business Technologies, Inc., Tokyo, Japan) set to D65 illuminant/10° observer. Film samples were measured against the surface of a standard white plate, and the CIELAB color space was used to obtain the color coordinates L^* (lightness) [black (0) to white (100)], a^* [green (−) to red (+)], and b^* [blue (−) to yellow (+)]. The color was expressed using the polar coordinates L^*, C^*, $h°$, and ΔE^*, where L^* is the same as above, C^* is chroma, $h°$ is hue angle, and $\Delta E^* = [(\Delta L^*)^2 + (\Delta a^*)^2 + (\Delta b^*)^2]^{1/2}$. Ten measurements were taken of each sample, and three samples of each film were measured.

2.3.4. Antimicrobial Activity of Coatings

Agar Diffusion

First, 100 µL of a bacterial suspension containing approximately 10^3 CFU/mL was spread on agar medium, and then a disk of film (2.5 cm diameter) was put over the agar with the coating surface facing the agar. Samples with and without chlorophyllins were irradiated with five 50 W LED lights for 15 min. At the same time, samples with and without active agent were treated in dark conditions by covering the Petri dishes with aluminum foil.

All the plates were incubated at 37 °C for 24 h, and the diameter of the resulting inhibition zone in the bacterial lawn was measured. The experiment was carried out in triplicate.

Surface Disinfection

The microorganisms to be tested were spread directly on the G, HPMC, PVOH, and PE coatings (with 1% E-140, and some without antimicrobial agent as controls). For this purpose, the coated films were placed on Petri dishes with the coating facing up and 100 µL of 10^5 CFU/mL of *L. monocytogenes* or *E. coli* was spread over them, ensuring direct contact between the bacteria and the coating. In the case of the *E. coli* bacteria, 20 or 40 mM of ethylenediaminetetraacetic acid (EDTA) was previously added to ease mass transport through the outer membrane. They were then irradiated with 5 LED lights for 15 min. The bacteria were collected with 10 mL of peptone water, and serial dilutions were made and plated on solid agar medium. The incubation was carried out at 37 °C for 24 h and CFU/mL was counted. The experiments were carried out in triplicate.

2.3.5. Release of Chlorophyllins

According to Council Directive 85/572/EEC, which determines the list of simulants that should be used to control the migration of components from materials and objects of plastic material intended to come into contact with food products, the simulant that best represents a fatty food such as bologna is vegetable oil. However, since the two chlorophyllins selected are water soluble, we decided to test the films with 50% ethanol, which is a valid simulant for oil-in-water emulsions. In this case, the exposure conditions were 10 mL of simulant and an exposure area of 12.5 cm^2 (5 cm × 2.5 cm rectangles) at a refrigeration temperature of 5 °C for 16 days. At 1 h, 2 h, 4 h, 6 h, 24 h, 4 days, 8 days, and 16 days, samples were analyzed at 655 nm by UV-vis spectrometry (Agilent UV-Visible 8453 Spectroscopy System, Agilent Technologies, Wilmington, DE, USA). To quantify the concentrations of migrated porphyrin, a previous calibration with known concentrations in 50% ethanol was prepared.

2.3.6. Application to Bologna Slices

Coatings on PET were used as separators of slices of a real food-bologna. In this test, the antibacterial activity of PVOH, G, PE, and HPMC with 1% porphyrin E-140 coatings was tested against the usual microbial load and against *L. monocytogenes*, which had been previously inoculated. An 800 g piece of classic bologna containing 50% pork and 15% turkey was purchased and sliced. Slices that were 3 mm thick and 6.8 cm in diameter were prepared, with an average weight per slice of 10.12 ± 0.13 g. Slices of bologna were inoculated with 100 µL of 10^5 CFU/mL of *L. monocytogenes* spread over their surface. Then, the active- or control-coated film (with/without porphyrin) was placed on the bologna slices and irradiated for 15 min with 5 LED lamps on a tray with ice to avoid sample overheating (Figure 1). In parallel, control samples with films with and without antimicrobial were kept in darkness for 15 min.

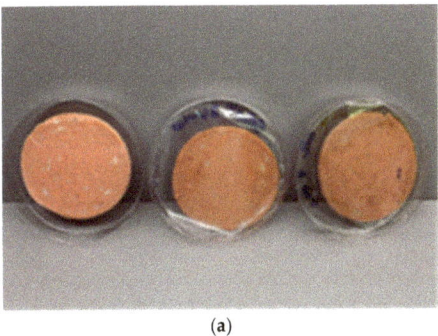 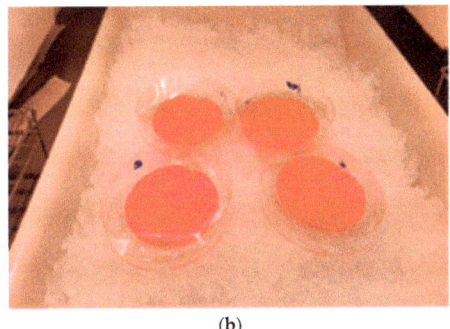

(a) (b)

Figure 1. Example of test: (**a**) from left to right: bologna slice without coating, sample with a control gelatin (G) coating on poly(ethylene terephthalate) (PET), and sample of G-coated PET containing E-140 porphyrin; (**b**) image of an experiment of treatment during the photoactivation process.

After the treatment, bologna slices were placed in a Stomacher bag together with 10 mL of peptone water and homogenized in a Stomacher (Bagmix, Interscience, St. Nom, France) for 4 min. Serial dilutions were carried out, and the enumeration of particular microbial groups was performed using the following culture conditions and media (from Scharlab, Barcelona, Spain): (a) PALCAM Listeria selective agar for *L. monocytogenes* incubated at 37 °C for 48 h; (b) plate count agar for total aerobic plate count, pour-plated and incubated at 30 °C for 48 h and plate count agar for total aerobic psychrotrophic count, pour-plated and incubated at 10 °C for 10 days; (c) Violet Red Bile Glucose agar for total enterobacteria, pour-plated and incubated at 37 °C for 48 h; (d) Eosin Methylene Blue agar for coliform bacteria, pour-plated and incubated at 37 °C for 48 h; (e) Brilliant Green agar for *Salmonella* isolation, pour-plated and incubated at 37 °C for 48 h; (f) King agar for *Pseudomonas* spp. count, pour-plated and incubated at 25 °C for 4 days; (g) De Man, Rogosa and Sharpe agar for lactic bacteria count, pour-plated and incubated at 25 °C for 4 days. The counts were performed in triplicate.

Once the antimicrobial activity had been evaluated, the possible impact on the color characteristics of the food due to active agent release was studied. The color of bologna samples before and after irradiation was determined using a Konica Minolta CM-3500d spectrophotometer and the same procedure as described in Section 2.3.3. Eight measurements were taken of each sample, and three samples of each film were measured.

2.3.7. Statistical Analysis

All the experiments were conducted at least in triplicate. The data were subjected to an analysis of main effects and interaction factors by multifactorial ANOVA using the StatGraphics Plus 5.1 program.

3. Results and Discussion

3.1. Minimum Inhibitory Concentration (MIC) and Minimum Bactericidal Concentration (MBC) of Chlorophyllins

Antimicrobial activity of E-140 and E-141 chlorophyllins was tested against *L. monocytogenes* and *E. coli* using two types of irradiation: halogen and LED lights. The results obtained are shown in Table 1 (halogen light) and Table 2 (LED light).

Table 1. Minimum inhibitory concentration (MIC in mg/mL) and minimum bactericidal concentration (MBC in mg/mL) of chlorophyllins E-140 and E-141 against *L. monocytogenes* and *E. coli* activated by halogen light (15 min).

Microorganism	E-140		E-141	
	MIC	MBC	MIC	MBC
L. monocytogenes	0.0031	0.10	2	10
E. coli	0.0031	0.63	3	11

Table 2. Minimum inhibitory concentration (MIC in mg/mL) and minimum bactericidal concentration (MBC in mg/mL) of chlorophyllins E-140 and E-141 against *L. monocytogenes* and *E. coli* activated by LED light (15 min).

Microorganism	E-140		E-141	
	MIC	MBC	MIC	MBC
L. monocytogenes	0.025	0.400	–	–
E. coli	0.100	0.630	–	–

The first important data were that the MIC values were higher against *E. coli* than against *L. monocytogenes*. In general, Gram-positive bacteria are more sensitive than Gram-negative bacteria to chemical compounds owing to the protection of the outer membrane surrounding the Gram-negative cell wall, which restricts the diffusion of antimicrobial substances [12,13]. Caires et al. [14] reported on the antimicrobial activity of two photosensitizers—eosin methylene blue and chlorophyllin sodium copper salt (E-141)—and showed that eosin methylene blue was able to photoinactivate *E. coli* and *S. aureus*, while E-141 was only active against *S. aureus*. In that test, the maximum concentration tested was 50 µM, which was much lower than the 4.15 mM obtained in this study. However, it is difficult to compare the published results because the important parameters of the photoactivation process—intensity, time, and type of radiation—were either different or were not provided. In this test, 15 min was enough to induce photoactivation, much less than the 120 min employed elsewhere [14].

It is also important to note that although both chlorophyllins presented high antimicrobial activity, porphyrin E-140 was much more effective (with much lower MIC and MBC values) than E-141. With regard to the type of photoactivation, halogen lights were more effective than LED lights. The use of LEDs slightly reduced the efficacy of E-140 against *L. monocytogenes* and *E. coli*. Other authors have reported the antimicrobial effectiveness of zinc complexes of tetrakis (*N*-methylpyridinium-4-yl) tetraiodide porphyrin and tetrakis (*N*-methylpyridinium-4-yl) tetraiodide phthalocyanine impregnated in paper using an inexpensive consumer LED lamp as activation mechanism [9].

In the case of E-141, porphyrin appeared to be inactive when exposed to the radiation of LED lamps, and neither the MBC nor the MIC of the two tested bacteria was found. Our hypothesis is that the LED lamps employed did not emit radiation of sufficient intensity in the wavelength range necessary for E-141 activation. Moreover, copper chlorophyllins have been reported to present low yield of singlet oxygen [15].

Finally, two control tests were carried out. In one, the two microorganisms were exposed to the chlorophyllins in darkness; in the other, they were exposed to radiation without porphyrin. The two microorganisms tested reflected the absence of antimicrobial effect, indicating that chlorophyllins were nontoxic for the two model bacteria, and light radiation alone did not produce antimicrobial activity.

3.2. Development of Coated Films

G, PE, PVOH, and HPMC coatings on PET film with and without 1% of E-140 or E-141 were successfully obtained. They were homogeneous, without discontinuities, flexible, transparent, had a light green color when the chlorophyllins were added, and with thicknesses that are shown in Table 3. As can be seen, the coating thicknesses varied according to the polymer material as a consequence of the different solid content of the film-forming solution and the polymer density; the PE-based materials were the thickest, while PVOH and HPMC were the thinnest. The incorporation of chlorophyllins in the coatings did not significantly affect thickness.

Table 3. Thicknesses (µm) of films and coatings with and without chlorophyllin.

Materials	Coatings	
	Control	E-140 and E-141
G	4.66 ± 1.63	5.16 ± 1.16
PVOH	2.28 ± 1.79	2.00 ± 1.03
PE	25.00 ± 1.87	25.17 ± 2.70
HPMC	2.14 ± 1.06	2.42 ± 1.61

The most significant effect of the addition of the chlorophyllins to the coatings was the yellowish (E-140) or greenish (E-141) color induced. Table 4 shows the color coordinates in the CIELAB system for different coatings, including chromaticity (C^*) and tone (h). As can be seen, in general, the coated films presented high luminosity as revealed by the L^* values ranging between 83 and 90. The only exceptions were the coatings based on PE-incorporated chlorophyllins, which presented a considerable reduction in luminosity. The addition of E-140 provided coatings with a yellowish color, which is characterized in CIELAB coordinates by positive b^* values and low negative a^* values. As a consequence of this, the tone of the coatings was characterized by values ranging between 95 and 110°. The incorporation of E-141 provided a greener color, with higher negative a^* values and positive b^* and tone values ranging between 107 and 135°. With respect to saturation or chromaticity, the values were greater for E-141 samples than for E-140. Comparing the polymeric materials, C^* values were greatest for the thickest material (PE).

Table 4. Color coordinates L^*, a^*, b^*, chromaticity (C^*) and tone (h in °) of films coated with gelatin (G), polyvinyl alcohol (PVOH), polyethylene (PE), (hydroxypropyl)methyl cellulose (HPMC) without (control) and with active agents (E-140 and E-141).

Coatings		a^*	b^*	L^*	C^*	h
G	Control	−0.23 ± 0.02 [c]	0.64 ± 0.02 [a]	88.04 ± 0.01 [c]	0.68 ± 0.05 [b]	109.84 ± 0.02 [c]
	E-140	−2.50 ± 0.01 [b]	8.18 ± 0.03 [a]	85.77 ± 0.01 [b]	8.55 ± 0.01 [a]	106.99 ± 0.02 [a]
	E-141	−3.02 ± 0.07 [a]	9.31 ± 0.01 [a]	83.61 ± 0.04 [a]	9.78 ± 0.01 [a]	107.97 ± 0.01 [b]
PVOH	Control	0.16 ± 0.04 [a]	0.32 ± 0.02 [b]	88.04 ± 0.02 [c]	0.36 ± 0.05 [a]	115.26 ± 0.01 [b]
	E-140	−2.89 ± 0.01 [b]	9.76 ± 0.03 [a]	85.26 ± 0.01 [b]	10.18 ± 0.02 [c]	106.54 ± 0.05 [a]
	E-141	−5.02 ± 0.01 [c]	7.89 ± 0.01 [a]	83.12 ± 0.01 [a]	9.35 ± 0.01 [b]	122.47 ± 0.02 [c]
PE	Control	−0.32 ± 0.01 [c]	−0.23 ± 0.02 [a]	89.23 ± 0.01 [c]	0.40 ± 0.06 [a]	216.41 ± 0.02 [c]
	E-140	−4.73 ± 0.01 [b]	38.06 ± 0.01 [b]	69.79 ± 0.02 [b]	38.35 ± 0.04 [b]	97.09 ± 0.01 [a]
	E-141	−26.24 ± 0.02 [a]	28.12 ± 0.02 [b]	67.51 ± 0.03 [a]	38.46 ± 0.04 [c]	133.01 ± 0.02 [b]
HPMC	Control	−0.26 ± 0.03 [c]	0.38 ± 0.04 [a]	88.64 ± 0.05 [c]	0.47 ± 0.01 [a]	124.54 ± 0.04 [b]
	E-140	1.11 ± 0.01 [b]	3.15 ± 0.02 [b]	87.69 ± 0.05 [b]	3.34 ± 0.04 [a]	109.33 ± 0.01 [a]
	E-141	−3.02 ± 0.04 [a]	3.41 ± 0.01 [c]	85.72 ± 0.03 [a]	4.56 ± 0.03 [a]	131.53 ± 0.01 [c]

[a-c] Different letters in the same column for a coating indicate significant differences (Tukey's adjusted analysis of variance, $p < 0.05$).

Figure 2 shows, as examples, photos of the different coatings. As can be seen, in the PE coatings there was a clearly visible alteration of the color of the material after incorporation of both chlorophyllins. On the other hand, the color of the HPMC coatings with E-140 or E-141 was hardly distinguishable from the control.

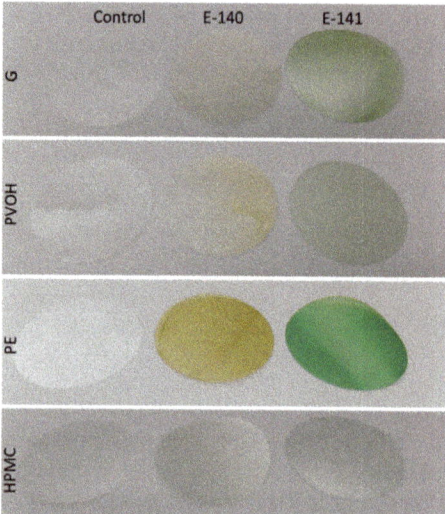

Figure 2. Images of the PET films coated with G, PVOH, HPMC, and PE for the control and the films containing chlorophyllins (E-140 and E-141).

3.3. Antimicrobial Activity of Films Coated with Chlorphyllins

From the results shown above, it was inferred that chlorophyllin E-140 was more efficient than E-141, provided less color change to the PET film, reduced energy consumption and, overall, was active when photoactivated with LED lights as it reduced the heating on the film and on the product. Therefore, coated films containing E-140 were selected to verify the effectiveness of porphyrin incorporated in various polymer matrices. For this purpose, two methods were used: the agar diffusion method and the surface disinfection method.

3.3.1. Agar Diffusion Method

The antimicrobial capacity of G, PVOH, PE, and HPMC films incorporating 1% of chlorophyllin E-140 was studied against *L. monocytogenes* and *E. coli* using the agar diffusion method. The results obtained showed that the G, PVOH, and HPMC films had a bactericidal effect against both microorganisms, while the PE film had a certain degree of inhibition, as can be seen in Table 5.

Table 5. Microbiological counts (log CFU/mL) on the surface of the agar in the area under the coated films and antimicrobial effect expressed as log reduction values (LRV).

Coatings	Dark Conditions		Light Conditions		LRV
	Control	E-140	Control	E-140	
Listeria monocytogenes					
G	6.90 ± 0.02	7.02 ± 0.01	7.09 ± 0.01	no growth	7.09
PVOH	7.13 ± 0.05	7.24 ± 0.04	7.25 ± 0.73	no growth	7.25
PE	7.05 ± 0.05	6.95 ± 0.01	7.38 ± 0.70	6.73 ± 0.05	0.65
HPMC	6.95 ± 0.01	6.90 ± 0.02	7.01 ± 0.02	no growth	7.01
Escherichia coli					
G	7.01 ± 0.02	7.02 ± 0.01	7.04 ± 0.01	no growth	7.04
PVOH	7.23 ± 0.02	7.01 ± 0.04	7.47 ± 0.67	no growth	7.47
PE	7.12 ± 0.03	6.95 ± 0.03	7.40 ± 0.70	6.79 ± 0.03	0.67
HPMC	7.01 ± 0.01	6.90 ± 0.02	7.04 ± 0.04	no growth	7.04

The results showed that the most hydrophilic films, i.e., with a higher degree of affinity for water, presented a greater antibacterial capacity. It is possible that the excited oxygen reacts with water molecules to produce hydroxyl radicals that have a longer lifetime and are more active against bacteria [16]. Accordingly, it was observed that photoexcited G, PVOH, and HPMC films with 1% chlorophyllin E-140 had a lethal effect on *L. monocytogenes* or *E. coli* in the area covered by the films. In the case of PE-coated films, a certain degree of inhibition was seen, but it was less than that with the other films, despite the higher amount of porphyrin present owing to the greater thickness of the coating. The absence of water in this hydrophobic polymer may limit the photoactivation to the formation of singlet oxygen, which rapidly reduces to triplet oxygen without interacting with the cells. Another explanation is that the low affinity between chlorophyllin and PE causes itsaggregation, and this results in a lower production of singlet oxygen. The films that developed acted mainly by contact, although in the case of the PVOH, G, and HPMC matrices, an inhibition halo of ca. 5 mm was observed around the films.

Bozja et al. (2003) tested the antimicrobial activity of porphyrins grafted onto nylon-6 fibers with polyacrylic acid. They found that the antimicrobial system developed was very effective against *S. aureus* but had no effect on *E. coli* bacteria at any light intensity [7]. There have been some similar studies, such as the one carried out in 2008 in which the effectiveness of E-140 porphyrin-bearing gelatin films as antibacterial agents [2] was demonstrated against *S. aureus* and *L. monocytogenes*, but was found to have no effect on *Salmonella* spp. and *E. coli* bacteria after photoactivation with 30,000 luxes for five minutes.

3.3.2. Surface Disinfection

The coatings were inoculated on the surface and then illuminated in order to verify whether they could be self-sanitized by photoactivation. As an example, Figure 3 shows the results obtained with gelatin films contaminated with *Listeria*. As can be seen, there is growth both in the control plate that was exposed to light and in the plate with porphyrin films that was kept in darkness. By contrast, irradiated samples containing porphyrin E-140 showed no microbial growth.

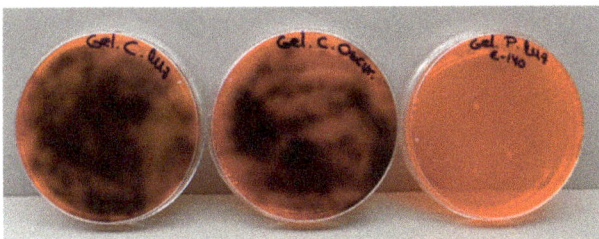

Figure 3. Microbial growth of *L. monocytogenes*. Left to right: control G film which had been irradiated, G film with E-140 which had been kept in darkness, and G film with E-140 which had been irradiated.

Similar results were obtained for the other films: HPMC, PVOH, and PE. This shows that the films developed can be self-sanitized in case of possible contamination simply by subjecting them to visible light irradiation for 15 min.

However, no clear bactericidal effect against *E. coli* was observed with any of the coated films that contained E-140. These results are in agreement with other authors who obtained an antimicrobial effect against Gram-positive bacteria using porphyrins grafted onto nylon fibers but found no effect on *E. coli* bacteria at any light intensity [7]. As explained previously, this may be due to the extra protection of the outer membrane in Gram-negative bacteria. The outer membrane of Gram-negative bacteria plays an important role, which is related to their resistance to many active agents that are very effective against Gram-positive bacteria [17]. The resistance of Gram-negative bacteria to photosensitization has been widely reported [18,19]. In order to increase the permeability of *E. coli*, we added 40 mM

EDTA (ethylenediaminetetraacetic acid) to the 100 μL dilution of *E. coli*. EDTA is a chelating agent that destabilizes the outer membrane and may facilitate access of free radicals generated with light to bacterial cells [20]. However, we did not observe significant differences between the controls and the treatment with light and porphyrin. Other authors have employed dimethyl sulfoxide (5%) to increase the permeability of *E. coli* to porphyrins without achieving the same results as with Gram-positive bacteria [14].

3.4. Migration of Chlorophyllins from Coatings to the Food Simulant

The release of chlorophyllin E-140 was studied in 50% ethanol as a fatty food simulant in accordance with Directive 85/572/EEC prior to application on bologna slices. The regulation stipulates that the results of overall migration obtained must be in accordance with the overall migration limit of 10 mg/dm^2 established in Directive 2002/72/EC and in Royal Decree 866/2008, which are both related to materials and plastic objects in contact with food products. The results are shown in Figure 4.

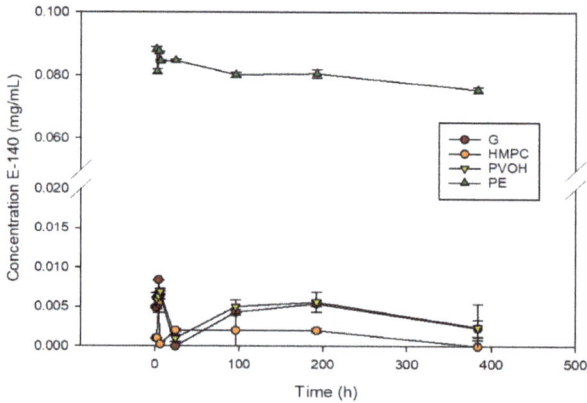

Figure 4. Concentration of E-140 released from G, PVOH, HPMC, and PE coatings on PET in fatty simulant as a function of time.

The release of chlorophyllins from G, PVOH, and HPMC coatings on PET was very low—practically zero—indicating that there was no substantial migration. This means the antimicrobial effect was produced by the generation of free radicals and it was not necessary for the chlorophyllins to migrate to the simulating medium. On the other hand, the PE coatings presented a maximum release of 6.4 mg/dm^2 (according to the surface of each coating, which was 0.125 dm^2), which was less than the overall migration limit set by legislation (10 mg/dm^2).

In the case of the other coatings, the low migration obtained makes it unnecessary to calculate migration limits, and therefore they can be used as packaging systems for fatty food. In this case, they were tested with bologna slices. The fact that the greatest release was observed in PE could be due to several factors. First, the thickness of the PE coating—well above 25 μm—results in a large amount of porphyrin in absolute values. Second, the poor affinity of LDPE for chlorophyllins might result in their separation, forming a two-phase matrix in which the agent is isolated in small regions dispersed in the pure LDPE matrix. When the material comes into contact with a solvent medium, the porphyrin located close to the surface is immediately released, as observed in Figure 4. In the other matrices there would be specific interactions between E-140 and the polymers, keeping the agent in the package where porphyrin molecules have a compatible chemical environment. Similar release results were observed for films containing E-141 (data not shown).

3.5. Application to Food

Once the antimicrobial and self-sanitizing capacity of the films had been determined, their effectiveness when applied to a real food was studied. The experiment was carried out with G, HPMC, PE, and PVOH coatings on PET films used as bologna slice separators.

This food product was chosen to test the coatings developed because it has a high water activity (a_w = 0.970) and is therefore very susceptible to spoilage by microorganisms. The antimicrobial effect of the films was studied against *L. monocytogenes*, which was inoculated on the bologna surface and against the usual microbial load of this meat product. The results are shown in Figure 5.

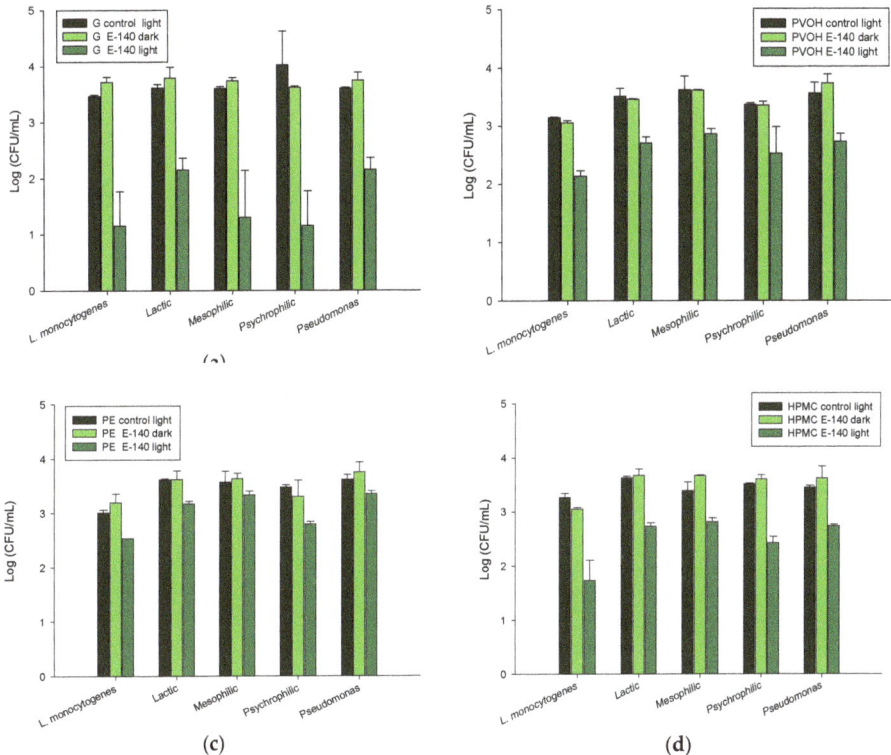

Figure 5. Antimicrobial activity of (**a**) G, (**b**) PVOH, (**c**) PE, and (**d**) HPMC coatings on PET films applied to bologna slices.

There was no microbial growth of *Salmonella* spp., enterobacteria, or coliform bacteria on any sample. Control coatings and coatings kept in darkness showed no antimicrobial effect, confirming the nontoxicity of the active agents without photoactivation. However, coatings with E-140 exposed to LED lights inhibited microbial growth of *L. monocytogenes* successfully. The greatest antimicrobial effect was observed with the G coating and the least with the PE coating. The HPMC and PVOH coatings were slightly less effective than G (Figure 5).

Growth of the microbial load was also inhibited in irradiated samples with chlorophyllin. Lactic, mesophilic, and psychrophilic bacteria and *Pseudomonas* were significantly reduced. Once again, the G coatings presented the greatest antimicrobial effect against all the microorganisms tested. The highest efficiency was observed against psychrophiles and mesophiles, with reductions of 2.87 and 2.30 log, respectively. In the case of lactic acid bacteria and *Pseudomonas*, although the degree of inhibition was lower, it was also considerable. The HPMC and PVOH coatings with chlorophyllin presented similar degrees of inhibition against all the microorganisms studied, ranging between 0.76 and 1 log. Finally, the antimicrobial effect of the PE coatings was lower, ranging between approximately 0.24 and 0.69 log.

Results showed that all the coatings developed in this study had antimicrobial activity, with gelatin being the most effective matrix. This is possibly due to its morphology that facilitates the release of free radicals produced by the porphyrin molecule, which is responsible for the antimicrobial action. It should also be taken into account that the G coatings were thicker, and therefore there was a greater amount of agent in absolute values than in the HPMC and PVOH coatings. The application of the gelatin coating would be very useful given gelatin is a component that is usually present in many foods and does not constitute a risk when it forms part of the surface of a package intended for food use.

To confirm the absence of effects on the organoleptic properties of the product and because the release of chlorophyllin could produce a green color, the color of the bologna slices was measured before and after the photoactivation treatment. The parameters L^* [black (0) to white (100)], a^* [green (−) to red (+)], and b^* [blue (−) to yellow (+)] were obtained, and the polar coordinates, chroma C^*, and hue angle h that were calculated are shown in Table 6.

Table 6. The color coordinates L^*, a^*, b^*, the chromaticity (C^*), and the tone (h in °) of bologna slices covered with G, PVOH, PE, and HPMC coatings on PET before and after photoactivation.

Coatings	Activation	a^*	b^*	L^*	C^*	h
G control	before	16.30 ± 0.04 [a]	10.30 ± 0.04 [a]	56.59 ± 0.04 [b]	19.28 ± 0.04 [c]	32.30 ± 0.17 [b]
	after	14.12 ± 0.02 [b]	12.70 ± 0.02 [b]	50.48 ± 0.02 [a]	18.99 ± 0.02 [c]	41.97 ± 0.06 [b]
G E-140	before	15.95 ± 0.05 [a]	9.24 ± 0.05 [a]	56.62 ± 0.05 [b]	18.44 ± 0.05 [a]	30.09 ± 0.13 [b]
	after	14.10 ± 0.03 [a]	14.07 ± 0.03 [b]	50.29 ± 0.03 [a]	19.92 ± 0.03 [b]	44.94 ± 0.10 [c]
PVOH control	before	17.70 ± 0.05 [a]	9.68 ± 0.05 [a]	57.10 ± 0.05 [b]	20.17 ± 0.05 [a]	28.66 ± 0.21 [b]
	after	13.53 ± 0.03 [b]	12.48 ± 0.03 [b]	50.47 ± 0.03 [a]	18.41 ± 0.03 [b]	42.69 ± 0.17 [a]
PVOH E-140	before	17.47 ± 0.03 [a]	9.42 ± 0.03 [a]	57.06 ± 0.03 [b]	19.85 ± 0.03 [a]	28.33 ± 0.10 [b]
	after	14.74 ± 0.01 [b]	13.34 ± 0.01 [b]	50.99 ± 0.01 [a]	19.88 ± 0.01 [a]	42.13 ± 0.09 [a]
HPMC control	before	16.20 ± 0.03 [b]	10.38 ± 0.03 [a]	55.27 ± 0.03 [b]	19.24 ± 0.03 [b]	32.65 ± 0.07 [b]
	after	13.48 ± 0.02 [a]	12.34 ± 0.02 [b]	50.23 ± 0.02 [a]	18.43 ± 0.02 [a]	41.41 ± 0.29 [a]
HPMC E-140	before	15.99 ± 0.04 [b]	10.56 ± 0.04 [a]	55.37 ± 0.04 [b]	19.17 ± 0.04 [b]	33.44 ± 0.21 [b]
	after	12.94 ± 0.07 [a]	12.56 ± 0.07 [b]	50.51 ± 0.07 [a]	18.03 ± 0.07 [a]	44.16 ± 0.07 [c]
PE control	before	17.59 ± 0.03 [a]	10.65 ± 0.03 [a]	54.36 ± 0.03 [b]	20.57 ± 0.03 [b]	31.20 ± 0.27 [b]
	after	13.26 ± 0.03 [b]	12.04 ± 0.03 [b]	50.78 ± 0.03 [a]	17.91 ± 0.03 [a]	42.26 ± 0.06 [a]
PE E-140	before	16.61 ± 0.02 [a]	10.53 ± 0.02 [a]	55.87 ± 0.02 [b]	19.67 ± 0.02 [b]	32.37 ± 0.08 [b]
	after	14.80 ± 0.01 [c]	12.39 ± 0.01 [b]	50.54 ± 0.01 [a]	19.30 ± 0.01 [b]	39.95 ± 0.10 [c]

[a–c] Different letters in the same column indicate significant differences (Tukey's adjusted analysis of variance, $p < 0.05$).

The bologna slices were displaced towards the + a coordinate, i.e., towards red and slightly towards the + b coordinate (yellow), which is logical given their rosy hue.

In general, irradiation produced a slight decrease in the luminosity of the bologna slices and a slight shift to the left of the b^* coordinates. However, these color modifications were not distinguishable to the naked eye. On the other hand, no color difference was observed between samples with and without porphyrin submitted to the same treatment (darkness or light). Therefore, from the data shown above, it can be confirmed that no substantial release of porphyrin to the bologna took place. Similar results were reported in a study of chlorophyllin gelatin films applied as wraps to frankfurter sausages, which found no substantial differences between uncoated and coated products [2].

Finally, it can be concluded that chlorophyllin-based photosensitization of coated films is an effective way of reducing the population of microorganisms naturally present in meat and poultry products and as a way of improving asepsis of packaging materials through their self-sanitizing function. Moreover, this efficiency has also been demonstrated when light is applied through the supporting material (in this case, PET) if the material is transparent to visible light. This aspect is of great importance because it would greatly facilitate carrying out its antimicrobial or self-sanitizing function in food products that have already been packaged. The process that has been developed in this study, which requires only white light, could become an alternative food process that is nonchemical, nonthermal, inexpensive, and environmentally friendly.

The films developed could be applied as part of a package intended for meat derivatives—either as an external protective cover or as separators of cold cuts—or in some dairy products, such as fresh cheese. In addition, the films developed could be used to wrap food that are sensitive to microbial contamination in daily use, such as pieces of cooked ham that are not immediately consumed completely but are rather manipulated (cut) on a number of occasions before being completely consumed, therefore making them potential vectors for the transmission of microorganisms. These films could be used to cover the cut surface and protect it from microbial spoilage thanks to the photoactivation induced by the white lights of refrigerated displays in grocery stores. Furthermore, this active packaging does not require any agent release, thereby increasing food shelf life without substantial changes in aroma, flavor, or color.

Author Contributions: Conceptualization, P.H.-M. and R.G.; Methodology, G.L.-C.; Validation, P.H.-M., R.G. and G.L.-C.; Formal Analysis, G.L.-C.; Investigation, P.H.-M., R.G. and G.L.-C.; Resources, P.H.-M. and R.G.; Writing, Original Draft Preparation, G.L.-C., R.G. and P.H.-M.

Funding: This research was funded by the Ministry of Economy and Competitiveness (No. AGL2015-64595-R).

Acknowledgments: The authors acknowledge the assistance of Karel Clapshaw (translation services) and the supply of materials by Paramelt B.V. (Heerhugowaard, The Netherlands) and the Nippon Synthetic Chemical Company (Osaka, Japan).

Conflicts of Interest: The authors declare no conflict of interest.

References

1. Flint, J.A.; Van Duynhoven, Y.T.; Angulo, F.J.; DeLong, S.M.; Braun, P.; Kirk, M.; Scallan, E.; Fitzgerald, M.; Adak, G.K.; Sockett, P.; et al. Estimating the burden of acute gastroenteritis, foodborne disease, and pathogens commonly transmitted by food: An international review. *Clin. Infect. Dis.* **2005**, *41*, 698–704. [CrossRef] [PubMed]
2. López-Carballo, G.; Hernández-Muñoz, P.; Gavara, R.; Ocio, M.J. Photoactivated chlorophyllin-based gelatin films and coatings to prevent microbial contamination of food products. *Int. J. Food Microbiol.* **2008**, *126*, 65–70. [CrossRef] [PubMed]
3. Luksiene, Z.; Paskeviciute, E. Novel approach to decontaminate food-packaging from pathogens in non-thermal and not chemical way: Chlorophyllin-based photosensitization. *J. Food Eng.* **2011**, *106*, 152–158. [CrossRef]
4. Evans, H.H.; DeMarini, D.M. Ionizing radiation-induced mutagenesis: Radiation studies in neurospora predictive for results in mammalian cells. *Mutat. Res. Rev. Mutat. Res.* **1999**, *437*, 135–150. [CrossRef]
5. Martins, D.; Mesquita, M.Q.; Neves, M.G.P.M.S.; Faustino, M.A.F.; Reis, L.; Figueira, E.; Almeida, A. Photoinactivation of *Pseudomonas syringae* pv. *Actinidiae* in kiwifruit plants by cationic porphyrins. *Planta* **2018**, *248*, 409–421. [CrossRef] [PubMed]
6. Paskeviciute, E.; Zudyte, B.; Luksiene, Z. Towards better microbial safety of fresh produce: Chlorophyllin-based photosensitization for microbial control of foodborne pathogens on cherry tomatoes. *J. Photochem. Photobiol. B Biol.* **2018**, *182*, 130–136. [CrossRef] [PubMed]
7. Bozja, J.; Sherrill, J.; Michielsen, S.; Stojiljkovic, I. Porphyrin-based, light-activated antimicrobial materials. *J. Polym. Sci. A Polym. Chem.* **2003**, *41*, 2297–2303. [CrossRef]
8. Krouit, M.; Granet, R.; Branland, P.; Verneuil, B.; Krausz, P. New photoantimicrobial films composed of porphyrinated lipophilic cellulose esters. *Bioorg. Med. Chem. Lett.* **2006**, *16*, 1651–1655. [CrossRef] [PubMed]
9. George, L.; Hiltunen, A.; Santala, V.; Efimov, A. Photo-antimicrobial efficacy of zinc complexes of porphyrin and phthalocyanine activated by inexpensive consumer led lamp. *J. Inorg. Biochem.* **2018**, *183*, 94–100. [CrossRef] [PubMed]

10. Smith-Palmer, A.; Stewart, J.; Fyfe, L. Antimicrobial properties of plant essential oils and essences against five important food-borne pathogens. *Lett. Appl. Microbiol.* **1998**, *26*, 118–122. [CrossRef] [PubMed]
11. LED Lights—Detrimental to Our Health and Environment. Available online: http://www.sunkissedsolar.com.au/led-light-globes-detrimental-health-environment (accessed on 27 July 2018).
12. Higueras, L.; López Carballo, G.; Hernández Muñoz, P.; Gavara, R.; Rollini, M. Development of a novel antimicrobial film based on chitosan with LAE (ethyl-N$^\alpha$-dodecanoyl-L-arginate) and its application to fresh chicken. *Int. J. Food Microbiol.* **2013**, *165*, 339–345. [CrossRef] [PubMed]
13. Delaquis, P.J.; Stanich, K.; Girard, B.; Mazza, G. Antimicrobial activity of individual and mixed fractions of dill, cilantro, coriander and eucalyptus essential oils. *Int. J. Food Microbiol.* **2002**, *74*, 101–109. [CrossRef]
14. Caires, C.S.A.; Leal, C.R.B.; Ramos, C.A.N.; Bogo, D.; Lima, A.R.; Arruda, E.J.; Oliveira, S.L.; Caires, A.R.L.; Nascimento, V.A. Photoinactivation effect of eosin methylene blue and chlorophyllin sodium-copper against *Staphylococcus aureus* and *Escherichia coli*. *Lasers Med. Sci.* **2017**, *32*, 1081–1088. [CrossRef] [PubMed]
15. Cavaleiro, J.A.S.; Görner, H.; Lacerda, P.S.S.; MacDonald, J.G.; Mark, G.; Neves, M.G.P.M.S.; Nohr, R.S.; Schuchmann, H.-P.; von Sonntag, C.; Tomé, A.C. Singlet oxygen formation and photostability of meso-tetraarylporphyrin derivatives and their copper complexes. *J. Photochem. Photobiol. A Chem.* **2001**, *144*, 131–140. [CrossRef]
16. Levine, J.S. Biomass burning: The cycling of gases and particulates from the biosphere to the atmosphere. In *Treatise on Geochemistry*; Holland, H.D., Turekian, K.K., Eds.; Pergamon: Oxford, UK, 2003; pp. 143–158.
17. Nikaido, H. Molecular basis of bacterial outer membrane permeability revisited. *Microbiol. Mol. Biol. Rev.* **2003**, *67*, 593–656. [CrossRef] [PubMed]
18. Nitzan, Y.; Gozhansky, S.; Malik, Z. Effect of photoactivated hematoporphyrin derivative on the viability of *Staphylococcus aureus*. *Curr. Microbiol.* **1983**, *8*, 279–284. [CrossRef]
19. Malik, Z.; Hanania, J.; Nitzan, Y. Bactercidal effects of photoactivated porphyrins-an alternative approach to antimicrobial drugs. *J. Photochem. Photobiol. B Biol.* **1990**, *5*, 281–293. [CrossRef]
20. Malik, Z.; Ladan, H.; Nitzan, Y. Phtodynamic inactivation of gram-negative bacteria-problem and possible solutions. *J. Photochem. Photobiol. B Biol.* **1992**, *14*, 262–266. [CrossRef]

© 2018 by the authors. Licensee MDPI, Basel, Switzerland. This article is an open access article distributed under the terms and conditions of the Creative Commons Attribution (CC BY) license (http://creativecommons.org/licenses/by/4.0/).

Review

Non-Conventional Tools to Preserve and Prolong the Quality of Minimally-Processed Fruits and Vegetables

Maria Rosaria Corbo *, Daniela Campaniello, Barbara Speranza, Antonio Bevilacqua and Milena Sinigaglia

Department of the Science of Agriculture, Food and Environment, University of Foggia, Via Napoli, 25, 71122 Foggia FG, Italy; daniela.campaniello@unifg.it (D.C.); barbara.speranza@unifg.it (B.S.); antonio.bevilacqua@unifg.it (A.B.); milena.sinigaglia@unifg.it (M.S.)
* Author to whom correspondence should be addressed; mariarosaria.corbo@unifg.it; Tel.: +39-881-589-232.

Academic Editor: Stefano Farris

Received: 7 October 2015; Accepted: 20 November 2015; Published: 26 November 2015

Abstract: The main topic of this paper is a focus on some non-conventional tools to preserve the microbiological and physico-chemical quality of fresh-cut fruits and vegetables. The quality of fresh-cut foods is the result of a complex equilibrium involving surface microbiota, storage temperature, gas in the headspace and the use of antimicrobials. This paper proposes a short overview of some non-conventional approaches able to preserve the quality of this kind of product, with a special focus on some new ways, as follows: (1) use of edible or antimicrobial-containing coatings (e.g., chitosan-based coatings) on fruits or vegetables; (2) alternative modified atmospheres (e.g., high O_2-modified atmosphere packaging (MAP)) or the use of essential oils in the headspace; (3) conditioning solutions with antimicrobials or natural compounds for fruit salad; and (4) biopreservation and use of a probiotic coating.

Keywords: fresh-cut; conditioning liquid; coatings; spoiling microorganisms; probiotics

1. Introduction

In recent years the attention towards a healthy diet is considerably growing; the consumer is paying more attention to the importance of the freshness and healthiness of foods. Consumption of fruits and vegetables is considered a successful way to contrast the current tendency toward obesity (especially in children). It is known that fruits and vegetables contain essential nutrients and are the major source of various antioxidants (vitamin C, vitamin E and β-carotene) that are necessary for the health, growth and development of the children. The World Health Organization (WHO), Food and Agriculture Organization (FAO), United States Department of Agriculture (USDA) and European Food Safety Authority (EFSA) recommended an increase of fruit and vegetable consumption to decrease the risk of cardiovascular diseases, cancer and ageing [1].

In addition, the consumer profile is changing as an effect of a stressed lifestyle: the number of working women and singles is gradually increasing; they are short on time and demand minimally-processed foods (also called ready-to-use) to save time on food preparation. To meet the consumer needs, the products should have an adequate shelf-life; it should be at least 4–7 days or even 21 days, depending on the products.

The International Fresh-cut Produce Association (IFPA) defines fresh-cut products as fruits or vegetables that have been trimmed and/or peeled and/or cut into a 100% usable product that is bagged or pre-packaged to offer consumers high nutrition, convenience and flavor, while still maintaining its freshness [2]. However, it is known that the processing of fruits and vegetables promotes physiological deterioration, biochemical changes and microbial degradation, beginning with raw materials, through processing methods and ending with packaging factors that affect the quality and shelf-life. These changes are due to tissue wounding, which includes browning, weight loss, accelerated respiration rate, off-flavor development, texture breakdown and increased susceptibility to microbial spoilage [3].

Thus, the search for new inexpensive and effective methods (such as chemical-based washing treatments, physical treatments, hurdle technology and packaging requirements) able to minimize these negative effects is of great interest to all of the stakeholders involved in the production and distribution of fresh fruits and vegetables.

The aim of this paper is to give insight into some non-conventional approaches able to ensure the microbiological stability and quality of minimally-processed fruits and vegetables with a special focus on some new ways currently available: (1) the use of edible or antimicrobial containing coatings (e.g., chitosan-based coatings) on fruits or vegetables; (2) alternative modified atmospheres (e.g., high O_2-modified atmosphere packaging (MAP)) or the use of essential oils in the headspace; (3) conditioning solutions with antimicrobials or natural compounds for fruit-salad; and (4) biopreservation and the use of probiotic coatings.

2. Use of Edible or Antimicrobial-Containing Coatings (e.g., Chitosan-Based Coatings) on Fruits or Vegetables

Edible coatings (ECs) and edible films belong to the modern food protection system; over the past few years, interest in the use of edible coatings for perishable foods has considerably increased due to their advantages and potential applications [4]. Edible coatings and edible films are terms that are often used interchangeably; however, a distinction is necessary.

An edible coating is defined as a thin layer of edible material applied to the surface of foods in addition to or as a substitution for natural protective coatings, able to form a barrier to moisture, oxygen and solute movement for the food [5–10].

On the other hand, edible film is defined as a thin layer of edible material formed on a product surface as a coating or placed (pre-formed) on or between food components [11].

Thus, an edible film is a thin skin, which has been pre-formed (for example, by casting a biopolymer solution separately from the food to form a film later applied to the food), whilst an edible coating is a suspension or an emulsion, which is applied directly to the food surface and later forms a film [12].

Edible coatings and films do not replace traditional packaging materials, but provide an additional factor to be applied for food preservation; ECs are consumed along with the food, thus the composition must conform to the regulations applied to the food product.

One of the advantages in using edible coatings and films is the reduction of water loss, considered one of the main factors in the deterioration of perishable foods. In fact, this thin layer protects fruits and/or vegetables against moisture loss, maintaining the texture and extending the shelf-life of the product, forming a protective barrier. On the other hand, when edible coatings are poor in water vapor barrier properties, a weight or moisture loss of the product could be recovered.

Numerous benefits result when edible coatings are applied, and these are summarized in Figure 1.

In addition, edible coatings may enhance sensory characteristics, can be consumed along with the food, provide additional nutrients and include quality-enhancing antimicrobials. Furthermore, they may reduce the cost and also the amount of traditional packaging used [13].

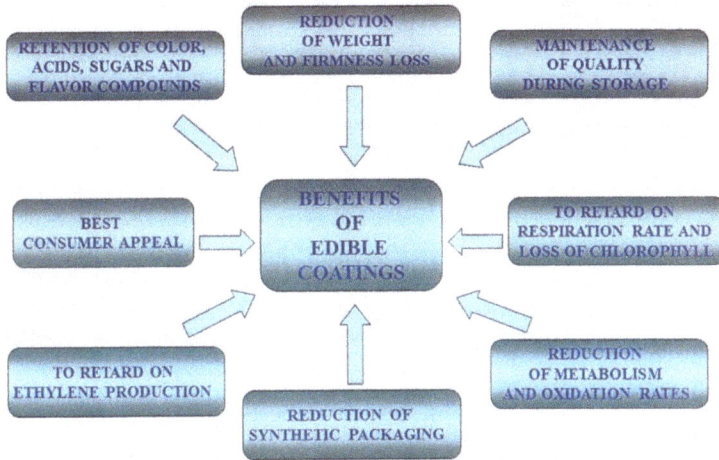

Figure 1. Main benefits of edible coatings.

2.1. Edible Coating Materials

Numerous varieties of fruits and vegetables are characterized by cuticle, a natural waxy layer on the surface, which, generally, has a low permeability to water vapor. To enhance the barrier properties of cuticle and/or substitute it when processing operations remove it, edible coatings could be applied.

Water and ethanol (or a blend of these compounds) are the solvents generally used for edible coating production.

Various coating materials could be added, but it is necessary that they possess some specific requirements (water solubility, hydrophobic of hydrophilic nature, easy formation of coatings, good sensorial properties) to obtain the resulting coating. In addition, an essential requirement is the plasticizing capacity to provide films with good flexibility properties; for this purpose, glycerol is generally used, followed by sorbitol, polyethylene glycol (PEG) and sugars [13].

Hydrocolloids have a good aptitude for forming continuous and cohesive matrices thanks to the hydrogen bonding of their polymeric chains.

Polysaccharides, proteins and lipids, alone or in combination, can also be used to produce edible films and coating. Lipids (together with waxes and fatty acids) do not have a suitable stand-alone filmmaking nature; they are usually opaque and relatively inflexible; the resulting films could also be quite fragile and unstable (rancidity) [9]. For this reason, lipids are generally incorporated into hydrocolloids-based films formulations [13].

The main polysaccharide, protein and lipid compounds used in edible coatings are reported in Table 1.

Table 1. Main polysaccharides, proteins and lipid compounds used for edible coatings.

	Description
Polysaccharides	**Starch** contains amylose ((1→4)-α-D-glucopyranosyl) and amylopectin (amylose branched with side units of D-glucopyranosyl linked by α-1,6-glycosidic bonds). Amylose has a film-forming ability, rendering strong, isotropic, odorless, tasteless and colorless films. It is inexpensive, widely available and easy to handle. Due to its good oxygen barrier, starch is used for coating fruits and vegetables characterized by high respiration rates, thus suppressing respiration and retarding the oxidation of the coated products.
	Dextrins derive from starch and are characterized by a smaller molecular size. Coatings resulting from dextrins provide a better water vapor resistance than starch coatings [14].
	Pullulan derives from starch and is edible and biodegradable; relative films are transparent, elastic, odorless and tasteless [15]. Pullulan-based coatings have shown potential for preserving fresh strawberries and kiwifruits because of their barriers to moisture, O_2 and CO_2 [16]. Pullulan films themselves do not demonstrate antimicrobial activity [17].
	Cellulose is the structural material of plant cell walls. It is composed of linear chains of (1→4)-β-D-glucopyranosyl units. Cellulose ethers (carboxymethylcellulose (CMC); methylcellulose (MC); hydroxypropyl cellulose (HPC); hydroxypropyl methylcellulose (HPMC)) are obtained by partial substitution of hydroxyl groups in cellulose by ether functions. Cellulose-derived films are tough, flexible, totally transparent, water soluble presence and are resistant to fats and oils [15,18], but are too expensive. Crosslinking treatments decrease the water solubility of cellulose ethers [19]. Cellulose derivative-based coatings were applied to some fruits and vegetables for providing barriers to oxygen, oil or moisture transfer [15]. CMC coatings help: (i) to retain the original firmness and crispness of apples, berries, peaches, celery, lettuce and carrots when used in a dry coating process; (ii) to preserve flavor components of some fresh fruits and vegetables; and (iii) to reduce oxygen uptake without increasing the carbon dioxide level in the internal environment of coated apples and pears by simulating a controlled atmosphere environment [15]. Li and Barth [20] observed that, after 3 weeks of storage, cellulose-based edible coating-treated carrots maintained a fresh appearance and higher carotene retention compared to controls.
	Alginates are extracted from brown seaweeds. They are sodium salts of alginic acid, which is a linear (1→4) linked polyuronic acid containing poly-β-D-manopyranosyluronic acid (M) blocks, poly-α-L-gulo pyranosyluronic acid (G) blocks and M–G blocks containing both polyuronic acids. Alginates possess good film-forming properties, producing uniform, transparent films with poor water resistance. Alginates react irreversibly with polyvalent metal cations (for example, calcium ions) to produce water-insoluble polymers. Alginate coatings are good oxygen barriers and retard lipid oxidation in fruits and vegetables. They show good ability to reduce weight loss and natural microflora counts in minimally-processed carrots [21]. Calcium alginate coatings reduce shrinkage, oxidative rancidity, moisture migration, oil absorption and sealing-in volatile flavors, improving appearance and color and reducing the weight loss of fresh mushrooms in comparison with uncoated ones [22].
	Carrageenan is extracted from red seaweeds. It is a complex mixture of several water-soluble galactose polymers. There are 3 different carrageenans (lambda-, iota- and kappa-carrageenan), differing in sulfate ester content and the distribution of 3,6-anhydro-α-D-galactopyranosyl residues [13]. Carrageenan-based coatings were applied to fresh apples for reducing moisture loss, oxidation or disintegration of the apples [23,24]. In combination with ascorbic acid, carrageenan-based coatings resulted in positive sensory results and reduction of microbial levels on minimally-processed apple slices [24].

Table 1. *Cont.*

	Description
	Gums include exudate gums (arabic, tragacanth and karaya), seed gums (locust bean and guar) and microbial fermentation gums (xanthan and gellan). Xanthan gum provides uniform coatings. Gum arabic was used for coating pecan nut halves to eliminate a moist and oily appearance [25].
	Pectins are polymers mainly composed of (1→4) α-D-galactopyranosyluronic acid units naturally esterified with methanol. According to their content of methyl esters or the degree of esterification (DE), pectins are divided into high methoxyl (HM, DE > 50%) or low methoxyl (LM, DE < 50%). The DE has a decisive effect on pectin solubility and gelation properties. The literature on pectin-based coatings is poor. Maftoonazad *et al.* [26] evaluated the protective effect of a pectin-based edible emulsion coating on the activity of *Lasiodiplodia theobromae*, a plant pathogen, in avocados, finding that the pectin-based coating was effective at controlling the spread and severity of stem end rot in avocados.
	Chitosan is a natural polysaccharide prepared by the alkaline deacetylation of chitin β-1,4-*N*-acetylglucosamine (found in fungi, arthropods and marine invertebrate); commercially, it is produced from exoskeletons of crustacean, such as crab, shrimp and crawfish. Chitosan is used for its film-forming ability; it is environmentally friendly, due to its biodegradability, biocompatibility, antimicrobial activity, non-toxicity and versatile chemical and physical properties. When applied on fruit and vegetables, chitosan-based coatings have shown effectiveness in delaying ripening and decreasing respiration rates [27,28].
Proteins	**Casein** forms transparent, flavorless and flexible films. Casein-based emulsion films (emulsified with lipid-based materials) were more effective than pure caseinate films in controlling moisture loss of fruits and vegetables [29].
	Whey proteins produce transparent, flavorless and flexible films, similar to caseinate films. Whey protein-based films possess excellent oxygen barrier properties comparable to the synthetic polymer films [30]; they also are good grease barriers [31,32].
	Zein is the principal protein of corn. It is characterized by water vapor permeabilities. Applied on vegetable and fruits, zein-based coatings are able to retard the ripening of tomatoes, to maintain the color and firmness of broccoli florets, to provide an adhesive and stable coating, to reduce the growth of *Listeria monocytogenes* on cooked sweet corn and to maintain the gloss and other qualities of apples [15].
	Soy proteins (SP) are extracted from defatted protein meal. Soy protein coatings generally have poor moisture resistance and water vapor barrier properties due to the inherent hydrophilicity of the protein and the addition of hydrophilic plasticizers; whilst they are potent oxygen barriers. SP coatings are used to preserve the freshness of apple slices and/or to retard the senescence process of kiwifruit [15].
Lipids	Most **fatty acids** derived from vegetable oils are considered GRAS (generally recognized as safe). Lipid-based coatings are compatible with other coating-forming agents; in addition, they have high water vapor and gas barrier properties. However, lipid-based coatings present a greasy surface and undesirable organoleptic properties, such as waxy taste and lipid rancidity [33]. Finally, some lipid materials are unstable when subjected to temperature changes.
	Waxes (carnauba wax, beeswax, paraffin wax and others) have been used as protective coatings for blocking moisture transport, reducing surface abrasion during fruit handling, controlling the browning of the skin in fruits by improving mechanical integrity and controlling the internal gas composition of the fruits. Wax coatings are applied on fruits (citrus, apples, cucumbers) and vegetables (tomatoes, asparagus, beans, beets, carrots, celery, eggplant, peppers, potatoes, radishes, squash and turnips.
	Resin coatings are effective at reducing water loss, but are the least permeable to gases; thus, fruit can suffer undesirable changes, such as anaerobic respiration and flavor changes.

Polysaccharides include starch, dextrins, pullulan, cellulose and derivatives, alginate, carrageenan, gums, pectins and chitosan. Polysaccharides render transparent and homogeneous edible films; these films are oxygen, aroma and oil barriers (due to their tightly-packed, ordered hydrogen-bonded network structure and low solubility), but are not effective moisture barriers due to their hydrophilic nature. However, when applied in the form of high-moisture gelatinous coatings, they can retard the moisture loss of food [15].

Proteins can be obtained from animal sources, such as casein and whey protein (the main milk protein fractions: 80% and 20%, respectively), and from plant sources, such as zein, gluten and soy proteins [34].

Different proteins are able to form films and coatings; this ability depends on their molecular weight, conformations, electrical properties (charge *vs.* pH), flexibilities and thermal stabilities [18]. Nevertheless, proteins have been studied less extensively than polysaccharides.

Generally, protein-based films have good oxygen, carbon dioxide and lipid barrier properties and mechanical properties; on the other hand, the poor water vapor resistance limits their application; this can be attributed to the inherent hydrophilicity of proteins.

As protein films are generally brittle and susceptible to cracking (due to the strong cohesive energy density of the polymers), an improvement of their properties could be attained by adding other components; for example, the addition of compatible plasticizers could improve their extensibility [35].

Edible lipid coatings include neutral lipids, fatty acids, waxes and resins. These compounds are effective in providing a moisture barrier and improving the surface appearance. Triglycerides or neutral lipids form a continuous stable layer on the food surface, thanks to their high polarity.

The growing interest addressed toward edible coatings leads to the formulation of new compositions consisting of blends of polysaccharides, proteins and lipids. The combination could be proteins and carbohydrates, proteins and lipids and carbohydrates and lipids [13], and it is based on the fact that each polymer has characteristic functions, that, when combined with each other, enhance the final functionality of the coating [15], improving the mechanical properties and, with emulsifiers, stabilizing composite coatings and improving coating adhesion.

Edible coatings can incorporate several compounds, such as: plasticizers (glycerol, acetylated monoglyceride, polyethylene glycol and sucrose), antimicrobials (bacteriocins (nisin)), enzymes (lysozyme, peroxidase and lactoperoxidase), essential oils (cinnamon, oregano, lemongrass, clove, rosemary, tea tree, thyme and bergamot), nitrites and sulfites [36], as well as synthetic antioxidants (butylated hydroxyanisole, butylated hydroxytoluene, propyl gallate, octyl gallate, dodecyl gallate, ethoxyquin, ascorbyl palmitate and tertiary butyl hydroquinone) and natural antioxidants (tocopherols, tocotrienols, ascorbic acid, citric acid, carotenoids). In these cases, edible films and coatings act as carriers of active compounds that, applied to the surface of fruits and vegetables, lead to the extension of shelf-life, the reduction of the risk of foodborne pathogenic microorganisms' growth on cut surfaces [37] and the improvement of the quality, stability and safety of coated foods [15].

2.2. Edible Coating: A Focus on Chitosan

Physical and chemical damage accrued during minimal processing of fruits and vegetables causes disruption of the plant tissues, and the exudates become ideal substrates for the growth of several microorganisms (pathogens, molds and bacteria). Natural biodegradable compounds with antimicrobial activity are recognized as safe (generally recognized as safe (GRAS)) and environmentally friendly, and chitosan is one such compound. Chitosan is able to create a semi-permeable film on the fruit surface, which results in limiting respiration and/or transpiration and in reducing weight loss [38,39]. Furthermore, its compatibility with other substances and its capability to induce host resistance to pathogens [40] have prompted its application as a coating on fruits and vegetables [41].

Chitosan has been widely used in controlling weight-loss in fresh strawberries (*Fragaria x ananassa*) and raspberries (*Rubus idaeus*), mango (*Mangifera indica*), litchi, blueberries and other fruit and vegetables [42]. Meng *et al.* [43] and Romanazzi *et al.* [44] reported that postharvest application of chitosan coating has a good control effect on decay of grapes. Chitosan owes its antimicrobial activity to its polycationic nature, which allows the interaction and formation of polyelectrolyte complexes with acid polymers produced at the surface of microbial cells, increasing their permeability and causing cell death [45].

Several factors affect the antimicrobial activity: type of chitosan, degree of acetylation, molecular weight, concentration, target microorganism, pH of the medium and presence of other additives or food components [46].

Benhabiles et al. [47] demonstrated that by reducing the number of steps for the synthesis and chemical reagents, chitosan coating was effective at improving the quality of strawberries by delaying changes in weight loss and the appearance of molds. These authors used three different chitosan coatings (chitosan C1 obtained by the classical method, chitosan C2 without decoloration and chitosan C3 without the decoloration and deproteinization steps) on strawberries, and they observed the best quality for the strawberries coated with C3 (1%), which was obtained through a lesser number of steps.

In a recent work, chitosan (in acid and water solution) exhibited its antibacterial activity against *Burkholderia seminalis*, an apricot fruit rot pathogen [48]. Lou and his coworkers demonstrated that acid-solution chitosan at a concentration of 2 mg/mL inhibited *B. seminalis*, while water-solution chitosan showed limited inhibition activity.

Chitosan has been proven to be a natural compound with antifungal activity for a wide varieties of postharvest fruits [38,49]. As reported by Bautista-Banos et al. [38], the level of inhibition of fungi is highly correlated to chitosan concentration. It is known that the polycationic nature of this compound is the key to its antifungal properties and that the length of the polymer chain enhances its antifungal activity. An additional explanation includes the possible effect that chitosan might have on the synthesis of certain fungal enzymes. El Ghaouth et al. [50] studied the antifungal effect of chitosan against *Botrytis cinerea* and *Rhizopus stolonifer*. These authors hypothesized that the mechanisms by which chitosan coating reduced the decay of strawberries appear to be related to its fungistatic properties, rather than to its ability to induce defense enzymes. A further confirmation of chitosan's ability to control fungal growth was reported by Park et al. [51]. They obtained a reduction of 2.5 and 2 log CFU/g in the counts of *Cladosporium* sp. and *Rhizopus* sp., respectively, on strawberries coated with a chitosan-based edible film, just after the coating application. A reduction in the counts of aerobic and coliform microorganisms during storage has been also reported. Chien et al. [52] investigated the effects of coating with low and high molecular weight chitosan on the decay of citrus and the maintenance of its quality. A concentration of 0.2% low molecular weight chitosan (LMWC) exhibited its antifungal activity in controlling the growth of *Penicillium digitatum* and *Penicillium italicum*. LMWC coating resulted also in being able to improve firmness, titratable acidity, ascorbic acidity and the water content for citrus stored at 15 °C for 56 days.

González-Aguilar et al. [53] have also reported a reduction in mesophilic aerobic microorganism count when fresh-cut papayas were coated with an edible coating based on chitosan of low and medium molecular weight. These researchers also observed a complete inhibition of yeast and molds throughout the storage (14 days at 5 °C). As reported by Ali et al. [54], chitosan preserved papaya fruit, delaying the ripening process by reducing the respiration rate. These results could be the reason for the delayed senescence and reduced tendency to decay [40]. Chitosan had also made improvements in the taste, peel and pulp color, texture and flavor of treated papaya fruit, but the sensory features of the papaya fruits coated with a 1.5% chitosan concentration demonstrated overall superiority after five weeks of storage.

Pilon et al. [55] made an alternative use of chitosan. They obtained chitosan nanoparticles and demonstrated that chitosan, used as a coating based on nanoparticles, reduced the microbial growth in fresh-cut apples. The samples treated with chitosan-tripolyphosphate (CS-TPP) nanoparticles (10 nm) showed higher antimicrobial activity against mesophilic and psychrotrophic bacteria, as well as molds and yeasts than conventional chitosan coating and control [55].

Numerous papers have demonstrated that chitosan-based coatings inhibit microbial growth on fresh produce, increasing shelf-life.

In a recent paper, Benhabiles *et al.* [56] reported that chitosan coatings and a chitosan derivative (N,O-carboxymethyl chitosan (NOCC)) coating improved the quality of tomato fruits (through delaying ripening, reducing weight loss and retaining fruit firmness) and extended the shelf-life. No microbial growth was observed during storage.

Assis and Pessoa [57] and Han *et al.* [58] proposed chitosan for extending the shelf-life of sliced apples and fresh strawberries, respectively. Chien *et al.* [52] reported the effectiveness of chitosan coating (at a concentration of 0.5%, 1% and 2% (w/v)) for prolonging the quality and extending the shelf-life of sliced mango fruit through a delay in the growth of mesophilic aerobic bacteria.

Durango *et al.* [59] and Devlieghere *et al.* [60] used a chitosan-based coating to cover carrots and lettuce, respectively, observing a reduction in the respiration rate and ethylene production, as well as a decrease in firmness loss. In particular, Durango *et al.* [59] controlled the growth of mesophilic aerobes, yeasts, molds and psychrotrophs of minimally-processed carrots during the first five days of storage at 15 °C using an edible yam starch coating containing chitosan. Campaniello *et al.* [28] observed that chitosan coating in combination with low temperature and suitable packaging was able to control browning and decay in strawberry fruits. Chitosan affected the microbial growth (psychrotrophic, lactic acid bacteria and yeasts) and did not affect the visual appearance. pH and thickness values remained unchanged by chitosan coating, whereas color was positively influenced.

Pushkala *et al.* [61] investigated chitosan-based powder coating on radish shreds, demonstrating the favorable effects of two different forms of chitosan (purified chitosan (CH) and chitosan lactate (CL)) on shelf-life extension of radish shreds by a minimum of 3 d over the control. Both CH- and CL-coated samples enhanced the microbial quality and sensory acceptability of the radish shreds, exhibiting a lower degree of weight loss, respiration rate, titrable acidity, % of soluble solids, a higher content of phytochemicals, moisture and pH, compared to control samples. The treated samples also exhibited lower exudate volume, lesser browning and lower microbial load compared to the control.

Sometimes, a chitosan-only coating demonstrated certain defects (including limited inhibition to some microorganisms that led fruit to decay and a poor coating structure); thus, chitosan was combined with other substances to improve its performance [62]. Chitosan coatings combined with organic acids are easy to handle, biodegradable and cause no harm to the coated fruit and/or vegetable [62]. For example, Yu *et al.* [63] combined 1% phytic acid (known as inositol hexakisphosphate (IP6), inositol polyphosphate) with 1% chitosan to preserve fresh cut lotus root. This composite coating decreased the weight loss rate and malondialdehyde (MDA) content of fresh-cut lotus root, postponed browning, restrained the activities of peroxidase (POD), polyphenol oxidase (PPO) and phenylalanine ammonia-lyase (PAL) and maintained the content of vitamin C and polyphenol at a relatively high level.

Chitosan combined with natamycin significantly decreased fresh melon decay and weight loss caused by *Alternaria alternata* and *Fusarium semitectum*, two strains of spoilage fungi [64].

Zhang *et al.* [65] reported that chitosan was able to inhibit the growth of *Botrytis cinerea* and *Rhizopus* sp. by increasing the activities of various defense enzymes, such as β-1,3-glucanase (in orange, strawberries and raspberries) and phenylalanine ammonia-lyase (PAL) (in strawberries and table grapes).

As expected, the antimicrobial activity of chitosan is also dependent on the food matrix; generally, the antimicrobial activity of chitosan is higher at a low pH because more of its amino groups are protonated; thus, it is able to interact with the negatively-charged surfaces inhibiting bacterial growth. Since the antimicrobial activity of chitosan is dependent on the charges on chitosan and the electrostatic forces, each food component could influence these interactions, affecting its activity.

Regarding this issue, Devlieghere *et al.* [60] published an interesting work on the interaction of chitosan and food components. The authors evaluated the effect of different food components (starch, proteins, NaCl and fat) on the antimicrobial activity of chitosan, following the growth of *Candida lambica*, a spoiling yeast strain, in a laboratory medium.

The authors reported that the higher concentrations of starch (30% (w/v)) inhibited the antimicrobial activity of chitosan, hypothesizing that it was due to a protective effect of starch or to electrostatic interactions when the starch was charged by modification. Proteins influenced the antimicrobial activity of chitosan depending on the pH of the medium, as their charges depend on the combination of the iso-electric point (IEP) of the proteins and the pH of the medium. If the pH is lower than the IEP, proteins and chitosan are positively charged; thus chitosan can exert its antimicrobial activity, as the interactions between both will be restricted. If the pH is higher than the IEP of the protein, the antimicrobial activity of chitosan is inhibited: proteins are negatively charged and neutralize most of the positive charges on the chitosan; thus, it cannot interact with the negatively-charged microbial surfaces [60].

NaCl reduces the antimicrobial activity of chitosan, because it interferes with the electrostatic forces between chitosan: Cl^- ions neutralize the positive charges on the chitosan, and on the other hand, the Na^+ ions compete with chitosan for the negative charges on the cell surface. Devlieghere et al. [60] observed an improved solubility of chitosan by adding NaCl at different concentrations to the medium. The authors explained that this behavior was probably due to a shielding effect of NaCl against the positive charges, which led to a coiled structure of chitosan and less interactions with components in the media. Finally, the influence of fat on the antimicrobial activity of chitosan was found negligible.

The addition of the essential oils to enhance chitosan antimicrobial action is common.

The active agents embedded into composite films may be released into or absorb substances from the packaged food or its surrounding environment. The interactions that can occur between the food product, the film or coating used as packaging and the surrounding environment are governed by different mass transfer processes: migration, adsorption, absorption and permeation [66]. For example, through these processes, the essential oils included in the coating could be transferred to the product and modify its organoleptic and nutritional characteristics by interacting with peptides, vitamins, etc. It is recognized that some important food components, such as vitamins, minerals and other nutrients, can also be sequestered with the consequent modification of food properties and functionality. In addition, when transferred from the coating to the food to pursue their protective action (antioxidants, antimicrobials, etc.), these compounds could also lose their effectiveness. It is, in fact, well recognized that generally, the efficacy of various antimicrobial compounds may be reduced by food components; for example, essential oils (EOs) and/or their components have a significant antimicrobial activity *in vitro*, but higher amounts are required (1%–3%) to achieve the same effect in foods. The presence of fat, protein, carbohydrate, water, salt, antioxidants, preservatives, other additives and pH reaction strongly influence the effectiveness of the natural compounds in foods.

High levels of fat and/or protein in foodstuffs protect bacteria from the action of EOs in some way. Some authors suggested that the fat provides a protective layer around the bacteria, or the lipid fraction could absorb the antimicrobial agent, thus decreasing its concentration and effectiveness in the aqueous phase [67]. Patrignani et al. [68] reported that the low fat content of vegetables may contribute to the successful use of EOs. Due to their lipophilic nature, EOs could share fat, missing the microbial target.

The pH is an important factor affecting the activity of oils. At a low pH, the hydrophobicity of some EOs increases due to their ability to dissolve more easily in the lipid phase of the bacterial membrane, thus enhancing the antimicrobial action.

Carbohydrates in foods do not protect bacteria from this action, whilst high water and/or salt level seems to facilitate the action of EOs.

The antibacterial activity of chitosan-based films combined with lemon (LO), tea tree (TTO) or bergamot essential oils (BO) was tested against two Gram-positive bacteria (*Staphylococcus aureus* and *L. monocytogenes*) and one Gram-negative bacteria (*Escherichia coli*) [69]. CH-EO composite films exhibited a significant antimicrobial activity against the three pathogens tested. The nature and concentration of the EOs, the film matrices and the interactions between CH and EOs affected the antimicrobial activity of the films. When TTO was added, CH exhibited the highest antimicrobial activity.

Chitosan-cinnamon oil coating extended the shelf-life of sweet pepper: after storage at 8 °C for 35 days, samples treated with chitosan-oil coatings showed a lower percentage of infected peppers and good sensory acceptability [70]. Sessa *et al.* [71] studied a novel approach to preserve vegetable products through modified chitosan edible coatings containing nanoemulsified natural antimicrobial compounds (lemon, mandarin, oregano or clove essential oils). A modified chitosan edible coating combined with lemon essential oil resulted in a remarkable increase in antimicrobial activity, with respect to other essential oils. This combination prolonged the shelf-life of rucola leaves from 3 to 7 days, in comparison to the untreated samples.

Moreover, the modified chitosan containing the nanoemulsified antimicrobial caused a significantly longer shelf-life also in comparison to a coating made of modified chitosan or essential oil alone. Thanks to this novel treatment, it was possible to prolong the shelf life of rucola leaf vegetables to about 10–14 days, without alteration of the organoleptic properties of the product, preventing the loss of firmness and color changes and preserving palatability during storage.

More recently, Randazzo *et al.* [72] used chitosan-based and methylcellulose-based films added to several EOs derived from citrus fruits (orange, mandarin and lemon) to perform the antilisterial assay and concluded that chitosan films containing essential oil from lemon were the most effective at reducing *L. monocytogenes* counts.

A chitosan-methyl cellulose-based film was used as a coating on cantaloupe fruit. The application of a chitosan (1.5% w/v)/methyl cellulose (0.5% w/v) film on fresh-cut cantaloupe reduced populations of *E. coli* inoculated on fresh-cut cantaloupe by more than 5 log CFU/piece in 8 days at 10 °C. Furthermore, a reduction of 3 log CFU/piece of *Saccharomyces cerevisiae* in 4 days of storage at 10 °C was reported [73].

Krasaekoopt and Mabumrung [74] observed that the incorporation of 1.5% and 2% chitosan in the methylcellulose coating, applied on fresh-cut cantaloupe, produced a better microbiological quality in the final product. This coating reduced the growth of mesophilic aerobes, psychrotrophs, lactic acid bacteria, yeast and molds and prevented the multiplication of *E. coli* and *Salmonella* strains.

3. Alternative Modified Atmospheres (e.g., High O_2-MAP) or the Use of Essential Oils in the Headspace

The final operation in producing minimally-processed fruits and vegetables is packaging.

Packaging performs several functions: it protects fresh-cut fruits and vegetables against deteriorative effects, contains the products, communicates to the consumers as a marketing tool and provides consumers products with ease of use and convenience. Fresh fruits and vegetables continue to respire, consuming oxygen and producing carbon dioxide and water vapor; thus, the atmosphere surrounding the package changes into another composition before sealing in vapor-barrier materials [75].

Modified atmosphere packaging (MAP) is a technique used for prolonging the shelf-life of fresh or minimally-processed foods [76]. MAP reduces the unwanted metabolic reactions and helps to protect processed products against contamination by microorganisms, thus slowing down the process of ageing [77]. MAP can be vacuum packaging (VP) and controlled atmosphere packaging (CAP). VP removes most of the air, or air is removed by vacuum or flushing and replaced with another gas mixture, before the product is packaged; the headspace atmosphere and product could change during storage, without additional manipulation of the internal environment. Differently, CAP continuously controls the environment to maintain a stable gas atmosphere within the package, as well as the temperature and humidity are monitored; it is generally used to control the ripening and spoilage of fruits and vegetables [76]. The composition, microflora, pH and the organoleptic characteristics of fruits and vegetables are extremely variable; therefore, it is impossible to define a univocal mixture of gases. It is necessary to maintain an optimum balance of oxygen and carbon dioxide; thus, an appropriate MAP must be studied for each product. Many factors affect the MAP of fresh produce and are summarized in Figure 2.

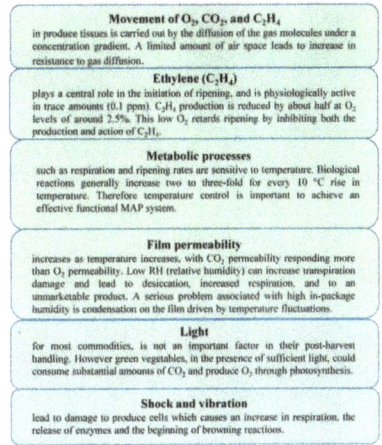

Figure 2. Factors affecting MAP of fresh products [76].

3.1. Gases Used in MAP

The main gases used in MAP are CO_2, O_2 and N_2; carbon monoxide (CO) and sulfur dioxide (SO_2) are also used. Noble gases, generally, are used for some products, such as coffee and snacks, but recently have been used also for minimally-processed apples and kiwi fruits [76]. The choice of gas depends on the food product being packed. Singly or in combination, these gases are commonly used to balance safe and shelf-life extension with optimal sensorial properties of the food. Table 2 reports the main gases used in MAP.

Low levels of oxygen and high levels of carbon dioxide reduce the produce respiration rate, delay senescence and, consequently, extend the shelf-life of fruits and vegetables. Once the package is closed, the composition of the gases inevitably changes due to produce respiration and the gas permeability of the film. If the oxygen levels are too low, fermentative processes are favored.

It is known that fresh-cut processing involves several forms of damage:

- Increases respiration rates;
- Causes major tissue disruption: enzymes and substrates, normally contained within the vacuole, are mixed with other cytoplasmic and nucleic substrates and enzymes;
- Increases wound-induced C_2H_4, water activity and surface area per unit volume, which lead to an accelerated water loss and favored microbial growth;
- Causes flavor loss;
- Causes cut surface discoloration or color loss;
- Causes decay;
- Causes increased rate of vitamin loss;
- Causes rapid softening;
- Causes shrinkage and a shorter storage life.

Thus, MAP facilitates the maintenance of fresh-cut products.

Actually, the safety of minimally-processed fruits and vegetables is mainly based on the correct chilling chain and hygienic practices, which seem to not be able to guarantee a sufficient degree of safety; in fact, the number of documented outbreaks of human infections associated with the consumption of these products has considerably increased. On these bases, traditional MAP is not enough to ensure the improvement of the quality and safety characteristics. In this context, many researchers have studied alternative tools, for improvement of MAP for minimally-processed fruits and vegetables; for example, the use of non-conventional atmospheres, including active packaging containing some natural antimicrobial compounds or MAP with high O_2, has been proposed.

Table 2. Main gases used in MAP.

Gases	Characteristics and Functions
Carbon Dioxide (CO_2)	It is a colorless gas with a slight pungent odor at very high concentrations. It is an asphyxiant and slightly corrosive in the presence of moisture. CO_2 is lipid soluble and also dissolves readily in water to produce carbonic acid (H_2CO_3), which causes an increase of the acidity of the solution, reducing the pH; this significantly affects the microbiology of packed foods. CO_2 has bacteriostatic and fungistatic properties. This bacteriostatic effect depends on the concentration and partial pressure of CO_2, the volume of headspace gas, the type of microorganism, the age and load of the initial bacterial population, the microbial growth phase, the growth medium used, the storage temperature, the acidity, the water activity and the type of product being packaged [78–81]. As MAP contains CO_2, it is not an advisable option for these products where spoilage yeasts are stimulated by high levels of CO_2. In addition, CO_2 is not effective towards *Clostridium perfringens* and *Clostridium botulinum*. Carbon dioxide is generally effective in fresh fruits and vegetables where the normal spoilage organisms consist of aerobic, Gram-negative psychrotrophic bacteria. To guarantee the maximum antimicrobial effect, low storage temperatures for MAP are recommended, because CO_2 is most soluble at low temperatures. The fat and moisture of the product affect the absorption of CO_2. Fruits and vegetables can suffer physiological damage due to high CO_2 levels.
Oxygen (O_2)	Oxygen is a colorless and odorless gas and is highly reactive. It has a low solubility in water and promotes several deteriorative reactions in foods (fat oxidation, browning reactions and pigment oxidation). Most of the common spoilage bacteria and fungi require oxygen for growth; therefore, if the pack atmosphere contains a low concentration of oxygen, the shelf-life of foods is prolonged. As an alternative to low oxygen concentrations, superatmospheric O_2 concentrations (≥ 70 kPa) have been proposed.
Nitrogen (N_2)	Nitrogen is an inert, tasteless, colorless and odorless gas and is relatively un-reactive. It has a lower density than air, is non-flammable and has a low solubility in water and other food constituents. N_2 is used to balance CO_2 and O_2 gases during food packaging, thus preventing packaging collapse that could occur when high concentrations of CO_2 are used. In addition, N_2 is able to delay oxidative rancidity and inhibit the growth of aerobic microorganisms without affecting the growth of anaerobic bacteria. Nitrogen can also indirectly influence the microorganisms in perishable foods by retarding the growth of aerobic spoilage microorganisms.
Carbon Monoxide (CO)	CO is a colorless, tasteless and odorless gas; it has a low solubility in water, but it is relatively soluble in some organic solvents. CO has been licensed for use in the USA to prevent browning in packed lettuce. It is highly reactive and very inflammable, thus its commercial application has been limited. Finally, CO has little inhibitory effect on microorganisms.
Noble Gases	These include helium (He), argon (Ar), xenon (Xe) and neon (Ne). These gases are not very reactive and are used in numerous food applications, e.g., potato-based snack products.

3.2. MAP and Natural Antimicrobial Compounds

Plants and plant products are generally considered natural alternatives to improve the shelf-life and the safety of foods, since they are characterized by a wide range of GRAS volatile compounds, which are used as food flavoring agents [76]. They are able to inhibit numerous microorganisms;

thus they are used as components of biological means for prolonging the shelf-life of post-harvest or minimally-processed fruits and vegetables [76].

Literature data indicate that volatile compounds can represent a useful tool to increase the shelf-life of plant products. Corbo *et al.* [82] evaluated the effects of hexanal and *trans*-2-hexenal on the shelf-life of fresh sliced apples. The authors added *trans*-2-hexenal to the gas mixture containing 70% N_2 and 30% CO_2 and found a significant extension of the shelf-life also when *Pichia subpelliculosa* (a spoilage yeast) was inoculated and abuse storage temperatures were used, although it had a weak negative effect on color retention.

Lanciotti *et al.* [83], postulated that future trends in the use of natural compounds, such as hexanal, 2-(E)-hexenal and hexyl acetate, would be focused on the use of specific active packaging able to release the active molecules in the head space slowly over time.

The authors reported that 150 ppm of hexanal, 20 ppm of 2-(E)-hexenal and 150 ppm of hexyl acetate displayed a bactericidal effect on *L. monocytogenes* and caused a significant extension of the lag phase of *E. coli* and *Salmonella* Enteritidis inoculated at levels of 10^4–10^5 CFU/g in fresh sliced apples packaged in ordinary or modified atmosphere.

Campaniello *et al.* [84] investigated the possibility of combining hexanal and MAP (65% N_2, 30% CO_2 and 5% O_2) on minimally-processed cactus pear fruits. The hexanal showed an antimicrobial effect against *Enterobacteriaceae*, normally contaminating minimally-processed fruits, both in the control and the modified atmosphere. The inclusion of the antimicrobial compound in the atmosphere determined an improvement of the original color retention; as well as a mesophilic selection favoring *Pantoea* spp., which have antagonistic activity against molds responsible for the decay of fruits during post-harvest phase.

Siroli *et al.* [85] proposed the use of several antimicrobial compounds (citron EO, hexanal, 2-(E)-hexenal, citral and carvacrol) alone or in combination in order to increase the shelf-life and quality parameters (texture and color) of sliced apples packaged in active modified atmosphere (7% O_2 and 0% CO_2), into medium permeability bags and stored at 6 °C. In all of the samples, the spoilage yeast threshold was not attained within the 35 days of storage. When treated with citral/2-(E)-hexenal and hexanal/2-(E)-hexenal, the sample showed a good color retention. This latter combination also improves on the retention of firmness, which was the best throughout 35 days of storage.

Furthermore, plant essential oils, constituted mainly by terpenoids, are used for their antimicrobial activity against many microorganisms. Most of the essential oils are GRAS; nevertheless, their use is often limited because of a high impact on the organoleptic characteristics of food products.

The activity of oils from *Labiatae* and citrus fruits, as well as the action of single constituents have been studied in order to better understand the cell targets of the molecules. Due to their antimicrobial effect, citrus essential oils could represent good candidates to improve the shelf-life and the safety of minimally-processed fruits. Essential oils, such as citrus, mandarin, cider, lemon and lime, were able to increase the shelf-life and safety of minimally-processed fruit salads without any impact on the sensory characteristics, even when the product was inoculated with spoilage or pathogenic bacterial species [76].

Citrus essential oils (EOs) exhibited their antimicrobial effect against a range of food poisoning-causing bacteria. A blend of citrus EO vapor against vancomycin-resistant (VRE) and vancomycin-susceptible (VSE) *Enterococcus faecium* and *Enterococcus faecalis* on lettuce and cucumber was assessed. Food samples were subjected to the vapor for 45 s in a 600-L vapor chamber at 25 °C. Results showed that microbial cell load was reduced and that no significant changes in taste were observed [86].

Tian *et al.* [87] tested the antifungal activity of essential oil extracted from the fruits of *Cicuta virosa* L. var. *latisecta* Celak (CVEO) against *Aspergillus niger*, *Aspergillus flavus*, *Aspergillus oryzae* and *Al. alternata* strains inoculated on cherry tomatoes. The samples were pre-treated with ethanol and wounded with a sterilized cork borer; then, each fruit was separately inoculated with 10 µL of a spore suspension containing 1×10^6 spores/mL of each fungal strain. CVEO (dissolved separately in 0.5 mL of 5% of Tween-20) was pipetted aseptically onto filter paper discs respectively placed into individual weighing bottles (without lids) to produce the concentrations of 200, 100 and 50 µL/mL. The essential oil was vaporized inside the containers spontaneously at 18 °C. The authors reported that at 200 µL/mL, CVEO showed the lowest percentages of decayed cherry tomatoes for all fungi compared to the control, as well as the highest inhibition of fungal infection.

In a recent paper, Vitoratos et al. [88] studied the antifungal activity of several essential oils obtained from oregano (*Origanum vulgare* L. ssp. *hirtum*), thyme (*Thymus vulgaris* L.) and lemon (*Citrus limon* L.) plants against *Bo. cinerea* inoculated in tomatoes, strawberries and cucumbers. All of the fruits were pre-treated and packaged. Different concentrations of essential oils were placed in small glass containers placed in the bottom of the package. Oregano and lemon oils were very effective at controlling disease of infected fruit by *Bo. cinerea* in tomatoes, strawberries and cucumbers. In particular, in tomatoes, *Bo. cinerea* was inhibited by oregano essential oils at 0.30 µL/mL; moreover, lemon essential oils also induced a significant reduction of grey mold. An amount of 0.05 µL/mL of lemon essential oils leads to a complete inhibition of *Bo. cinerea* in strawberries, whilst in cucumber, it leads to a reduction (39%) of the infected fruits.

3.3. Potential Applications of High O_2

It is known that low O_2 modified atmosphere packaging is universally accepted to prolong the shelf-life of minimally-processed fruits and vegetables; however, the growth of some anaerobic pathogen strains might be allowed or even stimulated. In addition, when O_2 levels are too low, this induces anaerobic respiration, leading to fermentation processes and to the production of undesirable metabolites [89].

The application of high O_2 concentrations (>70% O_2) could overcome the disadvantages of low O_2 modified atmosphere packaging for some ready-to-eat fruits and vegetables. High O_2 was found to be particularly effective at inhibiting enzymatic discoloration, preventing anaerobic fermentation reactions and inhibiting microbial growth [90]. It is hypothesized that high O_2 levels may cause substrate inhibition of the enzyme polyphenol oxidase (PPO), which is the enzyme responsible for initiating discoloration on the cut surfaces of processed products [91].

High O_2-modified atmosphere packaging has been used in some vegetables and fruits, such as minimally-processed carrots, strawberry, minimally-processed baby spinach, fresh-cut mango, apple slices, and so on [89].

Amanatidou et al. [92] screened numerous microorganisms (Pseudomonas fluorescens, Enterobacter agglomerans, Aureobacterium strains 27, Candida guilliermondii, Candida sake, *Salmonella* Typhimurium, *Salmonella* Enteritidis, *E. coli*, *L. monocytogenes*, *Leuconostoc mesenteroides* var. *mesenteroides*, *Lactobacillus plantarum* and *Lactococcus lactis*) associated with the spoilage and safety of minimally-processed vegetables. High oxygen levels (80%–90%) caused a reduction in the growth rate of microbial targets at 8 °C, whilst the lag phase was prolonged (it was more evident at 90% O_2 concentrations). Amanatidou et al. [92] and Kader and Ben Yehoshua [93] hypothesized that high O_2 concentrations lead to intracellular generation of reactive oxygen radical species (ROS, O_2^-, H_2O_2, OH^-), causing damage to the vital cellular macromolecules, thus reducing cell viability when oxidative stresses overwhelm cellular protection systems.

Jacxsens et al. [90] studied the effect of high oxygen-modified atmosphere packaging (*i.e.*, >70% O_2) on microbial growth and the sensorial qualities of mushroom slices, grated celeriac and shredded chicory endive and observed that high O_2 atmospheres were effective at inhibiting enzymatic browning of the tested vegetables. In addition, an improvement of the microbial quality (as a reduction in yeast growth) was obtained.

In an interesting paper, Allende et al. [94] studied the effect of superatmospheric oxygen packaging on sensorial quality, spoilage and *L. monocytogenes* and *Aeromonas caviae* growth in fresh processed mixed salads. An initial O_2 concentration of 95 kPa was combined with two plastic films (low and high barrier impermeability for O_2). Packaged salads were stored up to eight days at 4 °C and at temperatures simulating the chilled distribution chain.

The authors reported that superatmospheric O_2 does not affect all microorganisms in the same way: high oxygen levels affected lactic acid bacteria and *Enterobacteriaceae*, which were inhibited in both plastic films (with low and high permeability). On the contrary, the growth of yeast and *A. caviae* seemed to be stimulated, whereas the growth of psychrotrophic bacteria and *L. monocytogenes* was not affected. The general appearance was maintained for longer, and the shelf-life of the mixed salads was prolonged by using O_2 concentrations higher than 60 kPa throughout the storage period.

Chunyang *et al.* [89] studied the effect of high oxygen-modified atmosphere packaging on fresh-cut onion quality at room temperature. Onion slices were packaged in a high-barrier film package of 70 μm in thickness and stored in five different modified atmospheres (100% O_2; 95% O_2/5% CO_2; 80% O_2/20% CO_2, 75% O_2/25% CO_2; and air). Results showed that the fresh-cut onions packaged in air had a short shelf-life, because the respiration rate was quickened due to mechanical damage, and microorganisms were not inhibited. High O_2 reduced the weight loss, the respiration rate, the total reducing sugar loss and the total titrable acidity increase of the fresh-cut onions. The best modified atmosphere seemed to be 80% O_2/20% CO_2, where total bacteria counts increased slowly. High oxygen affected also the sensory characteristics: sensory quality was acceptable up to five days, while air-packaged fresh-cut onions were not acceptable after two days of storage at room temperature.

Day [91] also confirmed that the most effective high O_2 gas mixtures were found to be 80%–85% O_2/15%–20% CO_2. This had the most noticeable sensory quality and antimicrobial benefits on a range of freshly-prepared produce items.

In fact, recommended optimal headspace gas levels immediately after freshly-prepared produce package sealing are: 80%–95% O_2 and 5%–20% N_2. After package sealing, headspace O_2 levels will decline, whereas CO_2 levels will increase during storage, due to the intrinsic respiratory nature of fresh product.

The levels of O_2 and CO_2 within packages are influenced by numerous variables:

- Intrinsic produce respiration rate (which itself is affected by temperature, atmospheric composition, produce type, variety, cultivar, maturity and the severity of preparation);
- Packaging film permeability;
- Package volume;
- Surface area and fill weight;
- Produce volume/gas volume ratio;
- Degree of illumination.

It would be preferable to maintain headspace levels of O_2 > 40% and CO_2 between 10% and 25% to maximize the benefits of high O_2 MAP. This can be achieved by:

- Lowering the temperature of storage;
- Selecting produce having a lower intrinsic respiration rate;
- Minimizing cut surface tissue damage;
- Reducing the ratio of produce volume/gas;
- Using a packaging film able to maintain high levels of O_2 whilst allowing excess CO_2 to go out
- By incorporating an innovative active packaging sachet that can absorb excess CO_2 and emit an equal volume of O_2.

A further measure to maintain a gas mixture O_2 > 40%/CO_2 10%–25% could be obtained by introducing the highest level of O_2 (balanced with N_2) possible just prior to prepared produce package sealing. Generally, it is not necessary to have any levels of CO_2 in the initial gas mixture, since it will increase rapidly during chilled storage.

Numerous products (iceberg lettuce, sliced mushrooms, broccoli florets, lettuce, baby-leaf spinach, Lollo Rossa lettuce, flat-leaf parsley, cubed swede, coriander, raspberries, strawberries, grapes and oranges) processed with high O_2 MAP reported beneficial effects on sensory quality if compared to industry-standard air packing and low O_2 MAP [91].

Finally, the recommended packaging material for high O_2-modified atmosphere is 30-μm orientated polypropylene (OPP), which has sufficient O_2 barrier properties (to maintain high in-pack O_2 levels >40%), and it is sufficiently permeable to ensure that CO_2 did not rise above 25% after 7–10 days of storage at 5–8 °C.

Other packaging materials suitable for high O_2 MAP of fresh prepared produce are:

- Laminations or extrusions of OPP with low density polyethylene (LDPE);
- Ethylene-vinyl acetate (EVA);
- Polyvinyl chloride (PVC).

or other medium to very high O_2 permeability films.

4. Conditioning Solutions with Antimicrobials or Natural Compounds for Fruit Salad

Coatings, dipping, spraying, *etc.*, are the main applications of antimicrobial compounds, as natural alternatives to chemical additives, whilst little is known about the use of antimicrobials as filling liquids.

An initial work by Senesi *et al.* [95] on the use of rectified apple juice as a filling liquid to increase the quality and shelf-life of fresh-cut apples was followed, more recently, by D'Amato *et al.* [96]. The authors evaluated the possibility of using a chitosan, honey and pineapple juice solution as filling liquids to prolong the microbiological shelf-life of a fruit-based salad.

"Granny Smith" apples, "Gialla" first crop cactus pear fruits and "Regina" table grapes were washed; then, apples and cactus pears were peeled and sliced, while grapes were cut without seeds. Mixed fruits were placed in conical frustum-shaped cups, filled with four different solutions (sterile distilled water as the control; 30% of acacia honey and 70% of distilled water; 50% of pineapple juice and 50% of distilled water; 50% of low molecular weight chitosan solution and 50% of sterile distilled) and stored at 4, 8 and 12 °C.

The authors observed that the use of the natural antimicrobial compounds, as a filling liquid, affected the microbiological shelf-life of salad.

At 4 °C, mesophilic and psychrotrophic bacteria did not reach the established limit (1×10^6 CFU/g) during the whole storage period (14 days), in samples with honey and chitosan solutions, whereas it ranged from about 6.34–7.81 days and from 9.38 to 11.49 for the control and samples with pineapple juice, respectively.

At 8 °C, all of the samples were stored for 10 days. For the samples treated with honey and chitosan solutions, psychrotrophic bacteria did not reach the established limit during the storage, while in control and pineapple juice, it was of 5–6 days. Concerning mesophilic bacteria, microbiological shelf-life was strongly reduced in the control samples and in pineapple juice (about 3.7–4.3 days), whilst in chitosan and honey, it was over six days. In addition, in the samples with chitosan solution, the microbiological shelf-life, calculated by using the growth of yeasts, was over seven days and, by using lactic acid bacteria, did not reach the established limit.

At 12 °C, the samples were stored for six days. The microbiological shelf-life of fruit salad without antimicrobial compounds and with pineapple juice was always very low (1.5 and 4 days, respectively).

In the samples honey added, mesophilic and psychrotrophic bacteria did not reach the limit of 1×10^6 CFU/g. For lactic acid bacteria and yeasts, microbiological shelf-life reached 3.82 and 1.81 days, respectively.

Chitosan exhibited its high antimicrobial activity towards psychrotrophic bacteria, which did not reach the established limit, while no advantage resulted for mesophilic bacteria and lactic acid bacteria and yeasts.

The authors concluded that:

- Honey inhibited mesophilic and psychrotrophic bacterial growth; the antimicrobial activity of this natural antimicrobial compound was less effective on lactic acid bacteria and yeasts at all storage temperatures;
- Chitosan exerted a high antimicrobial activity for all microbial groups considered, particularly at a low temperature of storage;
- Pineapple juice was not effective towards all microbial groups, probably due to the high amount of nutrients.

Further investigations are needed to improve the antimicrobial efficacy of honey and chitosan for potential commercial applications. The literature is poor in articles addressing this topic; thus, it would be advisable to explore the use of antimicrobial compounds as filling liquids.

5. Biopreservation and the Use of Probiotic Coatings

In recent years, biological preservation has emerged as a promising strategy to extend the shelf-life and to improve the microbiological safety of foods [97], since it fits well with the diffuse desire to preserve foods by natural means. Several bacteria and yeasts have been already identified as bioprotective agents [98], and different studies have been carried out on their application to fresh-cut fruits and vegetables [99]. Table 3 summarizes the most recent studies performed on this topic.

In general, lactic acid bacteria (LAB) have shown the greatest potential as biocontrol agents of several minimally-processed foods, because they are widely used in fermented foods and have a long history of safe use [100]. However, several other bacteria and yeasts, often selected among the naturally-occurring microflora, including strains of *Pseudomonas syringae*, *Pseudomonas graminis*, *Gluconobacter asaii*, *Candida* spp., *Dicosphaerina fagi*, *Metschnikowia pulcherrima* and *C. sake*, have been proposed as biocontrol agents in fresh-cut fruits and vegetables [101–104].

Table 3. Overview of the most recent studies performed on the use of biocontrol microorganisms to biopreserve minimally-processed products.

Minimally-Processed Product	Proposed Microorganisms	Reference
Biopreservation		
Apples	*Leuconostoc mesenteroides, Leuconostoc citreum*	[103]
Apples	*Pseudomonas graminis*	[102,105]
Apples	*Candida sake*	[101]
Apples	*Candida* sp., *Gluconobacter asaii, Dicosphaerina fagi, Metschnikowia pulcherrima*	[106]
Apples	*Enterobacteriaceae*	[107]
Apples	*Lactobacillus plantarum*	[99]
Apples	*Lactococcus lactis*	[108]
Iceberg lettuce	*Leuconostoc mesenteroides, Leuconostoc citreum*	[103]
Iceberg lettuce	*Leuconostoc* spp.	[104]
Lamb's lettuce	*Lactobacillus plantarum, Lactobacillus casei*	[99]
Melon	*Pseudomonas graminis*	[109]
Melon	*Pseudomonas graminis*	[110]
Peaches	*Enterobacteriaceae*	[107]
Scarola salad	*Lactobacillus casei*	[111]
Probiotic Biopreservation		
Apple	*Lactobacillus rhamnosus* GG	[112]
Apple	*Lactobacillus rhamnosus* GG	[113]
Apple	*Bifidobacterium lactis* Bb-12	[114]
Papaya	*Bifidobacterium lactis* Bb-12	[114]

The success of LAB in preventing the growth and activity of undesirable microorganisms is due to a large diversity of mechanisms of action related to the production of antimicrobial compounds, organic acids, hydrogen peroxide, bacteriocins and diacetyl [103,104,115]. The combination of low pH values and antibacterial activities of organic molecules produced by LAB remains the main mechanism for biopreservation [97]. Several bacteriocin-producing LAB have been shown to be effective against spoilage and pathogenic microorganisms in minimally-processed fruits and vegetables [116,117]. However, the direct application of bacteriocins to fresh-cut products did not provide completely satisfactory results probably due to the adsorption or inactivation of the added compound into the food product [116]; on the other hand, the direct use of living bacteriocinogenic bioprotective strains could lead to a more effective protection of the food product by circumventing the mentioned problems because of the localized and constant delivery of the antibacterial compound, which will add to other advantages, like space colonization by the strain [116,117]. The inhibitory properties against contaminating foodborne pathogens and spoilage microorganisms could also consist of a mere competition for nutrients (vitamins, minerals, trace elements and peptides), and therefore, via competition or antibiosis, LAB are able to function as a hurdle to pathogen growth and survival [118].

In the last decade, different LAB species were proposed as biocontrol agents in minimally-processed fruits and vegetables (see Table 3). In 2004, Scolari and Vescovo [111] performed several challenge tests on salad leaves by simultaneously inoculating *Lactobacillus casei* and various pathogens (*S. aureus, Aeromonas hydrophila, E. coli* and *L. monocytogenes*). A significant inhibitory effect by the LAB towards all of the pathogenic strains was observed and confirmed by the same authors during a subsequent study about the influence of *Lb. plantarum* on the growth of *S. aureus* [119]. Trias *et al.* [103] found five strains of LAB (some *Leuconostoc* spp. strains) that were able to inhibit *L. monocytogenes* and *Salmonella* Typhimurium in cut iceberg lettuce leaf, but were not effective in reducing the amount of *E. coli*. More recently, Siroli *et al.* [108] showed the good performance of a nisin-producing strain of *Lc. lactis*, which was able to inhibit *L. monocytogenes*, *E. coli* and the total mesophilic species when added at a level of 7 log CFU/mL in the washing solution of minimally-processed lamb's lettuce and combined or not with thyme essential oil. In 2015, the same authors proposed two other LAB strains (*Lb. plantarum* V7B3 and *Lb. casei* V4B4) to be used as biocontrol agents alone or in combination with thyme essential oil (EO) in lamb's lettuce [99]. In this work, the use of the *Lb. plantarum* V7B3 strain (6 log CFU/mL) during the washing phase of fresh-cut lettuce increased product shelf-life and safety; *L. monocytogenes* and *E. coli* viabilities were significantly reduced over the nine days of refrigerated storage. Promising results were also obtained for biopreservation of minimally-processed golden delicious apples packaged in a modified atmosphere alone or in combination with natural antimicrobials (2-(E)-hexenal/hexanal and 2-(E)-hexenal/citral) [99]: a strain of *Lb. plantarum* (CIT3) was able to increase the safety of sliced apples, when inoculated at levels of 6–7 log CFU/g in the washing dipping solution, both alone or in combination with natural antimicrobials.

Other non-lactic acid bacteria and yeasts were also proposed as biocontrol agents (Table 3) For example, the growth of *L. monocytogenes* and *Salmonella* Enterica in fresh-cut apples has been prevented using fungal antagonists [106]. When inoculated at a low level, *L. monocytogenes* cell loads were greatly reduced (from 5.7 to 6.0 log units after seven days) by strains of *G. asaii, Candida* spp., *D. fagi* and *M. pulcherrima*. At high pathogen inoculum levels, only *G. asaii* and *Candida* spp. reduced the cell load of *L. monocytogenes* population to non-detectable levels. Abadias *et al.* [101] found that the application of the fungal postharvest antagonist *C. sake* CPA-1 reduced the growth of a mixture of *E. coli* strains in fresh-cut apples at 25 °C, whereas Alegre *et al.* [107] isolated a new strain of *Enterobacteriaceae* that reduced the growth of *Salmonella, L. monocytogenes* and *E. coli* O157:H7 on fresh-cut apples and peaches. *P. graminis* was found be able to reduce or slow down the development of foodborne pathogens on minimally-processed fresh-cut apples and peaches [105]; in this case, the inhibitory effect of the antagonists on the foodborne pathogens was not instantaneous and became apparent after 6 days at 5 °C. This strain was also proposed for fresh-cut melon, reducing *Salmonella* and *L. monocytogenes* growth during storage at 5, 10 and 20 °C [109]. The strain effectiveness depended on the pathogens' concentration and on storage temperature. At a low pathogen concentration and 20 °C, *L. monocytogenes* growth was reduced between 2.1 and 5.3 log CFU/g after two days of storage and *Salmonella* growth between 2.0 and 7.3 log CFU/g.

Although research on the use of biocontrol agents in minimally-processed fruits and vegetables has increased in recent decades, the standardization of a biopreservative approach is still difficult to realize. Independent of the species and strains proposed, in fact, the various studies available in the literature clearly highlight that the efficacy of biocontrol agents is affected by different factors, such as the inoculation level, the presence of other bacteria, the physico-chemical and compositional features of the products and the storage conditions. Further investigations are required, especially considering that process conditions have to be taken into account, during the scaling up at the industrial level.

An interesting modern challenge is to incorporate probiotic bacteria into coated processed fruits and vegetables in order to improve their shelf-life (biopreservation), while providing new non-dairy functional foods. As is well accepted, edible films/coatings may serve as carriers of food additives, such as anti-browning agents, antimicrobials, colorants, flavors, nutrients and spices [24]. Some authors [114,120] proved that the coatings were also good carriers for antioxidant agents, such as cysteine, glutathione and ascorbic and citric acids; thus, the addition of LAB to obtain functional edible films and coatings could have successful implications. Several studies were already conducted on different fruits, as is shown in Table 3. This new trend has arisen from more considerations. As just mentioned, the inclusion into coatings of cultures with inhibitory properties could improve the shelf-life and safety of minimally-processed products, while reducing the need to use increasing levels of chemical additives [118]. In addition, the steps of minimally-processing vegetables, such as peeling and cutting, promote the release of cellular content rich in minerals, sugars, vitamins and other nutrients, creating ideal conditions for microbial growth: this characteristic allows the use of fruit and vegetable food products as probiotic carriers. According to Soccol *et al.* [121], in fact, minimally-processed fruits and vegetables are very good matrices providing ideal substrates for probiotics, since they contain minerals, vitamins, antioxidants and fibers. Moreover, due to their cellulose content, fruits, such as apples and pears, may also exert a protective effect on the probiotic microorganisms during passage through the intestinal tract [122], allowing these microorganisms to reach the colon and benefit the host. It has been reported that the optimum probiotic growth temperature is between 35 and 40 °C, and the best pH is between 6.4 and 4.5, ceasing when a pH of 4.0–3.6 is reached [123]. This situation could be solved by using some supports, such as agar, polyacrylamide, calcium pectate gel, chemically-modified chitosan beads and alginates, to provide a physical barrier against unfavorable conditions [124–126].

One of the first reports about this issue is the study of Tapia *et al.* [114]; fresh-cut apple and papaya cylinders were coated with 2% (w/v) alginate or gellan film-forming solutions containing viable Bifidobacteria, namely *Bifidobacterium lactis* Bb-12. The bifidus-containing coatings were more permeable to water vapor than the corresponding films without probiotics, and gellan coatings were more resistant to water transfer than the alginate ones. The most important result was that the edible coatings were efficient in supporting *Bb. lactis* Bb-12 on fresh-cut apple and papaya. In 2010, Rößle *et al.* [112] applied a probiotic microorganism (*Lactobacillus rhamnosus* GG; LGG) to fresh-cut apple wedges (cultivar Braeburn). All samples were able to maintain a probiotic load of about *ca.* 10^8 CFU/g over 10 storage days, which is sufficient for a probiotic effect, and this is comparable to the counts of probiotic bacteria in commercially-available dairy products. It is also important to underline that the physico-chemical properties of the apple wedges containing LGG compared to the control (without probiotics) remained stable over the observation period. In a subsequent work [113], the effectiveness as a biocontrol agent of the same probiotic strain against *Salmonella* and *L. monocytogenes* on minimally-processed apples throughout storage, as well as its effect on apple quality and natural microflora was evaluated. The obtained results showed that *Salmonella* was not affected by co-inoculation with LGG, whereas the *L. monocytogenes* population was 1-log unit lower in the presence of probiotic population maintained over recommended levels for probiotic action (10^6 CFU/g) along 14 days.

Although still unexplored, this new challenge appears to be highly advantageous, since minimally-processed fruits and vegetables are a food category rich in nutrients, intended for consumption by all individuals and widely accepted among consumers [127]. Thus, edible coatings, including probiotic bacteria, when applied to minimally-processed products, open new possibilities to improve their shelf-life and safety while providing innovative functional foods.

6. Concluding Remarks

The use of non-conventional tools to preserve and prolong the quality of minimally-processed fruits and vegetables seems to be the future trend to obtain improved final products. In particular, the application of the antimicrobial compounds into coatings and/or filling liquids should be further investigated due to their environmental friendliness and versatility. A combined approach of these technologies, through additive and synergistic interactions, could be useful to obtain innovative and safe minimally-processed products.

Author Contributions: The authors contributed equally to this paper.

Conflicts of Interest: The authors declare no conflict of interest.

References

1. Allende, A.; Tomas-Barberan, F.A.; Gil, M.I. Minimal processing for healthy traditional foods. *Trends Food Sci. Tech.* **2006**, *7*, 179–186. [CrossRef]
2. Lamikanra, O. Preface. In *Fresh-Cut Fruits and Vegetables*; Lamikanra, O., Ed.; CRC Press: Boca Raton, FL, USA, 2002.
3. Gonzales-Aguilar, G.A.; Ayala-Zavala, J.F.; de La Rosa, L.A.; Ivarez-Parrilla, E.A. Preserving quality of fresh-cut product using safe technology. *J. Verbrauch. Lebensm.* **2010**, *5*, 65–72. [CrossRef]
4. Boesso Oriani, V.; Molina, G.; Chiumarelli, M.; Pastore, G.M.; Dupas Hubinger, M. Properties of cassava starch-based edible coating containing Essential Oils. *J. Food Sci.* **2014**, *79*, E189–E194. [CrossRef] [PubMed]
5. McHugh, T.H.; Senesi, E. Apple wraps: A novel method to improve the quality and extend the shelf life of fresh-cut apples. *J. Food Sci.* **2000**, *65*, 480–485. [CrossRef]
6. Nisperos-Carriedo, M.O.; Baldwin, E.A.; Shaw, P.E. Development of an edible coating for extending postharvest life of selected fruits and vegetables. *Proc. Fla. State Hort. Soc.* **1991**, *104*, 122–125.
7. Lerdthanangkul, S.; Krochta, J.M. Edible coating effects on post harvest quality of green bell peppers. *J. Food Sci.* **1996**, *61*, 176–179. [CrossRef]
8. Avena-Bustillos, R.J.; Krochta, J.M.; Saltveit, M.E. Water vapor resistance of red delicious apples and celery sticks coated with edible caseinate-acetylated monoglyceride films. *J. Food Sci.* **1997**, *62*, 351–354. [CrossRef]
9. Guilbert, S.; Gontard, N.; Gorris, L.G.M. Prolongation of the shelf-life of perishable food products using biodegradable films and coatings. *LWT—Food Sci. Technol.* **1996**, *29*, 10–17. [CrossRef]
10. Smith, S.; Geeson, J.; Stow, J. Production of modified atmospheres in deciduous fruits by the use of films ad coatings. *Hort. Sci.* **1987**, *22*, 772–776.
11. Krochta, J.M.; Mulder-Johnston, C.D. Edible and biodegradable polymer films: Challenges and opportunities. *Food Technol.* **1997**, *51*, 61–74.
12. Souza, B.W.S.; Cerqueira, M.A.; Teixeira, J.A.; Vicente, A.A. The use of electric fields for edible coatings and films development and production: A review. *Food Eng. Rev.* **2010**, *2*, 244–255. [CrossRef]
13. Campos, C.A.; Gerschenson, L.N.; Flores, S.K. Development of edible films and coatings with antimicrobial activity. *Rev. Food Bioprocess. Technol.* **2010**, *4*, 849–875. [CrossRef]
14. Allen, L.; Nelson, A.I.; Steinberg, M.P.; McGill, J.N. Edible corn-carbohydrate food coatings. Development and physical testing of starch-algin coating. *Food Technol.* **1963**, *17*, 1437–1442.
15. Lin, D.; Zhao, Y. Innovations in the development and application of edible coatings for fresh and minimally processed fruits and vegetables. *Compr. Rev. Food Sci. Food Saf.* **2007**, *6*, 60–75. [CrossRef]

16. Diab, T.; Biliaderis, C.G.; Gerasopoulos, D.; Sfakiotakis, E. Physicochemical properties and application of pullulan edible films and coatings in fruit preservation. *J. Sci. Food Agric.* **2001**, *81*, 988–1000. [CrossRef]
17. Synowiec, A.; Gniewosz, M.; Kraśniewska, K.; Przybył, J.L.; Bączek, K.; Węglarz, Z. Antimicrobial and antioxidant properties of pullulan film containing sweet basil extract and an evaluation of coating effectiveness in the prolongation of the shelf life of apples stored in refrigeration conditions. *Inn. Food Sci. Emer. Technol.* **2014**, *23*, 171–181. [CrossRef]
18. Vargas, M.; Pastor, C.; Chiralt, A.; McClements, D.J.; González-Martínez, C. Recent advances in edible coatings for fresh and minimally processed fruits. *Crit. Rev. Food Sci.* **2008**, *48*, 496–511. [CrossRef] [PubMed]
19. Coma, V.; Deschamps, A.; Martial-Gros, A. Bioactive packaging materials from edible chitosan polymer-antimicrobial assessment on dairy related contaminants. *J. Food Sci.* **2003**, *68*, 2788–2792. [CrossRef]
20. Li, P.; Barth, M.M. Impact of edible coatings on nutritional and physiological changes in lightly processed carrots. *Postharvest Biol. Tec.* **1998**, *14*, 51–60. [CrossRef]
21. Amanatidou, A.; Slump, R.A.; Gorris, L.G.M.; Smid, E.J. High oxygen and high carbon dioxide modified atmospheres for shelf life extension of minimally processed carrots. *J. Food Sci.* **2000**, *65*, 61–66. [CrossRef]
22. Hershko, V.; Nussinovitch, A. Relationships between hydrocolloid coating and mushroom structure. *J. Agric. Food Chem.* **1998**, *46*, 2988–2997. [CrossRef]
23. Bryan, D.S. Prepared citrus fruit halves and method of making the same. U.S. Patent 3,707,383, 26 December 1972.
24. Lee, J.Y.; Park, H.J.; Lee, C.Y.; Choi, W.Y. Extending shelf-life of minimally processed apples with edible coatings and antibrowning agents. *Lebens. Wissen. Technol.* **2003**, *36*, 323–329. [CrossRef]
25. Arnold, F.W. Infrared Roasting of Coated Nutmeats. U.S. Patent 3,383,220, 14 May 1968.
26. Maftoonazad, N.; Ramaswamy, H.S.; Moalemiyan, M.; Kushalappa, A.C. Effect of pectin-based edible emulsion coating on changes in quality of avocado exposed to *Lasiodiplodia theobromae* infection. *Carbohyd. Polym.* **2007**, *68*, 341–349. [CrossRef]
27. Vargas, M.; Albors, A.; Chiralt, A.; Gonzalez-Martinez, C. Quality of cold-stored strawberries as affected by chitosan-oleic acid edible coatings. *Postharvest Biol. Tec.* **2006**, *41*, 164–171. [CrossRef]
28. Campaniello, D.; Bevilacqua, A.; Sinigaglia, M.; Corbo, M.R. Chitosan: Antimicrobial activity and potential applications for preserving minimally processed strawberries. *Food Microbiol.* **2008**, *25*, 992–1000. [CrossRef] [PubMed]
29. Krochta, J.M.; Pavlath, A.E.; Goodman, N. Edible films from casein-lipid emulsions for lightly-processed fruits and vegetables. In *Engineering and food, Preservation Processes and Related Techniques*; Spiess, W.E., Schubert, H., Eds.; Elsevier Science Publishers: New York, NY, USA, 1990; pp. 329–340.
30. Trezza, T.A.; Krochta, J.M. Application of edible protein coatings to nut and nut-containing food products. In *Protein-Based Films and Coatings*; Gennadios, A., Ed.; CRC Press: Boca Raton, FL., USA, 2002; pp. 527–549.
31. Chan, M.A.; Krochta, J.M. Grease and oxygen barrier properties of whey-protein isolate coated paperboard. *Solutions* **2001**, *84*, 57.
32. Lin, S.Y.; Krochta, J.M. Plasticizer effect on grease barrier properties of whey protein concentrate coatings on paperboard. *J. Food Sci.* **2003**, *68*, 229–233. [CrossRef]
33. Guilbert, S. Technology and application of edible protective films. In *Food Packaging and Preservation*; Mathlouthi, M., Ed.; Elsevier Applied Science Publishers: London, UK, 1986; pp. 371–394.
34. Gennadios, A.; McHugh, T.H.; Weller, G.L.; Krochta, J.M. Edible coatings and films based on proteins. In *Edible Coatings and Films to Improve Food Quality*; Krochta, J.M., Baldwin, E.A., Nisperos-Carriedo, M.O., Eds.; Technomic Publishing Co., Inc.: Lancaster, PA, USA, 1994; pp. 201–277.
35. Lim, L.T.; Mine, Y.; Britt, I.J.; Tung, M.A. Formation and properties of egg white protein films and coatings. In *Protein-Based Films and Coatings*; Gennadios, A., Ed.; CRC Press: Boca Raton, FL, USA, 2002; pp. 233–252.
36. Franssen, L.R.; Krochta, J.M. Edible coatings containing natural antimicrobials for processed foods. In *Natural Antimicrobials for Minimal Processing of Foods*; Roller, S., Ed.; CRC Press: Boca Raton, FL, USA, 2003; pp. 250–262.
37. Martín-Belloso, O.; Rojas-Graü, M.A.; Soliva-Fortuny, R. Delivery of flavor and active ingredients using edible films and coatings. In *Edible Films and Coatings for Food Applications*; Embuscado, M.E., Huber, K.C., Eds.; Springer: New York, NY, USA, 2009; pp. 295–313.

38. Bautista-Banos, S.; Hernandez-Lauzardo, A.N.; Velazquez-del Valle, M.G.; Hernandez-Lo, M.; Ait Barka, E.; Bosquez-Molina, E.; Wilson, C.L. Chitosan as a potential natural compound to control pre and postharvest diseases of horticultural commodities. *Crop. Prot.* **2006**, *25*, 108–118. [CrossRef]
39. Jiang, Y.M.; Li, J.R.; Jiang, W.B. Effects of chitosan coating on shelf life of cold stored litchi fruit at ambient temperature. *LWT-Food Sci. Technol.* **2005**, *38*, 757–761. [CrossRef]
40. Shiekh, R.A.; Malik, M.A.; Al-Thabaiti, S.A.; Shiekh, M.A. Chitosan as a novel edible coating for fresh fruits. *Food Sci. Technol. Res.* **2013**, *19*, 139–155. [CrossRef]
41. Shahidi, F.; Arachchi, J.K.V.; Jeon, Y. Food application of chitin and chitosan. *Trends Food Sci. Tech.* **1999**, *10*, 37–51. [CrossRef]
42. Moncayo, D.; Buitrago, G.; Algeciras, N. The surface properties of biopolymer-coated fruit: A review. *Ing. Invest.* **2013**, *33*, 11–16.
43. Meng, X.H.; Li, B.Q.; Liu, J.; Tian, S.P. Physiological responses and quality attributes of table grape fruit to chitosan preharvest spray and postharvest coating during storage. *Food Chem.* **2008**, *106*, 501–508. [CrossRef]
44. Romanazzi, G.; Nigro, F.; Ippolito, A.; di Venere, D.; Salerno, M. Effects of pre- and postharvest chitosan treatments to control storage grey mold of table grapes. *J. Food Sci.* **2002**, *67*, 1862–1867. [CrossRef]
45. Ziani, K.; Fernandez Pan, I.; Royo, M.; Maté, J. Antifungal activity of films and solutions based on chitosan. *Food Hydrocoll.* **2009**, *23*, 2309–2314. [CrossRef]
46. Aider, M. Chitosan application for active bio-based films production and potential in the food industry: Review. *LWT Food Sci. Technol.* **2010**, *43*, 837–842. [CrossRef]
47. Benhabiles, M.S.; Drouiche, N.; Lounici, H.; Pauss, A.; Mameri, N. Effect of shrimp chitosan coatings as affected by chitosan extraction processes on postharvest quality of strawberry. *Food Meas.* **2013**, *7*, 215–221. [CrossRef]
48. Lou, M.M.; Zhu, B.; Muhammad, I.; Li, B.; Xie, G.L.; Wang, Y.L.; Li, H.Y.; Sun, G.C. Antibacterial activity and mechanism of action of chitosan solutions against apricot fruit rot pathogen *Burkholderia seminalis*. *Carbohydr. Res.* **2011**, *346*, 1294–1301. [CrossRef] [PubMed]
49. Liu, L.; Liu, C.K.; Hicks, K.B. Composite films from pectin and fish skin gelatin or soybean flour protein. *J. Agric. Food Chem.* **2007**, *55*, 2349–2355. [CrossRef] [PubMed]
50. El Ghaouth, A.; Arul, J.; Grenier, J.; Asselin, A. Antifungal activity of chitosan on two post-harvest pathogens of strawberry fruits. *Phytopathology* **1992**, *82*, 398–402. [CrossRef]
51. Park, S.I.; Stan, S.D.; Daeschel, M.A.; Zhao, Y. Antifungal coatings on fresh strawberries (*Fragaria* × *ananassa*) to control mold growth during cold storage. *J. Food Sci.* **2005**, *70*, 202–207. [CrossRef]
52. Chien, P.J.; Sheu, F.; Lin, H.R. Coating citrus (*Murcott tangor*) fruit with low molecular weight chitosan increases postharvest quality and shelf life. *Food Chem.* **2007**, *100*, 1160–1164. [CrossRef]
53. Gonzalez-Aguilar, G.A.; Valenzuela-Soto, E.; Lizardi-Mendoza, J.; Goycoolea, F.; Martinez-Tellez, M.A.; Villegas-Ochoa, M.A.; Monroy-Garcia, I.N.; Ayala-Zavala, J.F. Effect of chitosan coating in preventing deterioration and preserving the quality of fresh-cut papaya "Maradol". *J. Sci. Food Agric.* **2009**, *89*, 15–23. [CrossRef]
54. Ali, A.; Muhammad, M.T.M.; Sijam, K.; Siddiqui, Y. Effect of chitosan coatings on the physicochemical characteristics of *Eksotika*. II papaya (*Carica papaya* L.) fruit during cold storage. *Food Chem.* **2011**, *124*, 620–626. [CrossRef]
55. Pilon, L.; Poliana, C.; Spricigo, M.M.; de Moura, M.R.; Assis, O.B.G.; Mattoso, L.H.C.; Ferreira, M.D. Chitosan nanoparticle coatings reduce microbial growth on fresh-cut apples while not affecting quality attributes. *Int. J. Food Sci. Technol.* **2015**, *50*, 440–444. [CrossRef]
56. Benhabiles, M.S.; Tazdait, D.; Abdi, N.; Lounici, H.; Drouiche, N.; Goosen, M.F.A.; Mameri, N. Assessment of coating tomato fruit with shrimp shell chitosan and N,O-carboxymethyl chitosan on postharvest preservation. *Food Meas.* **2013**, *7*, 66–74. [CrossRef]
57. Assis, O.B.; Pessoa, J.D. Preparation of thin films of chitosan for use as edible coatings to inhibit fungal growth on sliced fruits. *Braz. J. Food Technol.* **2004**, *7*, 7–22.
58. Han, C.; Lederer, C.; McDaniel, M.; Zhao, Y. Sensory evaluation of fresh strawberries (*Fragaria ananassa*) coated with chitosan-based edible coatings. *J. Food Sci.* **2005**, *70*, S172–S178. [CrossRef]
59. Durango, A.M.; Soares, N.F.F.; Andrade, N.J. Microbiological evaluation of an edible antimicrobial coating on minimally processed carrots. *Food Control* **2006**, *17*, 336–341. [CrossRef]

60. Devlieghere, F.; Vermeulen, A.; Debevere, J. Chitosan: Antimicrobial activity, interactions with food components and applicability as a coating on fruit and vegetables. *Food Microbiol.* **2004**, *21*, 703–714. [CrossRef]
61. Pushkala, R.; Raghuram, P.K.; Srividya, N. Chitosan based powder coating technique to enhance phytochemicals and shelf life quality of radish shreds. *Postharvest Biol. Tec.* **2013**, *86*, 402–408. [CrossRef]
62. Jianglian, D.; Shaoying, Z. Application of chitosan based coating in fruit and vegetable preservation: A review. *J. Food Process. Technol.* **2013**, *4*, 227:1–227:4. [CrossRef]
63. Yu, Y.W.; Li, H.; Jinhua, D.; Ren, Y.Z. Study of natural film with chitosan combining phytic acids on preservation of fresh-cutting lotus root. *J. Chin. Inst. Food Sci. Technol.* **2012**, *12*, 131–136.
64. Cong, F.; Zhang, Y.; Dong, W. Use of surface coatings with natamycin to improve the storability of Hami melon at ambient temperature. *Postharvest Biol. Technol.* **2007**, *46*, 71–75. [CrossRef]
65. Zhang, H.; Li, R.; Liu, W. Effects of chitin and its derivative chitosan on postharvest decay of fruits: A Review. *Int. J. Mol. Sci.* **2011**, *12*, 917–934. [CrossRef] [PubMed]
66. Salgado, P.R.; Ortiz, C.M.; Musso, Y.S.; di Giorgio, L.; Mauri, A.N. Edible films and coatings containing bioactives. *Curr. Opin. Food Sci.* **2015**, *5*, 86–92. [CrossRef]
67. Speranza, B.; Corbo, M.R. Essential Oils for Preserving Perishable Foods: Possibilities and Limitations. In *Application of Alternative Food Preservation Technologies to Enhance Food Safety and Stability*; Bevilacqua, A., Corbo, M.R., Sinigaglia, M., Eds.; Bentham Publisher: Saif Zone, Sharjah, 2010; pp. 35–57.
68. Patrignani, F.; Siroli, L.; Serrazanetti, D.I.; Gardini, F.; Lanciotti, R. Innovative strategies based on the use of essential oils and their components to improve safety, shelf-life and quality of minimally processed fruits and vegetables. *Trends Food Sci. Technol.* **2015**. [CrossRef]
69. Sánchez-González, L.; Cháfer, M.; Hernández, M.; Chiralt, A.; González-Martínez, C. Antimicrobial activity of polysaccharide films containing essential oils. *Food Control.* **2011**, *22*, 1302–1310. [CrossRef]
70. Xing, Y.; Li, X.; Xu, Q.; Yun, J.; Lu, Y.; Tang, Y. Effects of chitosan coating enriched with cinnamon oil on qualitative properties of sweet pepper (*Capsicum annuum* L.). *Food Chem.* **2011**, *124*, 1443–1450. [CrossRef]
71. Sessa, M.; Ferraria, G.; Donsìb, F. Novel edible coating containing Essential Oil nanoemulsions to prolong the shelf life of vegetable products. *Chem. Eng. Trans.* **2015**. [CrossRef]
72. Randazzo, W.; Jimenez-Belenguer, A.; Settanni, L.; Perdones, A.; Moschetti, M.; Palazzolo, E.; Guarrasi, V.; Vargas, M.; Germanà, M.A.; Moschetti, G. Antilisterial effect of citrus essential oils and their performance in edible film formulations. *Food Control.* **2016**, *59*, 750–758. [CrossRef]
73. Sangsuwan, J.; Rattanapanone, N.; Rachtanapun, P. Effect of chitosan/methyl cellulose films on microbial and quality characteristics of fresh-cut cantaloupe and pineapple. *Postharvest Biol. Technol.* **2008**, *49*, 403–410. [CrossRef]
74. Krasaekoopt, W.; Mabumrung, J. Microbiological evaluation of edible coated fresh-cut cantaloupe. *Kasetsart. J. Nat. Sci.* **2008**, *42*, 552–557.
75. Corbo, M.R.; Speranza, B.; Campaniello, D.; D'Amato, D.; Sinigaglia, M. Fresh-cut fruits preservation: Current status and emerging technologies. In *Current Research, Technology and Education topics in Applied Microbiology*; Mendez-Villas, A., Ed.; Formatex Research Center: Badajoz, Spain, 2011; pp. 1143–1154.
76. Corbo, M.R.; Bevilacqua, A. Alternative Modified Atmosphere for Fresh Food Packaging. In *Application of Alternative Food Preservation Technologies to Enhance Food Safety and Stability*; Bevilacqua, A., Corbo, M.R., Sinigaglia, M., Eds.; Bentham Publisher: Saif Zone, Sharjah, 2010; pp. 196–204.
77. Siddiqui, M.W.; Chakraborty, I.; Ayala-Zavala, J.F.; Dhua, R.S. Advances in minimal processing of fruits and vegetables: A review. *J. Sci. Ind. Res.* **2011**, *70*, 823–834.
78. Church, N. Developments in modified-atmosphere packaging and related technologies. *Trends Food Sci. Technol.* **1994**, *5*, 345–352. [CrossRef]
79. Farber, J.M. Microbiological Aspects of Modified-Atmosphere Packaging Technology—A Review. *J. Food Protect.* **1991**, *54*, 58–70.
80. Phillips, C.A. Review: Modified Atmosphere Packaging and Its Effects on the Microbiological Quality and Safety of Produce. *Int. J. Food Sci. Technol.* **1996**, *31*, 463–479. [CrossRef]
81. Church, I.J.; Parsons, A.L. Modified atmosphere packaging technology: A Review. *J. Sci. Food Agric.* **1995**, *67*, 143–152. [CrossRef]

82. Corbo, M.R.; Lanciotti, R.; Gardini, F.; Sinigaglia, M.; Guerzoni, M.E. Effects of hexanal, (E)-2-hexenal, and storage temperature on shelf life of fresh sliced apples. *J. Agric. Food Chem.* **2000**, *48*, 2401–2408. [CrossRef] [PubMed]
83. Lanciotti, R.; Gianotti, A.; Patrignani, F.; Belletti, N.; Guerzoni, M.E.; Gardini, F. Use of natural aroma compounds to improve shelf-life and safety of minimally processed fruits. *Trends Food Sci. Technol.* **2004**, *15*, 201–208. [CrossRef]
84. Campaniello, D.; D'Amato, D.; Corbo, M.R.; Sinigaglia, M. Combined action of hexanal and MAP on minimally processed cactus pear fruit. *Int. J. Food Sci.* **2005**, 1–10.
85. Siroli, L.; Patrignani, F.; Serrazanetti, D.I.; Tabanelli, G.; Montanari, C.; Tappi, S.; Rocculi, P.; Gardini, F.; Lanciotti, R. Efficacy of natural antimicrobials to prolong the shelf-life of minimally processed apples packaged in modified atmosphere. *Food Control.* **2014**, *46*, 403–411. [CrossRef]
86. Fisher, K.; Phillips, C.; McWatt, L. The use of an antimicrobial citrus vapour to reduce *Enterococcus* sp. on salad products. *Int. J. Food Sci. Technol.* **2009**, *44*, 1748–1754. [CrossRef]
87. Tian, J.; Ban, X.; Zeng, H.; He, J.; Huang, B.; Wang, Y. Chemical composition and antifungal activity of essential oil from *Cicuta virosa* L. var. *latisecta* Celak. *Int. J. Food Microbiol.* **2011**, *145*, 464–470. [CrossRef] [PubMed]
88. Vitoratos, A.; Bilalis, D.; Karkanis, A.; Efthimiadou, A. Antifungal Activity of Plant Essential Oils against *Botrytis cinerea*, *Penicillium italicum* and *Penicillium digitatum*. *Not. Bot. Horti Agrobot.* **2013**, *41*, 86–92.
89. Chunyang, H.; Xiqing, Y.; Fei, L.; Binxin, S. Effect of High Oxygen Modified Atmosphere Packaging on Fresh-Cut Onion Quality. In Proceedings of the 17th IAPRI World Conference on Packaging, Tianjin, China, 12–15 October 2010.
90. Jacxsens, L.; Devlieghere, F.; Van der Steen, C.; Debevere, J. Effect of high oxygen modified atmosphere packaging on microbial growth and sensorial qualities of fresh-cut produce. *Int. J. Food Microbiol.* **2001**, *71*, 197–210. [CrossRef]
91. Day, B.P.F. New modified atmosphere techniques fresh prepared fruit and vegetables. In *Fruit and Vegetables Processing*; Jongen, W., Ed.; Woodhead Publishing Limited: Cambridge, UK, 2002; pp. 310–330.
92. Amanatidou, A.; Smid, E.J.; Gorris, L.G.M. Effect of elevated oxygen and carbon dioxide on the surface growth of vegetable-associated micro-organisms. *J. Appl. Microbiol.* **1999**, *86*, 429–438. [CrossRef] [PubMed]
93. Kader, A.; Ben-Yehoshua, S. Effects of superatmospheric oxygen levels on postharvest physiology and quality of fresh fruits and vegetables. *Postharvest Biol. Technol.* **2000**, *20*, 1–13. [CrossRef]
94. Allende, A.; Jacxsens, L.; Devlieghere, F.; Debevere, J.; Artés, F. Effect of Superatmospheric Oxygen Packaging on Sensorial Quality, Spoilage, and *Listeria monocytogenes* and *Aeromonas caviae* Growth in Fresh Processed Mixed Salads. *J. Food Protect.* **2002**, *65*, 1565–1573.
95. Senesi, E.; De Regibus, P. Use of rectified apple juice as filling liquid to increase quality and shelf-life of fresh-cut apples (Succo di mela rettificato come liquido di governo per prolungare la curabilita' di spicchi di mela di IV gamma). *Ind. Aliment.* **2002**, *41*, 139–146.
96. D'Amato, D.; Sinigaglia, M.; Corbo, M.R. Use of chitosan, honey and pineapple juice as filling liquids for increasing the microbiological shelf life of a fruit-based salad. *Int. J. Food Sci. Technol.* **2010**, *45*, 1033–1041. [CrossRef]
97. Galvez, A.; Abriouel, H.; Benomar, N.; Lucas, R. Microbial antagonists to food-borne pathogens and biocontrol. *Curr. Opin. Biotechnol.* **2010**, *21*, 142–148. [CrossRef] [PubMed]
98. Vermeiren, L.; Devlieghere, F.; Debevere, J. Evaluation of meat born lactic acid bacteria as protective cultures for the biopreservation of cooked meat products. *Int. J. Food Microbiol.* **2004**, *96*, 149–164. [CrossRef] [PubMed]
99. Siroli, L.; Patrignani, F.; Serrazanetti, D.I.; Tabanelli, G.; Montanari, C.; Gardini, F.; Lanciotti, R. Lactic acid bacteria and natural antimicrobials to improve the safety and shelf-life of minimally processed sliced apples and lamb's lettuce. *Food Microbiol.* **2015**, *47*, 74–84. [CrossRef] [PubMed]
100. Carr, F.J.; Chill, D.; Maida, N. The lactic acid bacteria: A literature survey. *Crit. Rev. Microbiol.* **2002**, *28*, 281–370. [CrossRef] [PubMed]
101. Abadias, M.; Usall, J.; Alegre, I.; Torres, R.; Vinas, I. Fate of *Escherichia coli* in apple and reduction of its growth using the postharvest biocontrol agent *Candida Sake* CPA-1. *J. Sci. Food Agric.* **2009**, *89*, 1526–1533. [CrossRef]

102. Alegre, I.; Vinas, I.; Usall, J.; Anguera, M.; Altisent, R.; Abadias, M. Antagonistic effect of *Pseudomonas graminis* CPA-7 against foodborne pathogens in fresh-cut apples under simulated commercial conditions. *Food Microbiol.* **2013**, *33*, 139–148. [CrossRef] [PubMed]
103. Trias, R.; Badosa, E.; Montesinos, E.; Baneras, L. Bioprotective *Leuconostoc* strains against *Listeria monocytogenes* in fresh fruits and vegetables. *Int. Microbiol.* **2008**, *127*, 91–98. [CrossRef] [PubMed]
104. Trias, R.; Baneras, L.; Badosa, E.; Montesinos, E. Bioprotection of Golden Delicious apples and Iceberg lettuce against foodborne bacterial pathogens by lactic acid bacteria. *Int. J. Food Microbiol.* **2008**, *123*, 50–60. [CrossRef] [PubMed]
105. Alegre, I.; Vinas, I.; Usall, J.; Anguera, M.; Figge, M.J.; Abadias, M. Control of foodborne pathogens on fresh-cut fruit by a novel strain of *Pseudomonas graminis*. *Food Microbiol.* **2013**, *34*, 390–399. [CrossRef] [PubMed]
106. Leverentz, B.; Conway, W.S.; Janisiewicz, W.; Abadias, M.; Kurtzman, C.P.; Camp, M.J. Biocontrol of the food-borne pathogens *Listeria monocytogenes* and *Salmonella* Enterica serovar poona on fresh-cut apples with naturally occurring bacterial and yeast antagonists. *Appl. Environ. Microb.* **2006**, *72*, 1135–1140. [CrossRef] [PubMed]
107. Alegre, I.; Vinas, I.; Usall, J.; Anguera, M.; Figge, M.J.; Abadias, M. An *Enterobacteriaceae* species isolated from apples controls foodborne pathogens on fresh-cut apples and peaches. *Postharvest Biol. Technol.* **2012**, *74*, 118–124. [CrossRef]
108. Siroli, L.; Patrignani, F.; Salvetti, E.; Torriani, S.; Gardini, F.; Lanciotti, R. Use of a nisin-producing *Lactococcus lactis* strain, combined with thyme essential oil, to improve the safety and shelf-life of minimally processed lamb's lettuce. In Proceeding of the 11th International Symposium on lactic acid bacteria, Egmond aan Zee, The Netherlands, 31 August–4 September 2014.
109. Abadias, M.; Altisent, R.; Usall, J.; Torres, R.; Oliveira, M.; Vinas, I. Biopreservation of fresh-cut melon using the strain *Pseudomonas graminis* CPA-7. *Postharvest Biol. Technol.* **2014**, *96*, 69–77. [CrossRef]
110. Plaza, L.; Altisent, R.; Alegre, I.; Vinas, I.; Abadias, M. Changes in the quality and antioxidant properties of fresh-cut melon treated with the biopreservative culture *Pseudomonas graminis* CPA-7 during refrigerated storage. *Postharvest Biol. Technol.* **2016**, *111*, 25–30. [CrossRef]
111. Scolari, G.; Vescovo, M. Microbial antagonism of *Lactobacillus casei* added to fresh vegetables. *Int. J. Food Sci.* **2004**, *16*, 465–475.
112. Rößle, C.; Auty, M.A.E.; Brunton, N.; Gormley, R.T.; Butler, F. Evaluation of fresh-cut apple slices enriched with probiotic bacteria. *Innov. Food Sci. Emerg.* **2010**, *11*, 203–209. [CrossRef]
113. Alegre, I.; Viñas, I.; Usall, J.; Anguera, M.; Abadias, M. Microbiological and physicochemical quality of fresh-cut apple enriched with the probiotic strain *Lactobacillus rhamnosus* GG. *Food Microbiol.* **2011**, *28*, 59–66. [CrossRef] [PubMed]
114. Tapia, M.S.; Rojas-Graü, M.A.; Rodríguez, F.J.; Ramirez, J.; Carmona, A.; Martín-Belloso, O. Alginate and Gellan-based edible films for probiotic coatings on fresh-cut fruits. *J. Food Sci.* **2007**, *72*, 190–196. [CrossRef] [PubMed]
115. Cleveland, J.; Montville, T.J.; Nes, I.F.; Chikindas, M.L. Bacteriocins: Safe, natural antimicrobials for food preservation. *Int. J. Food Microbiol.* **2001**, *71*, 1–20. [CrossRef]
116. Allende, A.; Martınez, B.; Selma, V.; Gil, M.I.; Suarez, J.E.; Rodrıguez, A. Growth and bacteriocin production by lactic acid bacteria in vegetable broth and their effectiveness at reducing *Listeria monocytogenes in vitro* and in fresh-cut lettuce. *Food Microbiol.* **2007**, *24*, 759–766. [CrossRef] [PubMed]
117. Randazzo, C.L.; Pitino, I.; Scifo, G.O.; Caggia, C. Biopreservation of minimally processed iceberg lettuces using a bacteriocin produced by *Lactococcus lactis* wild strain. *Food Control.* **2009**, *20*, 756–763. [CrossRef]
118. Schuenzel, K.M.; Harrison, M.A. Microbial antagonists of foodborne pathogens on fresh, minimally processed vegetables. *J. Food Protect.* **2002**, *65*, 1909–1915.
119. Scolari, G.; Vescovo, M.; Zacconi, C.; Bonade, A. Influence of *Lactobacillus plantarum* on *Staphylococcus aureus* growth in a fresh vegetable model system. *Eur. Food Res. Technol.* **2004**, *218*, 274–277.
120. Rojas-Graü, M.A.; Sobrino-López, A.; Tapia, M.S.; Martín-Belloso, O. Browning inhibition in fresh-cut Fuji apple slices by natural antibrowning agents. *J. Food Sci.* **2006**, *71*, 59–65. [CrossRef]
121. Soccol, C.R.; Vandenberghe, L.P.S.; Spier, M.R.; Medeiros, A.B.P.; Yamaguishi, C.T.; Lindner, J.D.; Pandey, A.; Soccol, V.T. The potential of probiotics. *Food Tech. Biotechnol.* **2010**, *48*, 413–434.

122. Kourkoutas, Y.; Kanellaki, M.; Koutinas, A.A. Apple pieces as immobilization support of various microorganisms. *LWT-Food Sci. Technol.* **2006**, *39*, 980–986. [CrossRef]
123. Shah, N.P. Functional cultures and health benefits. *Int. Dairy J.* **2007**, *17*, 1262–1277. [CrossRef]
124. Kailasapathy, K. Microencapsulation of probiotic bacteria: Technology and potential application. *Curr. Issues Intest. Microbiol.* **2002**, *3*, 39–49. [PubMed]
125. Guerin, D.; Vuillemard, J.C.; Subirade, M. Protection of *Bifidobacteria* encapsulated in polysaccharide-protein gel beads against gastric juice and bile. *J. Food Protect.* **2003**, *66*, 2076–2084.
126. Kourkoutas, Y.; Xolias, V.; Kallis, M.; Bezirtzoglou, E.; Kanellaki, M. *Lactobacillus casei* cell immobilization on fruit pieces for probiotic additive, fermented milk and lactic acid production. *Process. Biochem.* **2005**, *40*, 411–416. [CrossRef]
127. Saad, S.M.I.; Bedani, R.; Mamizuka, E.M. Benefícios à Saude dos Probióticos e Prebióticos. In *Probióticos e Prebióticos em Alimentos: Fundamentos e Aplicações Tecnológicas*; Saad, S.M.I., Cruz, A.G., Faria, J.A.F., Eds.; Editora Varela: São Paulo, Brazil, 2011; pp. 51–84, In Portuguese.

© 2015 by the authors. Licensee MDPI, Basel, Switzerland. This article is an open access article distributed under the terms and conditions of the Creative Commons Attribution (CC BY) license (http://creativecommons.org/licenses/by/4.0/).

Review

Gelatin-Based Films and Coatings for Food Packaging Applications

Marina Ramos *, Arantzazu Valdés, Ana Beltrán and María Carmen Garrigós

Analytical Chemistry, Nutrition & Food Sciences Department, University of Alicante, San Vicente del Raspeig, 03690 Alicante, Spain; arancha.valdes@ua.es (A.V.); ana.beltran@ua.es (A.B.); mc.garrigos@ua.es (M.C.G.)
* Correspondence: marina.ramos@ua.es; Tel.: +34-965-903-400 (ext. 3117); Fax: +34-965-903-697

Academic Editor: Jari Vartiainen
Received: 31 July 2016; Accepted: 20 September 2016; Published: 28 September 2016

Abstract: This review discusses the latest advances in the composition of gelatin-based edible films and coatings, including nanoparticle addition, and their properties are reviewed along their potential for application in the food packaging industry. Gelatin is an important biopolymer derived from collagen and is extensively used by various industries because of its technological and functional properties. Nowadays, a very wide range of components are available to be included as additives to improve its properties, as well as its applications and future potential. Antimicrobials, antioxidants and other agents are detailed due to the fact that an increasing awareness among consumers regarding healthy lifestyle has promoted research into novel techniques and additives to prolong the shelf life of food products. Thanks to its ability to improve global food quality, gelatin has been particularly considered in food preservation of meat and fish products, among others.

Keywords: edible films; food coatings; food preservation; biopolymers; antioxidant and antimicrobial agents

1. Introduction

In the last few decades, there has been a marked increase in the use of natural polymer-based film materials and coatings in packaging for food industry, which protect food from external contamination, retarding its deterioration by extending its shelf-life and maintaining its quality and safety [1]. In addition to consumer requirements and in order to substitute petroleum-based plastic packaging, a wide variety of biopolymers that come from agro-food industrial wastes and renewable low cost natural resources have emerged [2]. In this context, the formulation of films and coatings for food packaging applications must include at least one component capable of forming a cohesive three-dimensional matrix. Biopolymers directly extracted from biomass mainly used for edible films in food packaging are proteins, polysaccharides and lipids, and the physical and chemical properties of the biopolymer used determine the final properties of the developed films [3].

Proteins can be defined as natural polymers able to form amorphous three-dimensional structures stabilized mainly by non-covalent interactions. The functional properties of the final materials are highly dependent on the structural heterogeneity, thermal sensitivity, and hydrophilic behaviour of proteins. Different vegetable and animal proteins are commonly used as biodegradable polymers, such as corn zein, wheat gluten, soy protein, collagen and gelatin, casein and caseinates, and whey proteins, among others [1,4,5].

Regarding polysaccharides for material applications, the main ones used are cellulose and starch, but increasing attention is being given to more complex carbohydrate polymers produced by bacteria and fungi, especially to polysaccharides such as xanthan, curdlan, pullulan and hyaluronic acid [6]. In addition, the incorporation of lipid materials such as animal and vegetable oils and fats, waxes and natural resins, into polysaccharide and protein matrices to form edible composite films and coatings has the potential to improve film moisture barrier [7].

The use of biopolymers, especially gelatin, in packaging of highly perishable food products such as meat and fish is based on some particular properties such as cost, availability, functional attributes, mechanical (flexibility, tension) and optical (brightness and opacity) properties, barrier effect against gas flow, structural resistance to water and microorganisms and sensory acceptability. In this article, the latest advances in gelatin-based films and coatings including composition (additives to be used in the gelatin matrix, including nanoparticles addition) and properties are reviewed, as well as new research trends for different food applications.

Gelatin and Film-Forming Properties

Gelatin is a natural water soluble protein characterized by the absence of an appreciable odour and the random configuration of polypeptide chains in aqueous solution. It is obtained from the partial hydrolysis of collagen; a fibrous protein mainly found in certain parts of vertebrate and invertebrate animals as bones, skins, connective tissues and tendons [8]; and its structure consists of rigid bar-like molecules that arranged in fibres inter-connected by covalent bonds.

Pig skin was used as raw material to manufacture gelatin in the 1930s and continues to be the most important material for large-scale food industrial production; whereas for more expensive uses, such as pharmaceuticals, gelatin is generally obtained from cattle bones, which is considered a more complex and costly extraction process. However, in a move to get away from porcine and bovine gelatin, the production of fish gelatin has increased in the last decade, accounting for more than 1.5% of total gelatin production [9].

In recent years, by-products obtained from the fishing industry, such as heads, skin, bones, fins, muscle pieces, scales, viscera and others, are considered potential sources of exploration, rather than disposable waste. However, one of the main drawbacks of fish gelatin is its rheological properties, being less stable than the obtained from mammalian sources [10]. Moreover, since the production of gelatin from fish and poultry is still limited, the obtained products are less competitive in price than those from mammalian gelatins.

Soluble gelatin is produced by the destabilization of the collagen triple-helix. In general, gelatin properties are influenced by two main factors: the characteristics of the initial collagen and the extraction process. In this sense, the degree of collagen conversion into gelatin is dependent on the pre-treatment with warm-water extraction, temperature, pH, and extraction time. Interstitial collagen molecules are composed of three polypeptide α-chains intertwined and stabilized by hydrogen bonding and hydrophobic interactions. The destabilization is produced by breaking hydrogen and covalent bonds as a result of the heat treatment, resulting in helix-to-coil transition and subsequent conversion into soluble gelatin. Previously, the insoluble native collagen must be pre-treated to break non-covalent bonds so as to disorganize the protein structure, thus producing adequate swelling and collagen solubilisation, suitable for extraction [9].

Gelatin is a heterogeneous polypeptide mixture of α-chains (one polymer/single chain), β-chains (two α-chains covalently crosslinked) and γ-chains (three covalently crosslinked α-chains). Figure 1 shows a typical amino acid composition of gelatin: Ala-Gly-Pro-Arg-Gy-Glu-4Hyp-Gly-Pro-; with an elemental composition of 50.5% carbon, 25.2% oxygen, 17% nitrogen and 6.8% hydrogen [11].

Depending on the processing method, gelatin can be classified into two types: (1) type A: with an isoelectronic point at pH ~8–9, obtained from acid treated collagen; and (2) type B: with an isoelectronic point at pH ~4–5, derived from an alkali treated precursor which converts asparagine and glutamine residues into their respective acids, resulting in higher viscosity. Gelatin derived from pig skin is normally referred as type A and that derived from beef skin or pig cattle hides and bones is referred as type B [8].

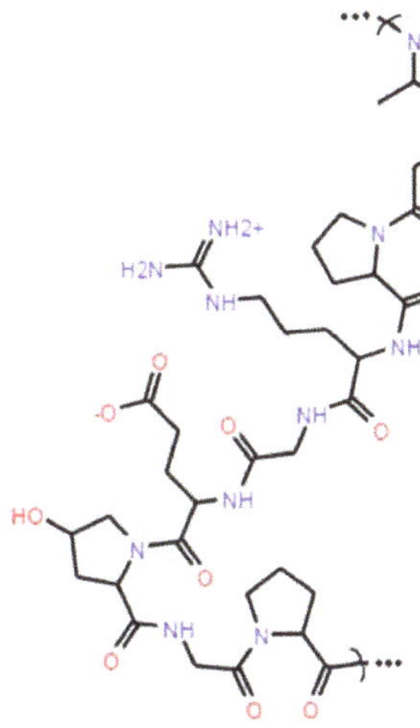

Figure 1. Representative gelatin structure according to its typical amino acid composition (adapted from N. Hanani et al. [11]).

Basic physico-chemical properties, such as solubility, composition parameters, colour, transparency, odourless and tasteless, are the main attributes that best define the overall commercial quality of gelatin. Also, gel strength (expressed in the normalized "bloom value") and viscosity should be considered as the most important physical properties since they influence gelatin quality and potential applications. In addition, gelatin can be produced in powdered or granulated form.

Gelatin has also distinctive functional properties that can be divided into two groups: (i) properties associated to surface behaviour such as protective colloid function, emulsion and foam formation and stabilization, adhesion and cohesion and film-forming capacity and (ii) properties related to gelling behaviour like gel formation, thickening, texturizing and water binding capacity [12]. Therefore, a wide number of final applications and uses can be obtained, in food, packaging, pharmaceutical, cosmetic and photographic industries (Figure 2) [13]. In particular, gelatin is used to provide gelling, stabilization, texturization and emulsification for bakery, beverages, confectionary and dairy products in food industry [10]. However, the limited thermal stability and mechanical properties of gelatin especially during processing, limit its potential applications.

Film-forming properties have been extensively used to protect food during its shelf life, as an outer film, from dryness, exhibition to light and/or exposure to oxygen. Due to the highly hygroscopic nature of gelatin, it has a tendency to swell or be dissolved when putting in contact with the surface of foodstuffs with high moisture content. Several research studies have been conducted to evaluate the overall effect of the addition of different substances, such as crosslinkers, strengthening agents, plasticizers or additives with antimicrobial or antioxidant properties, in gelatin-based products to improve the functional properties of gelatin and the shelf-life of food products [14,15]. The improvement in these properties occurs when intermolecular forces of protein chains are reduced by the action of molecular structures modifying their hydrophilic character or promoting the formation of strong covalent bonds in the protein network of the film [10]. Zhao et al. demonstrated the viability of using a natural extract as a new natural crosslinker for the modification of gelatin (type B, from bovine bone) by hydrogen bonding formation between water and free hydroxyl groups of amino or polyphenol groups. The results showed that the incorporation of this extract into gelatin significantly increased gel strength compared to the untreated gelatin [16]. The combination of gelatin with other biopolymers with different characteristics, such as whey proteins [17], starch [2], chitosan [18–20] or pectin [21], could be a good strategy for the development of films with improved mechanical and water resistance properties.

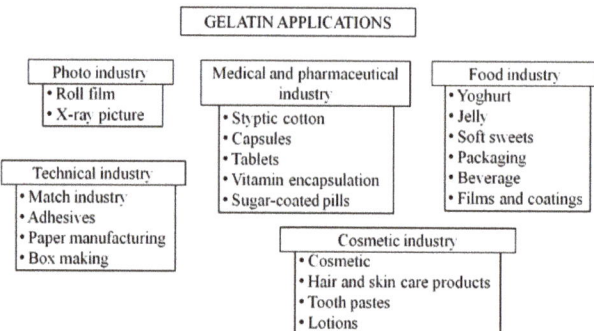

Figure 2. Some final applications and uses of gelatin.

The film properties of gelatin are also determined by the gelatin used, since the molecular weight distribution and amino acid composition vary greatly between gelatins obtained from different sources. In this sense, fish gelatin extracted at high temperature exhibits lower molecular weight profiles that gelatin from extraction at lower temperatures [10]. Muyonga et al. used sodium dodecyl sulphate gel electrophoresis (SDS–PAGE) to determine the molecular weight distribution of Nile perch gelatins as function of α and β chains, concluding that it could vary with the collagenous tissue used as raw material at high and low temperatures. Indeed, when gelatins from the same raw materials were compared, they found more peptides (molecular weight less than α chain) and lower proportion of high molecular weight (greater than β) fractions at high temperatures compared to those found at low temperature extractions [22]. Some authors have reported that these parameters also play a key role in the mechanical and barrier properties of the resulting films [23]. A recent study compared films made from fish gelatin derived from the bones of red snapper and grouper, generated as wastes by the fish processing industries, and mammalian gelatin showing fish gelatin films 17%–21% lower tensile strength than mammalian gelatin films [24].

In recent years, the interest in agro-industrial by-products has gradually increased. The revalorization of these materials is an upward trend, being considered as a potential source of resources for exploration, rather than as disposable waste. For example, the use of residues from gelatin capsules generated by the nutraceutical field is increasing despite the treatment and disposal of this residue imply economic and environmental issues [25]. Thus, the residues coming from gelatin, mainly used as encapsulating materials to deliver bioactive food compounds with active principles, can be revalorized as a potential source for the development of biodegradable films mainly composed of gelatin, glycerin and water [26,27].

Gelatin can be considered as a competitive alternative biopolymer in the market, being its use directly correlated to novel technological developments in order to improve its functional properties.

2. Gelatin-Based Films and Coatings for Food Packaging

Gelatin-based edible films and coatings have already been proposed to protect, maintain or extend the shelf-life of food products. Factors that should be considered when designing this type of system include the chemical nature of food, controlled release mechanisms, food organoleptic characteristics and additive toxicity, storage and distribution, physical and mechanical properties of packaging materials and regulations to be applied in this framework [14]. Consequently, different types of additives could be added to improve or modify the final properties in order to achieve suitable gelatin-based films or coatings for food packaging.

Recent studies have focused on interesting techniques to develop active packaging films and coatings, including antimicrobial, antioxidant and other agents which can enhance the biological features of food [8,14]. These components are usually essential oils or extracts obtained from plants and spices which exhibit antimicrobial and antioxidant properties, and most of them are considered to be Generally Recognized as Safe (GRAS) [28]. In order to reduce the use of synthetic chemical additives in the food industry, the use of natural food additives with antimicrobial and/or antioxidant properties without negative effects on human health has increased in the last years. These natural additives are able to prevent or reduce the deterioration of food caused by oxidation or microbiological effects, thus helping to preserve and extend food shelf-life [29].

2.1. Antimicrobial Agents

The use of antimicrobial additives in gelatin-based films or coatings for food packaging applications is a promising area, with the main goal being the prolongation of food shelf-life based on retarding deterioration mechanisms inside the package by using natural additives. A wide range of agents with antimicrobial properties has been proposed, e.g., organic acids, bacteriocins, spice extracts, thiosulphates, enzymes, proteins, isothiocyanates, antibiotics, fungicides, chelating agents, parabens and metals [30]. The research in this field is focused on the search for natural compounds to be used in active packaging formulations as substitutions for synthetic additives. As a result, many studies have been performed to propose the use of compounds obtained from natural sources with antimicrobial characteristics (Table 1). These additives can be obtained from different sources, including plants, animals, bacteria, algae, fungi and by-products generated during fruit and vegetable processing.

Essential oils have been extensively used in edible films. Martucci et al. developed gelatin-based films by using lavender or oregano essential oils and a mixture of them (50:50) at concentrations ranging between 0 and 6000 ppm. *Escherichia coli* (*E. coli*) as Gram-negative and *Staphylococcus aureus* (*S. aureus*) as Gram-positive bacteria were selected to evaluate the antimicrobial activity of the new films. The results showed that both microorganisms exhibited sensitivity to all active films, showing lower values of inhibition zone for *S. aureus* compared to *E. coli*, being 10.6 ± 1.5 mm and 13.7 ± 0.5 mm, respectively, at 4000 ppm [31]. Similar results were obtained by Alparslan et al. when studying the antimicrobial activity of gelatin-based films with orange leaf essential oil against five food-borne bacteria by the agar well-diffusion method. The gelatin film including 2% essential oil showed the highest antimicrobial effect against all microorganisms with inhibition zones of 14.5 ± 0.7 mm for *S. aureus* and 19.0 ± 1.4 mm for *E. coli* [32]. Antibacterial activity of fish skin gelatin films incorporating peppermint and citronella oils at 10%, 20% and 30% were studied by Yanwong et al., obtaining a growth inhibition of *E. coli* and *S. aureus* higher than 80% at 10% loading of each oil [33]. The obtained differences between Gram-negative and Gram-positive bacteria might be due to the presence of a thin peptidoglycan layer in Gram-negative bacteria that makes them more resistant against essential oils [34].

Regarding the mechanisms of action of this type of antimicrobial agent against bacteria, they have not been clearly detailed, since each compound present in the essential oil composition exhibits a unique mechanism of action that is specific to a particular range of food and microorganisms [35]. Different mechanisms have been identified: damage to the cell wall, interaction with and disruption of the cytoplasmic membrane, damage of membrane's proteins, leakage of cellular components, coagulation of cytoplasm and depletion of the proton motive force. All these effects produce microorganism death by the modification of the structure and composition of the bacteria cells [34].

Another important feature related to the use of this type of additive in food packaging systems is their poor stability at high temperatures and the need to control their release into the food sample over time. In fact, the release rate is a key parameter to allow for good and suitable microbial inhibition. Recent works have reported the use of alternative techniques for the incorporation of these additives into gelatin by using micro- or nano-encapsulations with the purpose of improving and controlling their release rate. Wu et al. developed fish gelatin films incorporated with cinnamon essential oil nanoliposomes. An evaluation of the antimicrobial stability of the films by using the disc diffusion method in the third and thirtieth days of storage was carried out. The results showed a higher inhibition zone for the obtained film with cinnamon essential oil nanoliposomes compared to that of gelatin with cinnamon essential oil, demonstrating an improvement in antimicrobial stability along with a decrease in release rate after storage for one month [36].

Table 1. Different compounds used as active additives in gelatin-based films and coatings.

Gelatin	Active Additive	Application	Main Benefits	Ref.
Fish gelatin	*Origanum vulgare* L. essential oil	Films	Enhancement in WVP, solubility, barrier capability to ultraviolet light Enhancement in antimicrobial properties	[37]
Fish gelatin	Nanoencapsulated *Origanum vulgare* L. essential oil	Films	Maintenance of initial thermal stability Less resistant and more flexible films Decrease in WVP Exhibited antimicrobial activity	[38]
Bovine gelatin	Bacteriocins and flavonoid ester prunin laurate	Films	Maintenance of functional properties Enhancement in antimicrobial properties and synergistic effect	[39]
Gelatin	Silver nanoparticles	Films	Enhancement in hydrophobicity, water vapour and UV barrier Compact surface structure Strong antibacterial activity	[40,41]
Gelatin	Zinc oxide nanoparticles	Films	Crystalline structure Enhancement in thermal stability, moisture content, water contact angle, WVP and elongation at break Strong antibacterial activity	[42]
Skate skin gelatin	Thyme essential oil	Chicken tenderloin (wrap)	Enhancement in antimicrobial properties Extend shelf-life of chicken tenderloin Increase in elongation at break	[43]
Grouper bone gelatin	Chitosan, clove and pepper essential oils	Fish steaks (coating)	Enhancement in antimicrobial properties Extend the shelf-life of fish steaks	[44]
Fish gelatin	Cinnamon essential oil nanoliposomes	Films	Decrease in tensile strength, water soluble, water content and WVP Sustained release effect and improvement in antimicrobial stability	[36]
Fish skin gelatin	Peppermint and citronella essential oils	Films	Enhancement in antimicrobial properties	[33]
Fish gelatin	Green tea, grape seed, ginger or gingko leaf	Films	Enhancement in antioxidant properties	[45]
Bovine gelatin	Brown seaweed *Ascophyllum nodosum*	Films	Increase in hydrophilicity Enhancement in antioxidant properties	[46]
Residues of gelatin capsules	Beet root residue powder	Films	Enhancement in antioxidant properties Maintenance of initial thermal stability	[47]
Bovine gelatin residue	Carrot residue fibre derived from minimally processed carrots	Films	High barrier, optical and thermal properties Capacity for protecting sunflower oil from primary rancidity reactions	[27]
Pork gelatin	Ethanolic hop extract	Films	Enhancement in antioxidant properties	[48]
Gelatin	Free/encapsulated tea polyphenols	Sunflower oil packaging	No significant differences in visual aspect Enhancement in antioxidant properties Good oxidation inhibitory effect over 6 weeks of storage	[49]
Gelatin	Tea polyphenols	Films	Enhancement in antioxidant properties	[50]
Pig skin gelatin	Hydrolysable chestnut tannin	Films	Enhancement in antimicrobial and antioxidant properties	[51]
Beef gelatin	Articoat DLP 02, Artemix Consa 152/NL, Auranta FV and sodium octanoate	Films	Enhancement in antimicrobial and antioxidant properties at different degrees Enhancement in oxygen transmission rate	[52]
Food grade gelatin	Orange leaf essential oil	Shrimps (coating)	Shelf-life extension Enhancement in antimicrobial and antioxidant properties	[32]
Bovine hide gelatin	Oregano and lavender essential oils	Films	Enhancement in antimicrobial and antioxidant properties	[31]

Metallic nanofillers have recently been considered in packaging technologies for the production of active gelatin-based films with potential antimicrobial effects since these additives are able not only to enhance barrier and mechanical properties when they are incorporated into the matrix, but also to improve food preservation and shelf-life through their antimicrobial performance [40,53]. Silver and zinc oxide nanoparticles are examples proposed by some authors in different studies. P. Kanmani et al. introduced different amounts of silver nanoparticles (0, 10, 20, 30, and 40 mg) into gelatin using a solution casting method. The antibacterial activities of films were evaluated using the agar well diffusion and colony count methods at time intervals of 2 h during 12 h by using *E. coli*, *Listeria monocytogenes* (*L. Monocytogenes*), *Salmonella typhimurium*, *S. aureus* and *Bacillus cereus*. The results showed that all films significantly decreased the cell viability of food-borne bacteria except the control and gelatin films with a lower amount of silver nanoparticles (10 mg), where no inhibition was observed. The film containing 40 mg of silver nanoparticles solution exhibited excellent antimicrobial effects against bacteria with values lower than 10^1 CFU/mL compared to the other films [40]. The antimicrobial mechanism suggested by several authors is supported by the morphological and structural changes found in the bacterial cells and the possibilities for silver nanoparticles to penetrate inside the bacterial structure due to their attachment to the cell membrane [54]. In line with this study, Shankar et al. prepared composite films based on gelatin incorporated with four different types of zinc oxide nanoparticles obtaining strong antibacterial activity against both Gram-positive and Gram-negative bacteria, *L. monocytogenes* and *E. coli*, respectively, for films with nanoparticles, with values of cell viability lower than 10^2 CFU/mL after 12 h of study. These results could be related to the release of Zn^{2+} ions, which could penetrate through the cell wall of bacteria and react with the cytoplasmic content, leading to microorganism death [42].

2.2. Antioxidant Agents

Nowadays, research in the field of active packaging is also focused on the development of novel food packaging materials with antioxidant agents from natural sources such as plant and spices extracts instead of synthetic antioxidants such as butylated hydroxytoluene (BHT) or butylated hydroxyanisole (BHA), since synthetic antioxidants are suspected of causing some safety concerns and have been restricted in their use as food additives [55]. In this context, some studies have reported that natural antioxidants show enough capacity to control lipid oxidation inside the food package since oxidative processes can cause the degradation of proteins, pigments and lipids, which limits food shelf-life [28,56,57]. Table 1 summarizes some research studies performed to enhance the final properties and applicability of food packaging and to extend the shelf-life of food products based on gelatin films and coatings incorporated with antioxidant additives.

Extracts obtained from green tea, grape seed, ginger or gingko leaf have been studied for their excellent antioxidant properties due to the presence of some compounds in their compositions, such as polyphenolic compounds in the case of green tea extract; flavones glycosides in ginkgo leaf extract; gingerol, gingerdiol, gingerdione and other antioxidant compounds for ginger extract; or tannins and monomeric flavonoids such as catechin and epicatechin for grape seed extracts. Li et al. incorporated natural extracts into fish skin gelatin at three different concentrations, 0.01, 1.0 and 5.0 mg/mL, by using the casting technique. In this work, physical and mechanical properties of films were studied and antioxidant activity was evaluated by using three commonly methods: DPPH radical scavenging assay, reducing power and peroxide value analysis. Results showed the strongest scavenging activity against DPPH radicals (around 90%) for the formulations with 1.0 mg/mL of each extract used except for the ginger one, whose value was around 17%. In a similar way, films mixed with natural antioxidants had high absorbance values, indicating an increase in the reducing powder compared to the control except for the film added with ginger. The obtained antioxidant capacity was mainly determined by the phenolic compounds present in the extract composition. As an example, regarding green tea and grape seed extracts, the amount of epicatechin, caffeic acid and catechin was relatively high. In addition, films mixed with these extracts had greater reducing power compared to the control except for the film with ginger extract. Authors also reported a reduction of around 30% in water vapour permeability (WVP) for gelatin-based films incorporated with green tea extract. This fact can be attributed to the presence of polyphenols which could be able to form hydrogen and covalent bonds with the polar groups of polypeptide in gelatin modifying the structure [45].

Edible gelatin coating solutions enriched with orange leaf essential oil obtained from orange (*Citrus sinensis* (L.) *Osbeck*) leaves were used as a coating for shrimp. The antioxidant activity evaluation of the obtained film forming solutions showed an optimum DPPH scavenging activity around 52% with 2% of essential oil. In addition, in this study, it was demonstrated that the coating improved the quality of shrimp during the storage period in terms of chemical indices determined in shrimp meat, preserving shrimp quality during cold storage with a shelf-life extension of 10 days [32].

The mechanisms of action of these natural antioxidants in contact with food are related to lipid oxidation reactions. In addition, they are focused on phenols and other compounds with hydroxyl groups present in the essential oils composition. Hydrogen atoms from phenol hydroxyl groups could react with peroxyl radicals produced in the early stages of the oxidation mechanisms to yield stable phenoxyl radicals and, consequently, resulting in the termination of the lipid peroxidation chain reactions. However, understanding the antioxidant activity mechanisms of these phenolic compounds is a hard task since this activity depends on the electronic and steric effects of their ring substituents, the strength of hydrogen-bonding interactions between the phenol and the solvent in the essential oil, and the interactions with matrix and food [56].

In order to protect additives from temperature or light, encapsulation is a promising technique that can be used during film processing [58,59]. Liu et al. investigated the applicability of gelatin-based films packaged with sunflower oil with different free/encapsulated tea polyphenol ratios through the synthesis of chitosan nanoparticles at 3 different encapsulation efficiencies (50%, 80% and 100%). The results showed a reduction in the oxidation of sunflower oil obtaining lower peroxide (PV) and thiobarbituric acid reactive substance (TBARS) values for oils exposed to the new films. In addition, an improvement in antioxidant activity when using an optimum partition of free and encapsulated (20:80, respectively) additives was demonstrated over a long period of storage (6 weeks) as well as the preservation of the functional properties of the new films [49].

2.3. Other Agents

As it has been mentioned before, as a consequence of gelatin's highly hygroscopic nature, it tends to swell or dissolve easily in contact with food despite its good barrier properties to oxygen and carbon dioxide [11]. Also, gelatin films show lower mechanical strength compared to synthetic ones [60]. To avoid these drawbacks, gelatin can be blended with different substances and/or polymers to obtain bio-composite films and coatings that combine the advantages of each component [11].

Hydrophobic substances such as lipids and oils have been used to improve the water vapour barrier properties of gelatin films. Limpisophon et al. [61] introduced stearic and oleic fatty acids into edible films based on blue shark skin gelatin by the casting technique. Stearic and oleic acid content in film solution were 25%, 50% and 100% (w/w) of the protein content. As stearic acid content increased, a reduction in WVP from 1.04 ± 0.09 to $0.70 \pm 0.06 \times 10^{-10}$ g·m^{-1}·Pa^{-1}·s^{-1} was reported, which was higher than that obtained for oleic acid at the same concentration (from 1.02 ± 0.06 to $0.91 \pm 0.06 \times 10^{-10}$ g·m^{-1}·Pa^{-1}·s^{-1}). Tongnuanchan et al. recently studied the physical, barrier, structural and thermal properties of fish skin gelatin films containing palm oil at 25%, 50%, 75% and 100% (based on protein) showing a reduction in WVP of 35.83%, 53.54%, 56.30% and 72.52%, respectively [62]. In other study, Bertan et al. evaluated the incorporation of Brazilian elemi oil (1%, 2.5%, 5%, 10%, 15% and 20%, w/w of dry gelatin) into bovine hide type A gelatin to obtain films by the casting technique, using a blend of palmitic and stearic acids (1:1 stearic/palmitic acid, 10%, w/w of dry gelatin) [63]. As a result, the addition of 10% elemi oil reduced WVP by about 57% compared to the film containing only plasticizer and the fatty acid blend. Similarly, Ma et al. developed composite films from bovine hide gelatin type B with olive oil (olive oil/ protein weight ratios of 5%, 10%, 15% and 20%) by the microfluidic emulsification technique [64]. A decrease in WVP from 5.610 ± 0.068 to $4.194 \pm 0.044 \times 10^{-10}$ g·m^{-1}·Pa^{-1}·s^{-1} when 20% of oil was incorporated was obtained. Xiao et al. reported the development of new bio-films by the casting technique based on the addition of palm oil at different degrees (8°, 18°, 24°, 33° and 44°) with a significant reduction in WVP at 36 wt % of gelatin content compared to the control gelatin film [65]. In particular, the lowest WVP value was achieved for the film containing the palm oil with 24° (2.19 ± 0.07 × 10^{-11} g·m^{-1}·s^{-1}·Pa^{-1}) in contrast to the control (1.55 ± 0.05 × 10^{-11} g·m^{-1}·s^{-1}·Pa^{-1}). Finally, Nilsuwan et al. investigated the influence of palm oil concentration (250, 500, 750 and 1000 g·kg^{-1} protein) and soy lecithin surfactant (500 g·kg^{-1} palm oil) on the stability of film-forming dispersion and properties of fish tilapia skin gelatin films obtained by the casting technique [66]. In general, films showed an improvement in WVP properties with increasing palm oil concentration, obtaining values of WVP of 24.52 ± 0.51 for the control and 6.37 ± 0.30 × 10^{-11} g·m^{-1}·s^{-1}·Pa^{-1} for 1000 g·kg^{-1} for the films added with palm oil. As a conclusion, from these studies it could be suggested that oils added into a gelatin matrix could act as hydrophobic and nonpolar substances, increasing hydrophobicity with a decrease in the permeation of moisture through the films [65].

Oils have also been used to enhance the structural and mechanical properties of gelatin. Tongnuanchan et al. developed new bio-films based on fish gelatin obtained from tilapia skin and 25% (w/w) of basil and citronella essential oils at a ratio of 1:1 (w/w) by the casting technique [67]. In general, higher opaqueness, lower T_g and thermal degradation temperatures were reported with essential oils incorporation. Wang et al. evaluated the effect of pH and corn oil addition on the mechanical properties of porcine skin gelatin films [68], obtaining an optimum film-forming solution with 55.18% of corn oil with a pH of 10.54 and a predictive value of tensile strength of 17.58 MPa, elongation at break of 305.90% and WVP of 44.21 g mm kPa^{-1}·d^{-1}·m^{-2}. These results suggest the interruption of protein–protein interactions with an increase in chain mobility in gelatin. A simplified illustration of film matrix interactions with and without oil incorporation after the casting technique is showed in Figure 3. When oil is incorporated into the gelatine matrix, the protein–protein interactions by hydrogen bonds are reduced and a different orientation of the gelatine matrix takes place. Then, two different phases coexist in the matrix, the hydrophobic phase of the oil and the hydrophilic phase, characteristic of the protein which is stabilized by hydrogen and hydrophobic interactions among them.

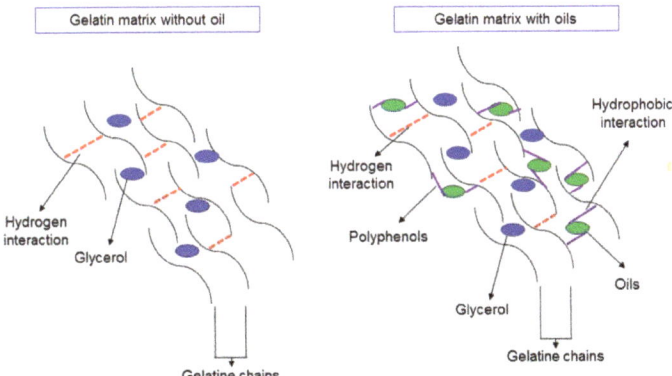

Figure 3. Scheme of gelatin matrix interactions with and without oil incorporation.

Regarding polymer blends, shellac, a special natural polymer obtained from purified resinous secretion of lac insects, *Laccifer Lacca*, has been widely studied as an edible film coating. Puncture strength and percentage elongation of composite films based on shellac (6% w/w) and gelatin type A (10%, 20%, 30%, 40% and 50% w/w into the shellac solution), obtained by the casting technique, increased from 3.61 to 15.58 MPa and from 3.80% to 32.47% as the gelatin concentration increased to 50% w/w, respectively, indicating an enhancement in strength and flexibility of the shellac film [69]. Regarding other blends, sago starch and fish gelatin at different ratios (1:0, 2:1, 3:1, 4:1, and 5:1) plasticized with glycerol or sorbitol (25%, w/w) were developed by the casting technique [70]. In this study, fish gelatin was extracted from fish waste provided by a local surimi processing plant.

By varying the ratio of the two polymers, the strength and extensibility of the composite films can be modified. Starch/gelatin solutions at 4:1 ratio formed good flexible films with tensile strength values of 9.87 ± 0.64 MPa for the control film containing only gelatin matrix compared to 18.06 ± 0.55 MPa for the 4:1 film. Also, elongation at break and Young modulus decreased from $17.11\% \pm 6.11\%$ and 6.17 ± 0.01 N·m$^{-2} \times 10^7$ for the control sample to $5.53\% \pm 0.42\%$ and 1.71 ± 0.05 N·m$^{-2} \times 10^7$ for the 4:1 film, respectively. In addition, Tao et al. studied the effect of pH (3, 7 and 10) on the physical properties of surimi-gelatin composite films at different blending ratios (10:0, 8:2, 6:4, 5:5, 4:6, 2:8 and 0:10) by the casting technique [71]. The composite films of surimi and gelatin could be formed irrespective of pH, and they became stronger under acidic or alkaline conditions. In general, as a higher content of gelatin was used in the blend, higher values of tensile strength and lower elongation at break were obtained.

Other biopolymers have been blended with gelatin with an improvement in gelatin film properties such as chitosan [72,73], lignin [74], lignosulphonates isolated from spent sulphite liquors [75] or fish protein isolate [76].

Nowadays, agricultural by-products are normally incinerated or dumped, causing environmental problems such as air pollution, soil erosion and decreasing soil biological activity [77]. The incorporation of agricultural residues into polymer matrices is currently a trending topic in research due to the relatively high strength, stiffness and low density of natural fibres present in these residues [78–80]. Coconut husk, the fibrous external portion of the fruit of coconut palms, is a by-product of the copra extraction process and is generally considered waste [81]. To revalorize this by-product, the effect of ethanolic extracts from coconut husks (0%–0.4% w/w, on protein basis) on properties of tilapia skin gelatin films obtained by casting were reported [82]. Gelatin film with 0.05% of ethanolic extract from coconut husk showed an improvement in mechanical properties with Young modulus, tensile strength and elongation at break of 1048.03 ± 31.40 MPa, 41.93 ± 0.49 MPa and $7.90\% \pm 0.03\%$ for the control sample compared to 1129.63 ± 25.58 MPa, 43.65 ± 0.68 MPa and $7.63\% \pm 0.01\%$ for the formulation with 0.05%. These positive results could be explained due to higher interactions between functional groups of gelatin and phenolic compounds.

Soy protein isolate is an abundant, inexpensive, biodegradable, and nutritious raw material, whereas microcrystalline cellulose is a commercially available material prepared by acid hydrolysis of wood fiber, back-neutralization with alkali, and spray-drying [83]. The effect of these two compounds in a gelatin matrix was studied after casting preparation of films. In particular, microcrystalline cellulose content of 2.5% in soy protein isolate and gelatin matrix significantly improved mechanical and barrier to water properties with values of Young modulus of 45.32 ± 3.28 MPa for the control to 107.35 ± 6.13 MPa for the formulation at 2.5%. Also, moisture content decreased from $20.28\% \pm 9.07\%$ for control to $16.81\% \pm 8.75\%$ for the 2.5% film.

In recent years, innovative food packaging technologies using biopolymer-based nanocomposites have emerged in response to increasing global waste disposal problems caused by non-biodegradable petroleum-based plastic packaging materials [84]. In this context, chitosan nanoparticles have been used as reinforcement agents in gelatin matrix [85]. The application of chitosan nanoparticles synthesized by the ionic cross-linking method modified the crystalline structure of gelatin mainly due to the nucleating effect of nanoparticles as detected by X-ray diffraction assays. This addition also decreased the T_g of gelatin increasing chains mobility. Finally, the thermal stability of the nanocomposite films increased up to 7 °C for the onset degradation temperature. Other authors reported a remarkable increase in tensile strength caused by the addition of chitosan nanoparticles from 7.44 ± 0.17 MPa for the control sample to 11.28 ± 1.02 MPa for films with a chitosan nanoparticles content of 8% [86]. Also, an increase in elastic modulus with additive content from 287.03 ± 14.25 MPa for control to 467.2 ± 49.63 MPa for the addition of 8% of the additive was reported.

Recent trends in the use of nanoclays as reinforcement agents in gelatin have been reported in the literature, such as montmorillonites [87–89] and laponites [90]. However, there is still some controversy regarding the current legislation about the use of these compounds for food packaging applications despite the fact that the research in this area is becoming more and more relevant.

For food packaging applications, it is necessary to maintain the gelatin structure to guarantee the film stability under humid conditions. Up to now, glutaraldehyde has been used as a crosslinker agent, but it has one great disadvantage, since it is a systemic and cell toxic compound [4]. To avoid the use of this substance, Biscarat et al. developed alternative matrices based on gelatin type A with three different crosslinkers: *N*-hydroxysuccinimide, Bis(succinimidyl)nona(ethylene glycol) and ferulic acid to prepare films by dry-casting [4]. Among them, the use of ferulic acid allowed flexible films to be obtained without using toxic agents. Also, it was shown that gelatin films with ferulic acid supported humid conditions (98% RH at 20 °C) for 15 days without breaking, making them a promising, environmentally friendly packaging system for food applications.

Other agents have been used in gelatin matrices to develop innovative food packaging technologies that include smart materials to extend the safety and quality of food products during their shelf-life. As an example, Musso et al. developed bio-films by casting based on bovine gelatin, glycerol and three acid-base indicators (methyl orange, neutral red and bromocresol green) capable of sense pH changes [91]. Results showed that colour was reversibly modified in all samples when they were put in contact with liquid, semisolid and gaseous media at different pHs, making them an environmentally friendly alternative to replace synthetic indicators.

3. Edible Film and Coating Applications

As has been mentioned in the present review, gelatin has several advantages when used as edible film and coating in different food products. The present section is focused on current applications of gelatin in food packaging reported in recent years (Table 2).

Table 2. Different gelatin matrices used in edible films and coatings for meat and fishery products.

Food Applicability	Product	Matrix	Processing Method	Final Product	Ref.
Meat products	Beef steaks	Bovine gelatin type B mixed with chitosan	Dipping into matrix solution	Coating	[92]
	Pork sausages	Gelatin, pectin and sodium alginate blends	Extrusion	Film	[93]
	Pork loin	Porcine gelatin	Dipping into gelatin matrix	Coating	[94]
	Kabanosy dry sausages	Pork gelatin, kappa-carrageenan and glycerol	Dipping into matrix solution	Coating	[95]
	Chicken tenderloin	Skate skin gelatin with thyme essential oil	Casting	Film	[96]
	Raw beef	Gelatin, Tween 80 and essential oils of *Thymus vulgaris* and *Rosmarinus officinalis*	Dipping into matrix solution	Coating	[97]
	Turkey bologna	Gelatin, glycerol and Nisaplin and Guardian CS1-50 antimicrobial additives	Casting	Film	[98]
	Bacon	Gelatin	Casting	Film	[99]
Fishery products	Rainbow trout	Cold water fish skin, chitosan and glycerol	Casting and dipping into matrix solution	Film and coating	[100]
	Cod fillets	Bovine hide gelatin, chitosan, sorbitol and glycerol with clove essential oil	Casting	Film	[99]
	Minced trout fillets	Cold water fish skin gelatin, chitosan, glycerol, red grape seed extract and *Ziziphora linopodioides* essential oil	Casting	Film	[101]
	Rainbow trout fillets	Food grade gelatin, glycerol, sorbitol, Tween 20 and laurel leaf essential oil	Casting	Film	[102]
	Tuna meat	Gelatin, red pepper seed meal protein and several plasticizers (glycerol, sorbitol, fructose and sucrose)	Casting	Film	[96]
	Fish sausages	Warm-water fish gelatin, chitosan, shrimp concentrate, Tween 80 and glycerol	Casting	Film	[19]
	Atlantic Salmon	Warm-water fish gelatin, lignin, sorbitol and glycerol	Casting	Film	[103]
	Salmon	Porcine skin gelatin, barley bran protein, sorbitol and grapefruit seed extract	Casting	Film	[104]
	Cold smoked Salmon	Pork gelatin, chicken feather protein, sorbitol and clove oil	Casting	Film	[105]
	Shrimps	Gelatin, glycerol, sorbitol, Tween 20 and orange leaf extract	Dipping into matrix solution	Coating	[32]

3.1. Meat Products

Regarding meat products, gelatin has been used blended with chitosan as a coating to reduce colour deterioration from red to brown as a consequence of a gradual accumulation of metmyoglobin in the meat's surface, mainly due to oxygen exposition and lipid oxidation of beef steaks [92]. The gelatin-chitosan coating successfully maintained the organoleptic properties of beef steaks during 5 days of retail display increasing their shelf-life. In other work, the potential use of blend films from pectin, gelatin and sodium alginate for breakfast pork sausages was reported [93].

According to Davis and Lin [106], approximately 50% of the worldwide daily protein intake is from pork. The application of a gelatin coating (0%, 10% and 20%) led to an improvement in preserving the quality and shelf-life of refrigerated pork meat during a storage period of 7 days. No significant colour changes due to the retard of metamyoglobin formation and lipid oxidation were observed, underlying the potential of gelatin as a bio-based material to be used as a coating to extend the shelf-life of meat products. Similarly, Tyburcy and Kozyra studied the effect of coating dry sausages with pork gelatin, kappa-carrageenan and glycerol as an alternative to vacuum packaging to avoid weight loss which was directly related to profit loss [95]. As a result, coating meat reduced its weight loss and, therefore, financial benefits could be achieved by the application of this type of packaging. However, according to the authors, more studies are necessary to reduce coating thickness.

Gelatin extracted from natural sources such as skate skin was used with thyme essential oil to prepare antimicrobial edible films for chicken tenderloin packaging [96]. The film containing 1% thyme oil reduced the population of *L. monocytogenes* and *E. coli* on chicken tenderloin during storage. The contamination of meat products with *L. monocytogenes* has been considered a serious public health problem [107]. Oliveira et al. studied the antimicrobial effect of the addition of *Thymus vulgaris* and *Rosmarinus officinalis* essential oils to gelatin solution in raw bovine meat pieces [97]. The effectiveness and viability of this coating were proven with a reduction in *L. monocytogenes* proliferation accompanied by acceptable sensory properties of the packaged meat. The use of antimicrobial coatings based on gelatin to reduce *L. monocytogenes* growth was also reported in other ready-to-eat poultry meats such as turkey bologna [98]. In this case, two commercial antimicrobial agents were added (Nisaplin and Guardian CS1-50) into gelatin to obtain films using glycerol as a plasticizer by the casting technique. The incorporation of the antimicrobial additives reduced the tensile strength of films whereas increased the elongation at break. Despite these structural changes, active films effectively inhibited *L. monocytogenes* on bologna at 4 °C up to 8 weeks.

The development of edible films and coatings and their applications on meat food products have been subject of a great number of scientific publications during the last decade, but several patents have also been commercialized. As an example, the development of gelatin and carboxymethylcellulose films with potassium sorbate were proven to be effective in extending the shelf-life of bacon (CN 102487988B patent) [108].

3.2. Fishery Products

Fish is one of the most perishable food products mainly due to chemical reactions, enzymatic response, and microbial spoilage [109]. As a consequence of its reduced shelf-life, the freshness and quality of fish have always gained the attention of Food Regulatory Agencies and Food Processing Industries. Proper handling, pre-treatment and preservation techniques can improve the quality of fish products. Much research in this field has been focused on the development of edible films and coatings to increase fish products shelf-life maintaining their quality parameters. Coating and films based on cold water fish-skin gelatin and chitosan blends were reported for rainbow trout fillet packaging, showing antioxidant properties. However, higher protective effect against lipid oxidation was obtained for coatings compared to films, due to higher chitosan migration in solution as an active additive [99]. Gelatin and chitosan blends were also used to obtain antimicrobial films for cod fillet packaging by adding clove essential oil to the matrix, resulting in a drastic reduction in microorganism growth for gram-negative bacteria [100]. In a different study, the development of chitosan-cold water fish skin gelatin films incorporated with grape seed extract (1% and 2%) and *Ziziphora clinopodioides* essential oil (1% and 2%), separately and in combination, led to a decrease of *L. monocytogenes* and shelf-life extension of minced trout fillets during refrigerated storage at 4 °C over a period of 11 days [101]. Also, trout fillets wrapped with 8% gelatin films containing laurel essential oil (0%, 0.1% and 1%, *w/w*) and vacuum packaged were evaluated to study the quality of fish during refrigerated storage at 4 °C over a period of 26 days [102].

In order to revalorize a form of agricultural food waste, fatty tuna meat was packaged into blend films of gelatin and red pepper seed meal isolated protein by the casting technique [96]. The results showed the potential of this material as an antimicrobial and antioxidant packaging, being the optimal formulation that containing 1% of gelatin and 4% of red pepper seed protein. As a result, tensile strength and elongation at break were improved and *L. monocytogenes* and *Salmonella Typhimurium* growths were reduced on tuna meat compared to the control. In a different work, the effect of a formulation obtained from a shrimp concentrate waste from the seafood industry, as a coating and film, on the shelf-life and characteristics of fish sausages was studied [19]. An extension of the shelf-life of fish sausages to 15 days was observed with harder texture, lower pH and greater microbiological control.

Regarding oxidative and organoleptic degradation, salmon is probably the most studied fish since it is one of the most sensitive food products [110]. In fact, various studies have applied gelatin in combination with other biopolymers or active additives such as lignin [103], and barley bran protein and grapefruit seed extract [104] to protect salmon against cooking processes. In addition, since cold-smoked salmon is generally consumed without cooking, it can cause serious health problems in consumers due to contamination with pathogenic bacteria, mainly *E. coli* and *L. monocytogenes* [111]. To avoid this problem, chicken feathers, a by-product of the poultry industry, were successfully used as a film base material after extraction of chicken feather protein in combination of gelatin and clove oil as an antioxidant and antimicrobial active agent to package smoked salmon [105].

3.3. Other Food Products

Aside from meat and fishery products, other food products are also susceptible to be coated or packaged into gelatin-based solutions or films. Potential applications of gelatin edible films in the food industry may include the transport of gases (O_2 and CO_2), water vapour, and flavours for fruits and vegetables [112]. As an example, refrigerated Red Crimson grapes were coated with gelatin type A, starch and glicerol films obtained by the casting technique [2]. As a result, an increase in gelatin concentration in the mixture provided an increase in thickness, WVP and mechanical resistance reducing the total weight loss without influencing consumers acceptance. Also, the incorporation of red bean powder as colorant and flavouring agent into gelatin films was studied for use in candies and brewing food as it was reported in CN 103589173A patent [113]. In another patent, gelatin and glucomannan films with garlic juice were described as antimicrobial and favouring agents (CN 103589168A patent) [114]. Other vegetables and fruits recently reported to be coated or wrapped with gelatin-based films and coatings are carrots [115], cherry tomatoes [116], calyx from physalis [117], oranges [118], banana and eggplant epicarps [119], fresh-cut melons [120], peppers [121], strawberries [122], blueberry fruit [123], pineapple fruit [124] and minimally processed persimmon [125].

Gelatin films prepared from cold-water fish show significantly lower WVP values than those from warm-water fish, due to their higher hydrophobicity directly related to lower amounts of two aminoacids, proline and hydroxyproline. As a consequence, cold-water fish gelatin films are particularly useful for applications related to reducing water loss from refrigerated or frozen food systems [126].

Residues generated by fruit and vegetable processing are well-studied sources of antioxidants, bio-polymers and dietary fibres [127]. Indeed, large amounts of oil nutraceutical capsule waste from coconut, chia, safflower and linseed, composed mainly of gelatin, are being generated with high waste treatment costs for industry. A reduction in the oxidative rancidity of sunflower oil exposed to films developed from chia oil nutraceutical capsule wastes by the casting technique after storage at 40 °C in the presence of light for 13 days was obtained [128]. Gelatin was extracted from capsule wastes composed of gelatin (48%), water (30%) and glycerol (22%). Blueberry pomace fibre and extract wastes were used as active additives. All gelatin films retarded oil lipid oxidation during the studied period. However, gelatin films added with fibre and extract significantly reduced lipid oxidation compared to the control film without antioxidants addition after 8 days of the oxidative treatment. Similarly, the effect of films based on beet root residue powder, obtained from peels, stalks, and shavings wastes derived from the production of linseed oil nutraceutical capsules, on the retardation of sunflower oil oxidation was recently reported [47]. Sunflower oil containing no artificial antioxidants was stored for 35 days at 35 °C and 54% RH and exposed to fluorescent light with an intensity of 900–1000 lux. Peroxide values were determined at different times. Films with antioxidants had a positive effect on the stability of sunflower oil during the entire storage period and, also, at the end of the experiment, presenting the packed oil peroxide values under the recommended limit of Codex Alimentarius (10 milliequivalent per oil kilogram). As a result, biodegradable films based on residues of beet root and gelatin capsules could be a potential tool to control and retard rancidity of different oils.

4. Conclusions

The use of gelatin-based edible films and coatings represents a stimulating route for creating new food packaging materials. Due to the hygroscopic properties of gelatin, some research studies have been conducted to evaluate the overall effect of the addition of different substances such as crosslinkers, strengthening agents, plasticizers or additives with antimicrobial or antioxidant properties to gelatin-based products to improve the functional properties of gelatin-based edible films and the shelf-life of food products. An increasing number of publications have reported the development of gelatin-based films for meat applications as coatings to reduce the colour deterioration from red to brown as a consequence of lipid oxidation. Regarding fish products, different studies have been focused on the application of gelatin in combination with other biopolymers or active additives to protect fresh fish against cooking processes and microbial/oxidation deterioration. In addition to fish and meat, some other food products such as fruits and vegetables can be coated with gelatin-based films in order to retard degradation processes due to the transport of gases (O_2 and CO_2) and water vapour. Extensive research is still needed on new methods for gelatin-based film formation to improve the final properties and potential applications.

Acknowledgments: Authors would like to thank Spanish Ministry of Economy and Competitiveness for financial support (MAT-2015-59242-C2-2-R).

Author Contributions: Marina Ramos and Ana Beltrán contributed to the definition of the review structure based on available literature survey, to writing, to integration of the different parts and to conclusions. Marina Ramos was also the editor of the manuscript. Arantzazu Valdés and María Carmen Garrigós contributed to writing and formatting. María Carmen Garrigós was also the main reviewer of content.

Conflicts of Interest: The authors declare no conflict of interest.

References

1. Malhotra, B.; Keshwani, A.; Kharkwal, H. Natural polymer based cling films for food packaging. *Int. J. Pharm. Pharm. Sci.* **2015**, *7*, 10–18.
2. Fakhouri, F.M.; Martelli, S.M.; Caon, T.; Velasco, J.I.; Mei, L.H.I. Edible films and coatings based on starch/gelatin: Film properties and effect of coatings on quality of refrigerated red crimson grapes. *Postharvest Biol. Technol.* **2015**, *109*, 57–64. [CrossRef]
3. Plackett, D. *Biopolymers—New Materials for Sustainable Films and Coatings*; John Wiley & Sons, Ltd.: Hoboken, NJ, USA, 2011.
4. Biscarat, J.; Charmette, C.; Sanchez, J.; Pochat-Bohatier, C. Development of a new family of food packaging bioplastics from cross-linked gelatin based films. *Can. J. Chem. Eng.* **2015**, *93*, 176–182. [CrossRef]

5. Arrieta, M.P.; Peltzer, M.A.; López, J.; Garrigós, M.D.C.; Valente, A.J.M.; Jiménez, A. Functional properties of sodium and calcium caseinate antimicrobial active films containing carvacrol. *J. Food Eng.* **2014**, *121*, 94–101. [CrossRef]
6. Ghanbarzadeh, B.; Almasi, H. Biodegradable polymers. In *Biodegradation—Life of Science*; Chamy, R., Rosenkranz, F., Eds.; InTech: Rijeka, Croatia, 2013.
7. Pérez-Gago, M.B.; Rhim, J.W. Edible coating and film materials: Lipid bilayers and lipid emulsions. In *Innovations in Food Packaging*, 2nd ed.; Elsevier: Amsterdam, The Netherlands, 2013; pp. 325–350.
8. Shankar, S.; Jaiswal, L.; Rhim, J.W. Gelatin-based nanocomposite films: Potential use in antimicrobial active packaging. In *Antimicrobial Food Packaging*; Elsevier: Amsterdam, The Netherlands, 2016; pp. 339–348.
9. Gómez-Guillén, M.C.; Giménez, B.; López-Caballero, M.E.; Montero, M.P. Functional and bioactive properties of collagen and gelatin from alternative sources: A review. *Food Hydrocoll.* **2011**, *25*, 1813–1827. [CrossRef]
10. Alfaro, A.T.; Balbinot, E.; Weber, C.I.; Tonial, I.B.; Machado-Lunkes, A. Fish gelatin: Characteristics, functional properties, applications and future potentials. *Food Eng. Rev.* **2014**, *7*, 33–44. [CrossRef]
11. Nur Hanani, Z.A.; Roos, Y.H.; Kerry, J.P. Use and application of gelatin as potential biodegradable packaging materials for food products. *Int. J. Biol. Macromol.* **2014**, *71*, 94–102. [CrossRef] [PubMed]
12. Gareis, H.; Schrieber, R. *Gelatine Handbook: Theory and Industrial Practice*; Wiley-VCH Verlag GmbH and Co. KGaA: Weinheim, Germany, 2007.
13. Galus, S.; Kadzińska, J. Food applications of emulsion-based edible films and coatings. *Trends Food Sci. Technol.* **2015**, *45*, 273–283. [CrossRef]
14. Mellinas, C.; Valdés, A.; Ramos, M.; Burgos, N.; Del Carmen Garrigós, M.; Jiménez, A. Active edible films: Current state and future trends. *J. Appl. Polym. Sci.* **2016**, *133*. [CrossRef]
15. Ortiz-Zarama, M.A.; Jiménez-Aparicio, A.R.; Solorza-Feria, J. Obtainment and partial characterization of biodegradable gelatin films with tannic acid, bentonite and glycerol. *J. Sci. Food Agric.* **2016**, *96*, 3424–3431. [CrossRef] [PubMed]
16. Zhao, Y.; Li, Z.; Yang, W.; Xue, C.; Wang, Y.; Dong, J.; Xue, Y. Modification of gelatine with galla chinensis extract, a natural crosslinker. *Int. J. Food Prop.* **2016**, *19*, 731–744. [CrossRef]
17. Taylor, M.M.; Lee, J.; Bumanlag, L.P.; Latona, R.J.; Brown, E.M. Biopolymers produced from gelatin and whey protein concentrate using polyphenols. *J. Am. Leather Chem. Assoc.* **2014**, *109*, 82–88.
18. Benbettaïeb, N.; Chambin, O.; Assifaoui, A.; Al-Assaf, S.; Karbowiak, T.; Debeaufort, F. Release of coumarin incorporated into chitosan-gelatin irradiated films. *Food Hydrocoll.* **2016**, *56*, 266–276. [CrossRef]
19. Alemán, A.; González, F.; Arancibia, M.Y.; López-Caballero, M.E.; Montero, P.; Gómez-Guillén, M.C. Comparative study between film and coating packaging based on shrimp concentrate obtained from marine industrial waste for fish sausage preservation. *Food Control* **2016**, *70*, 325–332. [CrossRef]
20. Benbettaïeb, N.; Chambin, O.; Karbowiak, T.; Debeaufort, F. Release behavior of quercetin from chitosan-fish gelatin edible films influenced by electron beam irradiation. *Food Control* **2016**, *66*, 315–319. [CrossRef]
21. Gupta, B.; Tummalapalli, M.; Deopura, B.L.; Alam, M.S. Preparation and characterization of in-situ crosslinked pectin-gelatin hydrogels. *Carbohydr. Polym.* **2014**, *106*, 312–318. [CrossRef] [PubMed]
22. Muyonga, J.H.; Cole, C.G.B.; Duodu, K.G. Extraction and physico-chemical characterisation of nile perch (lates niloticus) skin and bone gelatin. *Food Hydrocoll.* **2004**, *18*, 581–592. [CrossRef]
23. Nur Hanani, Z.A.; Roos, Y.H.; Kerry, J.P. Use of beef, pork and fish gelatin sources in the manufacture of films and assessment of their composition and mechanical properties. *Food Hydrocoll.* **2012**, *29*, 144–151. [CrossRef]
24. Jeya Shakila, R.; Jeevithan, E.; Varatharajakumar, A.; Jeyasekaran, G.; Sukumar, D. Comparison of the properties of multi-composite fish gelatin films with that of mammalian gelatin films. *Food Chem.* **2012**, *135*, 2260–2267. [CrossRef] [PubMed]
25. Valdés, A.; Mellinas, A.C.; Ramos, M.; Garrigós, M.C.; Jiménez, A. Natural additives and agricultural wastes in biopolymer formulations for food packaging. *Front. Chem.* **2014**, *2*, 1–10. [CrossRef] [PubMed]
26. Valdés, A.; Ramos, M.; García-Serna, E.; Carmen Garrigós, M.D.; Jiménez, A. Polymers extracted from biomass. In *Reference Module in Food Science*; Elsevier: Amsterdam, The Netherlands, 2016.
27. Iahnke, A.O.S.; Costa, T.M.H.; Rios, A.O.; Flôres, S.H. Residues of minimally processed carrot and gelatin capsules: Potential materials for packaging films. *Ind. Crops Prod.* **2015**, *76*, 1071–1078. [CrossRef]
28. Valdes, A.; Mellinas, A.C.; Ramos, M.; Burgos, N.; Jimenez, A.; Garrigos, M.C. Use of herbs, spices and their bioactive compounds in active food packaging. *RSC Adv.* **2015**, *5*, 40324–40335. [CrossRef]

29. Atarés, L.; Chiralt, A. Essential oils as additives in biodegradable films and coatings for active food packaging. *Trends Food Sci. Technol.* **2016**, *48*, 51–62. [CrossRef]
30. Sung, S.Y.; Sin, L.T.; Tee, T.T.; Bee, S.T.; Rahmat, A.R.; Rahman, W.A.W.A.; Tan, A.C.; Vikhraman, M. Antimicrobial agents for food packaging applications. *Trends Food Sci. Technol.* **2013**, *33*, 110–123. [CrossRef]
31. Martucci, J.F.; Gende, L.B.; Neira, L.M.; Ruseckaite, R.A. Oregano and lavender essential oils as antioxidant and antimicrobial additives of biogenic gelatin films. *Ind. Crops Prod.* **2015**, *71*, 205–213. [CrossRef]
32. Alparslan, Y.; Yapıcı, H.H.; Metin, C.; Baygar, T.; Günlü, A.; Baygar, T. Quality assessment of shrimps preserved with orange leaf essential oil incorporated gelatin. *LWT Food Sci. Technol.* **2016**, *72*, 457–466. [CrossRef]
33. Yanwong, S.; Threepopnatkul, P. Effect of Peppermint and citronella essential oils on properties of fish skin gelatin edible films. *IOP Conf. Ser. Mater. Sci. Eng.* **2015**, *87*, 012064. [CrossRef]
34. Calo, J.R.; Crandall, P.G.; O'Bryan, C.A.; Ricke, S.C. Essential oils as antimicrobials in food systems—A review. *Food Control* **2015**, *54*, 111–119. [CrossRef]
35. Bastarrachea, L.; Dhawan, S.; Sablani, S. Engineering properties of polymeric-based antimicrobial films for food packaging: A review. *Food Eng. Rev.* **2011**, *3*, 79–93. [CrossRef]
36. Wu, J.; Liu, H.; Ge, S.; Wang, S.; Qin, Z.; Chen, L.; Zheng, Q.; Liu, Q.; Zhang, Q. The preparation, characterization, antimicrobial stability and in vitro release evaluation of fish gelatin films incorporated with cinnamon essential oil nanoliposomes. *Food Hydrocoll.* **2015**, *43*, 427–435. [CrossRef]
37. Hosseini, S.F.; Rezaei, M.; Zandi, M.; Farahmandghavi, F. Bio-based composite edible films containing *Origanum vulgare* L. essential oil. *Ind. Crops Prod.* **2015**, *67*, 403–413. [CrossRef]
38. Hosseini, S.F.; Rezaei, M.; Zandi, M.; Farahmandghavi, F. Development of bioactive fish gelatin/chitosan nanoparticles composite films with antimicrobial properties. *Food Chem.* **2016**, *194*, 1266–1274. [CrossRef] [PubMed]
39. Ibarguren, C.; Céliz, G.; Díaz, A.S.; Bertuzzi, M.A.; Daz, M.; Audisio, M.C. Gelatine based films added with bacteriocins and a flavonoid ester active against food-borne pathogens. *Innov. Food Sci. Emerg. Technol.* **2015**, *28*, 66–72. [CrossRef]
40. Kanmani, P.; Rhim, J.W. Physicochemical properties of gelatin/silver nanoparticle antimicrobial composite films. *Food Chem.* **2014**, *148*, 162–169. [CrossRef] [PubMed]
41. Kanmani, P.; Rhim, J.W. Physical, mechanical and antimicrobial properties of gelatin based active nanocomposite films containing agnps and nanoclay. *Food Hydrocoll.* **2014**, *35*, 644–652. [CrossRef]
42. Shankar, S.; Teng, X.; Li, G.; Rhim, J.W. Preparation, characterization, and antimicrobial activity of gelatin/zno nanocomposite films. *Food Hydrocoll.* **2015**, *45*, 264–271. [CrossRef]
43. Lee, K.Y.; Lee, J.H.; Yang, H.J.; Song, K.B. Production and characterisation of skate skin gelatin films incorporated with thyme essential oil and their application in chicken tenderloin packaging. *Int. J. Food Sci. Technol.* **2016**, *51*, 1465–1472. [CrossRef]
44. Shakila, R.J.; Jeevithan, E.; Arumugam, V.; Jeyasekaran, G. Suitability of antimicrobial grouper bone gelatin films as edible coatings for vacuum-packaged fish steaks. *J. Aquat. Food Prod. Technol.* **2016**, *25*, 724–734. [CrossRef]
45. Li, J.-H.; Miao, J.; Wu, J.-L.; Chen, S.-F.; Zhang, Q.-Q. Preparation and characterization of active gelatin-based films incorporated with natural antioxidants. *Food Hydrocoll.* **2014**, *37*, 166–173. [CrossRef]
46. Kadam, S.U.; Pankaj, S.K.; Tiwari, B.K.; Cullen, P.J.; O'Donnell, C.P. Development of biopolymer-based gelatin and casein films incorporating brown seaweed ascophyllum nodosum extract. *Food Packag. Shelf Life* **2015**, *6*, 68–74. [CrossRef]
47. Iahnke, A.O.E.S.; Costa, T.M.H.; de Oliveira Rios, A.; Flôres, S.H. Antioxidant films based on gelatin capsules and minimally processed beet root (*Beta vulgaris* L. Var. Conditiva) residues. *J. Appl. Polym. Sci.* **2016**, *133*. [CrossRef]
48. Kowalczyk, D.; Biendl, M. Physicochemical and antioxidant properties of biopolymer/candelilla wax emulsion films containing hop extract—A comparative study. *Food Hydrocoll.* **2016**, *60*, 384–392. [CrossRef]
49. Liu, F.; Antoniou, J.; Li, Y.; Yi, J.; Yokoyama, W.; Ma, J.; Zhong, F. Preparation of gelatin films incorporated with tea polyphenol nanoparticles for enhancing controlled-release antioxidant properties. *J. Agric. Food Chem.* **2015**, *63*, 3987–3995. [CrossRef] [PubMed]
50. Wang, Y.; Liu, A.; Ye, R.; Li, X.; Han, Y.; Liu, C. The production of gelatin-calcium carbonate composite films with different antioxidants. *Int. J. Food Prop.* **2015**, *18*, 2442–2456. [CrossRef]

51. Peña-Rodriguez, C.; Martucci, J.F.; Neira, L.M.; Arbelaiz, A.; Eceiza, A.; Ruseckaite, R.A. Functional properties and in vitro antioxidant and antibacterial effectiveness of pigskin gelatin films incorporated with hydrolysable chestnut tannin. *Food Sci. Technol. Int.* **2015**, *21*, 221–231. [CrossRef] [PubMed]
52. Clarke, D.; Molinaro, S.; Tyuftin, A.; Bolton, D.; Fanning, S.; Kerry, J.P. Incorporation of commercially-derived antimicrobials into gelatin-based films and assessment of their antimicrobial activity and impact on physical film properties. *Food Control* **2016**, *64*, 202–211. [CrossRef]
53. Llorens, A.; Lloret, E.; Picouet, P.A.; Trbojevich, R.; Fernandez, A. Metallic-based micro and nanocomposites in food contact materials and active food packaging. *Trends Food Sci. Technol.* **2012**, *24*, 19–29. [CrossRef]
54. Reidy, B.; Haase, A.; Luch, A.; Dawson, K.; Lynch, I. Mechanisms of silver nanoparticle release, transformation and toxicity: A critical review of current knowledge and recommendations for future studies and applications. *Materials* **2013**, *6*, 2295–2350. [CrossRef]
55. Ramos, M.; Jiménez, A.; Garrigós, M.C. Chapter 26 Carvacrol-based films: Usage and potential in antimicrobial packaging. In *Antimicrobial Food Packaging*; Academic Press: San Diego, CA, USA, 2016; pp. 329–338.
56. Amorati, R.; Foti, M.C.; Valgimigli, L. Antioxidant activity of essential oils. *J. Agric. Food Chem.* **2013**, *61*, 10835–10847. [CrossRef] [PubMed]
57. Srinivasan, K. Antioxidant potential of spices and their active constituents. *Crit. Rev. Food Sci. Nutr.* **2012**, *54*, 352–372. [CrossRef] [PubMed]
58. Marques, H.M.C. A review on cyclodextrin encapsulation of essential oils and volatiles. *Flavour Fragr. J.* **2010**, *25*, 313–326. [CrossRef]
59. Ngamakeue, N.; Chitprasert, P. Encapsulation of holy basil essential oil in gelatin: Effects of palmitic acid in carboxymethyl cellulose emulsion coating on antioxidant and antimicrobial activities. *Food Bioprocess Technol.* **2016**, *9*, 1735–1745. [CrossRef]
60. Cuq, B.; Gontard, N.; Guilbert, S. Proteins as agricultural polymers for packaging production. *Cereal Chem. J.* **1998**, *75*, 1–9. [CrossRef]
61. Limpisophon, K.; Tanaka, M.; Osako, K. Characterisation of gelatin-fatty acid emulsion films based on blue shark (prionace glauca) skin gelatin. *Food Chem.* **2010**, *122*, 1095–1101. [CrossRef]
62. Tongnuanchan, P.; Benjakul, S.; Prodpran, T.; Nilsuwan, K. Emulsion film based on fish skin gelatin and palm oil: Physical, structural and thermal properties. *Food Hydrocoll.* **2015**, *48*, 248–259. [CrossRef]
63. Bertan, L.C.; Tanada-Palmu, P.S.; Siani, A.C.; Grosso, C.R.F. Effect of fatty acids and "brazilian elemi" on composite films based on gelatin. *Food Hydrocoll.* **2005**, *19*, 73–82. [CrossRef]
64. Ma, W.; Tang, C.-H.; Yin, S.-W.; Yang, X.-Q.; Wang, Q.; Liu, F.; Wei, Z.-H. Characterization of gelatin-based edible films incorporated with olive oil. *Food Res. Int.* **2012**, *49*, 572–579. [CrossRef]
65. Xiao, J.; Wang, W.; Wang, K.; Liu, Y.; Liu, A.; Zhang, S.; Zhao, Y. Impact of melting point of palm oil on mechanical and water barrier properties of gelatin-palm oil emulsion film. *Food Hydrocoll.* **2016**, *60*, 243–251. [CrossRef]
66. Nilsuwan, K.; Benjakul, S.; Prodpran, T. Emulsion stability and properties of fish gelatin-based films as affected by palm oil and surfactants. *J. Sci. Food Agric.* **2016**, *96*, 2504–2513. [CrossRef] [PubMed]
67. Tongnuanchan, P.; Benjakul, S.; Prodpran, T. Structural, morphological and thermal behaviour characterisations of fish gelatin film incorporated with basil and citronella essential oils as affected by surfactants. *Food Hydrocoll.* **2014**, *41*, 33–43. [CrossRef]
68. Wang, L.; Auty, M.A.E.; Rau, A.; Kerry, J.F.; Kerry, J.P. Effect of ph and addition of corn oil on the properties of gelatin-based biopolymer films. *J. Food Eng.* **2009**, *90*, 11–19. [CrossRef]
69. Soradech, S.; Nunthanid, J.; Limmatvapirat, S.; Luangtana-anan, M. An approach for the enhancement of the mechanical properties and film coating efficiency of shellac by the formation of composite films based on shellac and gelatin. *J. Food Eng.* **2012**, *108*, 94–102. [CrossRef]
70. Al-Hassan, A.A.; Norziah, M.H. Starch-gelatin edible films: Water vapor permeability and mechanical properties as affected by plasticizers. *Food Hydrocoll.* **2012**, *26*, 108–117. [CrossRef]
71. Tao, Z.; Weng, W.-Y.; Cao, M.-J.; Liu, G.-M.; Su, W.-J.; Osako, K.; Tanaka, M. Effect of blend ratio and pH on the physical properties of edible composite films prepared from silver carp surimi and skin gelatin. *J. Food Sci. Technol.* **2015**, *52*, 1618–1625. [CrossRef] [PubMed]
72. Liu, Z.; Ge, X.; Lu, Y.; Dong, S.; Zhao, Y.; Zeng, M. Effects of chitosan molecular weight and degree of deacetylation on the properties of gelatine-based films. *Food Hydrocoll.* **2012**, *26*, 311–317. [CrossRef]

73. Gómez-Estaca, J.; Gómez-Guillén, M.C.; Fernández-Martín, F.; Montero, P. Effects of gelatin origin, bovine-hide and tuna-skin, on the properties of compound gelatin–chitosan films. *Food Hydrocoll.* **2011**, *25*, 1461–1469. [CrossRef]
74. Núñez-Flores, R.; Giménez, B.; Fernández-Martín, F.; López-Caballero, M.E.; Montero, M.P.; Gómez-Guillén, M.C. Physical and functional characterization of active fish gelatin films incorporated with lignin. *Food Hydrocoll.* **2013**, *30*, 163–172. [CrossRef]
75. Núñez-Flores, R.; Giménez, B.; Fernández-Martín, F.; López-Caballero, M.E.; Montero, M.P.; Gómez-Guillén, M.C. Role of lignosulphonate in properties of fish gelatin films. *Food Hydrocoll.* **2012**, *27*, 60–71. [CrossRef]
76. Arfat, Y.A.; Benjakul, S.; Prodpran, T.; Osako, K. Development and characterisation of blend films based on fish protein isolate and fish skin gelatin. *Food Hydrocoll.* **2014**, *39*, 58–67. [CrossRef]
77. Pirayesh, H.; Khanjanzadeh, H.; Salari, A. Effect of using walnut/almond shells on the physical, mechanical properties and formaldehyde emission of particleboard. *Compos. Part B* **2013**, *45*, 858–863. [CrossRef]
78. Valdés, A.; Fenollar, O.; Beltrán, A.; Balart, R.; Fortunati, E.; Kenny, J.M.; Garrigós, M.C. Characterization and enzymatic degradation study of poly(ε-caprolactone)-based biocomposites from almond agricultural by-products. *Polym. Degrad. Stab.* **2016**, *132*, 181–190. [CrossRef]
79. Valdés García, A.; Ramos Santonja, M.; Sanahuja, A.B.; Del Carmen Garrigós Selva, M. Characterization and degradation characteristics of poly(ε-caprolactone)-based composites reinforced with almond skin residues. *Polym. Degrad. Stab.* **2014**, *108*, 269–279. [CrossRef]
80. Valdés, A.; Beltrán, A.; Garrigós, M.C. Potential Use of Nut Agricultural by-Products in Polymer Materials: A Review. In *Agricultural Wastes: Characteristics, Types and Management*; Nova Science Publishers, Inc.: New York, NY, USA, 2015; pp. 87–106.
81. Vázquez, H.; Canché-Escamilla, G.; Cruz-Ramos, C.A. Coconut husk lignin. I extraction and characterisation. *J. Appl. Polym. Sci.* **1992**, *45*, 633–644.
82. Nagarajan, M.; Benjakul, S.; Prodpran, T.; Songtipya, P. Properties and characteristics of nanocomposite films from tilapia skin gelatin incorporated with ethanolic extract from coconut husk. *J. Food Sci. Technol.* **2015**, *52*, 7669–7682. [CrossRef] [PubMed]
83. Moon, R.J.; Martini, A.; Nairn, J.; Simonsen, J.; Youngblood, J. Cellulose nanomaterials review: Structure, properties and nanocomposites. *Chem. Soc. Rev.* **2011**, *40*, 3941–3994. [CrossRef] [PubMed]
84. Rhim, J.-W.; Kim, Y.-T. Biopolymer-based composite packaging materials with nanoparticles. In *Innovations in Food Packaging*; Elsevier Academic Press: London, UK, 2014.
85. Hosseini, S.F.; Rezaei, M.; Zandi, M.; Farahmandghavi, F. Preparation and characterization of chitosan nanoparticles-loaded fish gelatin-based edible films. *J. Food Proc. Eng.* **2015**, *39*, 521–530. [CrossRef]
86. Hosseini, S.F.; Rezaei, M.; Zandi, M.; Farahmandghavi, F. Fabrication of bio-nanocomposite films based on fish gelatin reinforced with chitosan nanoparticles. *Food Hydrocoll.* **2015**, *44*, 172–182. [CrossRef]
87. Panzavolta, S.; Gioffrè, M.; Bracci, B.; Rubini, K.; Bigi, A. Montmorillonite reinforced type a gelatin nanocomposites. *J. Appl. Polym. Sci.* **2014**, *131*. [CrossRef]
88. Coronado Jorge, M.F.; Alexandre, E.M.C.; Caicedo Flaker, C.H.; Bittante, A.M.Q.B.; Sobral, P.J.D.A. Biodegradable films based on gelatin and montmorillonite produced by spreading. *Int. J. Polym. Sci.* **2015**, *2015*, 806791. [CrossRef]
89. Ge, L.; Li, X.; Zhang, R.; Yang, T.; Ye, X.; Li, D.; Mu, C. Development and characterization of dialdehyde xanthan gum crosslinked gelatin based edible films incorporated with amino-functionalized montmorillonite. *Food Hydrocoll.* **2015**, *51*, 129–135. [CrossRef]
90. Li, X.; Liu, A.; Ye, R.; Wang, Y.; Wang, W. Fabrication of gelatin-laponite composite films: Effect of the concentration of laponite on physical properties and the freshness of meat during storage. *Food Hydrocoll.* **2015**, *44*, 390–398. [CrossRef]
91. Musso, Y.S.; Salgado, P.R.; Mauri, A.N. Gelatin based films capable of modifying its color against environmental ph changes. *Food Hydrocoll.* **2016**, *61*, 523–530. [CrossRef]
92. Cardoso, G.P.; Dutra, M.P.; Fontes, P.R.; Ramos, A.D.L.S.; Gomide, L.A.D.M.; Ramos, E.M. Selection of a chitosan gelatin-based edible coating for color preservation of beef in retail display. *Meat Sci.* **2016**, *114*, 85–94. [CrossRef] [PubMed]

93. Liu, L.; Kerry, J.F.; Kerry, J.P. Application and assessment of extruded edible casings manufactured from pectin and gelatin/sodium alginate blends for use with breakfast pork sausage. *Meat Sci.* **2007**, *75*, 196–202. [CrossRef] [PubMed]
94. Herring, J.L.; Jonnalongadda, S.C.; Narayanan, V.C.; Coleman, S.M. Oxidative stability of gelatin coated pork at refrigerated storage. *Meat Sci.* **2010**, *85*, 651–656. [CrossRef] [PubMed]
95. Tyburcy, A.; Kozyra, D. Effects of composite surface coating and pre-drying on the properties of kabanosy dry sausage. *Meat Sci.* **2010**, *86*, 405–410. [CrossRef] [PubMed]
96. Lee, J.-H.; Yang, H.-J.; Lee, K.-Y.; Song, K.B. Physical properties and application of a red pepper seed meal protein composite film containing oregano oil. *Food Hydrocoll.* **2016**, *55*, 136–143. [CrossRef]
97. De Oliveira, M.M.M.; Brugnera, D.F.; Piccoli, R.H. Essential oils of thyme and rosemary in the control of listeria monocytogenes in raw beef. *Braz. J. Microbiol.* **2013**, *44*, 1181–1188. [CrossRef] [PubMed]
98. Min, B.J.; Han, I.Y.; Dawson, P.L. Antimicrobial gelatin films reduce listeria monocytogenes on turkey bologna. *Poult. Sci.* **2010**, *89*, 1307–1314. [CrossRef] [PubMed]
99. Nowzari, F.; Shábanpour, B.; Ojagh, S.M. Comparison of chitosan-gelatin composite and bilayer coating and film effect on the quality of refrigerated rainbow trout. *Food Chem.* **2013**, *141*, 1667–1672. [CrossRef] [PubMed]
100. Gómez-Estaca, J.; López de Lacey, A.; López-Caballero, M.E.; Gómez-Guillén, M.C.; Montero, P. Biodegradable gelatin-chitosan films incorporated with essential oils as antimicrobial agents for fish preservation. *Food Microbiol.* **2010**, *27*, 889–896. [CrossRef] [PubMed]
101. Kakaei, S.; Shahbazi, Y. Effect of chitosan-gelatin film incorporated with ethanolic red grape seed extract and ziziphora clinopodioides essential oil on survival of listeria monocytogenes and chemical, microbial and sensory properties of minced trout fillet. *LWT Food Sci. Technol.* **2016**, *72*, 432–438. [CrossRef]
102. Alparslan, Y.; Baygar, T.; Hasanhocaoglu, H.; Metin, C. Effects of gelatin-based edible films enriched with laurel essential oil on the quality of rainbow trout (oncorhynchus mykiss) fillets during refrigerated storage. *Food Technol. Biotechnol.* **2014**, *52*, 325–333.
103. Ojagh, S.M.; Núñez-Flores, R.; López-Caballero, M.E.; Montero, M.P.; Gómez-Guillén, M.C. Lessening of high-pressure-induced changes in atlantic salmon muscle by the combined use of a fish gelatin-lignin film. *Food Chem.* **2011**, *125*, 595–606. [CrossRef]
104. Song, H.Y.; Shin, Y.J.; Song, K.B. Preparation of a barley bran protein-gelatin composite film containing grapefruit seed extract and its application in salmon packaging. *J. Food Eng.* **2012**, *113*, 541–547. [CrossRef]
105. Song, N.-B.; Lee, J.-H.; Al Mijan, M.; Song, K.B. Development of a chicken feather protein film containing clove oil and its application in smoked salmon packaging. *LWT Food Sci. Technol.* **2014**, *57*, 453–460. [CrossRef]
106. Davis, C.G.; Lin, B.-H. *Factors Affecting US Pork Consumption*; United States Department of Agriculture Economic Research Service (USDA/ERS): Washington, DC, USA, 2005.
107. Mor-Mur, M.; Yuste, J. Emerging bacterial pathogens in meat and poultry: An overview. *Food Bioprocess Technol.* **2010**, *3*, 24–35. [CrossRef]
108. Chung, C.-K.; Zuo, L. Method for Refreshing Preserved Pork by Using Edible Composite Antibacterial Film. Patent CN 102487988B, 29 April 2013.
109. Vásconez, M.B.; Flores, S.K.; Campos, C.A.; Alvarado, J.; Gerschenson, L.N. Antimicrobial activity and physical properties of chitosan–tapioca starch based edible films and coatings. *Food Res. Int.* **2009**, *42*, 762–769. [CrossRef]
110. Yagiz, Y.; Kristinsson, H.G.; Balaban, M.O.; Welt, B.A.; Ralat, M.; Marshall, M.R. Effect of high pressure processing and cooking treatment on the quality of atlantic salmon. *Food Chem.* **2009**, *116*, 828–835. [CrossRef]
111. Rotariu, O.; Thomas, D.J.I.; Goodburn, K.E.; Hutchison, M.L.; Strachan, N.J.C. Smoked salmon industry practices and their association with listeria monocytogenes. *Food Control* **2014**, *35*, 284–292. [CrossRef]
112. Vidanarachchi, J.K.; Ranadheera, C.S.; Wijerathne, T.D.; Udayangani, R.M.C.; Himali, S.M.C.; Pickova, J. Applications of seafood by-products in the food industry and human nutrition. In *Seafood Processing by-Products: Trends and Applications*; Kim, S.-K., Ed.; Springer: New York, NY, USA, 2014; pp. 463–528.
113. Song, L. Edible Red Bean Film. Patent CN 103589173A, 19 February 2014.
114. Song, L. Edible Packaging Film for Garlic. Patent CN 103589168A, 19 February 2014.
115. Wang, X.; Kong, D.; Ma, Z.; Zhao, R. Effect of carrot puree edible films on quality preservation of fresh-cut carrots. *Irish J. Agric. Food Res.* **2015**, *54*, 64–71. [CrossRef]

116. Zhang, B.; Feng, X.; Han, P.; Duan, X. Effect of propolis/nano-silica composite coating on activities of ripening and senescence related enzymes in cherry tomato fruits. *J. Chin. Inst. Food Sci. Technol.* **2016**, *16*, 159–165.
117. Licodiedoff, S.; Koslowski, L.A.D.; Scartazzini, L.; Monteiro, A.R.; Ninow, J.L.; Borges, C.D. Conservation of physalis by edible coating of gelatin and calcium chloride. *Int. Food Res. J.* **2016**, *23*, 1629–1634.
118. Youssef, A.R.M.; Ali, E.A.M.; Emam, H.E. Influence of postharvest applications of some edible coating on storage life and quality attributes of navel orange fruit during cold storage. *Int. J. Chem. Technol. Res.* **2015**, *8*, 2189–2200.
119. Andrade, R.; Skurtys, O.; Osorio, F. Drop impact of gelatin coating formulated with cellulose nanofibers on banana and eggplant epicarps. *LWT Food Sci. Technol.* **2015**, *61*, 422–429. [CrossRef]
120. Poverenov, E.; Rutenberg, R.; Danino, S.; Horev, B.; Rodov, V. Gelatin-chitosan composite films and edible coatings to enhance the quality of food products: Layer-by-layer vs. Blended formulations. *Food Bioprocess Technol.* **2014**, *7*, 3319–3327. [CrossRef]
121. Poverenov, E.; Zaitsev, Y.; Arnon, H.; Granit, R.; Alkalai-Tuvia, S.; Perzelan, Y.; Weinberg, T.; Fallik, E. Effects of a composite chitosan-gelatin edible coating on postharvest quality and storability of red bell peppers. *Postharvest Biol. Technol.* **2014**, *96*, 106–109. [CrossRef]
122. Fakhouri, F.M.; Casari, A.C.A.; Mariano, M.; Yamashita, F.; Mei, L.H.I.; Soldi, V.; Martelli, S.M. Effect of a gelatin-based edible coating containing cellulose nanocrystals (CNC) on the quality and nutrient retention of fresh strawberries during storage. *IOP Conf. Ser. Mater. Sci. Eng.* **2014**, *64*. [CrossRef]
123. Feng, D.; Zhengguang, W.; Yimei, Z.; Xiang, Z.; Meng, G.X.; Xu, Y.; Bi, Y. Effect of chitosan composite coating on chinese blueberry fruit (*Vaccinium uliginosum* L.). *Acta Hortic.* **2014**, *1053*, 207–214. [CrossRef]
124. Bizura Hasida, M.R.; Nur Aida, M.P.; Zaipun, M.Z.; Hairiyah, M. Quality evaluation of fresh-cut "josapine" pineapple coated with hydrocolloid based edible coating using gelatin. *Acta Hortic.* **2013**, *1012*, 1037–1042. [CrossRef]
125. Neves, A.C.V., Jr.; Coneglian, R.C.C.; Soares, A.G.; Freitas, D.G.C.; Fonseca, M.J.O.; Barreira, F.R.; De Miranda, A.F.M. Physical and sensory characterization of edible coatings applied to minimally processed persimmon. *Acta Hortic.* **2012**, *934*, 537–542. [CrossRef]
126. Karim, A.A.; Bhat, R. Fish gelatin: Properties, challenges, and prospects as an alternative to mammalian gelatins. *Food Hydrocoll.* **2009**, *23*, 563–576. [CrossRef]
127. Galanakis, C.M. Recovery of high added-value components from food wastes: Conventional, emerging technologies and commercialized applications. *Trends Food Sci. Technol.* **2012**, *26*, 68–87. [CrossRef]
128. De Moraes Crizel, T.; Haas Costa, T.M.; de Oliveira Rios, A.; Hickmann Flôres, S. Valorization of food-grade industrial waste in the obtaining active biodegradable films for packaging. *Ind. Crops Prod.* **2016**, *87*, 218–228. [CrossRef]

© 2016 by the authors. Licensee MDPI, Basel, Switzerland. This article is an open access article distributed under the terms and conditions of the Creative Commons Attribution (CC BY) license (http://creativecommons.org/licenses/by/4.0/).

Review

Natural Pectin Polysaccharides as Edible Coatings

Arantzazu Valdés, Nuria Burgos, Alfonso Jiménez and María Carmen Garrigós *

Department of Analytical Chemistry, Nutrition & Food Sciences, University of Alicante, Campus San Vicente, 03690 San Vicente del Raspeig (Alicante), Spain; arancha.valdes@ua.es (A.V.); nuria.burgos@ua.es (N.B.); alfjimenez@ua.es (A.J.)

* Author to whom correspondence should be addressed; mc.garrigos@ua.es; Tel.: +34-96-590-1242; Fax: +34-96-590-3697.

Academic Editor: Stefano Farris
Received: 30 September 2015; Accepted: 11 November 2015; Published: 16 November 2015

Abstract: The most fashionable trends in food packaging research are targeted towards improvements in food quality and safety by increasing the use of environmentally-friendly materials, ideally those able to be obtained from bio-based resources and presenting biodegradable characteristics. Edible films represent a key area of development in new multifunctional materials by their character and properties to effectively protect food with no waste production. The use of edible films should be considered as a clean and elegant solution to problems related with waste disposal in packaging materials. In particular, pectin has been reported as one of the main raw materials to obtain edible films by its natural abundance, low cost and renewable character. The latest innovations in food packaging by the use of pectin-based edible films are reviewed in this paper, with special focus on the use of pectin as base material for edible coatings. The structure, properties related to the intended use in food packaging and main applications of pectins are herein reported.

Keywords: pectin; edible films; biopolymer coatings; fruits; vegetables; agricultural wastes; revalorisation

---

1. Introduction

Food packaging systems have been traditionally considered as simple containers to transport food from the place where they have been produced to the retail outlet and then to the consumer with no alteration of the food nutritional and organoleptic characteristics. Nevertheless, these systems are often unable to increase the shelf-life of fresh food resulting in a problem to producers, retailers and consumers. Since the main function of packaging is preservation of food from external contamination, other important features such as retardation of deterioration, extension of shelf-life, protection from transport impacts and maintenance of the food quality should be taken into account. Packaging materials should protect food from environmental influences such as heat, moisture, oxygen, enzymes, loss of aromas and unpleasant odor components, as well as from the attack from micro and macro-organisms. Furthermore, the global market is becoming more demanding and is continuously in need of novel and stable products which, at the same time, could retain the natural properties of food. In summary, the demand for new packaging materials and food packaging functionalities is increasing.

Among materials currently used in food packaging, polymers have taken a major share because of their versatility and advantageous performance/cost ratio. However, the most important polymers used for food packaging are obtained from non-renewable resources and they are not biodegradable or compostable, representing a global environmental problem. These drawbacks in the use of common polymers in food packaging are becoming an important issue in the design of attractive systems for distribution, making consumers aware of the problems related to the waste disposal. Although the stability of food packaging materials during the food shelf-life is an advantage, it turns into a disadvantage when the package enters the post-use phase. In summary, the use of polymers obtained from non-renewable resources and non-biodegradable in food packaging applications represents an important environmental impact and waste generation issue. In fact, packaging waste accounted for 32.5 million tons or 17.7% of the total municipal solid waste (MSW) in 2013 in the USA [1] and 19.3 million tons or around 25% in 2014 in Europe [2]. Currently, landfilling is the dominant method of packaging waste disposal, followed by recycling, incineration and composting. Even though recovery methods such as reuse, recycling and/or composting are encouraged as a way of reducing packaging waste disposal, there is still much work to do to substantially reduce the quantity of plastics present in MSW [3].

Some alternative packaging materials obtained from renewable resources, such as poly(lactic acid), PLA, poly(hydroxyalkanoates) (PHAs), starch or proteins, have been proposed as alternatives to non-biodegradable polymers in food packaging applications [4–7]. However, polysaccharides are gaining some space in their use as innovative packaging materials by their ubiquitous presence in Nature, as well as by their relative low cost compared to other biopolymers and the possibilities they offer to be used not only as polymer matrices but also in coatings. Polysaccharides are therefore strong environmentally-sound contenders in the food packaging market that fulfill all the environmental concerns (*i.e.*, derived from renewable raw materials and biodegradable) while being even possible to be metabolized by human body together with food, making them able to be used in edible films.

Polysaccharides are the main component of biomass, being the most abundant renewable polymer resources available from Nature. Among them, pectin is one of the most significant since the pectin world market demand is increasing, reaching a total production capacity around 45–50 Mton per year while the demand was around 140–160 Mton per year in 2011, showing that the industry interest on this complex polysaccharide is rising day by day.

One of the most important features in the still incipient commercial application in food packaging of bio-based polymers is the use of some of these materials (e.g., PLA, PHAs) for rigid containers [8], but flexible morphologies are still dependent on the use of additives. In addition, commercial biopolymers show some limitations in terms of performance (thermal resistance, poor barrier and brittleness) as well as high costs. In fact, the present situation is such that the development of polysaccharides-based materials in fields where unique performance or raw material characteristics show an added technological value (for instance, sourced from renewable materials, biodegradability, water barrier, antimicrobial properties, aroma barrier) or marketing-wise (e.g., green or sustainability image) is both a challenge and an opportunity for the food packaging industries.

The term edible coatings in food applications correspond to thin layers of edible materials applied onto surfaces of highly perishable foodstuff, such as fresh-cut fruits and vegetables. The aim of this paper is the full review of the properties and possible applications of pectin in the manufacture of edible films for food packaging. The most recent scientific and technological developments in this field will be highlighted and our goal would be to permit the reader getting a complete survey of the increasing use of this biopolymer in food industries.

2. Structure and Classification of Pectic Substances

Pectin is a white, amorphous and colloidal carbohydrate of high molecular weight occurring in ripe fruits, especially in apples, currants, *etc.*, and used in fruit jellies, pharmaceuticals and cosmetics for its thickening and emulsifying properties and ability to solidify to a gel. All these properties and applications have put pectin in the market of the biopolymers with great potential and possibilities for future developments. The structure of these polysaccharides is discussed in this section.

Pectic substances are present in the primary cell walls and middle lamellae of many plants and fruits, and they are frequently associated with cellulose, hemicellulose and lignin structures [9]. Their presence in the cell is important for some essential functions: (a) adhesion between cells; (b) mechanical strength of the cell wall; (c) ability to form stabilizing gels; and (d) they play a significant role in the growth of plant cells [10]. Pectin forms the most complex class of polysaccharides, mainly composed by high molecular weight heterogeneous groups of glycanogalacturonans and acidic structural polysaccharides with diverse structures. Pectin backbone consists of $(1\rightarrow 4)$-α-D-galacturonic acid molecules linked to a small number of rhamnose residues in the main chain and arabinose, galactose and xylose in the side chains [11]. Several authors stated that pectin polysaccharides can be classified in three types (Figure 1) [12–14].

Homogalacturonan (HG) is a linear polymer formed by D-galacturonic acid and it can be classified into three different families depending on the acetylation or methylation reactions suffered during polymerization: (i) pectin with more than 75% of methylated carboxyl groups; (ii) pectinic acid with less than 75% of methylation; and, finally, (iii) pectic acid or polygalacturonic acid without methyl-esterified carboxyl groups [12]. Is the bold/italics necessary?

Rhamnogalacturonan I (RGI) is composed by the repeating disaccharide rhamnose–galacturonic acid groups acetylated and linked to side chains of neutral sugars, such as galactose, arabinose and xylose. Finally, rhamnogalacturonan II (RGII) is also formed by homogalacturonan chains, but with complex side groups attached to 12 different types of glycosyl residues, such as rare sugar species (2-O-methyl xylose, 2-O-methyl fucose, aceric acid, 2-keto-3-deoxy-D-lyxo heptulosaric acid, and 2-keto-3-deoxy-D-manno octulosonic acid). RGI and RGII are called hairy regions in the pectin structure whereas HG is the smooth part of the molecule [13,14].

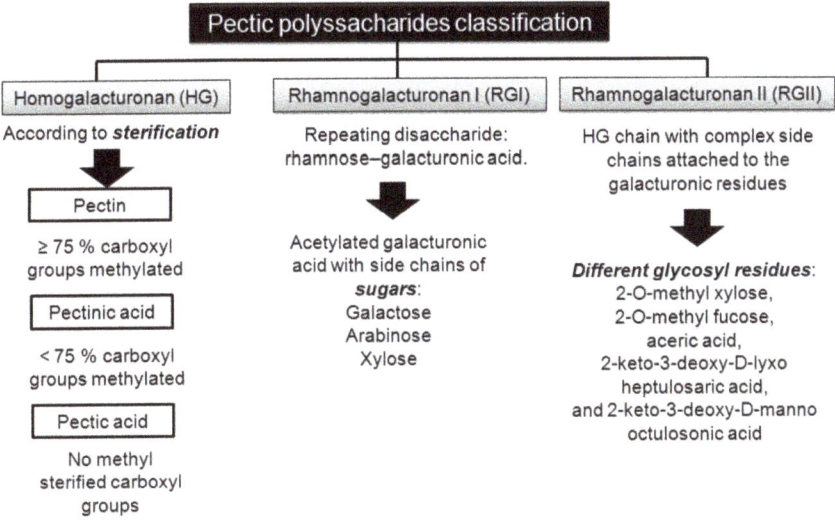

Figure 1. Classification of pectic polyssacharides based on D-galacturonic acid [12–14].

3. Properties and Current Applications in Food Industry

In general terms, pectins are used in food industry as stabilizers, thickening and gelling agents, crystallization inhibitors and encapsulating agents. But, in all cases, galacturonic residues should be previously modified by the addition of methyl groups to yield methoxides. This chemical modification is essential to improve pectins physical properties, which are dependent on their molecular masses and, primarily, on the degree of methyl-esterification (DM), calculated as moles of methanol per 100 moles of galacturonic acid [15]. In this context, two different DM degrees can be achieved resulting in different applications [10]:

- Pectins with high DM (HM pectins) containing 50% or higher of galacturonic residues. They are used as gelling component in heat-resistant bakery jams, fruit preservatives and juices,

confectionaries jellies, milky products, glazing cakes and soft drinks. HM pectins form gels by hydrophobic interactions under acidic conditions in aqueous media and high sugar content.
- Pectins with low DM (LM pectins) are obtained by de-esterification of HM pectins under controlled pH, temperature and time. LM pectins show DM degrees between 20% and 30% [16]. They are used to prepare gels at low pH values in the presence of divalent calcium. Applications in food industry include jams and jellies with low-sugar content, dairy desserts, ice cream with fruit gels, thickening agents of syrups for fruit and vegetable canning and food coatings.

In this context, edible coatings obtained from pectin and derivatives (pectate and amidated pectin) have been recently proposed in food-related applications by their excellent barrier to oxygen, aroma preservation, barrier to oil and good mechanical properties, but they are not effective against moisture transfer through films by their hydrophilic nature [17]. They are currently used in fresh and minimally processed fruits and vegetables, such as avocado, apple, apricot, chestnuts, berries, guava, melon, papaya, peach, walnuts, carrot and tomato [17]. Depending on the product, three different coating methods can be used, as shown in Figure 2 [18].

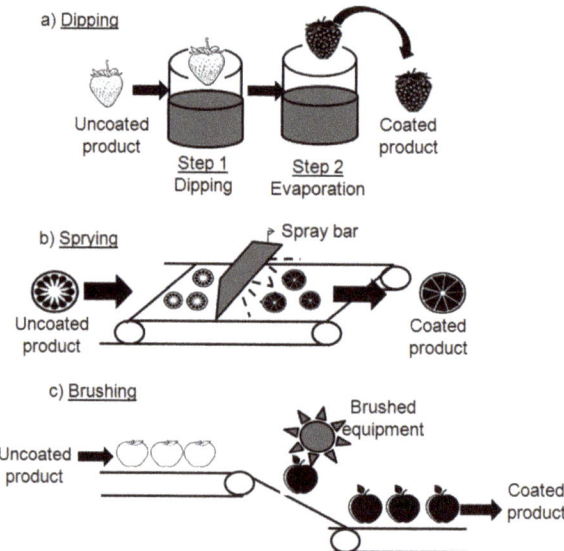

Figure 2. Main coating methods used in food industry. (Adapted from ref. [18]).

According to Dhanapal et al [18], dipping is the most common method to apply coatings to fruits and vegetables when the coating solution is highly viscous (Figure 2a). It is carried out by introducing the product for a time between 5 and 30 s in a coating solution under controlled conditions of density and surface tension.

However, when the coating solution is not highly viscous, spraying can be used (Figure 2b). The food product is introduced into the coating system and it is sprayed by controlling the final drop size of the spray solution, which depends on the thickness of the spray gun, nozzle temperature, air and liquid flow rates, humidity of incoming air and polymer solution, drying time and temperature [18]. The usual spraying instrument in this application can produce fine sprays, with drop-size distributions up to 20 µm. Finally, the brushing method is used in some products, such as fresh beans and strawberries, when the reduction of the moisture loss is an issue (Figure 2c). Thin coatings onto the surface of the product are obtained in all cases and they could act as semi-permeable membranes reducing gas transfer rates and creating new packaging materials to extend food shelf-life.

4. Extraction Methods for Pectins: New Trends for the Revalorization of Agricultural Residues

Pomace from apples, citric albedo, sugar beet pulp and sunflower rinds have been proposed as by-products from the agricultural and food industries to obtain pectic substances with contents over 15% on dry basis [19]. In fact, apple pomace (15% to 20% of pectin) and citrus peel (30% to 35% of pectin), obtained in fruit juice industries, are the two main raw materials used for the industrial extraction of pectins. However, other natural sources with potential in the pectins extraction have been described in literature [15]. For example, plants of *Lupinus* genus (pectins content 1.5%–7% depending on the variety, year, and climatic conditions), burdock from the *Arctium* genus (pectins content higher than 21%) or peach (*Persica vulgaris* L.) with pectins content up to 10% have been proposed.

The diverse nature of the pectic substances has resulted in different extraction methods. Depending on the raw material the extraction methods and further chemical purification treatments are different. This section focuses on natural pectin sources and the associated extraction methods used to produce high added-value products. The most frequent extraction techniques are: (i) conventional solvent extraction based on stirring and heating; (ii) microwave assisted extraction (MAE); (iii) electromagnetic induction (EMI); (iv) ultrasound-assisted extraction (UAE); and (v) enzymatic extraction.

In general terms, pectins are commonly extracted by solvent-liquid methods in acidic aqueous media (nitric, sulfuric, phosphoric or citric acid) at high temperatures due to their availability, applicability and ease of use [20]. Pectins are usually extracted by hot dilute mineral acids at pH 1.5–2.5, taking 2–4 h with further precipitation with ethanol or isopropyl alcohol, separation to remove impurities, drying, grinding and blending with other additives. The extracted pectins can be categorized into two major types, depending on their DM degree. Factors like pH, temperature, time and solvent:liquid (S:L) ratio are usually studied to optimize the extraction conditions. Table 1 shows the main extraction methods for natural pectins reported in literature.

Table 1. Natural pectins extracted by solvent extraction methods.

Natural Source	Extraction Parameters	Pectin Yield Extraction (%)	Ref.
Okra pods	Citric and phosphate buffer, pH 2 and 6, 60 min, 80 °C, 1:15 g·mL^{-1}	13.3 (pH 2.0); 15.7 (pH 6.0)	[21]
Durian rinds	HCl, pH 2.8, 43 min, 86 °C, S:L 1:10 g·mL^{-1}	9.1	[22]
Mango peel	Sulfuric acid in water, pH 1.5, 2.5 h, 90 °C	>70	[23]
Banana peel	Citric acid solution, pH 2.0, 160 min, 87 °C, 1:20 g·mL^{-1}	13.89	[24]
	Citric acid and HCl, pH 1.5, 4 h, 90 °C	16.54	[25]
Tropical fruit peels	Citric acid, pH of 2, 3.3 and 4.5, 120 min, 70 °C	12.56–14.24	[26]
Sugar beet pulp	Citric acid, pH 1, 166 min, 99 °C, 1:20 g·mL^{-1}	23.95	[27]
Bagasse and pomace lime fruit	Citric acid, 60 min, 90 °C, 1:20 g·mL^{-1}	13.31 (Bagasse); 15.1 (Pomace)	[28]
Valencia orange peel	Citric acid extraction, pH 1.5, 75 min, 90 °C	16.7	[29]
Watermelon seed	HCl, pH 2, 60 min, 85 °C, 1:15 g·mL^{-1}	19.75	[30]
Faba bean hulls	HCl, pH 1.5, 80 min, 85 °C, 1:25 g·mL^{-1}	15.75	[31]
Honeydew melon seeds and damaged skin	HCl at pH 1, 80 °C for 4 h	7.9	[32]
Tomato peel	Ammonium oxalate and oxalic acid, 90 °C in two extraction steps (24 and 12 h)	32.0	[33]
Sunflower head	Sodium citrate, 85 °C, 3.5 h, 1:40 g/mL	16.90	[34]

Alba, Laws and Kontogiorgos extracted okra pod pectins at different pH values (2.0 and 6.0 with 100 mM citric and phosphate buffer, respectively) and examined the effect of the extraction conditions on their molecular and compositional characteristics [21]. Pectins obtained could not be quantitatively recovered in a single extraction step and second extraction was required. They concluded that pectins extracted with citric buffer (pH 2.0) resulted in slightly lower yields (13.3% ± 0.3%) when compared to extraction at pH 6.0 (15.7% ± 0.2%). The highest extraction yields were obtained at high temperatures and low pH values due to the cleavage of pectin and other cell wall materials. Furthermore, okra pectins were found to be LM with DM 40.0% ± 1.6% and 24.6% ± 1.0% for pH 2.0 and 6.0, respectively. These results could be attributed to de-esterification process caused by the β-elimination of the esterified homogalacturonan backbone, removing methyl ester groups and resulting in pectins with lower DM and molecular mass. These pectins can be used as functional ingredients (thickeners, viscosity enhancers, gelling agents and texture modifiers) in the food industry.

Durian (*Durio zibethinus*) is considered as a high-value fruit due to its distinct flavor and unique taste, but its processing generates large amounts of waste and by-products rich in polysaccharides. A second-order polynomial equation using a three-level Box-Behnken response design to optimize the extraction conditions of pectins from Durian residues was recently proposed [22]. Those authors studied the influence of four variables: S:L ratio (5–15 g·mL^{-1}), pH (2–3), extraction time (20–60 min) and extraction temperature (75–95 °C) in the pectins extraction yield. The optimum selected conditions were pH 2.8, 43 min, 86 °C and 1:10 g·mL^{-1} S:L ratio, yielding 9.1% for the extraction efficiency.

Hubert *et al.* studied the impact of different Mango (*M. indica* L.) cultivars, technological procedures and ripeness degree on the pectin quality of their peels [23]. They concluded that grinding of mango peels (with particle sizes of 42 µm) significantly enhanced the pectins extraction yield (70%) without increasing the impurity contents.

Banana fruit has been considered an excellent source of pectins. A central composite design was used to determine the effect of pH (2.0–4.5), extraction temperature (70–90 °C) and time (120–240 min) in the extraction yield of pectins in citric acid solution (1:20 w/v) under stirring (150 rpm) [24]. High temperatures and pH values resulted in higher extraction yields with a decrease in DM from 79% to 43%. The optimum conditions for pectins extraction were 87 °C, 160 min and pH 2.0 with extraction yield 13.9%. Castillo-Israel *et al.* proved that banana peel waste could be a promising source of pectins with extraction yield 17.0% and DM 75.0% by using HCl (0.5 N, pH 1.5) and citric acid (0.5 N, pH 1.7) at 90 °C for 4 h [25]. Other agricultural residues used as pectins sources are peels of Yellow passion fruit (*Passiflora edulis* var. *flavicarpa*), red dragon fruit (*Hylocereus polyrhizus*) and soursop (*Annona muricata* L.) [26]. Peels of yellow passion fruit and dragon fruit gave pectins yield of 14.2% and 12.6% with DM 55.5% and 47.9% at pH = 2.4 and extraction times 58 and 65 min, respectively. It was concluded that the soursop peels are not suitable by their low pectins yield (lower than 6%). In contrast, the extraction of sugar beet pulp pectins in citric acid at 99 °C and 166 min at pH = 1.0 and S:L ratio 20 allowed the maximum extraction yield (23.9%) [27].

Bagasse and pomace of Mexican lime fruit have been proposed as pectins sources with S:L ratio 1:20, citric acid 1% at 90 °C for 60 min with yields of 13.3 and 15.1%, respectively [28]. Pectins yield extraction from Valencia orange peels was carried out by using citric acid solution at 90 °C for 75 min and pH 1.5 yielding acceptable pectins extraction (16.7%) [29]. Korish reported the use of the *Citrullus lanatus* var. Colocynthoides watermelon seeds as source of pectins [30]. The optimum yield was obtained with acidic water solution (HCl, pH 2.0, S:L ratio 1:15 and 85 °C), for 60 min yielding pectins extraction efficiency of 16.7%. This waste was proposed as raw material for the production of useful biomaterials and chemicals. These authors also studied the potential use of bean hulls as pectins source [31]. The maximum yield of the extracted pectins was not coincident with the highest esterification degree with the maximum yield (15.7%) recorded at pH 1.5 and 85 °C for 80 min and S:L ratio 1:25; while the highest degree of esterification (54.6%) occurred at pH 2.5 and 90 °C for 60 min. Therefore, it was concluded that the yield of pectins was variable with the extraction conditions.

Seeds and damaged skin of fresh honeydew melon (*Cucumis melo Inodorus*) were proposed as other pectins sources with acceptable efficiencies (7.9%) in HCl at pH = 1.0 [32]. Pectins from tomato (*Lycopersicum esculentum*) peels were also extracted under reflux at 90 °C with ammonium oxalate (16 g·L^{-1}) and oxalic acid (4 g·L^{-1}) in two extraction steps (24 and 12 h respectively) [33]. The highest pectins yields were 32.6% and 31.9% in two batches, but the pectin quality was low in the second extraction step. Kang *et al.* extracted pectins from sunflower heads (SFHP) with sodium citrate at 85 °C and S:L ratio 1:40 [34]. The content in uronic acid was used in this study as the indicator of the pectin extraction yield by using the carbazole colorimetric method with a maximum yield in uronic acid of 16.9%.

Giosafatto *et al.* recently reported the production of an edible hydrocolloid film made by using Citrus pectin and the protein phaseolin crosslinked by microbial transglutaminase, an enzyme able to covalently modify proteins by formation of isopeptide bonds between glutamine and lysine residues [35,36]. These authors included different biopolymer matrices in their formulations, including chitosan [37,38] and whey proteins [39].

In general terms, it can be concluded that the combination of high temperatures, times and solvents volume resulted in the recovery of pure pectins. However, two problems in the pectins extraction were described, since long heating times resulted in ruptures of the pectin chains leading to difficulties in pectin isolation, while acid extraction may result in corrosion and pollution whereas the alkaline extraction reduces the chain of galacturonic acid by beta-elimination [20]. Therefore, innovative methods such as MAE, EMI, UAE and enzymatic treatment have been proposed to counteract these effects. Table 2 summarizes the most recent trends on the application of these methods for pectins extraction.

Table 2. Innovative extraction methods for pectins.

Extraction Method	Natural Source	Extraction Parameters	Pectin Yield Extraction (%)	Ref.
MAE	Dragon fruit peel	400 W, 45 °C, 20 min, 24 g·mL^{-1}	7.5	[40]
		HCl, pH 2.07, 800 W, 65 s, 66.6 g·mL^{-1}	18.57	[41]
	Bagasse and pomace of Mexican lime fruit	Citric acid, 800 W, 120 °C, 5 min, 1:30 g·mL^{-1}	16.9 (Bagasse); 8.4 (Pomace)	[28]
	Pomelo peel	Tartaric acid solution, pH 1.5, 660 W, 9 min, 1:40 g·mL^{-1}	23.83	[42]
	Mango peel	Aqueous solution, pH 2.7, 413 W, 134 s, 1:18 g·mL^{-1}	28.86	[43]
	Papaya peel	Aqueous solution, 512 W, pH 1.8, 140 s, 1:15 g·mL^{-1}	25.41	[44]
EMI	Citrange albedos	Aqueous solution, pH 1.2 with H$_2$SO$_4$, 80 °C, 90 min	29	[45]
UAE	Pomegranate peel	Aqueous solution, pH 1.27, 61.9 °C, 28.31 min, 1:17.52 g·mL^{-1}	24.18	[46]
	Sisal fiber	Aqueous solution, 61 W, 50 °C, 26 min, 1:28 g·mL^{-1}	29.32	[47]
	Sugar beet pulp	Aqueous solution, 10.70 MPa, 120.72 °C, 30.49 min, 44.0 g·mL^{-1}	24.63	[48]
	Grapefruit peel	Deionized water, HCl, pH 1.5, 12.56 W/cm^2, 66.7 °C, 27.9 min, 1:50 g·mL^{-1}	27.34	[49]
Enzymes	Gold kiwifruit pulp, skin and seed	Purified water, pH 3.7, 25 °C, 30 min, Celluclast 1.5 L enzyme (1.05 mL·kg^{-1}), 1:3 g·mL^{-1}	2.14	[50]
	Gold kiwifruit pomace		4.5	[51]

MAE is an extraction technique that uses microwave power to heat samples in solution by enhancing the mass transfer rate of the solutes from the vegetable raw materials to solvents. Two main mechanisms allows the pectin extraction by MAE, *i.e.*, the rapid increase in temperature, which reduces the emulsion viscosity and breaks the chemical bonds in the raw material structure, and the molecular rotation due to the electrical charges of molecules, resulting in enhanced movement of ions which increase the efficiency of the extraction process [41]. MAE has been reported as the preferred extraction method of pectins from natural sources such as dragon fruit peels [40,41], bagasse and pomace obtained from Mexican lime fruit [28], pomelo [42], mango [43] and papaya peels [44] under different operation conditions. MAE conditions are dependent on different factors, such as microwave power, pH, time and S:L ratio. MAE methods show significant advantages over the conventional extraction techniques, such as the reduction in the amount of the extraction solvents, low energy consumption, high recoveries, good reproducibility, short extraction times (minutes rather than hours) and minimal sample manipulation.

Since these variables should be optimized for the development of every method, response surface methodologies (RSM) are used to study their interactive effects on pectins extraction with good correlation between experimental data and the predicted values. Table 2 summarizes the proposed MAE conditions to obtain maximum pectins extraction yields from natural sources. The most usual procedure includes the use of Box-Behnken experimental designs (BBD) or central composite designs (CCD). Good results were reported for the extraction of target compounds, such as antioxidants from almond skin [52], peanut skin [53] and batata leaves [54], highlighting the potential of MAE as a promising extraction method for pectins.

The use of EMI methods involving mixing, heating and maintenance of samples in an acidic medium to extract pectins has been also proposed. Zouambia *et al.* introduced a new approach for pectins extraction from citrange albedos [45]. In this study, the extraction yield, composition and physico-chemical characteristics of pectins extracted by electromagnetic induction heating were studied and compared with those obtained by direct boiling. EMI methods were also evaluated at different extraction times and power levels. A considerable reduction in time was observed by using EMI, since only 30 min were necessary to obtain pectins extraction yield (24%) similar to those obtained with conventional heating methods for 90 min. This innovative method can be suggested as an alternative for pectins extraction with short times, good yield and maintaining the main physico-chemical properties of pectins.

UAE is a process that uses acoustic energy and selected solvents to extract specific compounds from various matrices in relatively short times giving high yields. The mechanism of UAE is based on the interaction of the ultrasound waves and the solvent molecules allowing the acoustic cavitation and the disruption of cell walls with enhanced mass transfer across the cell membranes [20]. High efficiencies were recently reported for the UAE of pectins from several agricultural wastes, such as pomegranate peel [46] and sisal fiber [47]. In both cases, BBD was successfully used as valuable multivariate technique to evaluate the relationship between variables, helping to determine the accurate optimum values of all the experimental parameters with a reduced number of experiments [55]. In pomegranate peel, the effects of pH, S:L ratio, extraction time and temperature on the recovery of pectins were studied [46]. The maximum extraction yields were determined at pH = 1.27; 28.3 min 1:17.5 g·mL^{-1} S:L ratio and 61.9 °C resulting in yield extraction of 24.2%. In sisal fibers, the extraction conditions were also optimized [47]. The analysis of variance showed that the contribution of a quadratic model was significant for the pectins extraction yield. The optimal extraction condition were 61 W, 1:28 g·mL^{-1} S:L ratio, 26 min and 50 °C resulting in pectins yield of 29.3%. Therefore, UAE was proposed as an alternative method to extract pectins from sisal waste with the advantages of lower extraction temperatures and energies and shorter extraction times.

Several authors have combined UAE with other extraction techniques. Pectin-enriched materials obtained from sugar beet pulp wastes were extracted by UAE combined with subcritical water by using the RSM methodology [48]. The highest extraction yields (24.6%) were obtained at pressure 10.70 MPa, 120.7 °C, 30.5 min and S:L ratio 44.0 g·mL^{-1}.

The effect of the extraction temperature and sonication time is essential for UAE, but not much published work has focused on the optimization of UAE conditions and exploration of the combined effects of these variables in the pectins extraction. Wang *et al.* used RSM to study the extraction of grapefruit peel pectins by UAE, evaluating the combined effects of ultrasound power and extraction time and temperature on the extraction yield in HCl aqueous solutions (pH = 1.5) [49]. This study also compared UAE results with those obtained with conventional extraction methods using the same solvent at 80 °C for 90 min. UAE provided extraction yields of 16.3% at low temperatures and short times. In summary, UAE could be considered as an efficient and economic technique to extract pectins from plant resources.

All these methods have in common the use of diluted acid solutions, but in some cases non-solvent methods are preferred in food products. Enzymatic methods represent an innovative and clean alternative for the "green" extraction of pectins reducing water pollution and waste generation. These methods are based on the degradation of pectins by selective depolymerisation or their isolation by breaking the plants cell wall [20]. Only few works have reported the use of these methods by comparing the enzymatic extraction of pectins obtained from gold kiwifruit wastes at different stages of maturity with other extraction methods [50,51]. In both cases, the enzymatic treatment was conducted at 25 °C for 30 min with a Celluclast 1.5 L (1.05 mL·kg^{-1}) at pH 3.70 ± 0.05. These methods showed relative low yields (2.14% for peel, skin and seeds and 4.5% for pomace) when compared to acid/water extraction methods. Nevertheless, the advantages of enzymatic methods, *i.e.*, low temperatures that contribute to avoid degradation of pectins and low cost, make this method a promising alternative that deserves further study.

5. Recent Trends in the Use of Pectin Edible Coatings

Edible coatings are currently used in highly perishable foodstuff to protect their nutritional and organoleptic properties by their potential to extend shelf-life and reducing the negative effects caused by processing, such as enzymatic browning, texture breakdown and off-flavors development [56,57]. In addition, edible coatings may contribute to the reduction of gas exchange, moisture and solute migration as well as to the reduction of the respiration and oxidation reaction rates and the decrease of the risk of pathogens growth on food surfaces during storage [58].

The selection of the most adequate edible materials for this purpose depends on different factors, such as cost, availability, mechanical properties, transparency, brightness, gas barrier effects and resistance to water and microorganisms. The selection is also influenced by the coating processing conditions (pH, type of solvent, temperature), type and concentration of additives (plasticizers, emulsifiers or cross-linking agents) and by the presence of active compounds in the edible matrix (antimicrobials, antioxidants, texture enhancers or nutraceuticals) [56,58].

Plasticizers, such as glycerol or sorbitol, in edible coating formulations play the role of improving coatings flexibility by reducing the number of internal hydrogen bonds between polymer chains and by increasing the free volume in the matrix to permit oxygen and water vapor diffusion through the coating film. Emulsifiers are usually added into coating formulations to improve wettability, stability and adhesion to the food surface, while cross-linking agents are necessary to react with carbohydrates forming a solid polymeric matrix upon the food. In particular, the gelation of low-methoxyl pectins takes place in the presence of multivalent metal cations, such as calcium, due to their electrostatic interactions with the opposite charged cavities formed by polymer chains [59–61].

The most recent trends in the field of pectins coating applications include the shelf-life extension of fresh-cut highly perishable food, the application of pectin coatings as pre-frying treatment to reduce the oil consumption in deep-fat fried products and the use as pre-dried treatments to improve the retention of nutrients and quality characteristics of dehydrated and lyophilized food.

5.1. Shelf-life Extension of Fresh-cut Fruits and Vegetables

Consumers usually accept fresh food on the basis of their appearance and organoleptic characteristics, but their shelf-life is usually low, representing a clear problem for their commercial distribution. The increasing interest for fresh-cut fruits consumption has resulted in the development of physical technologies to extend their shelf-life [62]. The traditional method of storage at low temperatures to slow down the fruits metabolism can be used in combination with other strategies, being modified atmosphere packaging (MAP), thermal treatments (blanching and heat-shock), irradiation, UV light and edible films and coatings the most common preservation methods [63].

Among all these methods, the direct application of edible coatings onto the surface of fresh-cut horticultural products is one of the most studied possibilities. In particular, pectin coatings have been widely evaluated for this purpose by their good oxygen and carbon dioxide barrier, their ability to retard lipid migration and moisture loss while maintaining the sensory and quality of food. Low-methoxyl pectins are mostly used as edible coatings by their ability to form firm gels at low pH in the presence of calcium cations, promoting higher firmness and structural integrity while reducing water vapor permeability [64,65]. Table 3 summarizes the most recent reports on applications of edible pectin-based coatings to extend shelf-life in fresh fruits and vegetables.

In this context, low methoxyl edible pectins coated onto fresh-cut melon improved water vapor resistance, preventing dehydration and maintaining the initial firmness during storage (15 days at 4 °C) by using calcium chloride as cross-linking agent and sunflower oil as lipid source in the preparation of the film-forming emulsion. This edible coating did not reduce the microbiological growth, but could help to decrease the wounding stress of fresh-cut melon, providing antioxidant properties and maintaining quality attributes [66].

Table 3. Pectin-based coatings to extend shelf-life of fresh fruits and vegetables.

Coating	Active Agent	Food Matrix	Effect	Ref.
Low methoxyl pectin (2%), glycerol, sunflower oil, $CaCl_2$	–	Melons (*Cucumis melon* L.)	Antioxidant properties maintaining quality attributes (4 °C, 15 days)	[66]
Pectin (3%), sorbitol, beeswax	–	Mangoes (cv. Ataulfo)	Reduction of physiological changes, respiration rates and weight loss (15 days, room temperature)	[67]
Pectin (3%), sorbitol, beeswax	–	Avocados	Reduction of firmness, color, respiration rates and moisture loss (10 °C, 1 month)	[68]
Osmotic dehydration (with 0.5% calcium lactate) + pectin coating (1%)	–	Ripe Melons (*Cucumis melon* cv. *inodorus*)	Reduction of respiration rate maintaining sensory properties and quality parameters (5 °C, 80% RH, 14 days)	[69]
Pectin (2%) + glycerol combined with MAP	–	Nectarines (*Prunus persica* L. cv. Babygold)	Texture, color and hygienic quality (3 °C, 7 days)	[62]
Multi-layered alginate (1%)-β-CD-trans (2%)/pectin (2%) Calcium lactate	trans-cinnamaldehyde encapsulated in β-Cyclodextrins	Watermelon (*C. lanatus*)	Antimicrobial activity maintaining quality and sensory attributes (4 °C, 12 days)	[70]
Multi-layered chitosan (2%)-β-CD-trans (1%)/pectin (1%) $CaCl_2$	trans-cinnamaldehyde encapsulated in β-Cyclodextrins	Cantaloupe melon	Antimicrobial activity maintaining quality and sensory attributes (4 °C, 9 days)	[71]
Multi-layered chitosan (2%)-β-CD-trans (1%)/pectin (2%) $CaCl_2$	trans-cinnamaldehyde encapsulated in β-Cyclodextrins	Papaya fruits (*Carica papaya* L. cv Maradol)	Antimicrobial activity maintaining quality and sensory attributes (4 °C, 15 days)	[72]
Pectin (2%), glycerol, $CaCl_2$	N-acetylcysteine Glutathione	Pears (*Pyrus communis* L.)	Antibrowning, antimicrobial, antioxidant maintaining sensory attributes (4 °C, 14 days)	[73]
Pectin (3.5%), glycerol	Sodium benzoate; Potassium sorbate	Strawberries (*Fragaria ananassa* cv. Albion)	Antimicrobial activity maintaining physico-chemical and sensory attributes (4 °C, 15 days)	[74]
Pectin (2%), glycerol, $CaCl_2$	Apple fiber; Inulin; Ascorbic acid	"Golden delicious" apples	Nutritional value maintaining firmness, color and antioxidant activity (4 °C, 7 days)	[75]

Table 3. *Cont.*

Coating	Active Agent	Food Matrix	Effect	Ref.
Pectin (2%) CaCl$_2$	Eugenol Citral; Ascorbic acid	Strawberries	Antimicrobial, antioxidant, antibrowning activities maintaining sensory properties (0.5 °C, 7 days)	[76]
Pectin (1%) CaCl$_2$	Eugenol Citral; Ascorbic acid	Raspberries	Antimicrobial, antioxidant, antibrowning activities maintaining sensory properties (0.5 °C, 7 days)	[77]
Pectin (3%) + glycerol	Cinnamon leaf essential oil	Peach (*Prunus persica*)	Antimicrobial, antioxidant activity. Odor acceptability up to 10 days (5 °C)	[78]

Another study reported the use of pectin coatings to restrict the loss of moisture from avocados during storage, decreasing the O_2 absorption and slowing down the respiration rate [68]. The delay in changes on texture and color was observed, concluding that pectin coatings could extend the shelf-life of avocados to over a month at 10 °C. Similar results were reported by Moalemiyan *et al.* in fresh-cut mangoes coated with pectins, which remained in acceptable quality conditions for over two weeks by applying the optimized coating formulation: pectin (1.3% in water), sorbitol (28%) as plasticizer and beeswax (23%) based on pectins dry weight [67]. Authors reported that some optimization of the coating formulation was necessary to prevent the off-flavor development and the anaerobic respiration. A balance between gas and water vapor permeability should be achieved, but it is dependent on the respiration rate of each fruit or vegetable.

A recent report highlighted that the combination of osmotic dehydration (in 40 °Bx sucrose solution containing 0.5% calcium lactate) and pectins coating (1%) was a good preservation alternative for fresh-cut melon by reducing the fruit respiration rate and maintaining the organoleptic properties and quality parameters for 14 days of storage at 5 °C and 80% RH [69].

MAP is another popular technique to extend the shelf-life of fresh-cut fruits by decreasing O_2 and increasing CO_2 levels in the headspace of closed packages, reducing the respiration rate during storage with the positive effect of delaying ageing and senescence [63]. However, anaerobic respiration should be avoided to limit the proliferation of hazardous microorganisms and, consequently, some increase in ethanol production and off-flavors can be observed. In this sense, the combination of low-temperature storage, MAP and edible coatings have shown to be a promising alternative to control respiration of fruits [62]. It was reported that nectarine sections can be stored at 3 °C for 7 days maintaining texture, color and hygienic quality by using pectin, chitosan and sodium caseinate coatings combined with MAP.

Another current trend in food packaging research is the development of active systems with materials containing active compounds in their composition with the possibilities of release at controlled rate to the food surface during storage, resulting in shelf-life extension [79]. It was stated that pectins edible matrices show excellent performance as carriers for active compounds, such as antioxidants, antimicrobials and texture enhancers. The incorporation of active compounds into edible films and coatings has the main goal to reduce, inhibit or stop the microbial growth in food surfaces. The effect on the permeability and mechanical properties of active agents in coating formulations need further studies [58,80]. Herbs and spices are the most common sources of active additives, in particular phenolics, flavonoids and terpenoids [79]. Essential oils derived from plants and their components are also interesting bioactive agents in active packaging formulations. However, their application to packaging of fresh fruits and vegetables is still under investigation by their high volatility, reactivity and strong aroma that could affect the organoleptic characteristics of food. For example, the incorporation of cinnamon leaf essential oil into pectin edible coatings resulted in an effective antimicrobial activity and the enhancing of the antioxidant status of fresh-cut peach, with good acceptation by consumers [78].

The main components of essential oils obtained from herbs and spices have been evaluated as active agents in food packaging applications [79]. Eugenol and citral are examples of active compounds obtained from plant-derived essential oils, since they have been reported as good antimicrobial agents in pectin coatings to extend fresh strawberries and raspberries shelf-life [76,77]. In this context, N-acetylcysteine (0.75% w/v) and glutathione (0.75% w/v) were incorporated into pectin coatings (in the calcium chloride solution) to help in the enzymatic browning control while also retarding the microbial growth in fresh-cut pears and contributing to maintain the high antioxidant potential of this fruit without affecting firmness for 14 days at 4 °C [73]. Sodium benzoate and potassium sorbate are other active compounds successfully incorporated in pectin solutions to form edible coatings [74]. Results showed that pectin active coatings improved the physico-chemical, microbiological and sensory quality (color, flavor, texture and acceptance) of strawberries from 6 (control) to 15 days at 4 °C and 90% RH.

A limited number of studies on the effect of the incorporation of nutraceutical compounds into edible coatings to extend food shelf-life while providing nutritional value have been reported. Moreira *et al.* introduced apple fiber and inulin into pectins solutions as edible coatings in fresh-cut "Golden delicious" apples evaluating the effect on the quality attributes [75]. Ascorbic acid was also added as antioxidant. Results demonstrated that these edible active formulations increased the nutritional value of fresh-cut apples while maintaining the firmness and color and helping to keep their antioxidant potential for seven days of storage at 4 °C.

Multilayer coatings have received recent interest to overcome the problem of the poor adhesion between the commonly hydrophobic coatings and the hydrophilic surface of fruits. In the layer-by-layer technique fruits and vegetables are dipped into different solutions with opposite charged polyelectrolytes with the aim to obtain strong chemical bonds between them [71]. In this context, pectins have been studied in combination with other edible matrices, such as alginate or chitosan. Multilayer coatings were successfully prepared with the layer-by-layer technique by using a solution of sodium alginate containing one antimicrobial agent (trans-cinnamaldehyde encapsulated in β-cyclodextrins (β-CD-trans)) in combination with pectins and calcium lactate solutions to obtain stable and homogeneous coatings [70]. Authors reported that multilayer coatings with 1% alginate, 2% β-CD-trans and 2% pectins maintained the quality and organoleptic attributes of fresh-cut watermelon while extending the shelf life at 4 °C from 7 (control) to 12 days. In addition, microbiological analysis demonstrated the effectiveness of the encapsulated agents in the multilayer coatings against microbial growth. The application of multilayer coatings formed by chitosan (2%), 2% β-CD-trans and pectins (1%) (optimum formulation) was reported with the extension of shelf-life of fresh-cut cantaloupe at 4 °C up to 9 days [71]. Similar multilayer coatings were also successful in prolonging the shelf-life of fresh-cut papaya [72]. In both cases, calcium chloride (2%) was used as cross-linking agent.

5.2. Pectin Coatings in Pre-frying Treatments

Fried foods usually contain a significant amount of saturated and unsaturated fats, in some cases up to 1/3 of the total product mass. This high fats content has been related to obesity and coronary heart diseases. During deep-frying, the internal moisture of food is evaporated while oil is confined to the external surface. These processes are influenced by factors such as oil quality, frying conditions (temperature and time), food composition and pre-frying treatments (blanching and/or drying). The increase in the consumer's interest for more healthy food products has led to the trend of reducing lipid uptake in fried-food products while maintaining acceptable organoleptic characteristics. Edible coatings prepared from food hydrocolloids, such as cellulose and derivatives, gums, alginate, whey protein, albumin or pectins have been studied for such purpose. The high water absorption capacity and retention of the original food firmness of these coatings decrease the moisture loss by evaporation, reducing lipid uptake [56].

In the case of pectin coatings, few investigations have been reported. One of them studied the influence of sunflower head pectin and calcium chloride on the oil absorption in fried-potato chips [16]. Authors concluded that coatings prepared with 1% (w/v) pectins solutions induced by $CaCl_2$ (< 0.1 mol·L^{-1}) effectively decreased the lipid uptake by about 30%. These authors also observed that the

CaCl$_2$ concentration influenced significantly on the brittleness and crispiness of coated chips but the consumer's acceptation was still high.

In another study, the combination of carboxymethyl cellulose (CMC) and pectins (0.5 and 1% w/w respectively) was used as one layer in a double coating, formed by guar and xantham gums (1% w/w), or in a three-multilayer coating with CMC 1% (w/w) as the third layer [81]. Those authors recommended the CMC-pectin mixture for single-layer coatings in French-fries since this combination leads to the lowest lipid uptake. Further reductions in the oil absorption were observed in fried food coated with multilayer systems, but the high moisture content of the final products led to the decrease in the crunchiness and loss of acceptance by consumers.

Other authors observed that the application of pectin coatings (2% w/v) reduced significantly the acrylamide formation in banana chips (around 33%), while helping to slightly improve the organoleptic attributes and crunchiness of the final product [82]. In addition, a synergic effect of the combination of pectin edible coatings and blanching treatments at 90 or 100 °C for 1 min was observed, resulting in higher acrylamide reduction in fried-banana chips up to 91.9% and 90.8%, respectively. In summary, all these results indicate that the use of pectin coatings in pre-frying treatments is a promising process to reduce the lipid uptake and acrylamide content in potato and banana chips, respectively.

5.3. Pectin Coating as Pre-drying Treatment

Drying has been traditionally applied as a method of preservation in fruits and vegetables by reduction of their moisture content, decreasing the action of enzymes and preventing the microbial growth. However, this method implies important physical changes in color and texture, as well as chemical modifications related to the degradation of bioactive compounds sensitive to oxygen, light and heating, such as vitamins, antioxidants or minerals. Edible coatings have been proposed as a possibility to prevent the nutrients loss while improving the organoleptic quality of dried products after different pre-treatments, such as blanching and osmotic dehydration [83].

Some authors have recently reported that pectin coatings on papaya slices increased the moisture diffusion coefficient due to the hydrophilic character of pectins while enhancing the retention of vitamin C, resulting in a significant protective effect against nutritional losses [84]. Pectin coatings were also applied to osmotic dehydrated pineapple in sucrose (50%)/calcium lactate (4%)/ascorbic acid (2%) solutions before the application of the hot-air-drying method [85]. Results showed that pectin edible coatings were effective as oxygen barrier and did not affect the drying efficiency and the water diffusion coefficient. However, dried pineapples coated with pectins showed high retention of vitamin C and slightly changed their lightness. Pectin-based coatings (2% w/w) with and without vitamin C (1% w/w) were also applied to papaya slices before air-drying without reducing the efficiency of the dehydration process [86]. Authors observed that the incorporation of vitamin C on pectin coatings reduced its retention in dried samples after 30 days of storage, while increased significantly the vitamin C content in the final product. In addition, sensorial analysis confirmed that the pectin + vitamin C coating can be used as pre-drying treatment of papaya slices.

A combination of pectins (1.5%), CMC (1.5%) and ascorbic acid (0.6%) was used as edible coating of quince slices prior to their osmotic dehydration, with a subsequent hot-air-drying process [87]. In a first step, the composition of the osmotic solution was optimized (fructose 47.7%, CaCl$_2$ 4% and citric acid 3.5%) and authors reported the important effect of the fructose concentration on water loss. Pectin-CMC-based coatings increased the porosity of the quince dried structure, while the osmotic dehydration increased the collapse of their porous structure. These authors concluded that the decrease observed on the final product firmness could be due to the plasticizing effect of low molecular mass components, which were absorbed from the coating and osmotic solutions.

In summary, all these results suggest that pectin-based coatings can be effectively used as a valid pre-drying treatment constituting a promising alternative to improve the retention of nutrients and quality characteristics of dehydrated food, also opening the possibilities to carry other active additives and nutrients.

6. Conclusions

Much work has been recently carried out to propose new formulations of edible coatings based on pectins to be applied in the protection of fresh food, improving organoleptic and nutritional characteristics and extending shelf-life. Despite the fact that recent studies have reported significant improvements in specific applications of these formulations, there is still a large amount of work to be performed since some of the most remarkable improvements are not yet experimentally attained or reproducible at large scale. In general terms, there is still a need for a better understanding of the composition-structure-processing-properties relationships in edible films based on pectins for food packaging, both at the laboratory and industrial scale. Moreover, since many of the studies related to this issue have been carried out using some basic pectins already in the market, there is still a lot of room for variation and maturation in the development of edible coatings for application in food packaging.

Author Contributions: All authors performed the literatura survey and wrote the paper.

Conflicts of Interest: The authors declare no conflict of interest.

References

1. Advancing Sustainable Materials Magement: Facts and Figures 2013. Available online: http://www.epa.gov/wastes/nonhaz/municipal/pubs/2013_advncng_smm_rpt.pdf (accessed on 25 September 2015).
2. Collection and transport. Waste Management World Homepage. Available online: http://www.waste-management-world.com/collection-transport (accessed on 25 September 2015).
3. Jamshidian, M.; Tehrany, E.A.; Imran, M.; Jacquot, M.; Desobry, S. Poly-lactic acid: Production, applications, nanocomposites, and release studies. *Compr. Rev. Food Sci. Food Saf.* **2010**, *9*, 552–571. [CrossRef]
4. Martucci, J.F.; Ruseckaite, R.A. Biodegradable three-layer film derived from bovine gelatin. *J. Food Eng.* **2010**, *99*, 377–383. [CrossRef]
5. Kechichian, V.; Ditchfield, C.; Veiga-Santos, P.; Tadini, C.C. Natural antimicrobial ingredients incorporated in biodegradable films based on cassava starch. *LWT-Food Sci. Technol.* **2010**, *43*, 1088–1094. [CrossRef]
6. Arzu, A.B.; Tulay, O.; Oya, I.S.; Lutfiye, Y.E. The utilisation of microbial poly-hydroxy alkanoates (PHA) in food industry. *Res. J. Biotechnol.* **2010**, *5*, 76–79.
7. Ahmed, J.; Varshney, S.K. Polylactides—Chemistry, properties and green packaging technology: A review. *Int. J. Food Prop.* **2011**, *14*, 37–58. [CrossRef]
8. Auras, R.; Harte, B.; Selke, S. An overview of polylactides as packaging materials. *Macromol. Biosci.* **2004**, *4*, 835–864. [CrossRef] [PubMed]
9. Anuradha, K.; Padma, P.N.; Venkateshwar, S.; Reddy, G. Fungal isolates from natural pectic substrates for polygalacturonase and multienzyme production. *Indian J. Microbiol.* **2010**, *50*, 339–344. [CrossRef] [PubMed]
10. Lopes da Silva, J.A.; Rao, M.A. *Food Polysaccharides and Their Applications*, 2nd ed.; Taylor & Francis: Abingdon, UK, 2006.
11. Kohli, P.; Gupta, R. Alkaline pectinases: A review. *Biocatal. Agric. Biotechnol.* **2015**, *4*, 279–285. [CrossRef]
12. Pedrolli, D.B.; Monteiro, A.C.; Gomes, E.; Carmona, E.C. Pectin and pectinases: Production, characterization and industrial application of microbial pectinolytic enzymes. *Open Biotechnol. J.* **2009**, *3*, 9–18. [CrossRef]
13. Ridley, B.L.; O'Neill, M.A.; Mohnen, D. Pectins: Structure, biosynthesis, and oligogalacturonide-related signaling. *Phytochemistry* **2001**, *57*, 929–967. [CrossRef]
14. Caffall, K.H.; Mohnen, D. The structure, function, and biosynthesis of plant cell wall pectic polysaccharides. *Carbohydr. Res.* **2009**, *344*, 1879–1900. [CrossRef] [PubMed]
15. Ovodov, Y.S. Current views on pectin substances. *Russ. J. Bioorg. Chem.* **2009**, *35*, 269–284. [CrossRef]
16. Hua, X.; Wang, K.; Yang, R.; Kang, J.; Yang, H. Edible coatings from sunflower head pectin to reduce lipid uptake in fried potato chips. *LWT-Food Sci. Technol.* **2015**, *62*, 1220–1225. [CrossRef]
17. Ciolacu, L.; Nicolau, A.I.; Hoorfar, J. *Global Safety of Fresh Produce. A Handbook of Best Practice, Innovative Commercial Solutions and Case Studies*; Woodhead Publishing Limited: Sawston, UK, 2014.
18. Dhanapal, A.; Sasikala, P.; Rajamani, L.; Kavitha, V.; Yazhini, G. Edible films from polysaccharides. *Food Sci. Qual. Manag.* **2012**, *3*, 9–18.

19. Giovanetti, M.H.; Nogueira, A.; de Oliveira, C.L.; Wosiacki, G. *Chromatography—The Most Versatile Method of Chemical Analysis*; InTech: Rijeka, Croatia, 2012.
20. Sundari, N. Extraction of pectin from waste peels: A review. *Res. J. Pharm. Biol.* **2015**, *6*, 1841–1848.
21. Alba, K.; Laws, A.P.; Kontogiorgos, V. Isolation and characterization of acetylated lm-pectins extracted from okra pods. *Food Hydrocoll.* **2015**, *43*, 726–735. [CrossRef]
22. Maran, J.P. Statistical optimization of aqueous extraction of pectin from waste durian rinds. *Int. J. Biol. Macromol.* **2015**, *73*, 92–98. [CrossRef] [PubMed]
23. Geerkens, C.H.; Nagel, A.; Just, K.M.; Miller-Rostek, P.; Kammerer, D.R.; Schweiggert, R.M.; Carle, R. Mango pectin quality as influenced by cultivar, ripeness, peel particle size, blanching, drying, and irradiation. *Food Hydrocoll.* **2015**, *51*, 241–251. [CrossRef]
24. Oliveira, T.Í.S.; Rosa, M.F.; Cavalcante, F.L.; Pereira, P.H.F.; Moates, G.K.; Wellner, N.; Mazzetto, S.E.; Waldron, K.W.; Azeredo, H.M.C. Optimization of pectin extraction from banana peels with citric acid by using response surface methodology. *Food Chem.* **2015**, in press. [CrossRef]
25. Castillo-Israel, K.A.T.; Baguio, S.F.; Diasanta, M.D.B.; Lizardo, R.C.M.; Dizon, E.I.; Mejico, M.I.F. Extraction and characterization of pectin from Saba banana [*musa 'saba'(musa acuminata x musa balbisiana)*] peel wastes: A preliminary study. *Int. Food Res. J.* **2015**, *22*, 202–207.
26. Liew, S.Q.; Chin, N.L.; Yusof, Y.A.; Cheok, C.Y. Citric acid extraction of pectin from tropical fruit peels of passion fruit, dragon fruit and soursop. *J. Food Agric. Environ.* **2015**, *13*, 45–51.
27. Li, D.Q.; Du, G.M.; Jing, W.W.; Li, J.F.; Yan, J.Y.; Liu, Z.Y. Combined effects of independent variables on yield and protein content of pectin extracted from sugar beet pulp by citric acid. *Carbohyd. Polym.* **2015**, *129*, 108–114. [CrossRef] [PubMed]
28. Sánchez Aldana, D.; Contreras-Esquivel, J.C.; Nevárez-Moorillón, G.V.; Aguilar, C.N. Characterization of edible films from pectic extracts and essential oil from mexican lime. *CyTA J. Food.* **2014**, *13*, 17–25. [CrossRef]
29. Casas-Orozco, D.; Villa, A.L.; Bustamante, F.; González, L.M. Process development and simulation of pectin extraction from orange peels. *Food Bioprod. Process.* **2015**, *96*, 86–98. [CrossRef]
30. Korish, M. Potential utilization of citrullus lanatus var. Colocynthoides waste as a novel source of pectin. *J. Food Sci. Technol.* **2015**, *52*, 2401–2407. [CrossRef] [PubMed]
31. Korish, M. Faba bean hulls as a potential source of pectin. *J. Food Sci. Technol.* **2015**, *52*, 6061–6066. [CrossRef] [PubMed]
32. Denman, L.J.; Morris, G.A. An experimental design approach to the chemical characterisation of pectin polysaccharides extracted from cucumis melo inodorus. *Carbohydr. Polym.* **2015**, *117*, 364–369. [CrossRef] [PubMed]
33. Grassino, A.N.; Halambek, J.; Djaković, S.; Rimac Brnčić, S.; Dent, M.; Grabarić, Z. Utilization of tomato peel waste from canning factory as a potential source for pectin production and application as tin corrosion inhibitor. *Food Hydrocoll.* **2016**, *52*, 265–274. [CrossRef]
34. Kang, J.; Hua, X.; Yang, R.; Chen, Y.; Yang, H. Characterization of natural low-methoxyl pectin from sunflower head extracted by sodium citrate and purified by ultrafiltration. *Food Chem.* **2015**, *180*, 98–105. [CrossRef] [PubMed]
35. Giosafatto, C.V.L.; di Pierro, P.; Gunning, P.; Mackie, A.; Porta, R.; Mariniello, L. Characterization of citrus pectin edible films containing transglutaminase-modified phaseolin. *Carbohydr. Polym.* **2014**, *106*, 200–208. [CrossRef] [PubMed]
36. Giosafatto, C.V.L.; Di Pierro, P.; Gunning, A.P.; Mackie, A.; Porta, R.; Mariniello, L. Trehalose-containing hydrocolloid edible films prepared in the presence of transglutaminase. *Biopolymers* **2014**, *101*, 931–937. [CrossRef] [PubMed]
37. Porta, R.; Mariniello, L.; di Pierro, P.; Sorrentino, A.; Giosafatto, C.V.L. Transglutaminase crosslinked pectin- and chitosan-based edible films: A review. *Crit. Rev. Food Sci. Nutr.* **2011**, *51*, 223–238. [CrossRef] [PubMed]
38. Di Pierro, P.; Sorrentino, A.; Mariniello, L.; Giosafatto, C.V.L.; Porta, R. Chitosan/whey protein film as active coating to extend ricotta cheese shelf-life. *LWT Food Sci. Technol.* **2011**, *44*, 2324–2327. [CrossRef]
39. Rossi Marquez, G.; di Pierro, P.; Esposito, M.; Mariniello, L.; Porta, R. Application of transglutaminase-crosslinked whey protein/pectin films as water barrier coatings in fried and baked foods. *Food Bioprocess Technol.* **2014**, *7*, 447–455. [CrossRef]

40. Thirugnanasambandham, K.; Sivakumar, V.; Prakash Maran, J. Process optimization and analysis of microwave assisted extraction of pectin from dragon fruit peel. *Carbohyd. Polym.* **2014**, *112*, 622–626. [CrossRef] [PubMed]
41. Rahmati, S.; Abdullah, A.; Momeny, E.; Kang, O.L. Optimization studies on microwave assisted extraction of dragon fruit (*Hylocereus polyrhizus*) peel pectin using response surface methodology. *Int. Food Res. J.* **2015**, *22*, 233–239.
42. Quoc, L.P.T.; Huyen, V.T.N.; Hue, L.T.N.; Hue, N.T.H.; Thuan, N.H.D.; Tam, N.T.T.; Thuan, N.N.; Duy, T.H. Extraction of pectin from pomelo (*Citrus maxima*) peels with the assistance of microwave and tartaric acid. *Int. Food Res. J.* **2015**, *22*, 1637–1641.
43. Maran, J.P.; Swathi, K.; Jeevitha, P.; Jayalakshmi, J.; Ashvini, G. Microwave-assisted extraction of pectic polysaccharide from waste mango peel. *Carbohyd. Polym.* **2015**, *123*, 67–71. [CrossRef] [PubMed]
44. Maran, J.P.; Prakash, K.A. Process variables influence on microwave assisted extraction of pectin from waste *Carcia papaya* L. peel. *Int. J. Biol. Macromol.* **2015**, *73*, 202–206. [CrossRef] [PubMed]
45. Zouambia, Y.; Youcef Ettoumi, K.; Krea, M.; Moulai-Mostefa, N. A new approach for pectin extraction: Electromagnetic induction heating. *Arabian J. Chem.* **2014**. [CrossRef]
46. Moorthy, I.G.; Maran, J.P.; Surya, S.M.; Naganyashree, S.; Shivamathi, C.S. Response surface optimization of ultrasound assisted extraction of pectin from pomegranate peel. *Int. J. Biol. Macromol.* **2015**, *72*, 1323–1328. [CrossRef] [PubMed]
47. Maran, J.P.; Priya, B. Ultrasound-assisted extraction of pectin from sisal waste. *Carbohyd. Polym.* **2015**, *115*, 732–738. [CrossRef] [PubMed]
48. Chen, H.M.; Fu, X.; Luo, Z.G. Properties and extraction of pectin-enriched materials from sugar beet pulp by ultrasonic-assisted treatment combined with subcritical water. *Food Chem.* **2015**, *168*, 302–310. [CrossRef] [PubMed]
49. Wang, W.; Ma, X.; Xu, Y.; Cao, Y.; Jiang, Z.; Ding, T.; Ye, X.; Liu, D. Ultrasound-assisted heating extraction of pectin from grapefruit peel: Optimization and comparison with the conventional method. *Food Chem.* **2015**, *178*, 106–114. [CrossRef] [PubMed]
50. Yuliarti, O.; Matia-Merino, L.; Goh, K.K.T.; Mawson, J.; Williams, M.A.K.; Brennan, C. Characterization of gold kiwifruit pectin from fruit of different maturities and extraction methods. *Food Chem.* **2015**, *166*, 479–485. [CrossRef] [PubMed]
51. Yuliarti, O.; Goh, K.K.T.; Matia-Merino, L.; Mawson, J.; Brennan, C. Extraction and characterisation of pomace pectin from gold kiwifruit (actinidia chinensis). *Food Chem.* **2015**, *187*, 290–296. [CrossRef] [PubMed]
52. Valdés, A.; Vidal, L.; Beltrán, A.; Canals, A.; Garrigós, M.C. Microwave-assisted extraction of phenolic compounds from almond skin byproducts (prunus amygdalus): A multivariate analysis approach. *J. Agric. Food Chem.* **2015**, *63*, 5395–5402. [CrossRef] [PubMed]
53. Ballard, T.S.; Mallikarjunan, P.; Zhou, K.; O'Keefe, S. Microwave-assisted extraction of phenolic antioxidant compounds from peanut skins. *Food Chem.* **2010**, *120*, 1185–1192. [CrossRef]
54. Song, J.; Li, D.; Liu, C.; Zhang, Y. Optimized microwave-assisted extraction of total phenolics (TP) from ipomoea batatas leaves and its antioxidant activity. *Innov. Food Sci. Emerg. Technol.* **2011**, *12*, 282–287. [CrossRef]
55. Montgomery, D.C. *Design and Analysis of Experiments*; Wiley: Hoboken, NJ, USA, 2009.
56. Falguera, V.; Quintero, J.P.; Jiménez, A.; Muñoz, J.A.; Ibarz, A. Edible films and coatings: Structures, active functions and trends in their use. *Trends Food Sci. Technol.* **2011**, *22*, 292–303. [CrossRef]
57. Sánchez-Ortega, I.; García-Almendárez, B.E.; Santos-López, E.M.; Amaro-Reyes, A.; Barboza-Corona, J.E.; Regalado, C. Antimicrobial edible films and coatings for meat and meat products preservation. *Sci. World J.* **2014**. [CrossRef] [PubMed]
58. Rojas-Graü, M.A.; Soliva-Fortuny, R.; Martín-Belloso, O. Edible coatings to incorporate active ingredients to fresh-cut fruits: A review. *Trends Food Sci. Technol.* **2009**, *20*, 438–447. [CrossRef]
59. Rojas, M.A. Use of edible coatings for fresh-cut fruits and vegetables. In *Advances in Fresh-cut Fruits and Vegetables Processing*; CRC Press: Boca Raton, FL, USA, 2010; pp. 285–311.
60. Espitia, P.J.P.; Du, W.-X.; Avena-Bustillos, R.d.J.; Soares, N.d.F.F.; McHugh, T.H. Edible films from pectin: Physical-mechanical and antimicrobial properties—A review. *Food Hydrocolloid* **2014**, *35*, 287–296. [CrossRef]

61. Martín-Diana, A.B.; Rico, D.; Frías, J.M.; Barat, J.M.; Henehan, G.T.M.; Barry-Ryan, C. Calcium for extending the shelf life of fresh whole and minimally processed fruits and vegetables: A review. *Trends Food Sci. Technol.* **2007**, *18*, 210–218. [CrossRef]
62. Ramirez, M.E.; Timón, M.L.; Petrón, M.J.; Andrés, A.I. Effect of chitosan, pectin and sodium caseinate edible coatings on shelf life of fresh-cut *Prunus persica* var. Nectarine. *J. Food Process. Preserv.* **2015**. [CrossRef]
63. Rico, D.; Martín-Diana, A.B.; Barat, J.M.; Barry-Ryan, C. Extending and measuring the quality of fresh-cut fruit and vegetables: A review. *Trends Food Sci. Technol.* **2007**, *18*, 373–386. [CrossRef]
64. Zhang, Y.; Rempel, C.; McLaren, D. Edible coating and film materials: Carbohydrates. In *Innovations in Food Packaging*, 2nd ed.; Elsevier: Amsterdam, The Netherlands, 2014; pp. 305–323.
65. Ciolacu, L.; Nicolau, A.I.; Hoorfar, J. Edible coatings for fresh and minimally processed fruits and vegetables. In *Global Safety of Fresh Produce*; Hoorfar, J., Ed.; Woodhead Publishing: Sawston, UK, 2014; pp. 233–244.
66. Oms-Oliu, G.; Soliva-Fortuny, R.; Martín-Belloso, O. Using polysaccharide-based edible coatings to enhance quality and antioxidant properties of fresh-cut melon. *LWT Food Sci. Technol.* **2008**, *41*, 1862–1870. [CrossRef]
67. Moalemiyan, M.; Ramaswamy, H.S.; Maftoonazad, N. Pectin-based edible coating for shelf-life extension of ataulfo mango. *J. Food Process Eng.* **2012**, *35*, 572–600. [CrossRef]
68. Maftoonazad, N.; Ramaswamy, H.S. Effect of pectin-based coating on the kinetics of quality change associated with stored avocados. *J. Food Process. Preserv.* **2008**, *32*, 621–643. [CrossRef]
69. Ferrari, C.C.; Sarantópoulos, C.I.G.L.; Carmello-Guerreiro, S.M.; Hubinger, M.D. Effect of osmotic dehydration and pectin edible coatings on quality and shelf life of fresh-cut melon. *Food Bioprocess Technol.* **2013**, *6*, 80–91. [CrossRef]
70. Sipahi, R.E.; Castell-Perez, M.E.; Moreira, R.G.; Gomes, C.; Castillo, A. Improved multilayered antimicrobial alginate-based edible coating extends the shelf life of fresh-cut watermelon (citrullus lanatus). *LWT Food Sci. Technol.* **2013**, *51*, 9–15. [CrossRef]
71. Martiñon, M.E.; Moreira, R.G.; Castell-Perez, M.E.; Gomes, C. Development of a multilayered antimicrobial edible coating for shelf-life extension of fresh-cut cantaloupe (*Cucumis melo* L.) stored at 4 °C. *LWT-Food Sci. Technol.* **2014**, *56*, 341–350. [CrossRef]
72. Brasil, I.M.; Gomes, C.; Puerta-Gomez, A.; Castell-Perez, M.E.; Moreira, R.G. Polysaccharide-based multilayered antimicrobial edible coating enhances quality of fresh-cut papaya. *LWT-Food Sci. Technol.* **2012**, *47*, 39–45. [CrossRef]
73. Oms-Oliu, G.; Soliva-Fortuny, R.; Martín-Belloso, O. Edible coatings with antibrowning agents to maintain sensory quality and antioxidant properties of fresh-cut pears. *Postharvest Biol. Technol.* **2008**, *50*, 87–94. [CrossRef]
74. Treviño-Garza, M.Z.; García, S.; Flores-González, M.S.; Arévalo-Niño, K. Edible active coatings based on pectin, pullulan, and chitosan increase quality and shelf life of strawberries (*Fragaria ananassa*). *J. Food Sci.* **2015**, *80*, M1823–M1830. [CrossRef] [PubMed]
75. Moreira, M.R.; Cassani, L.; Martín-Belloso, O.; Soliva-Fortuny, R. Effects of polysaccharide-based edible coatings enriched with dietary fiber on quality attributes of fresh-cut apples. *J. Food Sci. Technol.* **2015**. [CrossRef]
76. Guerreiro, A.C.; Gago, C.M.L.; Faleiro, M.L.; Miguel, M.G.C.; Antunes, M.D.C. The use of polysaccharide-based edible coatings enriched with essential oils to improve shelf-life of strawberries. *Postharvest Biol. Technol.* **2015**, *110*, 51–60. [CrossRef]
77. Guerreiro, A.C.; Gago, C.M.L.; Faleiro, M.L.; Miguel, M.G.C.; Antunes, M.D.C. Raspberry fresh fruit quality as affected by pectin- and alginate-based edible coatings enriched with essential oils. *Sci. Hortic.* **2015**, *194*, 138–146. [CrossRef]
78. Ayala-Zavala, J.F.; Silva-Espinoza, B.A.; Cruz-Valenzuela, M.R.; Leyva, J.M.; Ortega-Ramírez, L.A.; Carrazco-Lugo, D.K.; Pérez-Carlón, J.J.; Melgarejo-Flores, B.G.; González-Aguilar, G.A.; Miranda, M.R.A. Pectin–cinnamon leaf oil coatings add antioxidant and antibacterial properties to fresh-cut peach. *Flavour Fragr. J.* **2013**, *28*, 39–45. [CrossRef]
79. Valdes, A.; Mellinas, A.C.; Ramos, M.; Burgos, N.; Jimenez, A.; Garrigos, M.C. Use of herbs, spices and their bioactive compounds in active food packaging. *RSC Adv.* **2015**, *5*, 40324–40335. [CrossRef]
80. Mellinas, C.; Valdés, A.; Ramos, M.; Burgos, N.; del Carmen Garrigós, M.; Jiménez, A. Active edible films: Current state and future trends. *J. Appl. Polym. Sci.* **2015**. [CrossRef]

81. Daraei Garmakhany, A.; Mirzaei, H.O.; Maghsudlo, Y.; Kashaninejad, M.; Jafari, S.M. Production of low fat french-fries with single and multi-layer hydrocolloid coatings. *J. Food Sci. Technol.* **2014**, *51*, 1334–1341. [CrossRef] [PubMed]
82. Suyatma, N.E.; Ulfah, K.; Prangdimurti, E.; Ishikawa, Y. Effect of blanching and pectin coating as pre-frying treatments to reduce acrylamide formation in banana chips. *Int. Food Res. J.* **2015**, *22*, 936–942.
83. Oliveira, S.M.; Brandão, T.R.S.; Silva, C.L.M. Influence of drying processes and pretreatments on nutritional and bioactive characteristics of dried vegetables: A review. *Food Eng. Rev.* **2015**. [CrossRef]
84. Garcia, C.C.; Caetano, L.C.; de Souza Silva, K.; Mauro, M.A. Influence of edible coating on the drying and quality of papaya (*Carica papaya*). *Food Bioprocess Technol.* **2014**, *7*, 2828–2839. [CrossRef]
85. Silva, K.S.; Garcia, C.C.; Amado, L.R.; Mauro, M.A. Effects of edible coatings on convective drying and characteristics of the dried pineapple. *Food Bioprocess Technol.* **2015**, *8*, 1465–1475. [CrossRef]
86. Canizares, D.; Mauro, M.A. Enhancement of quality and stability of dried papaya by pectin-based coatings as air-drying pretreatment. *Food Bioprocess Technol.* **2015**, *8*, 1187–1197. [CrossRef]
87. Akbarian, M.; Ghanbarzadeh, B.; Sowti, M.; Dehghannya, J. Effects of pectin-CMC-based coating and osmotic dehydration pretreatments on microstructure and texture of the hot-air dried quince slices. *J. Food Process. Preserv.* **2015**, *39*, 260–269. [CrossRef]

© 2015 by the authors. Licensee MDPI, Basel, Switzerland. This article is an open access article distributed under the terms and conditions of the Creative Commons Attribution (CC BY) license (http://creativecommons.org/licenses/by/4.0/).

Article

Technological Strategies to Preserve Burrata Cheese Quality

Cristina Costa [1], Annalisa Lucera [1], Amalia Conte [1,*], Angelo Vittorio Zambrini [2] and Matteo Alessandro Del Nobile [1]

[1] Department of Agricultural Sciences, Food and Environment, University of Foggia, Via Napoli, 25–71122 Foggia, Italy; cristina.costa@unifg.it (C.C.); annalisa.lucera@unifg.it (A.L.); matteo.delnobile@unifg.it (M.A.D.N.)
[2] Department of Quality, Innovation, Safety, Environment, Granarolo S.p.A., Via Cadriano, 27/2–40127 Bologna, Italy; vittorio.zambrini@granarolo.it
* Correspondence: amalia.conte@unifg.it; Tel.: +39-881-589-240

Received: 4 May 2017; Accepted: 4 July 2017; Published: 9 July 2017

Abstract: Burrata cheese is a very perishable product due to microbial proliferation and undesirable sensory changes. In this work, a step-by-step optimization approach was used to design proper processing and packaging conditions for burrata in brine. In particular, four different steps were carried out to extend its shelf life. Different headspace gas compositions (MAP-1 30:70 $CO_2:N_2$; MAP-2 50:50 $CO_2:N_2$ and MAP-3 65:35 $CO_2:N_2$) were firstly tested. To further promote product preservation, a coating was also optimized. Then, antimicrobial compounds in the filling of the burrata cheese (lysozyme and Na_2-EDTA) and later in the coating (enzymatic complex and silver nanoparticles) were analyzed. To evaluate the quality of the samples, in each step headspace gas composition, microbial population, and pH and sensory attributes were monitored during storage at 8 ± 1 °C. The results highlight that the antimicrobial compounds in the stracciatella, coating with silver nanoparticles, and packaging under MAP-3 represent effective conditions to guarantee product preservation, moving burrata shelf life from three days (control sample) to ten days.

Keywords: burrata cheese; shelf life; antimicrobial coating; packaging design

1. Introduction

Burrata is a fresh Italian cheese, typically produced in the Apulia region, made from fiordilatte paste and a cream called stracciatella. The outer shell is solid fiordilatte paste while the inside contains both fiordilatte pieces and cream, giving it a soft texture. Burrata is traditionally stored under refrigerated conditions, like fiordilatte and mozzarella cheese, and it can be packaged with or without brine in atmospheric conditions. Due to the high moisture and fat content, this dairy product results in rapid spoilage. It is well-known that the shelf life of fresh cheese is influenced by both microbial and sensory changes [1–3]. To preserve the characteristics of fresh cheese during storage several techniques have been suggested in the literature [4–6]. Several authors demonstrated the effects of modified atmosphere conditions (MAP) on extending the shelf life of fresh cheese [2,3,7].

In particular, the headspace gas composition characterized by a low concentration or lack of oxygen and high carbon dioxide improves the microbiological stability, reduces lipid oxidation and increases sensory acceptability [1,2,8]. A proper successful gas combination for burrata cheese packaged without brine was proposed by Conte et al. [9]. The application of antimicrobial compounds, such as essential oils, organic acids, bacteriocins and nanoparticles, was also found to be a strategic solution to improve the shelf life of different fresh cheeses [6,10–13]. To retard the microbial spoilage of dairy products, the antimicrobial compounds can be directly added into the product formulation, carried by coatings or loaded into the packaging materials. Among them, a great deal of attention has been devoted to inorganic nanoparticles (i.e., silver and copper) for their proven effects on the microbial stability of various foodstuffs [13–18]. In particular, scientific research has reported that silver nanoparticles immobilized on packaging materials or loaded in coatings improves the shelf life of mozzarella [12,17,19]. Enzymes can also play an interesting role in antimicrobial activity [6].

In particular, lysozyme, an enzyme found in many natural systems, shows antimicrobial activity against gram-positive bacteria more than gram-negative; however, the susceptibility of gram-negative bacteria can be enhanced by chelating agents such as ethylene diamine tetraacetic acid disodium salt (Na_2-EDTA). Different works have reported the antimicrobial activity of lysozyme alone or in combination with Na_2-EDTA [9,20,21]. The lactoperoxidase system—an enzymatic complex characterized by the lactoperoxidase enzyme, thiocyanate and hydrogen peroxide—is also active against gram-positive bacteria such as *Listeria monocytogenes* and *Staphylococcus aureus*, as well as gram-negative bacteria, including *Escherichia coli* [22,23]. The activity of the lactoperoxidase system has been verified on many bacterial species and pathogens in cow milk and dairy products [24,25].

To the best of our knowledge, no studies have reported on the packaging of burrata cheese with brine. The brine solution better preserves product hydration but it also concurs with the development of spoilage microorganisms. Therefore, the aim of this study was to evaluate step by step the influence of MAP, lysozyme and Na_2-EDTA in the burrata filling and the adoption of a proper active coating on burrata cheese shelf life. To assess the influence of the selected strategies, headspace gas composition, pH, spoilage microorganisms and sensory quality were monitored during the storage period.

2. Materials and Methods

2.1. Materials

Samples of burrata cheese (pieces of 150 g) were kindly provided by "Capurso Azienda Casearia SPA" (Gioia del Colle, Bari, Italy). The sodium alginic acid and the calcium chloride, as well as lysozyme, Na_2-EDTA and lactoperoxydase system (Sea-i® F75) were purchased from Perrin's Chemicals (Triggiano, Italy). Silver-montmorillonite nanoparticles were prepared by ion exchange reaction between Na^+-montmorillonite and silver nitrate solutions, as reported in detail by Incoronato et al. [12]. Polypropylene trays and multilayer film (oriented polyamide/polypropylen 75/15 µm) for tray tops were purchased from Orved (Musile di Piave, Venezia, Italy).

2.2. Sample Preparation

Table 1 briefly reports the different strategies applied to the burrata in each step. The details are reported below.

- Step 1: Two burrata samples were placed in a tray containing brine constituted by a NaCl solution (6 g·L^{-1}). Samples were packaged under MAP conditions (MAP-1 30:70 CO_2:N_2; MAP-2 50:50 CO_2:N_2 and MAP-3 65:35 CO_2:N_2) by means of a thermo-sealing machine (Orved, Musile di Piave, Venezia, Italy).
- Step 2: In the second step, the burrata cheese was coated by immersing samples first in a sodium alginic acid solution (2% w/v), then a solution of calcium chloride (5% w/v) was used to promote the alginate-gel-forming process by dipping the product for one min. Two coated samples were packaged with brine in air (Coat-Air) and under MAP (65:35 CO_2:N_2) (Coat-MAP).
- Step 3: In this step, the antimicrobial compounds were added to the burrata filling during the production process. In particular, lysozyme (500 mg·kg^{-1}) and Na_2-EDTA (50 mM) were dissolved in the cream and then mixed with the fiordilatte pieces. Samples of burrata with the antimicrobial compounds were packaged in air (LysEDTA-Air) and under MAP (65:35 CO_2:N_2) (LysEDTA-MAP). Moreover, burrata with the antimicrobial compounds was coated as described in Step 2 and packaged under MAP (65:35 CO_2:N_2) [Coat-LysEDTA].

- Step 4: In this step, burrata cheese with the antimicrobial compounds (lysozyme and Na_2-EDTA) was produced as described in Step 3. Subsequently, the samples were coated with alginic acid loaded with silver nanoparticles (250 mg·kg^{-1}) [NanoAg-A] or the lactoperoxidase system (10,000 mg·kg^{-1}) [LPX-A] and packaged under MAP (65:35 CO_2:N_2).

Table 1. Experimental step to improve burrata cheese shelf life.

Experimental Steps		Antimicrobial Compounds in the Burrata Filling			Coating with Antimicrobial Compound			Headspace Gas Composition			
		No. Antimicrobial	Lysozyme/ Na_2-EDTA	Absent	No. Antimicrobial	Silver Nanoparticles	Lactoperoxi- dase system	Air	MAP-1	MAP-2	MAP-3
Step-1	Cntr-1	✓		✓				✓			
	MAP-1	✓		✓					✓		
	MAP-2	✓		✓						✓	
	MAP-3	✓		✓							✓
Step-2	Cntr-2	✓		✓				✓			
	Coat-Air	✓			✓			✓			
	Coat-MAP	✓			✓						✓
Step-3	Cntr-3	✓		✓				✓			
	LysEDTA-Air		✓	✓				✓			
	LysEDTA-MAP		✓	✓							✓
	Coat-LysEDTA		✓		✓						✓
Step-4	Cntr-4	✓		✓				✓			
	Cntr-A		✓	✓				✓			
	NanoAg-A		✓			✓					✓
	LPX-A		✓				✓				✓

Due to the variability of the raw material, a control sample without coating and antimicrobial compounds and packaged in air was also prepared in each step (Cntr-1; Cntr-2; Cntr-3; Cntr-4). All the burrata samples were stored at 8 ± 1 °C.

2.3. Microbiological Analyses and Determination of pH

The burrata cheese (20 g) was diluted in a sterile saline solution (0.9%) and homogenized in a blender (Stomacher, International PBI, Milan, Italy). Decimal dilutions of homogenate cheese were made in a saline solution and plated on selective media for the determination of lactic acid bacteria, lactococci, total bacterial count, Enterobacteriaceae and *Pseudomonas spp*. The media and the conditions were the following: de Man Rogosa and Sharpe (MRS) agar, supplemented with cycloheximide (0.17 g·L^{-1}) (Sigma, Milan, Italy) incubated under anaerobic conditions at 37 °C for 48 h for lactic acid bacteria; M17 agar, incubated at 37 °C for 48 h for lactococci; plate count agar (PCA), incubated at 30 °C for 48 h for total bacteria count; Violet Red Bile Glucose Agar (VRBGA) incubated at 37 °C for 24 h for Enterobacteriaceae; Pseudomonas Agar Base (PAB), added with SR103 selective supplement (Oxoid, Milan, Italy) and incubated at 25 °C for 48 h for *Pseudomonas spp*. All media were purchased from Oxoid (Milan, Italy). All the analyses were performed in duplicate.

At each sampling time the pH of the burrata and brine solution were also measured by a pH meter (Crison, Barcelona, Spain) with an accuracy of 0.01 pH. The instrument was calibrated with 4.01 and 7.00 pH buffer solutions. The measurements were done in duplicate on two different samples.

2.4. Headspace Gas Composition

Oxygen and carbon dioxide headspace concentrations of packaged burrata were measured using a gas meter (PBI Dansensor, Checkmate 9900, Ringsted, Denmark), previously calibrated by the company at 23 ± 2 °C and a pressure of 1013 ± 50 hPa. The volume taken from the package headspace for gas analysis was about 10 cm^3. To avoid modifications in the headspace gas composition each package was used only for a single measurement of gas composition. Two packages were used at each sampling time.

2.5. Sensory Analysis

Sensory evaluation was carried out by seven trained panelists, members belonging to the food packaging laboratory. The selection of the panelists was made considering various aspects: interest and motivation, eating habits (consumption of fresh dairy products), ability to communicate sensations, time available for analysis sessions, ability to concentrate, and performance training. Eight sessions of 1 h each were required to define the sensory profile, with a frequency of three meetings a week. The sessions were used to familiarize the testers with the characteristics of burrata cheese samples in terms of odor, color, texture and overall quality [26,27]. The samples coated with alginic acid were presented with and without coating to the panelists in a random order [27]. The quality of the attributes (firmness, color, odor and overall quality) of the cheese samples were evaluated. The intensity of each attribute was quantified on a 7-point scale where 1 corresponded to fully poor quality, 2 to poor quality, 3 inadequate quality, 4 represented the threshold for product acceptability, 5 good quality, 6 very good quality, 7 excellent quality [2].

2.6. Shelf Life Calculation

For each step, to calculate the shelf life values of the burrata cheese, the microbial acceptability limit (MAL) and the sensory acceptability limit (SAL) were assessed with a re-parameterized Gompertz equation [2,26]. Shelf life was assumed to be the lowest value between MAL and SAL. In particular, the following equation was used to calculate the MAL:

$$Log(N(t)) = \log(N_{max}) - A \times e^{-e^{[\{[u_{max} \times 2.71 \frac{\lambda - MAL}{A}]+1\}]}} + A \times e^{-e^{[\{[u_{max} \times 2.71 \frac{\lambda - t}{A}]+1\}]}} \quad (1)$$

where $N(t)$ is the viable cell concentration (CFU/g) at storage time t; A is related to the difference between the decimal logarithm of maximum bacterial growth attained at the stationary phase and the decimal logarithm of the initial cell load concentration (CFU/g); μ_{max} is the maximal specific growth rate ($\Delta \log[\text{CFU/g}]/\text{day}$); λ is the lag time (day); t is the time (day); N_{max} is the cell load concentration threshold value (CFU/g); and MAL is the microbial acceptability limit (day) (i.e., the storage time at which the $N(t)$ equals N_{max}). The microbial threshold (N_{max}) was set to 10^6 CFU/g for total *Pseudomonas spp.*

A similar approach was used to quantitatively determine the efficacy of the tested variables (i.e., antimicrobial compounds, gas composition and coating) on sensory quality. To this aim, the re-parameterized [2,26] Gompertz equation was also fitted to the sensory data:

$$QSQ(t) = OSQ_{min} - A^Q \times e^{-e^{[\{[u^Q_{max} \times 2.71 \frac{\lambda^Q - SAL}{A^Q}]+1\}]}} + A^Q \times e^{-e^{[\{[u^Q_{max} \times 2.71 \frac{\lambda^Q - t}{A^Q}]+1\}]}} \quad (2)$$

where $OSQ(t)$ is the burrata cheese overall quality at time t; A^Q is related to the difference between the overall quality attained at the stationary phase and the initial value; u^Q_{max} is the maximal rate at which $OSQ(t)$ decreases; λ^Q is the lag time; OSQ_{min} is the threshold value; SAL is the sensory acceptability limit (i.e., the storage time at which the $OSQ(t)$ equals OSQ_{min}); and t is the storage time. The value of OSQ_{min} was set equal to 4.

2.7. Statistical Analysis

The MAL, SAL and shelf life values of all the investigated samples were compared respectively by one-way Anova analysis. A Duncan multiple range test, with the option of homogeneous groups ($p < 0.05$), was used to determine significance among differences. To this aim, Statistica 7.1 for Windows (StatSoft Inc., Tulsa, OK, USA) was used.

3. Results and Discussion

The first experimental step was aimed at assessing the optimal MAP composition to improve the microbial and sensory quality of burrata cheese packaged with brine. In particular, three different headspace gas compositions with increasing concentrations of carbon dioxide were tested (30:70; 50:50 and 65:35 $CO_2:N_2$). The data are reported in Figure 1, where a rapid reduction of carbon dioxide in all tested packages was detected due to the solubility and diffusivity of gas in brine and in burrata cheese [28]. After one week of storage, an increase of headspace carbon dioxide was observed above all for the Cntr-1 sample packaged in air, principally due to the metabolic reaction of microbial proliferation [29]. For the same reason, a gradual decrease of oxygen concentration (up to 0%) was found in Cntr-1. For all samples packaged under MAP conditions, a gradual oxygen increase (up to 3 ± 0.5%) was detected during the storage time, mainly attributable to film permeability.

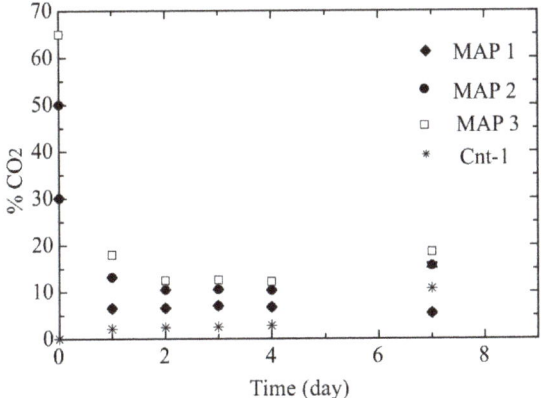

Figure 1. Carbon dioxide concentration in burrata samples packaged in air and under MAP.

Figure 2 shows the pH value of brine for samples packaged in air and under MAP. As expected, lower pH values were detected for samples packaged under MAPs compared to the control [28].

In particular, dissolved carbon dioxide is hydrated into carbonic acid, bicarbonate ion and carbonate ion, with the liberation of protons that cause pH change [28]. However, the pH of all the burrata samples remained constant for the entire storage period (6.50 ± 0.05) (data not shown).

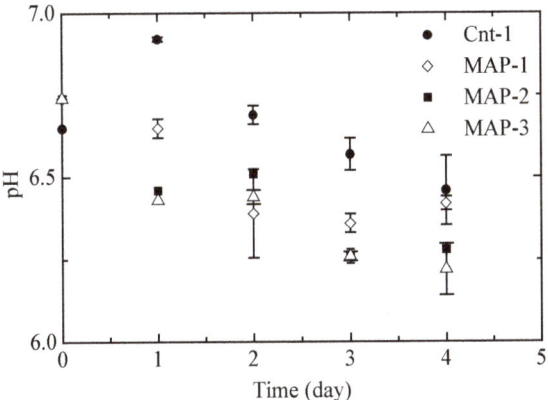

Figure 2. pH value of brine for samples packaged in air and under MAP.

From a microbiological point of view, spoilage microorganisms (Enterobacteriaceae and *Pseudomonas spp.*) were progressively delayed as carbon dioxide increased (data not shown). The increase in carbon dioxide also delayed sensory decay; therefore, among the MAP analyzed, MAP-3 (65:35 $CO_2:N_2$) was the headspace selected for the following steps due to its ability to prolong shelf life compared to the control sample in air (Table 2).

Table 2. Shelf life values (means ± sd) for each step calculated as the lowest value between *MAL* and *SAL*. The shelf life increase (%) compared to the control was also reported.

Experimental Steps		MAL (Day)	SAL (Day)	Shelf Life (Day)	Shelf Life Increase (%)
Step-1	Cntr-1	1.87 ± 0.11 [a]	2.11 ± 0.04 [a]	1.87 ± 0.11 [a]	–
	MAP-1	1.98 ± 0.05 [a]	3.81 ± 0.66 [b]	1.98 ± 0.05 [a]	5.88
	MAP-2	2.34 ± 0.04 [b]	3.84 ± 0.01 [b]	2.34 ± 0.04 [b]	25.13
	MAP-3	2.53 ± 0.09 [c]	5.04 ± 0.51 [c]	2.53 ± 0.09 [c]	35.29
Step-2	Cntr-2	1.68 ± 0.02 [a]	3.69 ± 0.01 [a]	1.68 ± 0.02 [a]	–
	Coat-Air	1.80 ± 0.18 [a]	4.19 ± 0.01 [b]	1.80 ± 0.18 [a]	7.15
	Coat-MAP	2.61 ± 0.06 [b]	5.76 ± 0.27 [c]	2.61 ± 0.06 [b]	55.35
Step-3	Cntr-3	1.67 ± 0.10 [a]	3.81 ± 0.02 [a]	1.67 ± 0.10 [a]	–
	LysEDTA-Air	2.10 ± 0.16 [a]	3.39 ± 0.18 [b]	2.10 ± 0.16 [a]	25.74
	LysEDTA-MAP	3.85 ± 0.05 [b]	4.36 ± 0.01 [c]	3.85 ± 0.05 [b]	130.53
	Coat-Lys EDTA	4.55 ± 0.76 [b]	5.34 ± 0.02 [d]	4.55 ± 0.76 [b]	172.45
Step-4	Cntr-4	3.00 ± 1.00 [a]	6.47 ± 0.14 [a]	3.00 ± 1.00 [a]	–
	Cntr-A	4.19 ± 0.13 [a]	5.62 ± 0.52 [a]	4.19 ± 0.13 [a]	39.66
	NanoAg-A	10 < day < 12	10.01 ± 0.64 [b]	10.01 ± 0.64 [b]	233.66
	LPX-A	9 < day < 10	8.21 ± 0.63 [c]	8.21 ± 0.63 [c]	173.66

Notes: For each step, data in the same column with different letters (a–d) are significantly different ($p < 0.05$). Step 1 (Cntr-1: burrata packaged in air; MAP-1: burrata packaged under 30:70 $CO_2:N_2$; MAP-2: burrata packaged under 50:50 $CO_2:N_2$; MAP-3 burrata packaged under 65:35 $CO_2:N_2$). Step 2 (Cntr-2: burrata packaged in air; Coat-Air: burrata coated with aginic acid and packaged in air; Coat-MAP: burrata coated with aginic acid and packaged under 65:35 $CO_2:N_2$). Step 3 (Cntr-3: burrata packaged in air; LysEDTA-Air: burrata with lysozyme and Na_2-EDTA dissolved in cream and packaged in air; LysEDTA-MAP: burrata with lysozyme and Na_2-EDTA dissolved in cream and packaged under 65:35 $CO_2:N_2$; Coat-LysEDTA: burrata with lysozyme and Na_2-EDTA dissolved in cream, coated and packaged under 65:35 $CO_2:N_2$). Step 4 (Cntr-4: burrata packaged in air; Cntr-A: burrata with lysozyme and Na_2-EDTA dissolved in cream, coated and packaged in air; NanoAg-A: burrata with lysozyme and Na_2-EDTA dissolved in cream, with active coating with nanoparticles and packaged under 65:35 $CO_2:N_2$; LPX-A: burrata with lysozyme and Na_2-EDTA dissolved in cream, with active coating with lactoperoxidase and packaged under 65:35 $CO_2:N_2$).

To further improve product quality, a coating was optimized prior to packaging the fresh cheese. The coating application combined with MAP retarded the spoilage, thus prolonging the microbial acceptability limit and consequently the shelf life (1.68 and 2.61 days for Cntr-2 and Coat-MAP, respectively—Table 2, Step 2). In particular, among spoilage microorganisms a rapid development of Enterobacteriaceae and *Pseudomonas spp.* was observed (Table 3); however, the development was better controlled by the application of combined strategies. As also confirmed by other scientific studies, the natural microflora (lactococci and lactobacilli) increased from 6 log CFU/g to about 8 log CFU/g during the storage period (data not shown) without being compromised by the coating application or the MAP conditions [11,17].

As regards sensory quality, Figure 3 reports the evolution of overall quality for the Step 2 burrata samples. The overall quality reflects the firmness, color, and odor of burrata. As can be seen, a rapid quality of decay was recorded for samples packaged in air (Cntr-2 and Coat-Air), even though the coating improved product acceptability, principally preserving cheese firmness [26,30,31]. The best results from the sensory point of view were obtained for samples coated and packaged under MAP.

Table 3. Log CFU/g ± sd of Enterobacteriaceae and *Pseudomonas spp.* for Step 2 burrata samples. (Cntr-2: burrata packaged in air; Coat-Air: burrata coated with aginic acid and packaged in air; Coat-MAP: burrata coated with aginic acid and packaged under 65:35 $CO_2:N_2$).

Microorganisms	Sampling Time (day)	Cntr-2 (log CFU/g)	Coat-Air (log CFU/g)	Coat-MAP (log CFU/g)
Enterobacteriaceae	0	1.78 ± 0.24	1.78 ± 0.24	1.78 ± 0.25
	3	5.55 ± 0.06	5.56 ± 0.08	5.84 ± 0.34
	4	7.44 ± 0.19	7.29 ± 0.30	6.62 ± 0.73
	6	–	–	7.43 ± 0.80
Pseudomonas spp.	0	3.41 ± 0.10	3.41 ± 0.10	3.41 ± 0.10
	3	6.54 ± 0.11	6.60 ± 0.01	6.29 ± 0.29
	4	6.57 ± 0.30	6.32 ± 0.16	5.93 ± 0.21
	6	–	–	6.73 ± 0.21

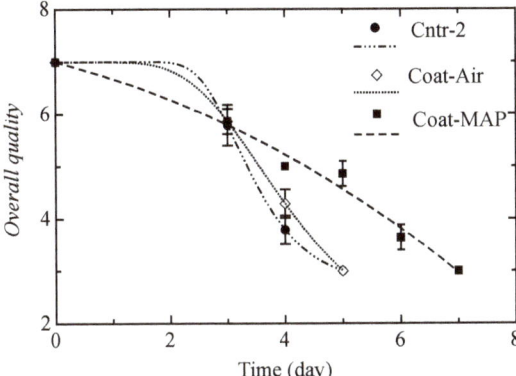

Figure 3. Overall Quality for Step 2 burrata samples (● Cntr-2: burrata packaged in air; ◇ Coat-Air: burrata coated with aginic acid and packaged in air; ■ Coat-MAP: burrata coated with aginic acid and packaged under 65:35 $CO_2:N_2$). The curves are the best fit of the re-parametrized Gompertz equation to the experimental data.

In Step 3, the antimicrobial activity of the lysozyme and the Na_2-EDTA added to the burrata filling and combined with coating and MAP were tested. The concentration of active compounds was taken from previously published data also dealing with burrata cheese [9]. As can be seen in Table 2, for Step 3 the shelf life was improved by the combination of these selected mild strategies. In fact, the active compounds exerted good effects in terms of both microbial and sensory quality, as is also confirmed in the literature [9,20], but their effect were more relevant when combined to coating or to MAP or to both. It is worth considering that the selected strategies did not influence the natural microflora of the burrata, as also confirmed by Conte et al. [9].

In the final experimental step (Step 4), the addition of lysozyme and Na_2-EDTA in the filling of the burrata was combined with the active coating containing the lactoperoxidase system or the silver nanoparticles and to MAP conditions during packaging. In Figure 4 the evolution of *Pseudomonas spp.* in samples from the last step can be observed. The control samples rapidly overlapped with the threshold within the first four days, whereas the burrata with the active coating maintained low microbial concentrations for much more than one week, above all when the silver nanoparticles were incorporated in the coating. The antimicrobial activity of the silver nanoparticles applied to dairy products is well documented in the literature [12,17,19], thus justifying the results also recorded in the current work.

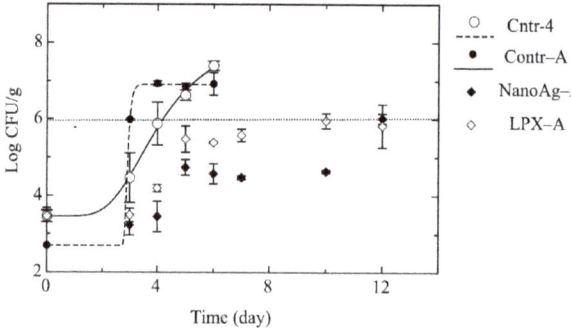

Figure 4. Evolution of *Pseudomonas spp.* for burrata samples of Step 4 (○ Cntr-4: Burrata packaged in air; • Cntr-A: Burrata with lysozyme and Na_2-EDTA dissolved in cream, coated and packaged in air; ♦ NanoAg-A: Burrata with lysozyme and Na_2-EDTA dissolved in cream, with active coating with nanoparticles and packaged under 65:35 $CO_2:N_2$; ◊ LPX-A: burrata with lysozyme and Na_2-EDTA dissolved in cream, with active coating with lactoperoxidase and packaged under 65:35 $CO_2:N_2$). The curves are the best fit of the re-parametrized Gompertz equation to the experimental data. The solid horizontal line represents the microbial threshold for *Pseudomonas spp.*

The LPX system greatly controlled the *Pseudomonas spp.* growth compared to the control samples, but it was less effective than nanoparticles. It is interesting to see from the data in Table 4 that the natural microflora proliferated without stress in the samples.

Table 4. Initial and final cell loads (log CFU/g ± sd) of lactic bacteria in Step 4.

Samples	Lactic Acid Bacteria		Mesophylic Lactococci	
	$\log_{(i)}$ CFU/g	$\log_{(f)}$ CFU/g	$\log_{(i)}$ CFU/g	$\log_{(f)}$ CFU/g
Cntr-4	5.11 ± 0.37	7.36 ± 0.35	6.08 ± 0.07	7.42 ± 0.11
Cntr-A	4.88 ± 0.03	6.99 ± 0.04	6.15 ± 0.09	7.34 ± 0.02
NanoAg-A	4.88 ± 0.03	7.12 ± 0.02	6.05 ± 0.03	7.49 ± 0.13
LPX-A	4.71 ± 0.05	7.27 ± 0.21	6.14 ± 0.17	7.62 ± 0.22

The active coating also preserved the sensory properties of the cheese, with slightly better effects for nanoparticles. The shelf life values reported in Table 2 highlight that a significant increase of more than 200% was recorded when the active compounds were used during burrata processing and an active coating with silver nanoparticles was applied prior to packaging the cheese under specific MAP conditions.

4. Conclusions

In this study, a step-by-step approach was used to optimize process and packaging for burrata cheese with brine. In particular, the shelf life was improved by combining the antimicrobial compounds to be applied during the processes of burrata filling, active coating and MAP. The results obtained in each step allowed the identification of the most appropriate conditions: the headspace gas composition, the coating application and the antimicrobial compounds in the burrata and in the coating. The results showed a significant increase in shelf life for samples under MAP compared to cheese packaged in air. The coating application above all improved the sensory quality. A relevant antimicrobial effect was recorded by applying active agents during the process and in the coating. Therefore, the combination of various mild strategies allowed us to reach a very significant extension of shelf life that accounted for a more than 200% increase, when compared to the control sample.

Acknowledgments: This study was financially supported by the project "Bioinnovazioni per produzioni lattiero casearie ad elevato contenuto salutistico" (Grant No. PON 01_00851). The authors also acknowledge the Institute for Composites and Biomedical Materials, CNR of Naples (Italy), in the person of Ing. G.G. Buonocore for kindly providing the active silver nanoparticles.

Author Contributions: For research articles with several authors, a short paragraph specifying their individual contributions must be provided. The following statements should be used Matteo Alessandro Del Nobile, Amalia Conte and Angelo Vittorio Zambrini conceived and designed the experiments; Cristina Costa and Annalisa Lucera performed the experiments; Matteo Alessandro Del Nobile and Amalia Conte analyzed the data; Angelo Vittorio Zambrini contributed reagents and materials; Cristina Costa and Amalia Conte wrote the paper. Authorship must be limited to those who have contributed substantially to the work reported.

Conflicts of Interest: The authors declare no conflict of interest.

References

1. Costa, C.; Conte, A.; Del Nobile, M.A. Use of Metal Nanoparticles for Active Packaging Applications. In *Antimicrobial Food Packaging*; Barros-Velázquez, J., Ed.; Academic Press: Cambridge, MA, USA, 2016; pp. 399–404.
2. Gammariello, D.; Conte, A.; Di Giulio, S.; Attanasio, M.; Del Nobile, M.A. Shelf life of stracciatella cheese under modified atmosphere packaging. *J. Dairy Sci.* **2009**, *92*, 483–490. [CrossRef] [PubMed]
3. Del Nobile, M.A.; Conte, A.; Incoronato, A.L.; Panza, O. Modified atmosphere packaging to improve the microbial stability of Ricotta. *Afr. J. Microbiol. Res.* **2009**, *3*, 37–142.
4. Selim, S. Antimicrobial activity of essential oils against Vancomycin-resistant enterococchi (VRE) and EscherichiaColi O157, H7 in feta soft cheese and minced beef meat. *Braz. J. Microbiol.* **2011**, *42*, 187–196. [CrossRef] [PubMed]
5. Di Pierro, P.; Sorrentino, A.; Mariniello, L.; Giosafatto, C.V.L.; Porta, R. Chitosan/whey protein film as active coating to extend Ricotta cheese shelf-life. *Food Sci. Technol.* **2011**, *44*, 2324–2327. [CrossRef]
6. Lucera, A.; Costa, C.; Conte, A.; Del Nobile, M.A. Food applications of natural antimicrobial compounds. *Front. Microbiol.* **2012**, *3*, 287–297. [CrossRef] [PubMed]
7. Papaioannou, G.; Chouliara, I.; Karatapanis, A.E.; Kontominas, M.G.; Savvaidis, I.N. Shelf-life of a Greek whey cheese under modified atmosphere packaging. *Int. Dairy J.* **2007**, *17*, 358–364. [CrossRef]
8. Dermiki, M.; Ntzimani, A.; Badeka, A.; Savvaidis, I.N.; Kontominas, N.G. Shelf-life extension and quality attributes of the whey cheese "Myzithra Kalathaki" using modified atmosphere packaging. *LWT* **2008**, *41*, 284–294. [CrossRef]
9. Conte, A.; Brescia, I.; Del Nobile, M.A. Lysozyme/EDTA disodium salt and modified-atmosphere packaging to prolong the shelf life of burrata cheese. *J. Dairy Sci.* **2011**, *94*, 5289–5297. [CrossRef] [PubMed]
10. Belewu, M.A.; Ahmed El-Imam, A.M.; Adeyemi, K.D.; Oladunjoye, S.A. Eucalyptus oil and lemon grass oil: Effect on chemical composition and shelf-life of soft cheese. *Environ. Nat. Resour. Res.* **2012**, *2*, 114–118. [CrossRef]
11. Martins, J.T.; Cerqueira, M.A.; Souza, B.W.S.; Avides, M.D.C.; Vicente, A.A. Shelf life extension of ricotta cheese using coatings of galactomannans from nonconventional sources incorporating nisin against Listeria monocytogenes. *Agric. Food Chem.* **2010**, *58*, 1884–1891. [CrossRef] [PubMed]
12. Incoronato, A.L.; Conte, A.; Buonocore, G.G.; Del Nobile, M.A. Agar hydrogel with silver nanoparticles to prolong the self life of Fiord di Latte cheese. *J. Dairy Sci.* **2011**, *94*, 1697–1704. [CrossRef] [PubMed]
13. Costa, C.; Lucera, A.; Lacivita, V.; Saccotelli, M.A.; Conte, A.; Del Nobile, M.A. Packaging optimisation for portioned Canestrato di Moliterno cheese. *Int. J. Dairy Technol.* **2016**, *69*, 401–409. [CrossRef]
14. Fernandez, A.; Soriano, E.; Lopez Carballo, G.; Picouet, P.; Lloret, E.; Gavara, R.; Hernandez-Mùnoz, P. Preservation of aseptic conditions in absorbent pads by using silver nanotechnology. *Food Res. Int.* **2009**, *42*, 1105–1112. [CrossRef]
15. De Azeredo, H.M.C. Nanocomposites for food packaging applications. *Food Res. Int.* **2009**, *42*, 1240–1253. [CrossRef]
16. Conte, A.; Longano, D.; Costa, C.; Ditaranto, N.; Ancona, A.; Cioffi, N.; Scrocco, C.; Sabbatini, L.; Contò, F.; Del Nobile, M.A. A novel preservation technique applied to fiordilatte cheese. *Innov. Food Sci. Emerg. Technol.* **2013**, *19*, 158–165. [CrossRef]

17. Mastromatteo, M.; Lucera, A.; Esposto, D.; Conte, A.; Faccia, M.; Zambrini, A.V.; Del Nobile, M.A. Packaging optimisation to prolong the shelf life of fiordilatte cheese. *J. Dairy Res.* **2015**, *1*, 43–51. [CrossRef] [PubMed]
18. Costa, C.; Conte, A.; Buonocore, G.G.; Lavorgna, M.; Del Nobile, M.A. Calcium-alginate coating loaded with silver-montmorillonite nanoparticles to prolong the shelf-life of fresh-cut carrots. *Food Res. Int.* **2012**, *48*, 164–169. [CrossRef]
19. Gammariello, D.; Conte, A.; Buonocore, G.G.; Del Nobile, M.A. Bio-based nanocomposite coating to preserve quality of Fior di latte cheese. *J. Dairy Sci.* **2011**, *94*, 5298–5304. [CrossRef] [PubMed]
20. Sinigaglia, M.; Bevilacqua, A.; Corbo, M.R.; Pati, S.; Del Nobile, M.A. Use of active compounds for prolonging the shelf life of mozzarella cheese. *Int. Dairy J.* **2008**, *18*, 624–630. [CrossRef]
21. Mastromatteo, M.; Lucera, A.; Sinigaglia, M.; Corbo, M.R. Synergic antimicrobial activity of lysozyme, nisin, and EDTA against Listeria monocytogenes in ostrich meat patties. *J. Food Sci.* **2010**, *75*, 422–429. [CrossRef] [PubMed]
22. Kennedy, M.; O'Rourke, A.L.; McLay, J.; Simmonds, R. Use of a ground beef model to assess the effect of the lactoperoxidase system on the growth of *Escherichia coli* O157:H7, *Listeria monocytogenes* and *Staphylococcus aureus* in red meat. *Int. J. Food Microbiol.* **2000**, *57*, 147–158. [CrossRef]
23. Vannini, L.; Lanciotti, R.; Baldi, D.; Guerzoni, M.E. Interactions between high pressure homogenization and antimicrobial activity of lysozyme and lactoperoxidase. *Int. J. Food Microbiol.* **2004**, *94*, 123–135. [CrossRef] [PubMed]
24. Seifu, E.; Buys, E.M.; Donkin, E.F. Significance of the lactoperoxidase system in the dairy industry and its potential applications: A review. *Trends Food Sci. Technol.* **2005**, *16*, 137–154. [CrossRef]
25. Seifu, E.; Buys, E.M.; Donkin, E.F.; Petzer, I.M. Antibacterial activity of the lactoperoxidase system against food-borne pathogens in Saanen and South African Indigenous goat milk. *Food Contr.* **2004**, *15*, 447–452. [CrossRef]
26. Conte, A.; Gammariello, D.; Di Giulio, S.; Attanasio, M.; Del Nobile, M.A. Active coating and modified-atmosphere packaging to extend the shelf life of Fior di Latte cheese. *J. Dairy Sci.* **2009**, *92*, 887–894. [CrossRef] [PubMed]
27. Faccia, M.; Angiolillo, L.; Mastromatteo, M.; Conte, A.; Del Nobile, M.A. The effect of incorporating calcium lactate in the saline solution on improving the shelf life of Fiordilatte cheese. *Int. J. Dairy Technol.* **2013**, *66*, 373–381. [CrossRef]
28. Chaix, E.; Guillaume, C.; Guillard, V. Oxygen and Carbon Dioxide Solubility and Diffusivity in Solid Food Matrices: A Review of Past and Current Knowledge. *Compr. Rev. Food Sci. Food Saf.* **2014**, *13*, 261–286. [CrossRef]
29. Cruz, R.S.; Soares, N.D.F.F.; de Andrade, N.J. Evaluation of oxygen absorber on antimicrobial preservation of lasagna—Type fresh pasta under vacuum packed. *Ciênc. Agroretec.* **2006**, *30*, 1135–1138. [CrossRef]
30. Kampf, N.; Nussinovitch, A. Hydrocolloid coating of cheeses. *Food Hydrocoll.* **2000**, *14*, 531–537. [CrossRef]
31. Laurienzo, P.; Malinconico, M.; Pizzano, R.; Manzo, C.; Piciocchi, N.; Sorrentino, A.; Volpe, M.G. Natural polysaccharide-based gels for dairy food preservation. *J. Dairy Sci.* **2006**, *89*, 2856–2864. [CrossRef]

© 2017 by the authors. Licensee MDPI, Basel, Switzerland. This article is an open access article distributed under the terms and conditions of the Creative Commons Attribution (CC BY) license (http://creativecommons.org/licenses/by/4.0/).

Article

Effects of Chitosan Coating Structure and Changes during Storage on Their Egg Preservation Performance

Dan Xu *, Jing Wang, Dan Ren and Xiyu Wu

College of Food Science, Food Storage and Logistics Research Center, Southwest University, Chongqing 400700, China; darenwl@foxmail.com (J.W.); rendan0709@swu.edu.cn (D.R.); xiyu.w@163.com (X.W.)
* Correspondence: xud@swu.edu.cn; Tel.: +86-23-6825-1298

Received: 8 August 2018; Accepted: 3 September 2018; Published: 7 September 2018

Abstract: To explore the influences of chitosan coating structure and structure changes during storage on egg preservation, eggs coated by chitosan solution for single time (CS1), two times (CS2), and three times (CS3) were prepared separately and stored with untreated eggs (CK1), eggs washed by water (CK2) and eggs treated by acetic acid solution (CK3) at 25 °C, 80% RH. The weight loss, Haugh unit, yolk index, albumen pH, eggshell morphologies and infrared (FTIR—Fourier Transform Infrared) spectra of all the samples were monitored. CS2 and CS3 presented the lowest weight loss, highest Haugh unit and yolk index, stabilized pH, and the highest thickness of chitosan coating layers (>2 μm) among all the groups, which extended egg shelf life for 20 days longer compared to CK1 and CK2. CS1 with very thin chitosan coating showed similar egg qualities with CK3, which are second only to CS2 and CS3. Furthermore, destructions were found on chitosan coatings during storage as revealed by the eggshell morphologies and FTIR spectra, which caused the quality deterioration of eggs. The results demonstrated that eggs with the thickest coating showed the best qualities during storage, while destructions on coating layers led to the quality drop of eggs.

Keywords: chitosan; coating; structural changes; egg preservation; shelf life

1. Introduction

Eggs are produced and consumed daily all over the world on a large scale, as excellent and inexpensive sources of high quality protein, certain minerals, and vitamins [1]. Nowadays, they have also been considered as ideal vehicles of bioactive components to enhance human nutrition [2]. Although the eggshells have been regarded as natural protective barriers of eggs, quantities of tiny pores presenting on the eggshells for gas exchange lead to the moisture and CO_2 loss of eggs, as well as the penetration of microbial, which would then cause the quality deterioration of eggs [3,4]. Moreover, eggshells are too fragile to retain their integrity during transport. Even a tiny crack in an eggshell would largely increase the risk of microbial contamination, particularly when eggs are stored at room temperature.

Attracted by the huge economic benefit, many strategies have been studied for egg preservation, which can be generally classified into two types based on the mechanism. The first one is to deactivate the microorganisms on the eggshells, including techniques such as ultrasonic treatment [5], ozone treatment [6], application of AgNPs-doped paper egg trays [7], and vacuum packaging [8]. The other one is to seal the pores on eggshells by coating, which serves as good barrier layers towards water vapor, gases, and microorganisms. The most widely studied coating materials are oil [9,10], proteins [11], biopolymers [3], etc. Biopolymers composed of biomass or derived polymers are intensively studied as edible coatings [12,13]. Among them, chitosan has attracted much attention of researchers due to its extraordinary performance, such as the excellent film forming property, low gas permeability, satisfied biocompatibility [14], and strong antimicrobial activities [15]. Therefore, coating of chitosan on eggshells can protect egg from both physicochemical changes and microbial contamination, leading to promising effects on quality preservation of eggs. Owing to the low cost, wide availability of chitosan and the operational feasibility of coating process, chitosan coating is considered an efficient and practical way to preserve eggs at room temperature. The influencing factors on the efficiency of chitosan coating such as the molecular weight [16], plasticizer types [17], sources of chitosan [18], and crosslinking agents [19] have been widely investigated. Combination coatings of chitosan with natural antimicrobial agents [20], pullulan [3], oil [21], and montmorillonite [22] have also been applied to eggs and showed improved performance.

However, although the coating materials have been intensively studied, the structure of the chitosan coating layer and their effects on egg preservation has not been carefully investigated so far. Suresh et al. [23] have coated eggs by chitosan a single time and three times, respectively. The coating layer on the eggshell with three-time coating was 20 μm thicker than that with single time coating, but no significant differences were found in weight loss, Haugh unit and albumin index of eggs between these two groups at 22 ± 1 °C. The preservation effects of chitosan coatings should be mainly dependent on their barrier properties and antimicrobial activities, which are highly related to the thickness and structures of coating layers on the eggshells.

Therefore, in this study, eggs were coated with chitosan one, two, and three times, separately, to observe their differences in coating structure and structural stability during storage in order to study the relationship between the coating structure and preservation effects. The quality of coated eggs were compared with those of the untreated, water washed and acetic acid treated eggs.

2. Materials and Methods

2.1. Materials and Reagents

The freshly laid eggs were provided by a local farm of Chongqing, China. Chitosan with molecular weight of 180 kDa and deacetylation degree of 90% was supplied by Weifang Haizhiyuan Biological Products Co., Ltd. (Weifang, China). Other chemicals were of analytical grade and were used as received.

2.2. Treatment of Eggs

Chitosan coating solution with a concentration of 0.5% (w/w) was prepared by dissolving certain amount of chitosan granules in 1 vol % acetic acid and stirring for 24 h. The freshly laid eggs were divided into 6 groups with 50 eggs in each. There were three coated groups, which were coated by chitosan solution for one, two and three times, separately, and were designated as CS1, CS2, and CS3, correspondingly. Each coating process was carried out as immersing eggs in the prepared coating solution for 2 min and drying under ambient conditions overnight with their small-ends down. The other three groups were set as control, including the untreated eggs (CK1), water washed eggs (CK2), and eggs immersed in 1 vol % acetic acid for 2 min (CK3). After the above treatments, all the eggs were placed in egg trays with their small-ends down, and were stored at 25 °C and 80% relative humidity (RH). Five marked eggs of each group were taken out to weigh every 5 days, and were returned back right after the measurement. Another five eggs of each group were picked up randomly to measure the Haugh unit, yolk index and pH of albumen every 5 days.

2.3. Characterizations of Eggs

2.3.1. Weight Loss

Weight loss (%) of a whole egg was calculated as the weight difference in percentage of an egg during storage compared to its whole weight at day 0 [24]. The weight of five marked eggs in each groups was measured using a balance every 5 days and given as average value ± standard deviation.

2.3.2. Haugh Unit

Haugh unit is a value related to egg weight and thick of albumen, which was calculated using Equation (1) [17]:

$$HU = 100 \log(H - 1.7 \times W^{0.37} + 7.6) \qquad (1)$$

where H is the albumen height (mm) and W is the weight of whole egg (g).

According to the United States Standards for Quality of Individual Shell Eggs (USDA 2000 [25]), eggs are classified in to AA, A and B grade, which require Haugh unit value to be above 72, 71 to 60, and below 60, correspondingly.

2.3.3. Yolk Index

Yolk index was calculated as the ratio of yolk height to yolk width [9]. The height and width of egg yolk were measured using a micrometer.

2.3.4. pH of Albumen

pH values of the albumen separated from the yolk was measured by a digital pH meter.

2.3.5. Scanning Electron Microscopy (SEM) of Eggshells

Morphologies of the surfaces and cross sections of eggshells were observed using a scanning electron microscope (JEM-2100, JEOL Ltd., Tokyo, Japan) after platinum sputtering of the samples.

2.3.6. Fourier Transform Infrared (FTIR) Spectra of Eggshells

FTIR spectra of the outer surface of eggshells were recorded using a FTIR spectrophotometer (Thermo-Nicolet 6700, Thermo Fisher Scientific, Waltham, MA, USA) with attenuated total reflectance (ATR) accessories. For each sample, 32 scans at 4 cm^{-1} resolution were used to collect the ATR spectra of the outer surface of eggshells.

2.3.7. Statistical Analysis

All the experiments were repeated at least three times to calculate the average values and standard deviations unless otherwise stated. All the data were statistically compared among groups by one-way analysis of variance (ANOVA). The significances of the mean values were determined by Duncan's multiple range testing with $p < 0.05$.

3. Results

3.1. Weight Loss

Generally, the weight loss of eggs gradually increased during storage, which was attributed to the escaping of CO_2 and water vapor in albumen through numerous pores on the eggshells [24]. The loss of CO_2 and water was responsible to many physical and chemical changes in albumen and yolk resulting in deterioration of eggs. Therefore, weight loss rate of eggs is an important index of egg quality. As shown in Table 1, the weight losses of CK1 and CK2 increased rapidly over time, which were significantly higher than others at day 36. CK3 and CS1 showed lower average values of weight loss, but were not significantly different from those of CK1 and CK2 ($p < 0.05$) until day 36. However, the weight losses of CS2 and CS3 were significantly lower than those of CK1 and CK2 since day 16. Up to day 36, the weight loss of CS2 was 29% less than that of CK2. The above results revealed that acetic acid treatment and one-time coating of chitosan showed limited effects to reduce the loss of CO_2 and water vapor of eggs, while coating two or three times largely slowed down the weight loss rate during storage. It is known that the main component of eggshell is $CaCO_3$ [26], which would react with acetic acid solution leading to some structural changes on the eggshell surface. Further coating with chitosan was more effective to seal the pores on the eggshells and to form dense and barrier outer layers, leading to considerably inhibited weight loss. Therefore, observations on the morphologies of eggshells would be helpful to further understand the weight loss change of each group.

Table 1. Changes in weight loss of different groups during storage.

Storage Time (day)	Weight Loss (%)					
	CK1	CK2	CK3	CS1	CS2	CS3
6	0.90 ± 0.07 [a]	1.01 ± 0.37 [a]	0.88 ± 0.17 [a]	0.78 ± 0.11 [a]	0.92 ± 0.17 [a]	0.80 ± 0.19 [a]
11	1.90 ± 0.15 [a]	2.12 ± 0.81 [a]	1.72 ± 0.47 [a]	1.66 ± 0.24 [a]	1.67 ± 0.45 [a]	1.68 ± 0.42 [a]
16	2.65 ± 0.23 [a]	2.76 ± 0.02 [a]	2.46 ± 0.45 [a,b]	2.31 ± 0.32 [a,b]	1.91 ± 0.27 [b]	2.01 ± 0.32 [b]
21	3.22 ± 0.24 [a,b]	3.27 ± 0.39 [a]	3.18 ± 0.58 [a,b]	2.95 ± 0.41 [a,b]	2.46 ± 0.35 [b]	2.61 ± 0.42 [a,b]
26	3.93 ± 0.33 [a,b]	4.21 ± 0.02 [a]	3.87 ± 0.69 [a,b,c]	3.58 ± 0.48 [a,b,c]	2.99 ± 0.42 [c]	3.19 ± 0.50 [b,c]
31	4.72 ± 0.36 [a]	4.74 ± 0.56 [a]	4.19 ± 0.07 [a,b]	3.96 ± 0.25 [a,b]	3.61 ± 0.52 [b]	3.84 ± 0.60 [b]
36	5.92 ± 0.48 [a,b]	6.18 ± 0.04 [a]	5.09 ± 0.09 [b,c]	5.02 ± 0.47 [b,c]	4.38 ± 0.63 [c]	4.64 ± 0.73 [c]

Note: Means in the same row with different superscripted letters (a–d) are significantly different ($p < 0.05$).

3.2. Haugh Unit

Haugh unit is an important index of egg quality [27], which is determined by the age-related changes of egg white proteins [28]. The reduced Haugh unit value during storage is a result of albumen thinning, which was caused by the increased concentration of clusterin and ovoinhibitor, as well as the disordering of ovalbumin structure [29]. These protein changes have been mainly attributed to the proteolysis of dense protein or to the increase of albumin pH [30], which are influenced by the losses of water and CO_2 during storage. As given in Table 2, the Haugh unit values of CK1 and CK2 decreased rapidly, which were significantly lower than those of other treated groups after stored for only 6 days. Three chitosan coated groups had similar Haugh unit values during the first 11 days. However, CS1 showed a sharp drop from day 11 to 16, making the value comparable to that of CK3. Moreover, the Haugh unit values of CS2 and CS3 were always significantly higher than those of CS1 and CK3 from day 16 to day 31. According to the egg grading standard given in 2.4, both CK1 and CK2 fell to grade B at day 16, followed by CK3 and CS1 both at day 21, while CS2 and CS3 were in grade B since day 36 and 31, respectively. Therefore, coating eggs with chitosan two times showed the best performance to significantly slow down the structural changes in albumen proteins, thus extending the shelf life by up to 20 days longer at 25 °C, 80% RH according to Haugh unit values.

Table 2. Changes in Haugh unit of different groups during storage.

Storage Time (day)	Haugh Unit					
	CK1	CK2	CK3	CS1	CS2	CS3
0	83.83 ± 0.81	83.83 ± 0.81	83.83 ± 0.81	83.83 ± 0.81	83.83 ± 0.81	83.83 ± 0.81
6	65.43 ± 2.85 [c]	62.34 ± 0.55 [c]	73.69 ± 4.56 [b]	74.90 ± 1.56 [a,b]	78.92 ± 3.98 [a,b]	80.44 ± 4.70 [a]
11	62.92 ± 1.67 [c]	61.72 ± 1.27 [c]	68.68 ± 1.89 [b]	73.37 ± 4.43 [a,b]	73.37 ± 4.43 [a,b]	73.54 ± 3.50 [a]
16	58.10 ± 2.14 [c]	57.89 ± 1.40 [c]	60.02 ± 3.46 [b,c]	62.04 ± 3.42 [b,c]	72.22 ± 4.58 [a]	73.09 ± 1.75 [a]
21	56.54 ± 3.93 [b]	54.39 ± 2.31 [b]	54.35 ± 2.69 [b]	54.03 ± 0.96 [b]	66.27 ± 1.41 [a]	68.34 ± 5.01 [a]
26	54.54 ± 5.25 [c]	57.04 ± 2.33 [b]	53.33 ± 1.79 [c]	55.16 ± 1.16 [b,c]	63.28 ± 3.81 [a]	65.79 ± 3.76 [a]
31	53.91 ± 2.89 [b]	52.81 ± 2.98 [b]	53.29 ± 3.89 [b]	54.55 ± 7.41 [b]	60.56 ± 5.26 [a]	56.95 ± 1.65 [a,b]
36	51.39 ± 2.74 [a]	50.53 ± 3.20 [a]	52.37 ± 1.08 [a]	53.96 ± 0.91 [a]	57.95 ± 8.11 [a]	56.53 ± 2.89 [a]

Means in the same row with different superscripted letters are significantly different ($p < 0.05$).

3.3. Yolk Index

Yolk index is another important measure of egg freshness besides Haugh unit, which is based on the yolk quality. As shown in Figure 1, CK1 and CK2 showed the similar and maximum decreasing rate of yolk index with storage time, followed by CK3 and CS1, while CS2 and CS3 presented the highest values during the whole storage. Moreover, it was worth noting that the yolk index of all the groups almost decreased linearly in two time periods, which were day 0 to 16, and day 20 to

36, respectively. To clearly demonstrate the changing trend, each curve in Figure 1 was linear fitted separately in these two periods. The slopes and correlation coefficients obtained from the linear fittings were listed in Table 3, which revealed some important results. First, the slope values of each curve were much higher in the first time period (day 0 to 16) than those in the second period (day 21 to 36), indicating that the yolk quality declined more seriously in the first time period compared to the second one. Second, the order of all the slope values in the first time period from small to large was CS2, CS3, CS1, CK3, CK1, and CK2. Particularly, CS2 and CS3 showed quite lower values compared to those of other groups. Third, the slope values of CS1, CS2, and CS3 were still lower than those of CK1, CK2, and CK3 in the second time period, but the differences among them became much smaller compared to those in the first period. Therefore, the above results showed that coating eggs with chitosan two or three times was very effective to slow down the quality drop of egg yolks, especially in the most important period of egg quality control, which is the first 16 days of storage.

The decreased values of yolk index with storage time was a result of liquefaction and flattening of yolk, which was attributed to the constant permeation of water from albumen to yolk through vitelline membrane driven by the osmotic pressure [11]. The osmotic pressure between albumen and yolk was associated to the albumen viscosity, which was reduced by the breakage of ovomucin-lyzyme complex. Therefore, when chitosan coating reduced the loss of CO_2 and water vapor and slowed down the structural changes in albumen, the increase in osmotic pressure between albumen and yolk would be slowed down resulting in improved yolk quality.

Furthermore, it was interesting to notice that the yolk index of CS1 showed a sharp decrease from day 11 to 16 as pointed by an arrow in Figure 1, which was in accordance with the rapid drop as mentioned above in Haugh unit value of CS1. As the changes in both yolk index and Haugh unit were associated to the barrier properties of eggshells, the sharp reductions of these values of CS1 might indicate some structural changes occurred on the chitosan coatings at this time period.

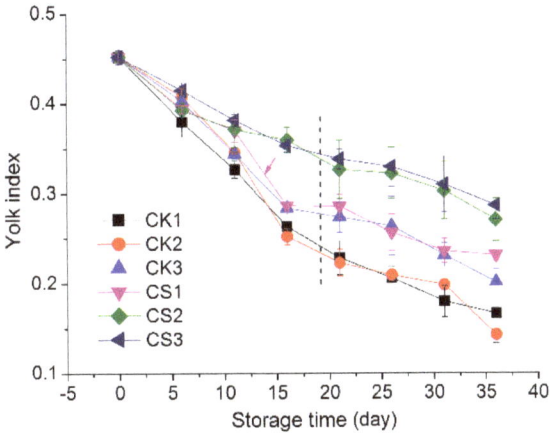

Figure 1. Changes in yolk index of eggs in different groups during storage.

Table 3. Slopes and correlation coefficients (R^2) of the yolk index curves of different groups at two time periods.

Samples	0–16 Days		21–36 Days	
	Slope	R^2	Slope	R^2
CK1	−0.0117	0.9990	−0.0042	0.9852
CK2	−0.0124	0.9585	−0.0050	0.8514
CK3	−0.0106	0.9916	−0.0050	0.9591
CS1	−0.0098	0.9536	−0.0037	0.9162
CS2	−0.0057	0.9064	−0.0038	0.8962
CS3	−0.0062	0.9991	−0.0035	0.9625

3.4. pH of Albumen

Since the albumen of fresh-laid eggs are saturated with CO_2, the evacuation of CO_2 through eggshells led to an increase in the albumen pH with increasing storage time. As given in Figure 2, the albumen pH of CK1 and CK2 increased to as high as 9.8 after storage for 31 days. The followed decrease at day 36 might be attributed to the breakdown of proteins in the albumen to fat and peptone. The albumen pH values of CK3 and CS1 presented similar changes, but were always lower than those of CK1 and CK2 at the same time. The pH value of CS1 also showed a sharp increase from day 11 to day 16. However, the albumen pH value of CS2 and CS3 was stabilized between 8.5 and 8.9 during the whole storage. Similar results have also been observed in eggs coated by proteins [11] and edible oils [31]. The stable albumen pH of CS2 and CS3 should be a result of the effective reductions of CO_2 loss in albumen, which further confirmed the considerably improved barrier properties of eggshells by chitosan coating two or three times.

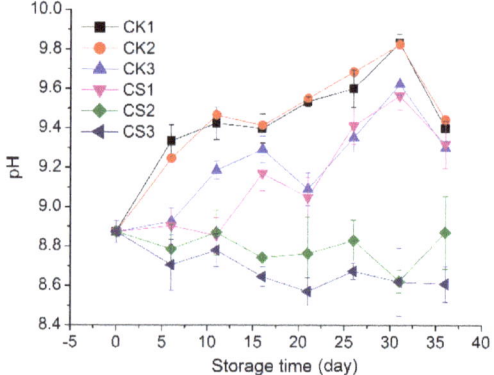

Figure 2. Changes in pH of albumen of eggs in different groups during storage.

3.5. Morphology of Eggshells

It has been generally agreed that coating eggs with polymers is able to seal the pores on eggshells and form barrier layers, thus providing effective protections to eggs to extend their shelf life. Therefore, it is necessary to examine and compare the eggshell morphologies of eggs in different groups, which is closely related to their barrier properties. The micrographs of the eggshell surfaces of all the groups are given in Figure 3. A large number of micro-cracks were observed on the surfaces of CK1 and CK2, which were probably produced during handling due to the fragile nature of eggshells. These micro-cracks would not only accelerate the permeation rate of CO_2 and water vapor from eggs to surroundings, but also provide pathways for the entry of bacteria. As a result, the protection function of eggshells would be largely compromised. However, after eggs were soaked in 1% acetic acid for 2 min, the micro-cracks on the eggshell surface of CK3 disappeared. It might be because the main component of eggshell, $CaCO_3$, reacts with acetic acid resulting in the filling up of micro-cracks on the eggshell surface. Therefore, the barrier properties of the eggshell of CK3 were superior to those of CK1 and CK2, which explained the improved quality of CK3. Moreover, the eggshell surfaces of CS1, CS2, and CS3 presented similar dense morphologies with that of CK3 when viewed at a low magnification (200×).

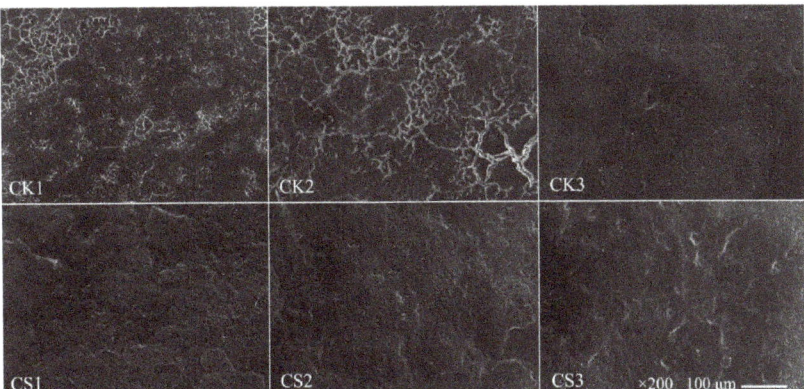

Figure 3. Scanning electron microscopy (SEM) micrographs of eggshell surfaces at day 0.

Therefore, the micrographs of the eggshell cross-sections were examined and given in Figure 4, chitosan coating layers with different thickness were found on the outside of the eggshells of CS1, CS2, and CS3. The chitosan coating layer on the eggshell of CS1 showed a thickness of approximately 0.7 μm, which was too thin to homogeneously cover the whole eggshell and to act as a dense and effective barrier layer. This might be the reason that CS1 showed comparable performance with that of CK3 during storage. The thicknesses of the coating layers of CS2 and CS3 were close to each other, and were larger than 2 μm. Although a dense eggshell surface was obtained by merely acetic acid treatment of eggs, eggshells of CS2 or CS3 coated by chitosan two or three times, which formed a dense polymer layer with a thickness of several micrometers on the outside of eggshells, provided more effective barriers compared to that of CK3.

Another thing that needed to be figure out is whether there are structural changes on the coating layers with increasing storage time, when they are constantly influenced by the environmental conditions and the gas transfer between eggs and surroundings. If the dense structure of coating layers was destructed during storage, the barrier properties of eggshells would certainly be compromised thus affecting the physical and chemical changes of eggs. Therefore, micrographs of eggshell surfaces were taken at different times during storage with a magnification of 2000× as given in Figure 5. It was clearly observed that the cracks on the eggshells of CK1 were gradually widened with increasing storage time, which indicated the weakened protection functions of the eggshell with time. For CK3, a stable structure with little change on the surface morphologies was observed on the eggshells treated by acetic acid. For CS1 and CS2, their eggshell surfaces were somewhat blurred at day 0 because of the soft texture of chitosan coating. As the storage time increased, some parts of the surfaces became clear with some micro-pits on it, which were similar to the morphology of CK3. Therefore, it might reveal that the dense structure of chitosan coating was gradually destructed during storage. Furthermore, this phenomenon was firstly observed for CS1 at day 11, while for CS2 at day 16 as observed in Figure 5. Since the coating layer on the eggshell of CS1 was thinner than that of CS2, it was reasonable that the coating on CS1 was more vulnerable to damages.

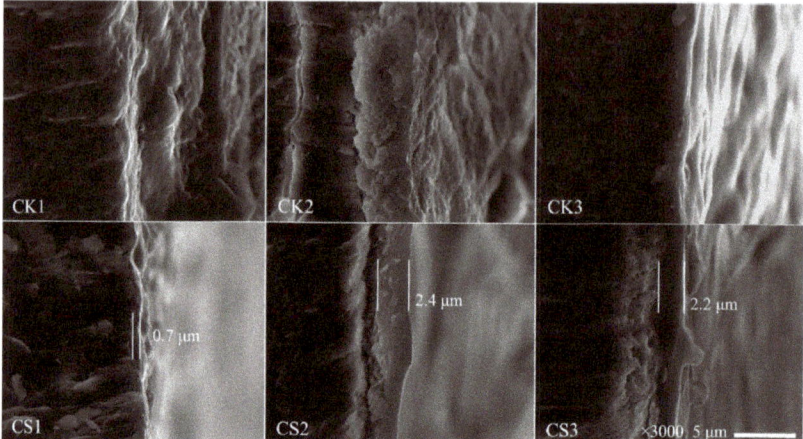

Figure 4. SEM micrographs of eggshell cross-sections at day 0.

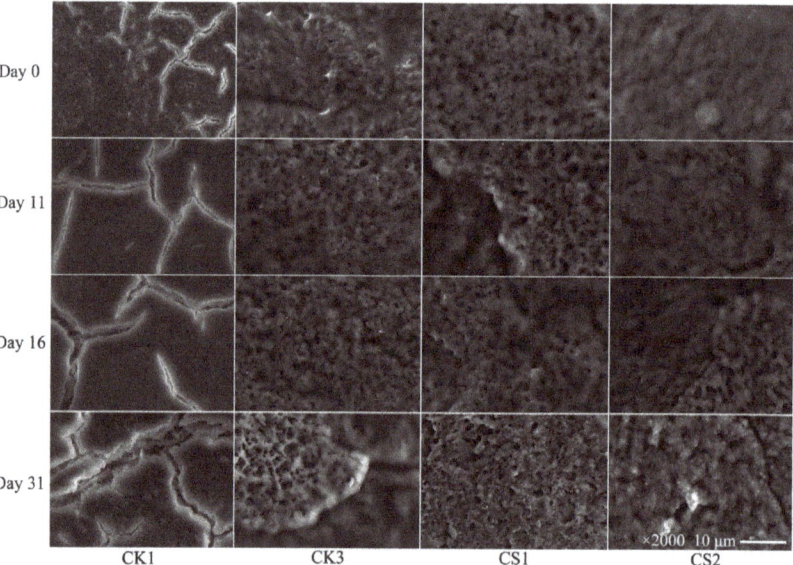

Figure 5. SEM micrographs of eggshell surfaces at day 0, 11, 16, and 31.

3.6. FTIR Spectra of Eggshells

To further confirm the structural changes on the coatings, the compositional changes of the eggshells were analyzed by the FTIR spectra, which was taken on the outer surface of eggshells at day 0 and day 36, respectively. As given in Figure 6, the spectra of CK1, CK2, and CK3 showed peaks at 872 cm^{-1} and 1397 cm^{-1}, which were due to the out-plane deformation and stretching vibration of carbonate groups, respectively [32], while peaks at 1026 cm^{-1} and 1647 cm^{-1} might be attributed to the C–O–C vibration of polysaccharides [33] and amide I band of proteins on the eggshell cuticles, respectively [34]. However, the intensity of peaks at 1026 cm^{-1} and 1647 cm^{-1} were relatively lower than that of peaks at 872 cm^{-1} and 1397 cm^{-1} in the spectra of CK3 compared to that of CK1 and CK2, which revealed the loss of polysaccharides and proteins on the eggshell cuticles after soaking in acetic acid solution. For CS2 and CS3, a new absorption at 1064 cm^{-1} appeared, which might be assigned to the asymmetric stretching of C–O–C of chitosan [35].

After storage for 36 days, the spectra of CK1, CK2, and CK3 showed little changes compared to those of them at day 0, which indicated that the compositions of their eggshell cuticles were stable. However, for CS1, CS2, and CS3, the intensity of peaks at 1064 cm^{-1} significantly decreased in day 36 compared to those in day 0, which further confirmed the destructions of chitosan coating layer in the coated groups during storage.

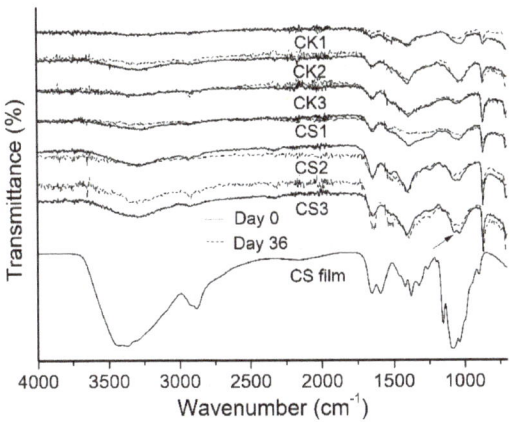

Figure 6. Fourier Transform Infrared (FTIR) spectra of CS film and eggshell surfaces at day 0 (solid line) and day 36 (dashed line).

4. Discussion

When changes in egg qualities are correlated to variations in coating morphologies, the relationship between them can be observed. CS2 and CS3 showed the highest thickness of coating layers and they were close to each other. At the same time, they both exhibited the best qualities during storage. CS1 and CK3 had thinner coatings presented poorer qualities but better than those of CK1 and CK2. It indicated that thicker coating layers led to better performance of coated eggs, and similar thickness of coating resulted in similar quality of eggs during storage. Therefore, it confirms that the thickness of chitosan coating layer, which determines their barrier properties, is responsible to the preservation performance of coatings on eggs.

Secondly, destructions on chitosan coating for different samples occurred at different time of storage. For CS1, destructions were observed on day 11, which was the same time that weight loss, Haugh unit, yolk index, and pH of CS1 showed sharp changes. Similarly, from day 16 to 21, more significantly changes on weight loss, Haugh unit, yolk index of CS2 were also observed compared to other time periods, which could correspond to the destructions on coatings of CS2 that appeared on day 16. Therefore, it indicated that egg quality was quite sensitive to the structural changes of coating layers during storage, which was dominant to the barrier properties of eggshells. However, thicker coatings are more resistant to the damages, thus exhibiting better preservation effects on eggs during the whole storage. The above findings inspired us to further extend the shelf life of eggs, efforts should be made not only to improve the initial barrier properties of the coatings, but also to enhance their structural stability during storage.

5. Conclusions

The results revealed that eggs coated by chitosan two or three times with a coating thickness of more than 2 µm showed a shelf life up to 36 days at 25 °C, 80% RH, which was 15 days longer than those of CS1 and CK3, and 20 days longer than those of CK1 and CK2. The thicker the coating layer, the better quality of the coated eggs. Therefore, the preservation effects of chitosan coating on eggs should be predominantly attributed to the barrier properties of coatings. However, destructions in the coating structure with increasing storage time were revealed by the micrographs and FTIR spectra of the outer surfaces of eggshells. Although the mechanism needed to be further explored, this study inspired us to effectively improve the storage qualities of eggs, coatings with stable structure and durable barrier properties should be developed.

Author Contributions: Conceptualization, D.X.; Methodology and Validation, J.W.; Formal Analysis, X.W.; Investigation, D.R.; Data Curation, J.W.; Writing—Original Draft Preparation, J.W.; Writing—Review & Editing, D.X.; Visualization, D.X.; Supervision, D.X.; Project Administration, D.X.; Funding Acquisition, D.X.

Funding: This research was funded by Fundamental Research Funds for the Central Universities of China (XDJK2016B012), National Natural Science Foundation of China (21306154), and Special Projects of Chongqing Social Undertakings and People's Livelihood Protection Technology innovation (cstc2015shmszx80011).

Conflicts of Interest: The authors declare no conflict of interest.

References

1. Kuroli, S.; Kanoo, T.; Itoh, H.; Ohkawa, Y. Nondestructive measurement of yolk viscosity in lightly heated chicken shell eggs. *J. Food Eng.* **2017**, *205*, 18–24. [CrossRef]
2. Kostogrys, R.B.; Filipiak-Florkiewicz, A.; Dereń, K.; Drahun, A.; Czyzynska-Cichon, I.; Cieślik, E.; Szymczyk, B.; Franczyk-Zarow, M. Effect of dietary pomegranate seed oil on laying hen performance and physicochemical properties of eggs. *Food Chem.* **2017**, *221*, 1096–1103. [CrossRef] [PubMed]
3. Morsy, M.K.; Sharoba, A.M.; Khalaf, H.H.; El-Tanahy, H.H.; Cutter, C.N. Efficacy of antimicrobial pullulan-based coating to improve internal quality and shelf-life of chicken eggs during storage. *J. Food Sci.* **2015**, *80*, M1066–M1074. [CrossRef] [PubMed]
4. Damaziak, K.; Marzec, A.; Riedel, J.; Szeliga, J.; Koczywas, E.; Cisneros, F.; Michalczul, M.; Łukasiewicz, M.; Gozdowski, D.; Siennicka, A. Effect of dietary canthaxanthin and iodine on the production performance and egg quality of laying hens. *Poult. Sci.* **2018**. [CrossRef] [PubMed]
5. Sert, D.; Aygun, A.; Demir, M.K. Effects of ultrasonic treatment and storage temperature on egg quality. *Poult. Sci.* **2011**, *90*, 869–875. [CrossRef] [PubMed]
6. Yüceer, M.; Aday, M.S.; Caner, C. Ozone treatment of shell eggs to preserve functional quality and enhance shelf life during storage. *J. Sci. Food Agric.* **2016**, *96*, 2755–2763. [CrossRef] [PubMed]
7. Viswanathan, K.; Priyadharshini, M.L.M.; Nirmala, K.; Raman, M.; Raj, G.D. Bactericidal paper trays doped with silver nanoparticles for egg storing applications. *Bull. Mater. Sci.* **2016**, *39*, 819–826. [CrossRef]
8. Aygun, A.; Sert, D. Effects of vacuum packing on eggshell microbial activity and egg quality in table eggs under different storage temperatures. *J. Sci. Food Agric.* **2013**, *93*, 1626–1632. [CrossRef] [PubMed]
9. Ryu, K.N.; No, H.K.; Prinyawiwatkul, W. Internal quality and shelf life of eggs coated with oils from different sources. *J. Food Sci.* **2011**, *76*, S325–S329. [CrossRef] [PubMed]
10. Figueiredo, T.C.; Assis, D.C.S.; Menezes, L.D.M.; Oliveira, D.D.; Lima, A.L.; Souza, M.R.; Heneine, L.G.D.; Cançado, S.V. Effects of packaging, mineral oil coating, and storage time on biogenic amine levels and internal quality of eggs. *Poult. Sci.* **2014**, *93*, 3171–3178. [CrossRef] [PubMed]
11. Caner, C.; Yüceer, M. Efficacy of various protein-based coating on enhancing the shelf life of fresh eggs during storage. *Poult. Sci.* **2015**, *94*, 1665–1677. [CrossRef] [PubMed]
12. Sagnelli, D.; Hooshmand, K.; Kemmer, G.C.; Kirkensgaard, J.J.K.; Mortensen, K.; Giosafatto, C.V.L.; Holse, M.; Hebelstrup, K.H.; Bao, J.; Stelte, W.; et al. Cross-linked amylose bio-plastic: A transgenic-based compostable plastic alternative. *Int. J. Mol. Sci.* **2017**, *18*, 2075. [CrossRef] [PubMed]
13. Giosafatto, C.V.L.; Di Pierro, P.; Gunning, P.; Mackie, A.; Porta, R.; Mariniello, L. Characterization of Citrus pectin edible films containing transglutaminase-modified phaseolin. *Carbohydr. Polym.* **2014**, *106*, 200–208. [CrossRef] [PubMed]
14. Kopacic, S.; Bauer, W.; Walzl, A.; Leitner, E.; Zankel, A. Alginate and chitosan as a functional barrier for paper-based packaging materials. *Coatings* **2018**, *8*, 235. [CrossRef]
15. Hu, D.; Wang, H.; Wang, L. Physical properties and antibacterial activity of quaternized chitosan/carboxymethyl cellulose blend films. *LWT—Food Sci. Technol.* **2016**, *65*, 398–405. [CrossRef]

16. Bhale, S.; No, H.K.; Prinyawiwatkul, W.; Farr, A.J.; Nadarajah, K.; Meyers, S.P. Chitosan coating improves shelf life of eggs. *J. Food Sci.* **2003**, *68*, 2378–2383. [CrossRef]
17. Kim, S.H.; No, H.K.; Prinyawiwatkul, W. Plasticizer types and coating methods affect quality and shelf life of eggs coated with chitosan. *J. Food Sci.* **2008**, *73*, S111–S117. [CrossRef] [PubMed]
18. *AMS-56 United States Standards, Grades, and Weight Classes for Shell Eggs*; USDA: Washington, DC, USA, 2000.
19. Pujols, K.D.; Osorio, L.; Carrillo, E.P.; Wardy, W.; Torrico, D.D.; No, H.K.; Corredor, J.A.H.; Prinyawiwatkul, W. Comparing effects of α- vs. β-chitosan coating and emulsion coatings on egg quality during room temperature storage. *Int. J. Food Sci. Technol.* **2014**, *49*, 1383–1390. [CrossRef]
20. Legendre, G.; Vallée-Réhel, K.; Linossier, I.; Faÿ, F. Evaluation of ionically cross-linked chitosan coating aimed at eggs' protection. *Int. J. Food Sci. Technol.* **2015**, *50*, 736–743. [CrossRef]
21. Jin, T.Z.; Gurtler, J.B.; Li, S.-Q. Development of antimicrobial coatings for improving the microbiological safety and quality of shell eggs. *J. Food Prot.* **2013**, *76*, 779–785. [CrossRef] [PubMed]
22. Torrico, D.D.; Wardy, W.; Carabante, K.M.; Pujols, K.D.; Xu, Z.; No, H.K.; Prinyawiwatkul, W. Quality of eggs coated with oil-chitosan emulsion: Combined effects of emulsifier types, initial albumen quality, and storage. *LWT—Food Sci. Technol.* **2014**, *57*, 35–41. [CrossRef]
23. Yi, S.-Z.; Liu, Q.; Xu, D.; Li, D. Preservation effects of chitosan/nano-montmorillonite coating on clean eggs. *Packag. Eng.* **2017**, *7*, 12. (In Chinese)
24. Suresh, P.V.; Raj, K.R.; Nidheesh, T.; Pal, G.K.; Sakhare, P.Z. Application of chitosan for improvement of quality and shelf life of table eggs under tropical room conditions. *J. Food Sci. Technol.* **2015**, *52*, 6345–6354. [CrossRef] [PubMed]
25. Jirangrat, W.; Torrico, D.D.; No, J.; No, H.K.; Prinyawiwatkul, W. Effects of mineral oil coating on internal quality of chicken eggs under refrigerated storage. *Int. J. Food Sci. Technol.* **2010**, *45*, 490–495. [CrossRef]
26. Munoz, A.; Dominguez-Gasca, N.; Jimenez-Lopez, C.; Rodriguez-Navarro, A.B. Importance of eggshell cuticle composition and maturity for avoiding trans-shell Salmonella contamination in chicken eggs. *Food Control* **2015**, *55*, 31–38. [CrossRef]
27. Pan, C.; Zhao, Y.X.; Liao, S.F.; Chen, F.; Qin, S.Y.; Wu, X.S.; Zhou, H.; Huang, K.H. Effect of selenium-enriched probiotics on laying performance, egg quality, egg selenium content, and egg glutathione peroxidase activity. *J. Agric. Food Chem.* **2011**, *59*, 11424–11431. [CrossRef] [PubMed]
28. Sheng, L.; Wang, J.; Huang, M.J.; Xu, Q.; Ma, M.H. The changes of secondary structures and properties of lysozyme along with the egg storage. *Int. J. Biol. Macromol.* **2016**, *92*, 600–606. [CrossRef] [PubMed]
29. Sheng, L.; Huang, M.J.; Wang, J.; Xu, Q.; Hammad, H.H.M.; Ma, M.H. A study of storage impact on ovalbumin structure of chicken egg. *J. Food Eng.* **2018**, *219*, 1–7. [CrossRef]
30. Omana, D.A.; Liang, Y.; Kav, N.N.; Wu, J. Proteomic analysis of egg white proteins during storage. *Proteomics* **2011**, *11*, 144–153. [CrossRef] [PubMed]
31. Nongtaodum, S.; Jangchud, A.; Jangchud, K.; Dhamvithee, P.; No, H.K.; Prinyawiwatkul, W. Oil coating affects internal quality and sensory acceptance of selected attributes of raw eggs during storage. *J. Food. Sci.* **2013**, *78*, S329–S335. [CrossRef] [PubMed]
32. Markovski, J.S.; Markovic, D.D.; Dokic, V.R.; Mitric, M.; Ristic, M.D.; Onjia, A.E.; Marinkovic, A.D. Arsenate adsorption on waste eggshell modified by goethite, α-MnO$_2$ and goethite/α-MnO$_2$. *Chem. Eng. J.* **2014**, *237*, 430–442. [CrossRef]
33. Pop, O.L.; Vodnar, D.V.; Suharoschi, R.; Mudura, E.; Socaciu, C. *L. plantarum* ATCC 8014 entrapment with prebiotics and lucerne green juice and their behavior in simulated gastrointestinal conditions. *J. Food Process Eng.* **2015**, *39*, 433–441. [CrossRef]
34. Dominguez-Gasca, N.; Muñoz, A.; Rodriguez-Navarro, A.B. Quality assessment of chicken eggshell cuticle by infrared spectroscopy and staining techniques: A comparative study. *Br. Poult. Sci.* **2017**, *58*, 517–522. [CrossRef] [PubMed]
35. Galvis-Sánchez, A.C.; Castro, M.C.R.; Biernacki, K.; Gonçalves, M.P.; Souza, H.K.S. Natural deep eutectic solvents as green plasticizers for chitosan thermoplastic production with controlled/desired mechanical and barrier properties. *Food Hydrocoll.* **2018**, *82*, 478–489. [CrossRef]

© 2018 by the authors. Licensee MDPI, Basel, Switzerland. This article is an open access article distributed under the terms and conditions of the Creative Commons Attribution (CC BY) license (http://creativecommons.org/licenses/by/4.0/).

Article

Elaboration and Characterization of Active Apple Starch Films Incorporated with Ellagic Acid

Juan Manuel Tirado-Gallegos [1], Paul Baruk Zamudio-Flores [1,*], José de Jesús Ornelas-Paz [1], Claudio Rios-Velasco [1], Guadalupe Isela Olivas Orozco [1], Miguel Espino-Díaz [1], Ramiro Baeza-Jiménez [2], José Juan Buenrostro-Figueroa [2], Miguel Angel Aguilar-González [3], Daniel Lardizábal-Gutiérrez [4], María Hernández-González [5], Francisco Hernández-Centeno [5] and Haydee Yajaira López-De la Peña [5]

[1] Coordinación de Tecnología de Alimentos de la Zona Templada, Centro de Investigación en Alimentación y Desarrollo, A.C., Avenida Río Conchos s/n, Parque Industrial, Cd. Cuauhtémoc, Chihuahua 31570, Mexico; jmtiradoga@gmail.com (J.M.T.-G.); jornelas@ciad.mx (J.d.J.O.-P.); claudio.rios@ciad.mx (C.R.-V.); golivas@ciad.mx (G.I.O.O.); aster3000@hotmail.com (M.E.-D.)
[2] Centro de Investigación en Alimentación y Desarrollo, A.C., Unidad Delicias, Av. Cuarta Sur 3820, Fraccionamiento Vencederos del Desierto, Cd. Delicias, Chihuahua 33809, Mexico; ramiro.baeza@ciad.mx (R.B.-J.); jose.buenrostro@ciad.mx (J.J.B.-F.)
[3] Centro de Investigación y de Estudios Avanzados del Instituto Politécnico Nacional, Unidad Saltillo, Av. Industria Metalúrgica 1062, Ramos Arizpe, Coahuila 25900, Mexico; mgzlz@hotmail.com
[4] Laboratorio Nacional de Nanotecnología, Centro de Investigación en Materiales Avanzados, Miguel de Cervantes 120, Chihuahua, Chihuahua 31109, Mexico; pauba99@yahoo.com
[5] Departamento de Ciencia y Tecnología de Alimentos, División de Ciencia Animal, Universidad Autónoma Agraria Antonio Narro, Calzada Antonio Narro 1923, Buenavista, Saltillo, Coahuila 2315, Mexico; maryhg12@yahoo.com (M.H.-G.); francisco.hdezc@gmail.com (F.H.-C.); Yajaira.lp@gmail.com (H.Y.L.-D.l.P.)
* Correspondence: pzamudio@ciad.mx; Tel.: +52-625-581-2920 (ext. 111)

Received: 18 September 2018; Accepted: 23 October 2018; Published: 27 October 2018

Abstract: Apple starch films were obtained from apples harvested at 60, 70, 80 and 90 days after full bloom (DAFB). Mechanical properties and water vapor permeability (WVP) were evaluated. The apple starch films at 70 DAFB presented higher values in the variables of tensile strength (8.12 MPa), elastic modulus (3.10 MPa) and lower values of water vapor permeability (6.77×10^{-11} g m^{-1} s^{-1} Pa^{-1}) than apple starch films from apples harvested at 60, 80 and 90 DAFB. Therefore, these films were chosen to continue the study incorporating ellagic acid (EA). The EA was added at three concentrations [0.02% (FILM-EA0.02%), 0.05% (FILM-EA0.05%) and 0.1% (FILM-EA0.1%) w/w] and compared with the apple starch films without EA (FILM-Control). The films were characterized by their physicochemical, optical, morphological and mechanical properties. Their thermal stability and antioxidant capacity were also evaluated. The FILM-Control and FILM-EA0.02% showed a uniform surface, while FILM-EA0.05% and FILM-EA0.1% showed a rough surface and insoluble EA particles. Compared to FILM-Control, EA modified the values of tensile strength, elasticity modulus and elongation at break. The antioxidant capacity increased as EA concentration did. EA incorporation allowed obtaining films with higher antioxidant capacity, capable of blocking UV light with better mechanical properties than film without EA.

Keywords: active films; thermogravimetric analysis; UV protection; X-ray diffraction

1. Introduction

Plastics are synthetic polymers derived from fossil fuels, which because of its production cost and versatility are one of the most popular food packaging materials [1,2]. Despite that, the non-biodegradable nature of these materials has generated serious pollution problems [3]. For these reasons, the interest on developing edible films and coating that can totally or partially replace the use of synthetic polymers derived from petroleum, has grown [4,5]. Edible films can be obtained from proteins, carbohydrates, glycoproteins, lipids or their mixtures [6,7]. One of the most widely used natural biopolymers with a great potential in the production of biodegradable materials is starch [1,8]. Starch is the main energy reserve of higher plants, a natural resource, renewable, abundant and low cost [4]. Starch films offer several advantages for use as food packaging, being colorless, odorless, tasteless, transparent, biodegradable, non-toxic and with low oxygen permeability [9]. On the other hand, compared with synthetic polymers, starch films are highly hydrophilic with poor mechanical properties [3,4]. It has been reported that the different properties of starch-based films vary with the botanical source [10–15]. These variations are mainly due to the amylose content in the native starches (18%–30%) [16]. The food industry is constantly looking for new sources of starches that may offer different or better functional properties [1,17]. Regarding to this, Stevenson et al. [18] evaluated the functional properties of starches isolated from immature apples of six cultivars, their results indicated that the amylose content was 26.0% to 29.3%. In this sense, apple starches may form films with acceptable properties (due to its amylose content) and comparable to those obtained from conventional sources such as corn and wheat. Apple is a climacteric fruit, whose ripening process is regulated by ethylene [19]. Within its growth and development, the fruit is characterized by significant concentrations of starch, which become almost zero at the time of commercial harvest [18,19]. One of the most common practices in apple orchards is the fruit thinning. The objective of this process is to achieve the development of high-quality fruits with regular yields and high blooming return [20,21]. Besides, it is important to mention that in order to obtain a good commercial quality harvest, it is required that only between 5% and 30% of the flowers end up as mature fruits [22]. When fruit thinning in apple orchards is done manually (hand-thinning), it is recommended to do it at 28 DAFB. However, late manual thinning (60 DAFB) is a common practice in the region [20], which may be extended up to 70 DAFB. Although its benefits are well known, this practice is usually delay due to the farmers fear, that the damages by frost in orchard whose fruits has been previously thinned, diminish the harvest yield. Apples removed during the fruit thinning are considered as a waste and in the best case, they are used as animal feed or as compost [21]. An alternative for the use of the fruits eliminated by this practice may be the isolation of their starch for the development of biodegradable films. The packaging plays an important role in the food industry, it must to contain and protect the food against external and internal factors. Biodegradable films can control the mass exchange between the food and the surrounding medium increasing the extension of shelf life and improving the quality of foods [23,24]. There is increasing interest in developing biodegradable packaging that can be used as active packaging [25]. According to these, starch-based films can develop activities as carriers and active substances-releasing agents (antioxidants, antimicrobials, flavorings, among others) in to the food matrix by diffusion process, which gives rise to the active films [26,27]. At present, there has been an increasing interest in incorporating antioxidants in starch-based films, mainly plants extracts, which are often characterized by their content of phenolic compounds [1,10,25,28,29]. Ellagic acid (EA) is a phenolic compound found in a wide variety of vegetables, usually in the form of ellagitannins (its precursors). EA is mainly found in fruits such as pomegranates, raspberries and blueberries [30], it shows antimicrobial and anticancer properties, besides inhibiting the formation and growth of tumors in animals [31]. Definitely, one of its most studied properties is its ability to capture free radicals [31–33], due to its high antioxidant capacity, since it is a stable molecule at high temperatures, which melting point is ~362 °C [34]. EA has previously been used to deal with melanoma cells in chitosan based-films, presenting anticancer activity at concentrations as low as 0.1% (w/v) [30]. It has also been applied on candelilla wax based coatings at a concentration of 0.01% (w/v) to minimize the changes of color and improve the texture of fresh-cut fruits [35] and to lengthen shelf life in avocados [36]. For all these reasons, it can be mentioned that the EA is a substance with varied biological activity, which has already been added to films for medical purposes; however, currently there are no reports on the incorporation on the properties of film formation in apple starches. The objective of this study consisted in evaluating the effect of incorporating EA, on the physicochemical, mechanical, thermal and antioxidant properties of apple starch-based films.

2. Materials and Methods

2.1. Sample and Reagents

Starch was obtained from Golden Delicious Smothee apples harvested at 60, 70, 80 and 90 days after full bloom (DAFB) during the production cycle of 2013 in the orchard "La Campana" at Ciudad Cuauhtemoc, Chihuahua, Mexico. Starch extraction was performed by wet milling according to the method reported by Tirado-Gallegos et al. [17].

2.2. Preparation of Coating Solutions and Films

Coating solutions and films were made with the plate casting method using the methodology proposed by Mali et al. [37]. Different aqueous dispersions were prepared with each of the starches. Starch was mixed with distilled water until reaching a concentration of 4% (w/w). The dispersion was heated on a heating plate (model 6795-220, CORNING, Monterrey, NL, Mexico) and kept under constant stirring (350 rpm) with overhead stirrer IKA (model RW 20 digital, WERKE, Wilmington, NC, USA), the dispersion was held at 85–90 °C for 15 min. Subsequently, the starch dispersion was cooled to 70 °C and glycerol (2%, w/w) as plasticizer was added [38]. The heating was maintained 15 min at this temperature under the same conditions of agitation. The filmogenic solutions with EA had the same starch and glycerol concentration and were prepared under the same conditions. However, glycerol was mixed with EA at three concentrations (0.02%, 0.05% and 0.1%, w/w) based on total mass of film forming solution. This mixture was kept under constant stirring for 12 h before being added to the starch solution. Once the process is complete, the filmogenic solution (film-forming solution) was cooled to 60 °C and emptied immediately in polystyrene Petri dishes (Ø = 15 cm) using 40 g/box. Boxes with the filmogenic solutions were dried under laboratory condition (RH ≈ 45% ± 5% and 25 ± 1 °C) until the formed film was peeled off easily from the plate (≈ 72 h) [24,39]. Subsequently, the films were conditioned for 48 h in desiccators containing a saturated salt solution of NaBr (RH = 55% ± 5%). After the conditioning, the properties of the films were characterized. The films obtained in the first stage from starch of apples harvested at 60, 70, 80 and 90 DAFB were abbreviated as: FILM-60, FILM-70, FILM-80 and FILM-90, respectively. The films made in the second stage, were obtained from starch of harvested apples at 70 DAFB. In this case, the film without EA was appointed as FILM-Control, while the films with 0.02%, 0.05% and 0.1% (w/w) of EA were identified as FILM-EA0.02%, FILM-EA0.05% and FILM-EA0.1%, respectively.

2.3. Films Characterization

2.3.1. Color

The films tri-stimulus color was evaluated with a Minolta colorimeter CR-300 (Minolta, Osaka, Japan) with a diffuse illumination/0° viewing geometry and a pulsed xenon arc lamp (PXA). Readings were recorded in the color space in accordance with the CIELAB (L^*, a^*, b^*) scale. Readings were taken in five random points on the surface of the films, using as background white standard used in the calibration of equipment. Moreover, the whiteness index (WI) of the films with EA and its control was measured according to the following equation [40]:

$$WI = 100 - \sqrt{(100 - L*)^2 + a*^2 + b*^2} \tag{1}$$

2.3.2. Optical Properties using Ultraviolet–visible (UV-vis) Spectroscopy

The light barrier properties of the films evaluated as transparency was determined according to the methodology proposed by Luchese et al. [39]. Sample rectangles were cut and placed in the cell of a spectrophotometer (Evolution 300, Thermo Scientific, Waltham, MA, USA) and readings of the absorbance of the cell were made with the film at 600 nm taking the filmless cell as blank. Films transparency (T) was calculated according to the following equation:

$$Transparency\ (T) = \frac{A_{600}}{e} \qquad (2)$$

where A_{600} is the absorbance of the cell with the film at 600 nm and e is the film's thickness. According to the above, high absorbance values mean less transparency in the films. In the case of the films with EA, a sweep of the absorbance from 200 to 800 nm was performed. Five replicates were performed.

2.3.3. Scanning Electron Microscopy

The surface morphology of the films was evaluated by scanning electron microscope JEOL (JEE400, Tokyo, Japan). The film sample was adhered to the sample holder and covered with gold to make it conductive. Finally, micrographs were taken at acceleration potential of 5 kV and a current intensity of 2 mA.

2.3.4. Thickness Measurement

The film thickness (e) was measured using a digital micrometer Mitutoyo (model Coolant Proof 293-348, Kanagawa, Japan) with an accuracy of 0.001 mm, the thickness was measured at 10 points randomly designated on the films.

2.3.5. Moisture Content

The moisture content was determined using the standard method of the International Association of Official Analytical Chemistry (AOAC) [41]. Samples of 0.5 g was weighed and dried in an oven Shel Lab (model 1370GM-2, Sheblon Manufacturing, Inc., Cornelius, OR, USA) at 105 °C to constant weight with a variation of 0.0001 g ($\approx$ 2.5 h). This analysis was carried out five times and the moisture content was expressed as percentage of water in the film according to following equation:

$$Moisture\ (\%) = \frac{W_i - W_f}{W_i} \times 100 \qquad (3)$$

where, W_i is the initial weight of the humid film and W_f is the final weight of the dry film.

2.3.6. Water Solubility

The films solubility was determined according to the methodology proposed by Colla et al. [42] with slight modifications. Films were cut with sizes of 2 × 3 cm², which were dried to constant weight at 105 °C during 2.5 h. Each sample was placed in a beaker with 80 mL of distilled water. The samples were kept under gentle stirring ($\approx$ 60 rpm) in a Corning plate (Model PC-620D, New York, NY, USA) at room temperature (23 ± 3 °C) within 24 h. Liquid together with the film was then filtered with filter paper (Whatman No. 1, pre-dried at 105 °C to constant weight). Filter papers with film residues were dried again at 105 °C to constant weight to obtain the dry matter. The solubility was recorded as the percentage of dry material of the solubilized film (%S) for 24 h and was determined using the following equation:

$$Solubility(\%S) = \frac{W_i - W_f}{W_i} \times 100 \qquad (4)$$

where, W_i is the weight of the dry film and W_f is the weight of the dry sample after immersion in water.

2.3.7. Water Vapor Permeability (WVP)

WVP was determined according to ASTM E-9680 [43] method, known as the cup method or test cell. The films were cut into circles and placed on cells containing approximately 12 g of dry silica gel (desiccant) to generate a relative humidity close to 0%. Subsequently, the cells were placed in a desiccator containing a saturated NaCl solution to generate a 75% RH at room temperature ($25 \pm 2\,°C$) solution. Weight variation of the cells over time was recorded, for which the cells every 60 min for at least 7 h were weighed. The recorded data were fit to a linear regression model and the transmission rate (TR) was calculated from the slope ($g\,s^{-1}$) obtained from the straight line and the effective permeation area (0.0031 m^2), while the WVP ($g\,Pa^{-1}\,s^{-1}\,m^{-1}$) was determined according to the equation:

$$TR = \left(\frac{\Delta w}{\Delta t}\right)\frac{1}{A} \quad (5)$$

$$WVP = \frac{(TR)(e)}{\Delta P} \quad (6)$$

where Δw is the weight change in the cell (g) at the time Δt (s), A is the exposed area of the film in the cell (m^2), e is the film thickness (m) and ΔP is the gradient of the partial pressure of water vapor (Pa) in the desiccator and inside the cell.

2.3.8. Mechanical Properties

Determinations were performed according to the methodology described by Mali et al. [44]. Ten strips (60 × 10 mm^2) were cut for each formulation; the thickness was measured at 10 random spots along each strip using a micrometer (Mitutoyo, Kobe, Japan). The films were subjected to a tensile stress in a TAXT-Plus texturometer (Stable Micro Systems, Surrey, UK) with a 30 kg load cell, following the guidelines of ASTM-882-95a [45]. Tensile tests were performed at a strain rate of 20 mm min^{-1} and a distance between the clamping pincers 4 cm. The tensile strength (TS) in MPa, the elongation at break (%E) in % were recorded and the elasticity modulus (EM) in MPa was determined.

2.3.9. X-ray Diffraction

The diffractograms of the films were obtained by means of an X-ray diffractometer (Panalytical Xpert PRO, Almelo, OV, The Netherlands) with Ni-filtered CuKα radiation at a voltage of 40 kV and a current of 30 mA (λ = 0.154 nm), equipped with an X´Celerator detector. The diffractograms were collected within the scanning angle 2θ from 5° to 40° with a scanning speed of 1° min^{-1}. The crystallinity of the films was calculated with the following equation:

$$Cristallinity\,(\%) = \frac{A_c}{A_c + A_a} \times 100 \quad (7)$$

where, A_c is the area of the crystalline region and A_a is the area of the amorphous region.

2.3.10. Fourier Transform Infrared Spectroscopy (FTIR)

The characterization of the films by FTIR was carried out with a Spectrum Two infrared spectrophotometer (Perkin Elmer Inc., Waltham, MA, USA) equipped with a universal module of attenuated total reflectance (ATR). The samples were placed in the sample holder for ATR and by means of a punch a pressure of 100 ± 1 N was exerted on the sample. The vibrational transition frequencies were reported in transmittance (%) according to the wave number (cm^{-1}) within the mid-infrared. An average of 34 sweeps per sample was recorded with a resolution of 4 cm^{-1} in the region from 450 to 4000 cm^{-1}. The spectra were analyzed with Spectrum Two software version 10.4.

2.3.11. Thermogravimetric Analysis

The thermal stability of the samples was assessed by thermogravimetric analysis according to procedure described by Teodoro et al. [46]. A TGA equipment was used (U600, TA Instruments, New Castle, DE, USA) under constant nitrogen flow (20 mL/min). The amount of approximately 8–10 mg was weighted. Sample was heated from 25 to 800 °C at a speed of 20 °C min^{-1}.

The following temperatures for each sample were determinate: Maximum degradation temperature (T_{max}), the percentage of moisture contained at 100 °C (MC100), the temperature at which occurs a weight loss of 10% (T_{90}), as well as the extrapolated onset temperature (T_{onset}) indicating the temperature at which weight loss begins and the residual mass at 800 °C extrapolated (MR800). The MC100, T_{onset} and MR800 were estimated from the weight loss with respect to temperature (DTG curves).

2.3.12. Antioxidant Capacity

The films were kept under stirring in alcohol (0.05 g/mL) for 12 h (in complete darkness). Subsequently, samples were centrifuged (16,000 × g, 10 min, 5 °C) and the supernatant was recovered and filtered (Whatman No. 1). The extracts obtained were used to evaluate antioxidant capacity. Evaluation of antioxidant capacity was performed by the 2,2-diphenyl-1-picrilhidracil method (DPPH) according to Brand-Williams et al. [47] with slight modifications. A methanolic solution of DPPH (25 mg/L) was prepared and adjusted its absorbance at 0.7 ± 0.02 a 490 nm since the molecule has its absorption maximum at this wavelength (according to a preliminary sweep). 20 µL of the diluted sample (dilution factor 1:10) were placed in wells of a microplate and 280 µL of the DPPH solution was added. Subsequently, the mixture was kept in the dark for 30 min and the absorbance was read on a microplate reader BIORAD at a wavelength of 490 nm. Results were expressed as TEAC values, which translate as µmol Trolox equivalent (TE)/g dries film.

The antioxidant capacity was also determined by the assay of 2,2'-azino-bis (3-ethylbenzthiazolin)-6-sulphonic acid (ABTS), for which the protocol described by Re et al. [48] was used. The ABTS cation radical was generated by mixing an aqueous solution of ABTS (7 mM) with 88 µL of a solution of potassium per sulfate (140 mM) (final concentration of 2.45 mM). The mixture was allowed to stand for 16 h under conditions of total darkness. Subsequently, the solution of ABTS radical was diluted with methanol to obtain an absorbance of 0.70 ± 0.05 at 734 nm. Ten µL of the diluted extract was placed in wells of a microplate and 290 µL of the ABTS solution was added and its absorbance was measured at 734 nm in a BIORAD microplate reader. Antioxidant capacity was reported in the same manner as in DPPH protocol.

2.4. Experimental Design

The experiment was planned in two stages. In the first stage, apple starch films plasticized with glycerol were formulated and obtained. Apple starches were obtained from unripe fruits harvested at 60, 70, 80 and 90 DAFB. Once the films were obtained, their different properties were evaluated. Based on the results of its mechanical and barrier properties, the best film was chosen. In the second stage the effect of the incorporation of EA on the films selected in stage 1 was evaluated, considering its physicochemical, mechanical, barrier and antioxidant properties. In both stages, a completely randomized blocks design was used. The data were reported as the average of three replicates ± standard error. The results were evaluated by analysis of variance one-way (ANOVA) and comparison of means was performed to detect significant differences ($p < 0.05$) by Tukey test. Data analysis was performed using SAS 9.0 (SAS Institute Inc., Cary, NC, USA) software.

3. Results and Discussion

3.1. Physical Properties of Apple Starch Films Harvested at DAFB

In Table 1 the physical, mechanical, barrier (WVP) and structural (crystallinity) properties of apple starch films harvested at 60, 70, 80 and 90 DAFB are shown. The thickness of the films did not show significant differences ($p > 0.05$), which was due to the fact that the preparation of films is standardized [38]. Moreover, this is an important variable that influences the mechanical and barrier properties. The moisture content of the films ranged significantly ($p < 0.05$) between 21.33% and 23.11%, with the highest moisture content FILM-60.

Table 1. Physical, mechanical, crystallinity and water vapor permeability (WVP) properties of apple starch films harvested at different days after full bloom (DAFB).

Analyzed Variable	Film			
	FILM-60	FILM-70	FILM-80	FILM-90
Moisture (%)	23.11 ± 0.49 [a]	21.56 ± 0.45 [b]	22.84 ± 0.39 [ab]	21.33 ± 0.33 [b]
Solubility (%)	18.55 ± 1.01 [ab]	19.89 ± 0.49 [a]	21.36 ± 0.89 [a]	15.62 ± 0.62 [b]
Thickness (μm)	102.30 ± 4.39 [a]	100.28 ± 2.74 [a]	99.33 ± 1.94 [a]	96.73 ± 1.48 [a]
Transparency	1.14 ± 0.02 [b]	1.18 ± 0.01 [b]	1.39 ± 0.04 [a]	1.06 ± 0.02b [c]
Color	—	—	—	—
L*	95.73 ± 0.09 [a]	95.81 ± 0.14 [a]	95.77 ± 0.11 [a]	95.46 ± 0.19 [a]
a*	0.21 ± 0.01 [ab]	0.19 ± 0.01 [b]	0.18 ± 0.01 [b]	0.25 ± 0.01 [a]
b*	2.04 ± 0.05 [a]	2.14 ± 0.05 [a]	2.15 ± 0.04 [a]	2.02 ± 0.04 [a]
Mechanical properties	—	—	—	—
TS (MPa)	4.70 ± 0.15 [b]	8.12 ± 0.36 [a]	7.37 ± 0.02 [a]	7.14 ± 0.35 [a]
%E (%)	54.99 ± 4.04 [a]	52.12 ± 4.30 [a]	56.59 ± 0.56 [a]	56.35 ± 2.79 [a]
EM (MPa)	0.92 ± 0.14 [c]	3.10 ± 0.27 [a]	2.03 ± 0.11 [ab]	1.71 ± 0.19 [bc]
Crystallinity (%)	28.29 ± 1.25 [a]	29.66 ± 2.52 [a]	33.50 ± 1.82 [a]	28.47 ± 1.20 [a]
WVP × 10^{-11} (g m^{-1} s^{-1} Pa^{-1})	11.97 ± 1.09 [a]	6.77 ± 0.85 [b]	7.36 ± 0.29 [b]	9.71 ± 0.25 [ab]

Values represent the average of three repetitions ± standard error; Different letters in each row are significantly different ($p < 0.05$); DAFB: days after full bloom; TS: tensile strength; %E: elongation at break; ME: elasticity modulus; WVP: water vapor permeability.

The solubility of the films ranged between 15.62% and 21.36%; while the moisture content of the films did not affect their solubility. These values were minor than reported in sago starch films (~25.2%) [49] and cassava starch films (28%) [50]. High values of solubility film have been related to a high biodegradability, this is an advantage in foods covered with edible coatings that will be cooked. However, low solubility is required in foods that are stored for a long time [49]. Moreover, the solubility is an important factor to take into account in the migration of active substances from the packaging to the food matrix [6].

The optical properties are another advantage of starch films, since being practically colorless allow consumers to examine the food inside the package. The transparency of the films ranged between 1.06 and 1.39, equivalent to values of internal transmittance of 70%–75%. The transparency of the starch films was similar to that of synthetic films such as low density polyethylene (3.05), oriented polypropylene (1.67) and polyester (1.51) [51], which suggests that they have enough transparency to be used as edible packaging. The tensile strength (TS) (4.70–8.12 MPa) and the elastic modulus (EM) (0.92–3.10 MPa) varied significantly and both mechanical properties presented the following descending order: FILM-70 > FILM-80 > FILM-90 > FILM-60. The elongation at break (%E) did not present significant differences ranging between 52.12% and 56.59% with the following decreasing order: FILM-80 > FILM-90 > FILM-60 > FILM-70. This was due to the fact that the mechanical properties are influenced by the microstructure of the films, as well as by the crystallinity of their components.

The crystallinity of the films evaluated fluctuated between 28.29% and 33.50% without presenting significant differences ($p > 0.05$) with the following descending order: FILM-80 > FILM-70 > FILM-90 > FILM-60. Just at the sample FILM-60 a feasible relation between the crystallinity and the lower numerical values was observed for TS and ME. Finally, the water vapor permeability (WVP) varied significantly among the tested films (Table 1). The highest values for WVP were observed in FILM-60 and FILM-90 with 11.97 and 9.71 × 10^{-11} g m^{-1} s^{-1} Pa^{-1}, respectively. On the other hand, the lowest values were for FILM-70 and FILM-80 with a WVP of 6.77 and 7.36 × 10^{-11} g m^{-1} s^{-1} Pa^{-1}, respectively. These values were lower than those reported in films of commercial corn´s starch (19.2 × 10^{-11} g m^{-1} s^{-1} Pa^{-1}) [52] and yam (18.10 × 10^{-11} g m^{-1} s^{-1} Pa^{-1}) [37] plasticized with glycerol. In relation to these results it is important to point out that an objective comparison of WVP can be difficult, because this variable depends on the botanical source (amylose/amylopectin ratio), amount of plasticizer and moisture gradient used in the determination of WVP [27]. The behavior of WVP is also influenced by the microstructure of the film, because the porosity and possible cracks increase the values of WVP [1,53]. When compared with synthetic plastics such as polyethylene, one of the great disadvantages of starch films is their high WVP, which limits their ability to maintain the quality of food during storage. Therefore, taking into account the highest values of TS, ME and the lowest value of WVP, the formulation FILM-70 was considered as the most appropriate to assess the effect of the addition of ellagic acid (EA) on its physicochemical, mechanical, barrier and antioxidants properties. It is important to note that there were no significant differences between the mechanical properties of FILM-70 and FILM-80. Nevertheless, late fruit hand-thinning is practiced at until 70 DAFB in our region, thus these immature apples are considered a waste that can be used as an unconventional source of starch.

3.2. Characterization of Films with Ellagic Acid (EA) in the FILM-70 Formulation

3.2.1. Scanning Electron Microscopy (SEM) Analysis

In Figure 1 the micrographs of the surface of apple's starch films without EA (FILM-Control) and EA at different concentrations (0.02%, 0.05% and 0.1%) are shown. The absence of EA yielded films with a compact and smooth surface, while the addition of EA to different concentrations promoted the formation of rough surfaces. Furthermore, the presence of aggregates of EA on the surface was evident, which indicated that there was no adequate dispersion of the EA in the polymeric matrix because the EA, besides being insoluble in water, is very poorly soluble in many organic solvents such as alcohol [34], so it probably requires the use of other types of solvents as the basic solutions [30]. This behavior was similar to that observed by Kim et al. [30] who formulated autoclaved chitosan films with EA. These authors observed that the surface became rough by increasing the concentration of EA. The morphology and microstructure of the films is of extreme importance, since it has a direct effect on the other properties [37].

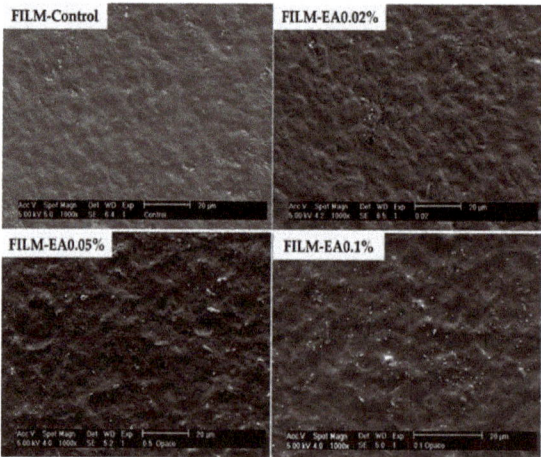

Figure 1. Surface morphology of apple starch films without ellagic acid (FILM-Control) and with ellagic acid (EA) added to 0.02% (FILM-EA0.02%), 0.05% (FILM-EA0.05%) and 0.1% (FILM-EA0.1%).

3.2.2. Color and UV-vis Spectroscopy

The optical properties (transparency and color) varied significantly ($p < 0.05$) when EA was added in films formulation FILM-70 (Table 2). The brightness of the films (L^*) decreased with the addition and the increase in the concentration of EA, which produced a slight darkening in the films, this was in agreement with the decrease in the whiteness index (WI). The variable a^* ranged from positive values (control film) to negative values (green tonality) when adding EA. The variable b^* was always maintained with positive values, fluctuating between 2.73 and 14.48, following the descending order FILM-EA0.1% > FILM-EA0.05% > FILM-EA0.02% > FILM-Control. This indicated that the films tended towards the yellow tonality (values + b^*) when adding and increasing the EA concentration (Figure 2a).

Table 2. Physical, mechanical, crystallinity and WVP properties of apple starch films without ellagic acid (FILM-Control) and with ellagic acid (EA) added to 0.02% (FILM-EA0.02%), 0.05% (FILM-EA0.05%) and 0.1% (FILM-EA0.1%).

Analyzed Variable	Film			
	FILM-Control	FILM-EA0.02%	FILM-EA0.05%	FILM-EA0.1%
Transparency	1.51 ± 0.09 [d]	3.73 ± 0.27 [c]	8.46 ± 0.194 [b]	17.45 ± 0.55 [a]
Color	—	—	—	—
L^*	95.87 ± 0.07 [a]	95.11 ± 0.13 [b]	94.13 ± 0.17 [c]	92.32 ± 0.14 [d]
a^*	0.22 ± 0.01 [a]	−0.82 ± 0.03 [b]	−1.24 ± 0.03 [c]	−1.31 ± 0.02 [c]
b^*	2.73 ± 0.06 [d]	6.77 ± 0.20 [c]	10.52 ± 0.29 [b]	14.48 ± 0.18 [a]
WI	95.04 ± 0.08 [a]	91.60 ± 0.24 [b]	87.89 ± 0.33 [c]	83.8 ± 83.56 [d]
Moisture (%)	23.83 ± 0.88 [a]	21.52 ± 0.49 [ab]	22.92 ± 0.49 [ab]	21.23 ± 0.32 [b]
Solubility (%)	20.51 ± 0.12 [b]	21.79 ± 0.43 [ab]	22.02 ± 0.93 [a]	23.79 ± 0.20 [a]
Thickness (μm)	102.79 ± 3.67 [a]	102.79 ± 3.67 [a]	104.42 ± 3.61 [a]	103.53 ± 4.17 [a]
Mechanical properties	—	—	—	—
TS (MPa)	6.51 ± 0.18 [b]	8.98 ± 0.28 [a]	9.63 ± 0.52 [a]	8.21 ± 0.45 [a]
%E (%)	65.11 ± 2.98 [a]	62.48 ± 2.98 [ab]	56.91 ± 3.01 [ab]	52.44 ± 1.21 [b]
EM (MPa)	1.79 ± 0.19 [c]	2.78 ± 0.13 [bc]	4.58 ± 0.42 [a]	3.63 ± 0.32 [ab]
Crystallinity (%)	28.81 ± 1.69 [a]	28.63 ± 0.64 [a]	28.99 ± 0.27 [a]	31.39 ± 1.66 [a]
WVP × 10^{-11} (g m^{-1} s^{-1} Pa^{-1})	6.59 ± 0.28 [b]	7.46 ± 0.34 [a]	6.65 ± 0.19 [ab]	6.35 ± 0.22 [b]

Values represent the average of three repetitions ± standard error; Different letters in each row are significantly different ($p < 0.05$); DAFB, days after full bloom; WI, whiteness index; TS: tensile strength; %E: elongation at break; EM: elasticity modulus; WVP: water vapor permeability.

On the other hand, transparency of the films increased between 1.51 and 17.45, according to the following order FILM-EA0.1% > FILM-EA0.05% > FILM-EA0.02% > FILM-Control. It is important to indicate that higher values in transparency result in materials that reduce transmission to light (more opaque). This characteristic can be observed in the Figure 2a, the increase in EA content promoted more opaque starch films than control film. Figure 2b shows the optical transparency of films evaluated a in the wavelength region of 200–800 nm, the FILM-Control virtually allowed light transmission between 60% and 70% under the visible light region. This light transmission decreased dramatically to 17% to 48%, 1% to 20% and 0.1% to 6% with addition of EA at 0.02 (FILM-EA0.02), 0.05 (FILM-EA0.05) and 0.1% w/w (FILM-EA0.1), respectively. UV-vis absorption curves measured from 200 to 800 are shown in Figure 2c, all the films could prevent the UV transmission compared to the film control, because of the addition of EA promoted significant ($p < 0.05$) high absorption of light. Moreover, in the UV-vis absorption spectrum, the presence of EA in the films promoted the apparition of absorption peaks at 370 and 400 nm and the intensity of peaks increased with increasing EA content in the films. The peak absorption band 370 nm is characteristic to the presence of flavonols, while the absorption band 400 nm exhibited the presence of the lactone ring in the EA structure [54]. Within UV radiation, the one that causes the most damage to sensitive components (pigments, vitamins, some enzymes, among others) present in food, is UV-A (ultraviolet light of long wavelength, 315–400 nm) [49]. The incorporation of EA significantly increased the absorption of UV-A, which was almost completely blocked with the addition of EA at the concentrations of 0.05% (FILM-EA0.05%) and 0.1% (FILM-EA0.1%) in films (Figure 2b). Generally, food packaging requires transparent packaging; however, mostly opaque materials can protect those components of food that are sensitive to light, thus maintaining their quality.

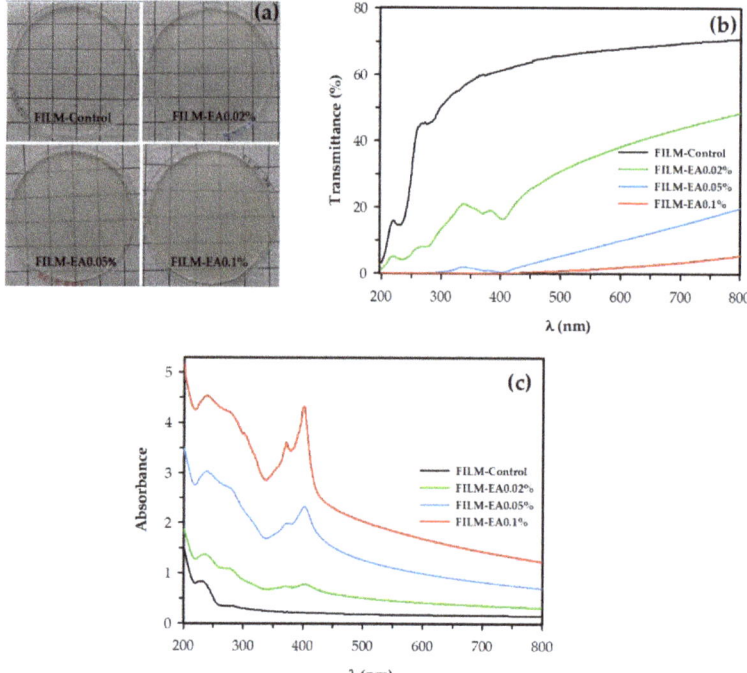

Figure 2. (a) Digital images, (b) UV-vis transmittance and (c) UV-vis absorbance spectra of apple starch films without ellagic acid (FILM-Control) and with ellagic acid (EA) added to 0.02% (FILM-EA0.02%), 0.05% (FILM-EA0.05%) and 0.1% (FILM-EA0.1%).

3.2.3. Moisture Content and Thickness of the Films

There were no significant differences ($p > 0.05$) in the thickness of the films (Table 2), so the addition of EA did not influence this determination. Thickness uniformity can be taken as an indicator that the process for preparing film was standardized, allowing control this feature. The film thickness is important because it influences the mechanical and barrier properties of these materials. Compared with the film without EA (FILM-Control), the moisture content only decreased ($p < 0.05$) when 0.1% of EA was added (FILM-EA0.1%), going from 23.83% to 21.23% (Table 2), with a decrease of 11%. These values were higher than those reported by Piñeros-Hernandez et al. [1] in starch films with rosemary extract (19.4%–19.8%). Moreover, they were considerably lower than those reported by Medina-Jaramillo et al. [50] in films from cassava starch with green tea (25.3%) and basil (28.6%) extract. Both researchers observed surface hydrophobicity after adding extracts rich in phenols.

3.2.4. Solubility and Water Vapor Permeability (WVP)

The solubility of the films is shown in Table 2, expressed as a percentage of dry matter soluble in water for 24 h. The solubility of the films ranged from 20.51% to 23.59%, with the following order FILM-EA0.1% > FILM-EA0.05% > FILM-EA0.02% > FILM-Control. In comparison with the film without EA, the solubility increased significantly by 16% when 0.05%–0.1% of EA was incorporated. This behavior could be explained based on the results reported by Kim et al. [30], who observed that increasing the concentration of EA in chitosan films also decreased the contact angle on its surface, promoting a more hydrophilic character. The solubility of the films is related to increased water resistance. The water vapor permeability (WVP) is a measure of the amount of vapor molecules of water passing through the film. Compared to FILM-Control, only FILM-EA0.05% increased the WVP from 6.59 to 7.46 × 10^{-11} g m^{-1} s^{-1} Pa^{-1}, showing an increase of 13%. WVP values reported in this paper are within those reported by Nouri and Mohammadi [49] and Piñeros-Hernandez et al. [1] in films of sago starch and cassava starch, respectively. As mentioned above, the incorporation EA in chitosan films promoted a character hydrophilic in their surface, so that based on the results of solubility shown in the Table 2, is very likely that such behavior is also present in the films obtained in this study. Therefore, the WVP values would be expected to increase with the EA concentration. It is possible that the presence of scattered particles in the films (Figure 1) will affect mass transfer in the contact area [25]; however, more studies are needed in this regard. The barrier properties of the films are related to their microstructure [37]. Figure 1 shows a mostly uniform surface in FILM-Control and only cracks were observed on the surface of the films with 0.02% of EA, which promoted an increase in the values of WVP. Piñeros-Hernandez et al. [1] observed cracks in the surface of starch films with 10% rosemary extract only when the magnification was 2000×. In addition, cross-sectional micrographs generate more information about the microstructure of these materials and impact properties (data not shown). In general, the solubility and values of WVP films should be low or minimum to preserve food during storage; however, high solubility would be an advantage if these materials are to be used as food coatings [49].

3.2.5. Mechanical Properties

The tensile strength (TS), elastic modulus (EM) and elongation at break (%E) of the starch films without and with EA are presented in Table 2. Compared with FILM-Control, the addition of EA affected significantly ($p < 0.05$) all the mechanical properties. Tensile strength increased to 38% from 6.51 MPa (FILM-control) to 8.98 MPa by adding 0.02% of EA in the formulation (FILM-EA0.02%). By increasing the concentration of EA were no significant differences between active films. In comparison with FILM-Control, elongation at break remained constant at EA concentrations of 0.02% (FILM-EA0.02%) and 0.05% (FILM-EA0.05%). However, increasing the concentration of EA to 0.1% (FILM-EA0.1%), %E experienced a decrease of 19%. EA concentration did not affect %E in active films. Moreover, the elasticity modulus unchanged ($p > 0.05$) when comparing FILM-Control and FILM-EA0.02%; however, the rigidity (EM) of the films increased by 155 and 102% by increasing the concentration of 0.05% EA (FILM-AE0.05%) and 0.1% (FILM-AE0.1%), respectively. On the other hand, the micrographs of the films (Figure 1) showed that FILM-Control presented, at least on the surface, a more compact structure (uniform and homogeneous) than the active films (films with EA). Theoretically, a more compact surface would provide better mechanical properties; however, the inclusion of EA in the structure generated rougher surfaces presenting mechanical properties (except %E) higher than those in FILM-Control. This implies that there was an interaction between the EA and the starch, such as the possible formation of ester bonds during the gelatinization process [1,55,56] or hydrogen bonding interactions [56,57]; however, further investigation needs to be done about this physicochemical phenomenon.

3.2.6. X-ray Diffraction and Crystallinity

The X-ray diffraction patterns of the developed films are shown in Figure 3. The film FILM-Control presented mostly pronounced peaks at approximately $2\theta = 17°$, 19.5° and 22.2°. Compared FILM-Control, films added with EA showed the same peaks of crystallinity. However, from concentrations higher than 0.02% of EA (i.e., in films FILM-EA0.05% and FILM-0.1%) a peak of crystallinity was observed at $2\theta = 12.3°$. In the film with the highest concentration of EA (FILM-0.1%), another peak of crystallinity appeared at $2\theta = 28.3°$. Kim et al. [30] obtained diffractograms of EA powder and added in chitosan films, the peaks of crystallinity in $2\theta = 13.2°$ and 28.3° were reported for EA powder and appeared in chitosan films with EA concentrations higher than 0.1%. However, in comparison with the control film, the presence of the peaks of crystallinity of the EA did not significantly affect ($p > 0.05$) the crystallinity of the films, ranging between 26.63% and 31.39% (Table 2) with the following order FILM-0.1% > FILM-EA0.05% > FILM-Control > FILM-EA0.02%.

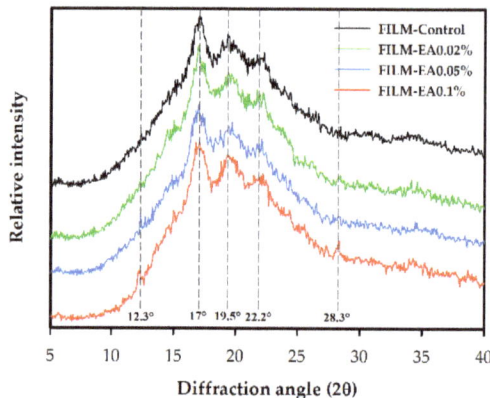

Figure 3. X-ray diffractograms of apple starch films without ellagic acid (FILM-Control) and with ellagic acid (EA) added to 0.02% (FILM-EA0.02%), 0.05% (FILM-EA0.05%) and 0.1% (FILM-EA0.1%).

3.2.7. Fourier Transform Infrared Spectroscopy

Fourier transform infrared (FTIR) spectra obtained from the apple starch film without EA (FILM-Control) and with EA (FILM-EA0.02%, FILM-EA0.05% and FILM-EA0.1%) are shown in Figure 4. These spectra were compared to determine the feasible chemical interactions between the components of the films (starch, glycerol, water and EA). It can be seen that in the wave number $(1/\lambda)$ from 4000 to 450 cm^{-1} (Figure 4a) all the spectra showed the characteristic peaks for the stretching of the OH bond (3300 cm^{-1}) and the stretching of the CH bonds associated with the glucose ring (2922 cm^{-1}) [58]. Within the region of the fingerprint $(1/\lambda = 400–1250$ cm$^{-1})$ [17] characteristic peaks were observed at 1149 cm^{-1}, 1078 cm^{-1} and 1004 cm^{-1}, representative of C–O–C and C–OH stretches in the glycosidic bonds of polysaccharides [59]. In the films with EA, starting at a concentration of 0.05% (FILM-EA0.05%) the appearance of a new peak was observed at 1507 cm^{-1}, which was intensified at a concentration of 0.1% (FILM-EA0.1%). Previously similar behaviors have been observed in chitosan films with different concentrations of EA (0.05, 0.1 and 0.5%) and this was related to the stretching of the C=C bonds present in the aromatic compounds [30]. These results are similar to those reported by Piñeros-Hernandez et al. [1], who obtained cassava starch films added with rosemary extract. FILM-Control and FILM-EA0.02% film had a clearly defined signal at 1645 cm^{-1}, which was the result of vibrations bending of the OH$^-$ groups of water absorbed in the amorphous region of the starch molecule [60]. However, in Figure 4b it can be seen that in films FILM-EA0.05% a new peak was developed at 1702 cm^{-1}, which was more intense in FILM-EA0.1%. This peak is related to the stretching of the ester bonds (C=O) [1,55,56]. Moreover, the new bands at 1620 and 1580 cm^{-1} in FILM-EA0.1% (Figure 4b) have been related with aromatic rings [61]. This suggests that there was possibly a chemical interaction between starch and ellagic acid, which could have caused the observed variations in the mechanical properties of films with EA.

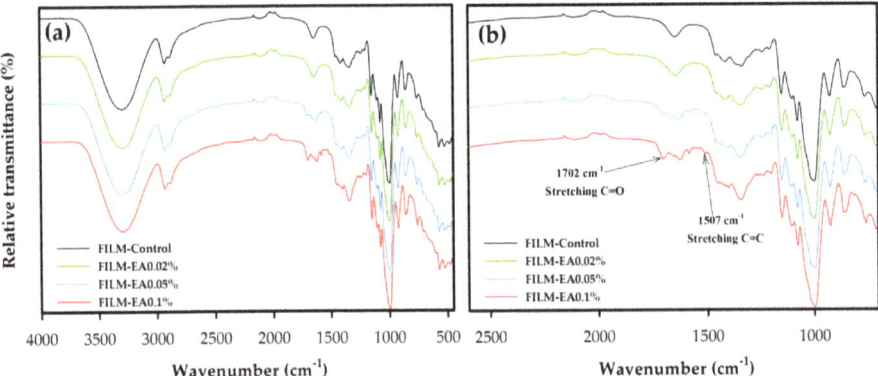

Figure 4. Fourier transform infrared (FTIR) spectra evaluated in: (a) 4000–650 cm^{-1} and (b) 2600–700 cm^{-1} of apple starch films without ellagic acid (FILM-Control) and with ellagic acid (EA) added to 0.02% (FILM-EA0.02%), 0.05% (FILM-EA0.05%) and 0.1% (FILM-EA0.1%).

3.2.8. Thermogravimetric Analysis

The effect of EA addition on the thermal stability of the FILM-70 films can be seen in the graphs of thermogravimetric analysis (TGA) shown in Figure 5. In the thermograms, three stages of weight loss versus temperature were noted, this behavior has been previously reported for starch films plasticized with glycerol [13,29]. The first stage occurred between 55 and 150 °C, this decrease in initial weight has been associated with the loss of water and other low molecular weight compounds [62]. Weight loss in the second stage (150–280 °C) has been associated with the decomposition of the glycerol-rich phase, which also contains starch [29]. Finally, the weight loss observed in the last stage (280–350 °C) was a result of the breakdown of carbohydrates [29,58].

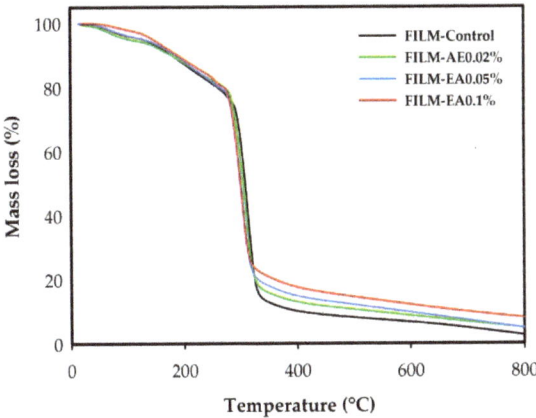

Figure 5. Thermogravimetric curves of apple starch films without ellagic acid (FILM-Control) and with ellagic acid (EA) added to 0.02% (FILM-EA0.02%), 0.05% (FILM-EA0.05%) and 0.1% (FILM-EA0.1%).

On the other hand, within the thermal variables evaluated (Table 3), it was shown that the temperature at which 10% of the weight of the sample was lost (T90) did not present a significant difference ($p > 0.05$) for all the films, which indicated that the thermal stability was not affected by the incorporation of AE in the evaluated concentrations [46]. No significant differences were observed ($p > 0.05$) in the percentage of humidity contained at 100 °C (MC100), with the lowest values corresponding to the films FILM-AE0.05% and FILM-AE0.1%, which was related to the moisture content of those films (Table 2). The T_{onset} and T_{max} temperatures were significantly higher in the control film, which has been linked to the low interaction between the matrix and the additive [63]. The residual mass (residue) is related to the nature of the additives, inorganic components and impurities further combustion in inert atmosphere (N_2) cannot even allow complete combustion of organic components [63].

Table 3. Thermal variables determined from film thermograms of apple starch films without ellagic acid (FILM-Control) and with ellagic acid (EA) added to 0.02% (FILM-EA0.02%), 0.05% (FILM-EA0.05%) and 0.1% (FILM-EA0.1%).

Film	T_{90} (°C)	MC100 (%)	T_{onset} (°C)	T_{max} (°C)	RM800 (%)
FILM-Control	176.83 ± 1.08 [a]	4.11 ± 0.22 [a]	291.20 ± 0.76 [a]	313.77 ± 0.83 [a]	1.63 ± 0.62 [a]
FILM-0.02%EA	170.86 ± 6.89 [a]	5.02 ± 0.47 [a]	280.63 ± 2.33 [b]	309.16 ± 1.27 [b]	4.61 ± 0.15 [b]
FILM-0.05%EA	180.30 ± 3.10 [a]	3.95 ± 0.21 [a]	278.37 ± 0.42 [b]	302.63 ± 0.33 [c]	5.31 ± 0.33 [b]
FILM-0.10%EA	179.86 ± 14.68 [a]	3.70 ± 1.22 [a]	278.50 ± 0.80 [b]	301.29 ± 0.01 [c]	7.58 ± 0.29 [c]

Values represent the average of three repetitions ± standard error. Different letters in each column are significantly different ($p < 0.05$).

3.2.9. Antioxidant Capacity

The antioxidant capacity of starch films with different concentrations of EA is presented in Figure 6. The antioxidant capacity of the films, determined by DPPH and ABTS, showed a similar behavior by both methods, with significant increases in antioxidant capacity with the increase of EA in the films. Based on films with 0.02% EA, with the DPPH protocol, antioxidant capacity increased 3.2 and 7.1 times in films with 0.05% and 0.1% EA, while with the ABTS protocol the increase was 3.02 and 5.7 times, respectively. The results obtained in this study were similar to those reported by López-Mata et al. [40] in chitosan films added with carvacrol; however, the concentrations used by these authors are up to 15 times larger, which is a clear evidence that the AE (compared to other substances) is a powerful antioxidant even at low concentrations as 0.02% (w/v). These results indicate that apple starch films added with EA may have antioxidant capacity even at the concentration of 0.02% (w/v). However, it is necessary to carry out studies on the kinetics of release of the antioxidant agent from the films, which is determined by the type of antioxidant and the nature of the food simulant [6].

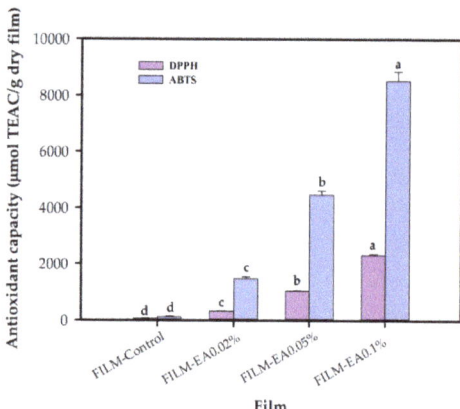

Figure 6. Antioxidant capacity (evaluated by DPPH and ABTS techniques) of apple starch films without ellagic acid (FILM-Control) and with ellagic acid (EA) added to 0.02% (FILM-EA0.02%), 0.05% (FILM-EA0.05%) and 0.1% (FILM-EA0.1%).

4. Conclusions

The mechanical properties of the films were affected depending on the concentration of EA, which was beneficial, because depending on the potential applications could adapt the mechanical properties. Incorporating EA promoted increases in opacity. This indicated that the films could be efficient in preventing the passage of UV light. The results of this study suggest that films based on apple starch added with ellagic acid (a nutraceutical agent with powerful antioxidant action) can be used for the development of active food packaging. Finally, it is necessary to explore strategies that increase the solubility of ellagic acid in the filmogenic solution and improve its incorporation in thermoplastic starch.

Author Contributions: Investigation and writing–original draft, J.M.T-G.; Methodology, J.d.J.O.-P., C.R.-V., G.I.O.O, M.A.A.-G. and D.L.-G.; Project administration, P.B.Z.-F.; Validation, H.Y.L.-D.l.P.; Visualization, F.H.-C.; Writing–original draft, M.E.-D. and M.H.-G.; Writing–review & editing, R.B.-J. and J.J.B.-F.

Funding: This study is supported by CONACYT-Mexico and the Research Center for Food and Development (CIAD, A.C.).

Acknowledgments: The first author, J.M. Tirado-Gallegos thanks the National Council of Science and Technology (CONACYT) of Mexico for the scholarship granted to carry out doctoral studies at the "Centro de Investigación en Alimentación y Desarrollo (CIAD), A.C., Unidad Cuauhtémoc." The authors thank Emilio Ochoa Reyes for their technical assistance. This study is the product of the Research Group in Carbohydrates, Packaging and Functional Foods (CEAF) of CIAD-Cuauhtemoc, Chihuahua, Mexico led by Paul Baruk Zamudio-Flores.

Conflicts of Interest: The authors declare no conflict of interest.

References

1. Piñeros-Hernandez, D.; Medina-Jaramillo, C.; López-Córdoba, A.; Goyanes, S. Edible cassava starch films carrying rosemary antioxidant extracts for potential use as active food packaging. *Food Hydrocoll.* **2017**, *63*, 488–495. [CrossRef]
2. Reis, L.C.B.; de Souza, C.O.; da Silva, J.B.A.; Martins, A.C.; Nunes, I.L.; Druzian, J.I. Active biocomposites of cassava starch: The effect of yerba mate extract and mango pulp as antioxidant additives on the properties and the stability of a packaged product. *Food Bioprod. Process.* **2015**, *94*, 382–391. [CrossRef]
3. Oniszczuk, T.; Wójtowicz, A.; Moácicki, L.; Mitrus, M.; Kupryaniuk, K.; Kusz, A.; Bartnik, G. Effect of natural fibres on the mechanical properties of thermoplastic starch. *Int. Agrophys.* **2016**, *30*, 211–218. [CrossRef]
4. Jiménez, A.; Fabra, M.J.; Talens, P.; Chiralt, A. Edible and biodegradable starch films: A review. *Food Bioprocess Technol.* **2012**, *5*, 2058–2076. [CrossRef]
5. López, O.V.; Castillo, L.A.; García, M.A.; Villar, M.A.; Barbosa, S.E. Food packaging bags based on thermoplastic corn starch reinforced with talc nanoparticles. *Food Hydrocoll.* **2015**, *43*, 18–24. [CrossRef]
6. Ashwar, B.A.; Shah, A.; Gani, A.; Shah, U.; Gani, A.; Wani, I.A.; Wani, S.M.; Masoodi, F.A. Rice starch active packaging films loaded with antioxidants—development and characterization. *Starch Stärke* **2015**, *67*, 294–302. [CrossRef]
7. Zamudio-Flores, P.B.; Bello-Pérez, L.A. Elaboración y caracterización de películas de glicoproteínas obtenidas mediante reacción de Maillard utilizando almidón acetilado y aislado proteico de suero lácteo. *Rev. Mex. Ing. Quím.* **2013**, *12*, 401–413. (In Spanish)
8. Sapper, M.; Chiralt, A. Starch-Based Coatings for Preservation of Fruits and Vegetables. *Coatings* **2018**, *8*, 152. [CrossRef]
9. Yan, Q.; Hou, H.; Guo, P.; Dong, H. Effects of extrusion and glycerol content on properties of oxidized and acetylated corn starch-based films. *Carbohydr. Polym.* **2012**, *87*, 707–712. [CrossRef]
10. Chandra mohan, C.; Rakhavan, K.R.; Sudharsan, K.; Radha krishnan, K.; Babuskin, S.; Sukumar, M. Design and characterization of spice fused tamarind starch edible packaging films. *LWT Food Sci. Technol.* **2016**, *68*, 642–652. [CrossRef]
11. Famá, L.; Flores, S.K.; Gerschenson, L.; Goyanes, S. Physical characterization of cassava starch biofilms with special reference to dynamic mechanical properties at low temperatures. *Carbohydr. Polym.* **2006**, *66*, 8–15. [CrossRef]

12. Mali, S.; Grossmann, M.V.E.; García, M.A.; Martino, M.N.; Zaritzky, N.E. Barrier, mechanical and optical properties of plasticized yam starch films. *Carbohydr. Polym.* **2004**, *56*, 129–135. [CrossRef]
13. Sanyang, M.L.; Sapuan, S.M.; Jawaid, M.; Ishak, M.R.; Sahari, J. Effect of plasticizer type and concentration on tensile, thermal and barrier properties of biodegradable films based on sugar palm (*Arenga pinnata*) starch. *Polymers* **2015**, *7*, 1106–1124. [CrossRef]
14. Sartori, T.; Menegalli, F.C. Development and characterization of unripe banana starch films incorporated with solid lipid microparticles containing ascorbic acid. *Food Hydrocoll.* **2016**, *55*, 210–219. [CrossRef]
15. Zamudio-Flores, P.B.; Vargas-Torres, A.; Pérez-González, J.; Bosquez-Molina, E.; Bello-Pérez, L.A. Films prepared with oxidized banana starch: Mechanical and barrier properties. *Starch Stärke* **2006**, *58*, 274–282. [CrossRef]
16. Daudt, R.M.; Avena-Bustillos, R.J.; Williams, T.; Wood, D.F.; Külkamp-Guerreiro, I.C.; Marczak, L.D.F.; McHugh, T.H. Comparative study on properties of edible films based on pinhão (*Araucaria angustifolia*) starch and flour. *Food Hydrocoll.* **2016**, *60*, 279–287. [CrossRef]
17. Tirado-Gallegos, J.M.; Zamudio-Flores, P.B.; Ornelas-Paz, J.d.J.; Rios-Velasco, C.; Acosta-Muñiz, C.H.; Gutiérrez-Meraz, F.; Islas-Hernández, J.J.; Salgado-Delgado, R. Efecto del método de aislamiento y el estado de madurez en las propiedades fisicoquímicas, estructurales y reológicas de almidón de manzana. *Rev. Mex. Ing. Quím.* **2016**, *15*, 391–408. (In Spanish)
18. Stevenson, D.G.; Domoto, P.A.; Jane, J.-L. Structures and functional properties of apple (*Malus domestica* Borkh) fruit starch. *Carbohydr. Polym.* **2006**, *63*, 432–441. [CrossRef]
19. Shi, Y.; Jiang, L.; Zhang, L.; Kang, R.; Yu, Z. Dynamic changes in proteins during apple (*Malus x domestica*) fruit ripening and storage. *Hortic. Res.* **2014**, *1*, 6. [CrossRef] [PubMed]
20. Berlanga Reyes, D.I.; Romo Chacón, A.; Martínez Campos, Á.R.; Guerrero Prieto, V.M. Apple fruit chemical thinning in Chihuahua, Mexico. *Rev. Fitotec. Mex.* **2008**, *31*, 243–250.
21. Rascón-Chu, A.; Martínez-López, A.-L.; Carvajal-Millán, E.; Martínez-Robinson, K.G.; Campa-Mada, A.C. Gelificación iónica de pectina de bajo grado de esterificación extraída de manzanas inmaduras de raleo. *Rev. Fitotec. Mex.* **2016**, *39*, 17–24. (In Spanish)
22. Berlanga-Reyes, D.I.; Rios-Velasco, C.; Romo-Chacón, A.; Guerrero-Prieto, V.M. Raleo químico de flores de manzano (*Malus x domestica* Borkh.) 'Golden Delicious' y 'RedChief Delicious'. *Tecnociencia Chihuah.* **2012**, *6*, 147–157. (In Spanish)
23. Romero, V.; Borneo, R.; Passalacqua, N.; Aguirre, A. Biodegradable films obtained from triticale (*x Triticosecale Wittmack*) flour activated with natamycin for cheese packaging. *J. Food Packag. Shelf Life* **2016**, *10*, 54–59. [CrossRef]
24. Cano, A.; Cháfer, M.; Chiralt, A.; González-Martínez, C. Development and characterization of active films based on starch-PVA, containing silver nanoparticles. *J. Food Packag. Shelf Life* **2016**, *10*, 16–24. [CrossRef]
25. Shah, U.; Gani, A.; Ashwar, B.A.; Shah, A.; Ahmad, M.; Gani, A.; Wani, I.A.; Masoodi, F.A.; Yildiz, F. A review of the recent advances in starch as active and nanocomposite packaging films. *Cogent Food Agric.* **2015**, *1*, 1115640. [CrossRef]
26. Mehyar, G.F.; Han, J.H. Physical and mechanical properties of high-amylose rice and pea starch films as affected by relative humidity and plasticizer. *J. Food Sci.* **2004**, *69*, E449–E454. [CrossRef]
27. De Araújo, G.K.P.; de Souza, S.J.; da Silva, M.V.; Yamashita, F.; Gonçalves, O.H.; Leimann, F.V.; Shirai, M.A. Physical, antimicrobial and antioxidant properties of starch-based film containing ethanolic propolis extract. *Int. J. Food Sci. Technol.* **2015**, *50*, 2080–2087. [CrossRef]
28. Corrales, M.; Han, J.H.; Tauscher, B. Antimicrobial properties of grape seed extracts and their effectiveness after incorporation into pea starch films. *Int. J. Food Sci. Technol.* **2009**, *44*, 425–433. [CrossRef]
29. Medina, J.C.; Gutiérrez, T.J.; Goyanes, S.; Bernal, C.; Famá, L. Biodegradability and plasticizing effect of yerba mate extract on cassava starch edible films. *Carbohydr. Polym.* **2016**, *151*, 150–159. [CrossRef] [PubMed]
30. Kim, S.; Liu, Y.; Gaber, M.W.; Bumgardner, J.D.; Haggard, W.O.; Yang, Y. Development of chitosan–ellagic acid films as a local drug delivery system to induce apoptotic death of human melanoma cells. *J. Biomed. Mater. Res. Part B Appl. Biomater.* **2009**, *90*, 145–155. [CrossRef] [PubMed]
31. Seeram, N.P.; Adams, L.S.; Henning, S.M.; Niu, Y.; Zhang, Y.; Nair, M.G.; Heber, D. In vitro antiproliferative, apoptotic and antioxidant activities of punicalagin, ellagic acid and a total pomegranate tannin extract are enhanced in combination with other polyphenols as found in pomegranate juice. *J. Nutr. Biochem.* **2005**, *16*, 360–367. [CrossRef] [PubMed]

32. Baek, B.; Lee, S.H.; Kim, K.; Lim, H.-W.; Lim, C.-J. Ellagic acid plays a protective role against UV-B-induced oxidative stress by up-regulating antioxidant components in human dermal fibroblasts. *Korean J. Physiol. Pharmacol.* **2016**, *20*, 269–277. [CrossRef] [PubMed]
33. Kilic, I.; Yeşiloğlu, Y.; Bayrak, Y. Spectroscopic studies on the antioxidant activity of ellagic acid. *Spectrochim. Acta Part A Mol. Biomol. Spectrosc.* **2014**, *130*, 447–452. [CrossRef] [PubMed]
34. Bala, I.; Bhardwaj, V.; Hariharan, S.; Kumar, M.R. Analytical methods for assay of ellagic acid and its solubility studies. *J. Pharm. Biomed. Anal.* **2006**, *40*, 206–210. [CrossRef] [PubMed]
35. Saucedo-Pompa, S.; Jasso-Cantu, D.; Ventura-Sobrevilla, J.; SÁEnz-Galindo, A.; RodríGuez-Herrera, R.; Aguilar, C.N. Effect of candelilla wax with natural antioxidants on the shelf life quality of fresh-cut fruits. *J. Food Qual.* **2007**, *30*, 823–836. [CrossRef]
36. Saucedo-Pompa, S.; Rojas-Molina, R.; Aguilera-Carbó, A.F.; Saenz-Galindo, A.; Garza, H.d.L.; Jasso-Cantú, D.; Aguilar, C.N. Edible film based on candelilla wax to improve the shelf life and quality of avocado. *Food Res. Int.* **2009**, *42*, 511–515. [CrossRef]
37. Mali, S.; Grossmann, M.V.E.; Garcia, M.A.; Martino, M.N.; Zaritzky, N.E. Microstructural characterization of yam starch films. *Carbohydr. Polym.* **2002**, *50*, 379–386. [CrossRef]
38. Zamudio-Flores, P.B.; Ochoa-Reyes, E.; Ornelas-Paz, J.D.J.; Tirado-Gallegos, J.M.; Bello-Pérez, L.A.; Rubio-Ríos, A.; Cárdenas-Felix, R.G. Caracterización fisicoquímica, mecánica y estructural de películas de almidones oxidados de avena y plátano adicionadas con betalaínas. *Agrociencia* **2015**, *49*, 483–498. (In Spanish)
39. Luchese, C.L.; Garrido, T.; Spada, J.C.; Tessaro, I.C.; de la Caba, K. Development and characterization of cassava starch films incorporated with blueberry pomace. *Int. J. Biol. Macromol.* **2018**, *106*, 834–839. [CrossRef] [PubMed]
40. López-Mata, M.A.; Ruiz-Cruz, S.; Silva-Beltrán, N.P.; Ornelas-Paz, J.D.J.; Zamudio-Flores, P.B.; Burruel-Ibarra, S.E. Physicochemical, antimicrobial and antioxidant properties of chitosan films incorporated with carvacrol. *Molecules* **2013**, *18*, 13735–13753. [CrossRef] [PubMed]
41. *Official Methods of Analysis*, 16th ed.; Association of Official Analytical Chemist: Washington, DC, USA, 1995.
42. Colla, E.; do Amaral Sobral, P.J.; Menegalli, F.C. Amaranthus cruentus flour edible films: Influence of stearic acid addition, plasticizer concentration and emulsion stirring speed on water vapor permeability and mechanical properties. *J. Agric. Food Chem.* **2006**, *54*, 6645–6653. [CrossRef] [PubMed]
43. *ASTM-E-96-80 Standard Methods of Test for Water Vapor Transmission of Materials in Sheet Form*; ASTM International: West Conshohocken, PA, USA, 2016.
44. Mali, S.; Sakanaka, L.S.; Yamashita, F.; Grossmann, M.V.E. Water sorption and mechanical properties of cassava starch films and their relation to plasticizing effect. *Carbohydr. Polym.* **2005**, *60*, 283–289. [CrossRef]
45. *ASTM-882-95a Standard Test Methods for Tensile Properties of Thin Plastic Sheeting*; ASTM International: West Conshohocken, PA, USA, 1995.
46. Teodoro, A.P.; Mali, S.; Romero, N.; de Carvalho, G.M. Cassava starch films containing acetylated starch nanoparticles as reinforcement: Physical and mechanical characterization. *Carbohydr. Polym.* **2015**, *126*, 9–16. [CrossRef] [PubMed]
47. Brand-Williams, W.; Cuvelier, M.E.; Berset, C. Use of a free radical method to evaluate antioxidant activity. *LWT Food Sci. Technol.* **1995**, *28*, 25–30. [CrossRef]
48. Re, R.; Pellegrini, N.; Proteggente, A.; Pannala, A.; Yang, M.; Rice-Evans, C. Antioxidant activity applying an improved ABTS radical cation decolorization assay. *Free Radic. Biol. Med.* **1999**, *26*, 1231–1237. [CrossRef]
49. Nouri, L.; Mohammadi, N.A. Antibacterial, mechanical and barrier properties of sago starch film incorporated with betel leaves extract. *Int. J. Biol. Macromol.* **2014**, *66*, 254–259. [CrossRef] [PubMed]
50. Medina-Jaramillo, C.; Ochoa-Yepes, O.; Bernal, C.; Famá, L. Active and smart biodegradable packaging based on starch and natural extracts. **2017**, *176*, 187–194. [CrossRef] [PubMed]
51. Shiku, Y.; HamaguchI, P.Y.; Tanaka, M. Effect of pH on the preparation of edible films based on fish myofibrillar proteins. *Fish. Sci.* **2003**, *69*, 1026–1032. [CrossRef]
52. García, M.A.; Martino, M.N.; Zaritzky, N.E. Lipid addition to improve barrier properties of edible starch-based films and coatings. *J. Food Sci.* **2000**, *65*, 941–944. [CrossRef]
53. Versino, F.; García, M.A. Cassava (*Manihot esculenta*) starch films reinforced with natural fibrous filler. *Ind. Crops Prod.* **2014**, *58*, 305–314. [CrossRef]
54. Arulmozhi, V.; Pandian, K.; Mirunalini, S. Ellagic acid encapsulated chitosan nanoparticles for drug delivery system in human oral cancer cell line (KB). *Colloids Surf. B Biointerface* **2013**, *110*, 313–320. [CrossRef] [PubMed]

55. Bikiaris, D.; Panayiotou, C. LDPE/starch blends compatibilized with PE-g-MA copolymers. *J. Appl. Polym. Sci.* **1998**, *70*, 1503–1521. [CrossRef]
56. Wu, Y.-P.; Ji, M.-Q.; Qi, Q.; Wang, Y.-Q.; Zhang, L.-Q. Preparation, structure and properties of starch/rubber composites prepared by co-coagulating rubber latex and starch paste. *Macromol. Rapid. Commun.* **2004**, *25*, 565–570. [CrossRef]
57. Chai, Y.; Wang, M.; Zhang, G. Interaction between amylose and tea polyphenols modulates the postprandial glycemic response to high-amylose maize starch. *J. Agric. Food Chem.* **2013**, *61*, 8608–8615. [CrossRef] [PubMed]
58. Pelissari, F.M.; Andrade-Mahecha, M.M.; do Amaral Sobral, P.J.; Menegalli, F.C. Isolation and characterization of the flour and starch of plantain bananas (*Musa paradisiaca*). *Starch Stärke* **2012**, *64*, 382–391. [CrossRef]
59. Mano, J.F.; Koniarova, D.; Reis, R.L. Thermal properties of thermoplastic starch/synthetic polymer blends with potential biomedical applicability. *J. Mater. Sci. Mater. Med.* **2003**, *14*, 127–135. [CrossRef] [PubMed]
60. Das, D.; Jha, S.; Kumar, K.J. Isolation and release characteristics of starch from the rhizome of Indian Palo. *Int. J. Biol. Macromol.* **2015**, *72*, 341–346. [CrossRef] [PubMed]
61. Hussein, M.Z.; Al Ali, S.H.; Zainal, Z.; Hakim, M.N. Development of antiproliferative nanohybrid compound with controlled release property using ellagic acid as the active agent. *Int. J. Nanomed.* **2011**, *6*, 1373–1383. [CrossRef] [PubMed]
62. Li, X.; Qiu, C.; Ji, N.; Sun, C.; Xiong, L.; Sun, Q. Mechanical, barrier and morphological properties of starch nanocrystals-reinforced pea starch films. *Carbohydr. Polym.* **2015**, *121*, 155–162. [CrossRef] [PubMed]
63. Perazzo, K.K.N.C.L.; de Vasconcelos Conceição, A.C.; dos Santos, J.C.P.; de Jesus Assis, D.; Souza, C.O.; Druzian, J.I. Properties and antioxidant action of actives cassava starch films incorporated with green tea and palm oil extracts. *PLoS ONE* **2014**, *9*, e105199. [CrossRef] [PubMed]

© 2018 by the authors. Licensee MDPI, Basel, Switzerland. This article is an open access article distributed under the terms and conditions of the Creative Commons Attribution (CC BY) license (http://creativecommons.org/licenses/by/4.0/).

Review

Recent Progress in Gas Barrier Thin Film Coatings on PET Bottles in Food and Beverage Applications

Masaki Nakaya [1,*], Akira Uedono [2] and Atsushi Hotta [3]

1   Research Laboratories for Packaging Technologies, R & D Department, Kirin Co., Ltd., Technovillage 3F, 1-17-1 Namamugi, Tsurumi-ku, Yokohama 230-8628, Japan
2   Division of Applied Physics, Faculty of Pure and Applied Science, University of Tsukuba, Tsukuba 305-8573, Japan; uedono.akira.gb@u.tsukuba.ac.jp
3   Department of Mechanical Engineering, Faculty of Science and Technology, Keio University 3-14-1, Hiyoshi, Kohoku-ku, Yokohama 223-8522, Japan; hotta@mech.keio.ac.jp
*   Author to whom correspondence should be addressed; masaki-nakaya@kirin.co.jp; Tel.: +81-45-521-0072; Fax: +81-45-505-3925.

Academic Editor: Alessandro Lavacchi
Received: 20 September 2015; Accepted: 20 November 2015; Published: 8 December 2015

Abstract: This article presents a short history and the recent advancement of the development of chemical vapor deposition technologies to form thin film gas barrier coatings on PET bottles and other plastic containers in food and beverage containers. Among different gas barrier enhancement technologies, coating can show unique performance where relatively high gas barrier enhancement is possible to various gas permeants. In this article, technologically common and different points of the current thin film coating methods in this field are summarized. This article also refers to recent market situations and technological challenges in the Japanese market.

Keywords: gas barrier; coating; thin film; PET bottle; DLC; SiO_x; SiOC

---

1. Background of Thin Film Coatings on Plastic Containers

Polymer materials have unique properties, such as being easy to shape, and are elastic to physical impacts compared to other types of materials, like metal, glass, and ceramics, and nowadays quite a wide variety of plastic containers are seen in the food and beverage industry. For example, PET (poly(ethylene terephthalate)) bottles are the most widely used package format in the soft drink segment and further use of PET bottles is expected both inside and beyond the soft drink segment [1,2].

From the view of package performance, light-weight, unbreakable, and transparent properties are favorable advantages of common plastic containers. To the contrary to these consumer benefits, gas permeability is a remarkable disadvantage of plastic containers compared to metal and glass containers [3], which virtually eliminate gas permeation, except sealing parts where polymer materials are usually used.

Especially, the permeation of oxygen and carbon dioxide molecules often limits the shelf-life of sensitive products. One of the most sensitive products to gas permeation is beer. Beer is quite sensitive to oxidation, and also sensitive to carbon dioxide release. From the view point of shelf-life extension, the degree of gas barrier improvement is often expressed by BIF (barrier improvement factor) [4]. The value of BIF can be calculated based on the gas transmission rate of normal container(s) divided by that of barrier improved container(s). PET bottles of single serve size require 10 or more times the oxygen barrier in BIF in order to achieve a realistic shelf-life of beer. Furthermore, they require seven or more times the carbon dioxide barrier in BIF if the equivalent shelf-life in glass bottles is demanded.

Since these sensitive products are seen quite often in our daily diet, like in juice, teas, seasoning, edible oil, and wine, as well as beer, significant effort has been made for improving the gas barrier performance of plastic containers. Among rigid containers used in the food and beverage industry, PET bottles are the most intensive category of plastic containers for gas barrier enhancement study because of their industrial scale of use. It should be stressed that the demand for high gas barrier PET bottles has been increasing because of the global trend in weight reduction, where thinner bottle walls show less gas barrier performance [5], and of a gradual increase of the applications of PET bottle formats.

Based on these backgrounds, this paper reviews the past and recent progress of gas barrier enhanced PET bottles, especially gas barrier thin film coated bottles.

2. Approach to the Gas Barrier Enhancement of PET Bottles Other Than Thin Film Coating

Major technologies to enhance the gas barrier property of PET bottles used in today's industry can be roughly classified into four categories, that is, (i) coating; (ii) multi-layer; (iii) blending; and (iv) oxygen scavengers, as illustrated in Table 1. It should be noted that different approaches can be combined together. For example, the core layer explained below in the multi-layer approach may include oxygen scavengers, or the blending additives explained below are added in PET layers of multi-layer walls.

Table 1. Rough classification of the current major gas barrier technologies for PET bottles [6,7].

Technology	Coating	Multilayer	Blending	O_2 Scavengers
Schematic Image				
O_2 Scavengers	High	Middle	Middle	High
Other gas barrier	High	Middle	Middle	Low
Recyclability	High	Middle	Middle	Middle
Installation cost	High	High	Low	–
Operation cost	Low	High	High	High

The multi-layer approach employs at least one core layer with higher gas barrier properties placed between PET layers. The core layer(s) provides the majority of the gas barrier property of the whole bottle. Some specific grades of polyamides are often used for core layer materials, even though other materials had been attempted [3,8]. While the multi-layer approach is widely used in many industrial fields and its process control has been well established, economics due to the use of specific injection machines for multilayer preforms and of relatively expensive core-layer materials are the barrier to further distribution in the PET bottle industry. From a technical standpoint of view, the core layers are usually adjusted to occupy several percentages of the whole bottle weight to shape the bottle properly, and the core layers of a bottle usually do not exist near the mouth part and the center of the bottom part. These factors limit the maximum oxygen barrier property of multi-layered bottles compared to oxygen scavengers and coating approaches. In Japan, the market share of barrier PET bottles based on this approach has been decreasing.

In the blending approach, higher gas barrier materials are added into melted PET resin before the shaping process. The additives increase the gas barrier property of the whole bottle depending on the concentration in the PET matrix. Some specific grades of polyamides are often used for additive materials [9], even though other materials had been attempted [3,10]. Due to the cost of additives and limited barrier performance compared to other approaches, the use of this approach is limited in these days in Japan. Additionally, in some countries such as Japan, possible mass use of polyamide additives is a concern to their recycling systems.

Oxygen scavengers are a type of additive which reacts with the oxygen permeant and results in restricting the passage of oxygen molecules through the bottle wall. The addition of a certain polyamide and transition metal complex into the PET matrix is an example of this approach [11]. In ideal conditions, this approach can inhibit the increase of dissolved oxygen in the liquid content of the bottle. However, it makes it difficult for bottle manufactures to control the quality of their products as additive concentration and shaping conditions of bottles affects to each other. Some application may not accept the tint and haze due to typical types of scavenger additives.

3. Thin Film Coating for the Gas Barrier Enhancement of PET Bottles

Coating forms thin films over the surface of PET bottles. Dense structures of the thin films, typically several tens of nanometers in thickness, behave like glass or ceramics, and block the passage of gas permeants. The current approach generally uses two types of thin film species, that is, (A) carbon thin films, often described as diamond-like carbon (DLC) or a-C:H [3,4], or (B) silicate oxide thin films, often described as SiO_x, where x is a number and often somewhere between 1.5 to 1.8 [3,12].

While each approach described in the previous chapter has its own advantages and disadvantages, the use of coating is an expanding trend, or is expected to expand [13]. At least in the Japanese market, the trend is remarkable in recent years [14]. One of the advantages in the coating approach is that relatively high gas barrier enhancement is possible to various gas components including oxygen, carbon dioxide, water vapor, and flavors. This favors the quality retention of beverages where quite complex combinations of flavors contribute to unique taste and mouth-feeling, for example, seen in wine and beer [15–17]. Another advantage lies in high recyclability. While other categories of the gas barrier enhancement approach of PET bottles usually require several percentages of foreign materials in the PET matrix in terms of weight, the foreign materials derived from coating amount to be, at most, several parts per million in terms of weight. As a result, coated bottles are usually no problem in recycling of normal PET bottles even in the case of mass use. From an economic point of view, relatively high capital cost to install coating machines is disadvantageous to coating, and this can explain the cause of the relatively slow increase of the use of coated bottles. On the other hand, relatively low operation cost is advantageous, and, in the case with high operational efficiency, coating is expected to require the lowest operation cost [6–8]. In brief, in the case where a remarkable increase of barrier PET bottles happens, especially involved with the mass use in beer and carbonated soft drinks, coating approaches are most likely to be accepted from the viewpoint of bottle performance, social systems, and economics. In other words, at present, coating can be considered to have the largest growth potential among the barrier enhancement technologies of PET bottles.

4. Current Methodology to Thin Film Formation onto PET Bottle Surface

While various techniques are known to form thin films on substrates, plasma assisted chemical vapor deposition (CVD) techniques are currently available for mass production machinery for gas barrier thin film coating of PET bottles. These techniques meet the requirements for food and beverage containers. At least several requirements are essential, as summarized in Table 2.

Table 2. Basic requirements for thin film coating to PET bottles.

No.	Property	Reason	Corresponding Process Design
1	High gas barrier	For the flavor quality of the bottle content	Special configuration in coating chambers
2	Flexible	To withstand bottle deformation	Limited coating thickness and/or use of adhesion layer(s)
3	Thin and clear	For recycling and bottle appearance	Limited coating thickness
4	Physically and chemically stable to the bottle content	For safety to human and the flavor quality of the bottle content	Choice of thin film species in case of inside coating
5	Short process time	Economics	Optimization between barrier enhancement and throughput

One of major conceivable reasons of the use of plasma-assisted CVD lies in low heat load to the substrate. The deformation of the containers is likely to occur when the temperature of the substrate increases above its glass transition temperature which, in the case of polyester-based plastic containers like PET and PLA bottles is 70–80 °C, and 60–70 °C, respectively [18].

A second conceivable reason is that plasma can relatively readily occur inside a bottle. While coating may be applied to the outer surface of a bottle, these types of technologies involve some difficulty to protect the physical damage to the coating during production in filling lines and transportation to retailers, and also to control coating conditions along with accumulating coating dusts inside vacuum chambers. On the other hand, in the case of coating on the inner surface of a bottle, the thin film is protected with the bottle wall from physical impacts from the outside of the bottle, and most coating dusts can be deposited inside the bottle and removed from the vacuum chamber. Physical impacts may be a concern even with the internal coating due to known "abuse", while typical production and transportation processes seem harmless to the barrier performance of the coating inside the bottle, as far as coated bottles were observed in Japanese market. Additionally, it should be noted that dust control is significantly important for continuous production which might last 20 h or longer. In the case of coatings over the inner surface of containers, thin films tend to come in contact with food and beverages, and are required to have physio-chemical stability which secures the safety to human diet.

The third reason is the relatively short process time for thin film formation. Usually, thin films of 10–100 nm in thickness are used in current technologies. Coating thickness is determined, depending on thin film species, based on economics and the optimal thickness for gas barrier properties [2,12]. It should be noted that an excessively thin film lacks in barrier property, and an excessively thick film decreases in visual and barrier quality due to the occurrence of cracks [2,19].

As a result, based on the deposition rates of roughly 2–60 nm per seconds, 1–5 s are taken for thin film deposition under vacuum conditions, such as 1–20 Pa before coating and 5–30 Pa during coating. The whole process time ranges from 6–30 s per one bottle coating, depending on coating conditions and machine configurations. These process conditions are summarized in Table 3.

Table 3. Summary of plasma assisted CVD techniques used for PET bottle coating.

No.	Coating Process/Device	Variations
1	Power frequency	2.45 GHz, 13.56MHz, or 6.0 MHz
2	Thin film species	Carbon (DLC) or SiO$_x$
3	Material gas	Acetylene, HMDSO, HMDSN
4	Coating chambers	With electrodes (capacitative systems), or without electrodes (inductive systems)
5	Vacuum pressure	Around 10 Pa
6	Coating time	Around 1–5 s
7	Coating surface	Inside of bottles

As a result, high throughput machines with a capacity of up to 40,000 bottles per hour have been in operation in soft drink and beer segments based on industriall-realistic economics. Figure 1 and Table 4 show an example of high throughput machine and details on coating process and performance, respectively, based on Kirin's DLC coating method [20].

Figure 1. Example of high throughput rotary coating machine for PET bottles (photo provided by courtesy of Mitsubishi Heavy Industry Food and Package Co., Ltd., Nagoya, Japan).

Table 4. Typical process conditions for DLC coating to PET bottles [20].

Process Parameter	Conditions
Power frequency	13.56 MHz, or 6.0 MHz
Power outlet	300–2500 W
Material gas	Acetylene
Gas flow rate	10–300 sccm
Vacuum pressure	5–10 Pa
Coating time	1–2 s
Resultant Properties	**Performance**
Deposition rate	Around 10 nm/s
Gas barrier improvement	Oxygen, Carbon dioxide, water vapor, and flavor components
Applicable container	1–5000 mL
Applicable filling manner	Aseptic to hot filling

Although differences in processes for coating bottles can be found among the current plasma-assisted CVD technologies, they have the basic process concept in common, that is, (i) to place a bottle into a vacuum chamber, and to vacuum the chamber; (ii) to supply material gas into the bottle; (iii) to apply electromagnetic wave to the inside of the bottle so that the material gas is decomposed into a plasma state; (iv) to allow the plasma to form a thin film on the inner surface of the bottle; and (v) to release the chamber to the atmospheric pressure, and to remove the coated bottle (as summarized in Figure 2). Obviously, these processes can be repeated continuously.

Figures 2 and 3 show an example of the coating processes of Kirin's DLC coating method and the coating system, respectively. In this system, an outer electrode functions as a part of vacuum chamber. Moreover, its internal shape similar to the bottle shape enables evenly distributed coating over the entire part of the bottle, based on that distance between the inner surface of the outer electrode and the bottle can control the voltage of the bottle surface and the resultant plasma distribution.

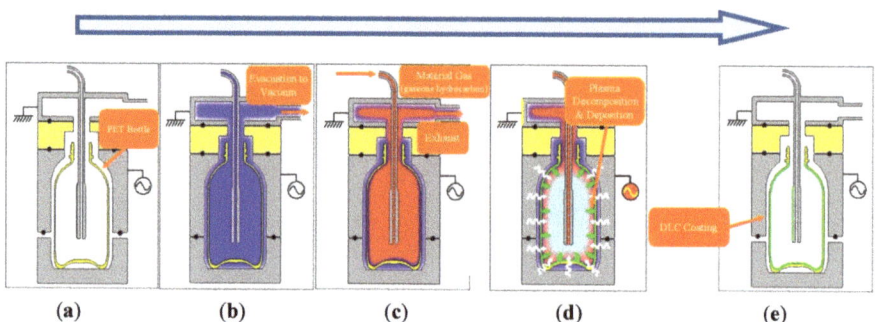

Figure 2. Schematic plasma CVD process for coating plastic bottles in case of Kirin's DLC coating. (**a**) Bottle placement into the coating chamber and vacuuming; (**b**) material gas supply; (**c**) power application to the coating chamber; (**d**) thin film deposition; and (**e**) pressure release and bottle removal from the coating chamber.

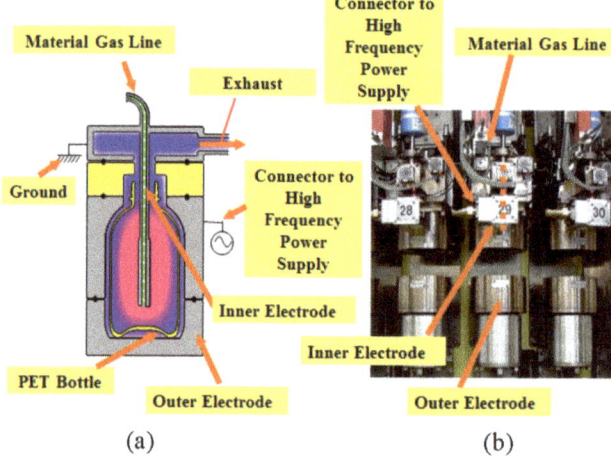

Figure 3. Example of the components of coating system for PET bottles: (**a**) schematic model; and (**b**) the corresponding part of the production machines (photo provided by courtesy of Mitsubishi Heavy Industry Food and Package Co., Ltd., Nagoya, Japan).

This basic process concept for hollow containers was seen at least as early as the 1980s, and some coating machines intended for commercial use were introduced early in 1990s [21,22], and various process conditions, including different material gas species, have been tried. As a result, the main difference of the processes among the current coating technologies for PET bottles, in general, lies in the material gas species and the frequency of power used to create plasma states.

Nowadays, types of metal oxides and nitrides, as well as carbons, are known to be possible to function as gas barrier thin films [23]. Carbon and silicate oxide thin films are, however, only two thin film materials available for mass production technologies for gas barrier enhanced PET bottles. The major reasons for the use of carbon and silicate oxide thin films for PET bottle applications lie in safety in food contact, the availability and relatively easy handling of material gas, and the economics to achieve sufficient gas barrier performance. Although aluminum and aluminum oxide thin films have a long history of use for the gas barrier enhancement of film and sheet applications [24], appropriate material gas species and coating processes for container applications have not yet been found.

In addition, the current plasma assisted CVD processes which are practical in the mass production can be found in vacuum conditions. Although it has been proved that certain atmospheric plasma-assisted CVD techniques can form gas barrier carbon and silicate oxide thin films based on dielectric barrier discharge techniques [25], their technical problems, such as dimensional limits, remain yet unsolved for the application of three-dimensional objects like PET bottles.

5. Difference in and between Carbon and Silicate Oxide Coatings

The current commercial carbon thin films have a slight, brownish to golden, tint [26]. Although this may restrict applicable product categories of carbon coated PET bottles, from the viewpoint of the visual quality of products, the degree of the tint appears to position within the scope of consumer acceptance, based on the commercial products of white wine (Figure 4) and edible oil categories in the Japanese market.

Figure 4. Example of DLC coated bottles for wine.

In case of beverage and liquor market in Japan, carbon coating is more often seen than silicate oxide coating in spite of the abovementioned disadvantage to carbon coating. The reason might be found in that carbon thin films are readily applicable to various product categories because carbon coating is inert to food and beverage solutions as long as the PET substrate is stable. On the other hand, some more remarkable limit in applicable product properties is known in typical silicate oxide coating. The gas barrier property of silicate oxide coatings may be decreased in contact with some solutions, for example, beverages of pH close to neutral [27].

C_2H_2 (acetylene) is the main material gas for carbon thin films for gas barrier-enhanced PET bottles. Derived from the hydrogen contained in acetylene molecules, the resultant carbon thin films contain hydrogen components up to 40% in atomic percentage. ERDA (Elastic Recoil Detection Analysis) analyses showed oxygen components up to 10% may be present in the carbon thin films [4,28], which is considered to be mainly derived from water vapor from PET substrates. The advantage of the use of acetylene lies in high deposition rates and economics, while CH_4 (methane) is used in many studies [4,26]. At least carbon thin films derived from acetylene contain the carbon bonding of sp^3, sp^2, and sp^1, based on XPS and FTIR studies [22]. In the Japanese market, carbon-coated PET bottles are derived from Kirin's DLC and Sidel's Actis™ technologies [14].

HMDSO (hexa-metyl-di-siloxan) and HMDSN (hexa-metyl-di-silazane) are the main material gases for silicate oxide thin films with aid of the controlled supply of oxygen. Based on the ratio in the mass flow of the material gas to oxygen and other conditions, the resultant thin films have different components consisting of silicate, carbon, oxygen, and hydrogen [29]. The components have impacts on gas barrier properties and stability in contact with beverage solutions, and sometimes also on tint. In the case with commercial gas barrier silicate oxide, thin films are totally colorless in visual observation. In the Japanese market, silicate oxide-coated PET bottles can be mainly seen in domestic edible oil and wine products, and rarely seen in imported carbonated water. Those bottles are derived from Toyo Seikan's Sibird™, Toppan's GL-C™, and KHS's Plasmax™ technologies [14].

From the viewpoint of the frequency of power used to cause the plasma states of material gas supplied, radio frequency (13.56 MHz) and microwave (2.45 MHz) are used in commercial technologies. The use of radio frequencies usually leads to a bi-electrode system, in other words, a type of capacitively-coupled plasma system, where sheath voltage and the resultant ion impact over the surface of the substrate can be controlled relatively precise manner [28]. It can be expressed that the use of these systems involves both merits and demerits to machine users. Examples of the merits are possible improvement in the performance of coating and stable application to relatively small or large containers, while those of the demerits are the possible increase of the change of mechanical parts for the application to containers of different shapes and sizes.

6. Recent Advancement in Commercialized Technologies for Coating Plastic Containers

In spite of the different nature of carbon and silicate oxide thin films as described above, it can be said that the difference between the two thin films is decreasing in the recent technical advancement.

It is obviously conceivable that the optimization of process conditions in parallel to the improvement in machinery has been performed in each technology, and as a result, deposition time has been shortened while the barrier properties of PET bottles coated are maintained or even improved. It is supposed that typical process conditions, including vacuum pressure, gas flow rate, and power application have been optimized. As a result, carbon coating has been less colored, and widened its applications (as shown in Figure 2). In the same way, silicate oxide coating has clarified and mitigated its limitation in applications, and widened its applications. In the case of the Japanese market, the use of coating technologies has been rapidly increased in recent years and, at present coating is the most abundant among technologies, compared to other gas barrier enhancement technologies applied to PET bottles [14].

An example of technological advancement has been found in the appropriate use of dielectric materials along electrodes in Kirin's DLC coating technology [30], and the modification of power frequency. Conventionally, this technology employed 13.56 MHz for power frequency and outer electrodes made of metal (conductive materials) parts only. Recently, power frequency was confirmed as one of the significant process parameters [28]. The use of 6 MHz for power frequency and outer electrodes fully or partially covered with dielectric material parts has been proposed in order to facilitate finding the appropriate process conditions for high gas barrier coatings (as shown in Figure 5) in addition to a decreased change of electrode parts for bottles of different shape and size. The results of the observation of coating thickness and plasma light emission indicate that the reason why 6 MHz power frequency showed the lowest gas barrier performance lies in the optimized spatial distribution of plasma. Compared to plasma produced with 13.56 MHz, where the plasma tends to concentrate around the neck part of the bottle, it seems that plasma with a lower frequency provides higher ion impacts to the PET substrate and the resultant secondary electrons modify the spatial distribution of the plasma to the direction of the bottom part of the bottle.

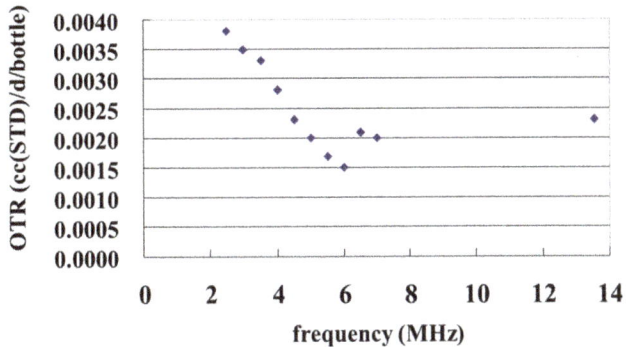

Figure 5. Example of the impact on power frequency to the performance of coated bottles. 500 mL PET bottles were coated with DLC using different power frequency ranging from 2.5 MHz to 13.56 MHz, and the oxygen transmission rate (OTR) of these bottles were measured. Optimized power frequency was found at 6.0 MHz in terms of OTR. The measurement of OTR was performed based on ASTM F1307 [31] method using an Oxran 2/21 device, Mocon Co., Ltd., (Brooklyn Park, MN, USA) under conditions of 23 °C and 90% relative humidity.

Another example has been found in the modification in the manner of material gas and oxygen supply during the coating process of KHS's Plasmax™. This technology is called Plasmax Plus™. Due to an extra carbon-rich layer formed on the conventional silicate oxide layer, the resultant coating can be stable in contact with solutions of pH close to neutral, which deteriorates the performance of coating based on the conventional process. Interestingly, the new coating manner requires no machinery modification [27].

7. Possible Near Future Advancement in This Field

The above description in this review mainly covered a brief history of gas barrier enhancement of PET bottles through plasma-assisted CVD techniques. On the other hand, a lot of effort has been made to other types of plastic containers and novel approaches to gas barrier enhancement.

Although the current era where PET bottles are the most abundant package format of rigid plastic containers is likely to last in this and the following decades because of their industrially-favorable balance between performance and economics, other plastic materials also have demands for functional thin film coating. Some polyolefins, such as PE (poly(ethylene)) and PP (poly(propylene)), are quite useful materials while the lack of oxygen and other barrier properties limits their benefits. For possible example, coated PP bottles or jars could keep the flavor quality of filled contents for certain extended periods of time in addition to high heat resistance, compared to PET containers, which are limited in applications below the boiling temperature of water.

The authors found a remarkable difference in the degree of oxygen gas barrier enhancement with DLC coating formed on various kinds of plastic film substrates, as shown in Table 5 and Figure 6. However, a positron annihilation [32] study by the authors indicates that, on these substrates, DLC coating can be formed homogeneously in terms of free volume, as shown in Figure 7. The positron annihilation method is based on a phenomenon where positrons implanted into a condensed matter annihilate with an electron and emits two 511-keV γ quanta. The spectra of γ energy, including the Doppler shift, are characterized by the S parameter, which mainly reflects changes due to the annihilation of positron-electron pairs with a low-momentum distribution. For amorphous materials, positronium (Ps: a hydrogen-like bound state between a positron and an electron) may form in open spaces (or free volumes). Figure 7 clearly shows that DLC films has a small S parameter, compared to polymer substrates, and that thin films of small free volume can function as barriers against gas permeation. Empirically, packages made of PE and PP tends to have relatively rough surface, and rough surface is considered to lead to significant defects in coating. When the surface of them and PET bottles is observed using an atomic force microscopy, the R_a of 1 μm square is usually 30–100 nm and less than 1 nm, respectively. A result of wet coating approach [33] supports this concept, where a specific type of organosilane materials placed between DLC thin films and PP substrates remarkably enhanced oxygen barrier property, even though the organosilane layer itself did not have a significant barrier property to the PP substrate. It should be noted that, when the surface of the organosilane layer is observed with an atomic force microscopy, the R_a of 1 μm square is usually around 1 nm. In this case, the smoothed surface with an increased anti-crack property due to the organosilane layer caused the enhancement of the coating. These results suggest the interface between thin films and substrates plays a crucial role on the enhancement of gas barrier property with dense thin film coating, and technologies for surface conditions are considered to be a key for the future commercialization of coated containers made of various plastics such as PE and PP.

In the other way, a novel approach to gas barrier thin film coating has been proposed, where a hot wire or catalytic CVD technique is applied to bottle coating in an attempt to achieve decreased installing expenditure based on the simple configuration of coating machines compared to that of conventional plasma assisted CVD machines. Furthermore, the application of this technique can produce unique gas barrier coating to PET bottles, like an intermediate between carbon and silicate oxide thin films [34]. Since the machine installation cost seems to be the bottleneck to further distribution of thin film coating technologies, significantly low cost machinery may be a remarkable breakthrough in this field.

Table 5. List of plastic materials used for comparing the degree of gas barrier enhancement with DLC coating.

Material	Abbreviation	Manufacturer	Type	Thickness
linear low-density poly(ethylene)	LLDPE	Toyobo Co., Ltd.	L6102	30 μm
low-density poly(ethtlene)	LDPE	–	Type S-1	40 μm
high-density poly(ethtlene)	HDPE	Mitsui-Toatsu Pleatec Co., Ltd.	Hiburon	25 μm
cast poly(propylene)	CPP	Toyobo Co., Ltd.	P1128	18 μm
retortable cast (propylene)	rCPP	Toyobo Co., Ltd.	P1153	40 μm
Oriented poly(propylene)	OPP	Toyobo Co., Ltd.	P2108	40 μm
poly(stylene)	OPS	Toyo Chemical Co., Ltd.	Hallen L	25 μm
poly(vinyl acetate)	EVA	Kaito Chemical Industry Co., Ltd.	Type E-30	30 μm
Poly(ethylene)-poly(vinyl acetate)	EVOH	Kuraray Co., Ltd.	Eval EF-F	30 μm
Oriented polyamid	ONY	Toyobo Co., Ltd.	N1100	30 μm
poly(acrylo nitril)	PAN	Mitsui-Toatsu Pleatec Co., Ltd.	Zecron	20 μm
poly(lactic acid)	PLA	Mitsubishi Plastics Co., Ltd.	–	50 μm
Oriented poly(ethylene terephthalate)	PET	Toyobo Co., Ltd.	ES100	12 μm

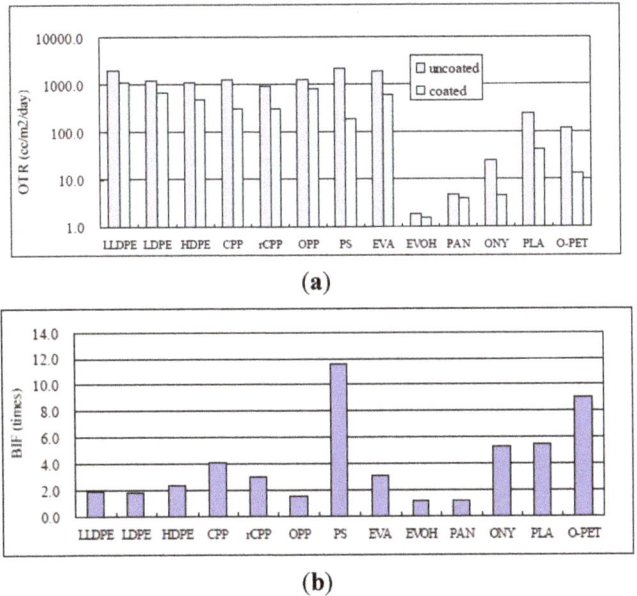

Figure 6. Comparison of (**a**) the oxygen transmission rate (OTR); and (**b**) BIF of DLC coating on different plastic films. Samples of 40 mm square size were placed on the center of body part of 500 mL PET bottles were coated and measured based on the ASTM D3985 method using Oxtran 2/21 devices, Mocon Co., Ltd. (Brooklyn Park, MN, USA), under conditions of 23 °C and 90% relative humidity, in the same manner in a previous study [28].

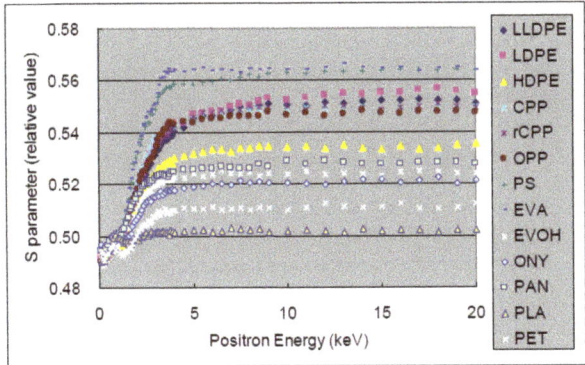

Figure 7. Depth profile of the positron annihilation of DLC coated samples. The S parameter of DLC coating (see the region of less than 1.5 keV) and plastic substrates (see the region of more than in 5.0 keV) was measured in the same manner in a previous study [28] for observing the relative free volume of DLC coating layers.

The above discussion on near future technologies indicates a high potential of further advancement in thin film coating technologies for hollow plastic containers in this field, including applications to food and beverage industry.

Acknowledgments: We would like to express our gratitude to Tetsuya Suzuki, and Taku Aoki, and Kazuhisa Tsuji, Keio University, Japan, for profound discussion. We are also thankful to all the staff related to this article in University of Tsukuba, Japan, and Kirin Co. Ltd., Tokyo, Japan.

Author Contributions: Masaki Nakaya conceived and designed the experiments on DLC coating to PET bottles in Figures 5 and 6, Table 5, and a part of Table 4. Masaki Nakaya and Akira Uedono conceived and designed the experiments in Figure 7. Masaki Nakaya and Atushi Hotta conceived and designed the experiments using organosilane. Masaki Nakaya performed the experiments in Figures 5 and 6, and a part of Table 4. Masaki Nakaya and Akira Uedono performed the experiments in Figure 7. Akira Uedono analyzed the data in Figure 7. Atsushi Hotta contributed PP substrate materials and analysis tools in the experiments using organosilane. Masaki Nakaya wrote the paper.

Conflicts of Interest: The authors declare no conflict of interest.

References

1. Downy, R. Evolving global demand and opportunities for PET. In Proceedings of the PETnology 2014, Nuernberg, Germany, 10 November 2014.
2. Kazminova, A.; Shelemin, A.; Pert, M.; Olylian, O.; Biederman, H. Barrier coatings on polymeric foils for food packaging. In Proceedings of the 22nd Annual Conference of Doctoral Students—WDS 2013-Proceedings of Contributed Papers, Part III, Plague, Czech Republic, 4–7 June 2013; pp. 128–133.
3. Lange, J.; Wyser, Y. Recent innovations in barrier technologies for plastic packaging—A review. *Packag. Technol. Sci.* **2013**, *16*, 149–158. [CrossRef]
4. Boutroy, N.; Pernel, Y.; Rius, J.M.; Auger, F.; Bardeleben, H.J.; Cantin, J.L.; Abel, F.; Zainert, A.; Casiraghi, C.; Ferrari, A.C.; et al. Hydrogenated amorphous carbon film coating of PET bottles for gas diffusion barriers. *Diam. Relat. Mater.* **2006**, *15*, 921–927. [CrossRef]
5. Shirakura, A.; Nakaya, M.; Koga, Y.; Kodama, H.; Hasebe, T.; Suzuki, T. Diamond-like carbon films for PET bottles and medical applications. *Thin Solid Films* **2006**, *494*, 84–91. [CrossRef]
6. *Barrier-Enhancing Technologies for PET & Polypropylene Containers & Closures*; Packaging Strategies; Business Development Associates, Inc.: Bethesda, MD, USA, 2002; pp. 8–57, 59–115, 124–130.
7. *PET Beverage Packaging Barrier Technologies*; Packaging Strategies; Robert Tremblay, Plastex Solutions, LLC: Fort Worth, TX, USA, 2007; pp. 1–14, 46–96, 108–128.
8. Brandan, O. Hidden issues in barrier technology adaptations for converters and brandowners. In Proceedings of the Novapack 2010, Orlando, FL, USA, 10 February 2010.
9. Hu, S.; Prattipati, V.; Mehta, S.; Schiraldi, D.A.; Hiltner, A.; Baer, E. Improving gas barrier of PET by blending with aromatic polyamides. *Polymer* **2005**, *46*, 2685–2698. [CrossRef]
10. Tsai, T.Y.; Li, C.H.; Chang, C.H.; Cheng, W.H.; Hwang, C.L.; Wu, R.J. Preparation of exfoliated polyster/clay nanocomposites. *Adv. Mater.* **2005**, *17*, 1769–1773. [CrossRef]
11. Galdi, M.R.; Nicolais, V.; Maio, L.D.; Incarnato, L. Production of active PET films: Evaluation of scavenging activity. *Packag. Technol. Sci.* **2008**, *21*, 257–268. [CrossRef]
12. Hartwig, K.; Behle, S.; Neuhäuser, M. The new plasma-coating-system for the PET-market. In Proceedings of the International Technical Conference 2002, Hamburg, Germany, 14–15 October 2002.
13. Platt, D. *The Future of Functional Additives and Barrier Coatings for Plastic Packaging to 2014*; Pira International Ltd.: Surrey, UK, 2010; pp. 41–42.
14. Aibara, K. *Container Market 2014*; Yano Research Institute Ltd.: Tokyo, Japan, 2015; pp. 109–110.
15. Casiraghi, C.; Robertson, J.; Ferrari, A.C. Diamond-like carbon for data and beer storage. *Mater. Today* **2007**, *10*, 44–53. [CrossRef]
16. Lopes, P.; Silva, M.A.; Pons, A.; Tominaga, T.; Lavigne, V.; Saucier, C.; Darriet, P.; Teissedre, P.L.; Dubourdieu, D. Impact of oxygen dissolved at bottling and transmitted through closures on the composition and sensory properties of a Sauvignon Blanc wine during bottle storage. *J. Agric. Food Chem.* **2009**, *57*, 10261–10270. [CrossRef] [PubMed]
17. Dombre, C.; Chalier, P. Evaluation of transfer of wine aroma compounds through PET bottles. *J. Appl. Polym. Sci.* **2015**, *132*, 41784–41794. [CrossRef]
18. Auras, R.; Harte, B.; Selke, S. An overview of polylactides as packaging materials. *Macromol. Biosci.* **2004**, *4*, 835–864. [CrossRef] [PubMed]

19. Abbas, G.A.; Roy, S.S.; Papakonstantinou, P.; McLaughlin, J.A. Structural investigation and gas barrier performance of diamond-like carbon based films on polymer substrates. *Carbon* **2005**, *43*, 303–309. [CrossRef]
20. Nagashima, K.; Shima, H. Carbon Film-Coated Plastic Container Manufacturing Apparatus and Method. Patent WO 1996005112 A1, 9 August 1996.
21. Felts, J.T. Thickness effects on thin film gas barriers: Silicon-based coatings. In Proceedings of 34th Annual Technical Conference, Society of Vacuum Coaters, Philadelphia, PA, USA, 17–22 March 1991; pp. 99–104.
22. Shirakura, A. DLC coating for high barrier plastic bottles. In Proceedings of 6th Pan Asian PET Markets' 99 Conference, Singapore, Singapore, 20 May 1999; p. 19.
23. Fahlteich, J.; Fahland, M.; Schönberger, W.; Schiller, N. Permeation barrier properties of thin oxide films on flexible polymer substrates. *Thin Solid Films* **2009**, *517*, 3075–3080. [CrossRef]
24. Kelly, R.S.A. Development of clear barrier films in Europe. In Proceedings of the 36th Annual Technical Conference, Society of Vacuum Coaters, Dallas, TX, USA, 1993; pp. 312–317.
25. Kodama, H.; Nakaya, M.; Shirakura, A.; Hotta, A.; Hasebe, T.; Suzuki, T. Synthesis of practical high-gas-barrier carbon films at low and atmospheric pressure for PET bottles. *New Diam. Front. Carbon Technol.* **2006**, *16*, 107–119.
26. Yamamoto, S.; Kodama, H.; Hasebe, T.; Shirakura, A.; Suzuki, T. Oxygen transmission rate of transparent diamond-like carbon films. *Diam. Relat. Mater.* **2005**, *14*, 1112–1115. [CrossRef]
27. Kuroiwa, T. Barrier technology—SiO_x coating to the inside of the PET bottle. *JPI J.* **2014**, *52*, 1006–1009.
28. Nakaya, M.; Shimizu, M.; Uedono, A. Impact of the difference in power frequency on diamond-like carbon thin film coating over 3-dimensional objects. *Thin Solid Films* **2014**, *564*, 45–50. [CrossRef]
29. Erlat, A.G.; Spontak, R.J.; Clarke, R.P.; Robinson, T.C.; Haaland, P.D.; Tropsha, Y.; Harvey, N.G.; Vogler, E.A. SiO_x gas barrier coatings on polymer substrates: Morphology and gas transport considerations. *J. Phys. Chem. B* **1999**, *103*, 6047–6055. [CrossRef]
30. Mitsubishi Heaby Industry Food Packaging and Kirin. Patent WO 2008114475 A1, 25 September 2008.
31. *ASTM F1307 Standard Test Method for Oxygen Transmission Rate Through Dry Packages Using a Coulometric Sensor*; ASTM International: West Conshohocken, PA, USA, 2014.
32. Uedono, A.; Ishibashi, S.; Ohdaira, T.; Suzuki, R. Point defects in group-III nitride semiconductors studied by positron annihilation. *J. Cryst. Growth* **2009**, *311*, 3075–3079. [CrossRef]
33. Tashiro, H.; Nakaya, M.; Hotta, A. Enhancement of the gas barrier property of polymers by DLC coating with organosilane interlayer. *Diam. Relat. Mater.* **2013**, *35*, 7–23. [CrossRef]
34. Shimizu, M.; Matsui, E.; Sato, A.; Nakaya, M.; Tabuchi, H. Gas-Barrier Plastic Molded Product and Manufacturing Process Therefor. Patent WO 2012091097 A1, 21 December 2011.

© 2015 by the authors. Licensee MDPI, Basel, Switzerland. This article is an open access article distributed under the terms and conditions of the Creative Commons Attribution (CC BY) license (http://creativecommons.org/licenses/by/4.0/).

Article

Alginate and Chitosan as a Functional Barrier for Paper-Based Packaging Materials

Samir Kopacic [1,*], Andrea Walzl [2], Armin Zankel [3], Erich Leitner [2] and Wolfgang Bauer [1]

1. Institute of Paper, Pulp and Fibre Technology, Graz University of Technology, Inffeldgasse 23, 8010 Graz, Austria; wolfgang.bauer@tugraz.at
2. Institute of Analytical Chemistry and Food Chemistry, Graz University of Technology, Stremayrgasse 9/2, 8010 Graz, Austria; andrea.walzl@tugraz.at (A.W.); erich.leitner@tugraz.at (E.L.)
3. Institute for Electron Microscopy and Nanoanalysis, NAWI Graz, Graz University of Technology and Centre for Electron Microscopy, Steyrergasse 17, 8010 Graz, Austria; armin.zankel@felmi-zfe.at
* Correspondence: samir.kopacic@tugraz.at; Tel.: +43-316-873-30788

Received: 30 May 2018; Accepted: 27 June 2018; Published: 3 July 2018

Abstract: Paper-based food packaging materials are widely used, renewable, and biodegradable. Because of its porous structure, paper has poor or no barrier performance against grease, water vapor, water, and volatile organic compounds. Moreover, recycled paperboard can be a source of organic residuals that are able to migrate into packed food. Two different types of paperboard produced from primary and secondary cellulosic fibers were coated using renewable materials, such as alginate and chitosan, and comprehensive barrier measurements showed multifunctional barrier properties of these two biomaterials. Both paper substrates were successfully coated using a draw-down coater, and the measured air permeability of the coated samples was 0 mL·min^{-1}. Grease resistance was improved, while it was possible to reduce water vapor transmission, the migration of mineral oil saturated hydrocarbons and mineral oil aromatic hydrocarbons (MOSH/MOAH), and the permeation of volatile compounds for both paper substrates when compared with uncoated substrates. Wettability and water absorptiveness of chitosan- and alginate-coated papers were found to be substrate-dependent properties, and could be significantly affected by bio-based coatings. In summary, industrially produced paperboard was upgraded by coating it with the naturally biodegradable biopolymers, alginate and chitosan, thus achieving extraordinary barrier performance for various applications within the packaging industry.

Keywords: barrier coating; paper-based food packaging material; chitosan; alginate; water vapor transmission rate; MOSH/MOAH migration; permeation; grease barrier; water absorptiveness; HPLC–GC coupled with a flame ionization detector (FID)

1. Introduction

Food packaging materials based on cellulosic fibers must keep their functionality under permanently changing conditions in the surrounding environment, such as temperature, storage time, or moisture, which are major influences on the shelf-life and quality of the packed food [1–3]. Paper-based packaging assures the strength and stability of the packaging, but due to its porous structure, paper lacks most of the important barrier functions needed nowadays. Furthermore, paperboard produced from so-called recycled or secondary cellulosic fibers can contain residues of mineral oils, better known as mineral oil saturated hydrocarbons (MOSH) and mineral oil aromatic hydrocarbons (MOAH), which represent a serious source of contaminants for packed food products. The sources of mineral oil residues in secondary fibers are ubiquitous, including mineral-oil-based printing inks in particular [4–8]. In order to control and prevent the permeation or migration of water vapor, mineral oils, grease, or liquids, paper must be further upgraded by a suitable barrier coating, in order to ensure the required packaging function. Therefore, the surface of paper-based packaging materials is treated either by extrusion, using thermoplastic petroleum-based polymers, or by dispersion coating, using synthetic water-based polymer dispersions [9–11]. In recent years, significant research efforts in academia, as well as in industry, focused on the replacement of oil-based polymer materials in the surface treatment of paperboard. Driving forces for these developments are not necessarily only coming from the producers, but also from the consumers [2]. Bio-based materials applied on paper could provide interesting barrier functionalities while still maintaining the environmentally friendly characteristics of the packaging material. The challenge with paper-based packaging materials is that, for different products, different barrier properties are needed. Multiple layers of barrier materials are sometimes the solution chosen in practice [3,12,13].

Our investigation focused on sodium alginate and chitosan, and their application as barrier materials for paper intended to be used as primary or secondary food-packaging materials. Alginate is a polysaccharide naturally present in brown algae, and is usually available as salts of sodium and calcium. Alginates and its derivatives are already used in large amounts in the food industry as additives, and therefore, are also considered to be safe for their use as functional barriers for food-contact materials. Various water-soluble alginate formulations are available on the market, which can be applied with conventional coating equipment used in the paper and packaging industries [14–16]. Chitosan is an abundant, natural polysaccharide derived from chitin, a substance in the exoskeletons of crustaceans and insects. Economically interesting quantities are already produced from fishing industry waste, mainly during the processing of crabs and shrimps [17–21].

Our work investigated the coatability and barrier properties of these two water-soluble biopolymers, chitosan and alginate, from renewable resources. Both were applied under the same conditions onto two different paper grades, with the aim of evaluating their potential for reducing the migration and permeation of mineral oil components (MOSH/MOAH) [11], aromatic components, and water vapor. Furthermore, the coating layer's quality was analyzed via scanning electron microscopy, and its resistance toward grease, water absorption, and air permeability was determined. Similar studies with chitosan and alginate were reported in the literature for some of above-stated barrier functions [22–30]. The novel aspect of this work involves the systematic comparison and quantitative study of the barrier properties of alginate and chitosan, and their interaction with two different industrially produced paper substrates. In particular, the effect of these two bio-based barrier materials against the migration of mineral oil fractions (MOSH and MOAH) contained in paper, measured and quantified with HPLC–GC coupled with a flame ionization detector (FID), is yet to be reported.

2. Materials and Methods

2.1. Coating Materials

Powdered chitosan used in the preparation of the coating solution was kindly supplied by BioLog Heppe GmbH (Landsberg, Germany). This industrially produced chitosan with a degree of deacetylation 88%–95% was made from the carapace skin of crustaceans. According to the product specifications, the chitosan powder consisted of particles with a diameter ≤200 µm, and an ash content <1% (w/w). The dynamic viscosity of a 1% (w/w) chitosan aqueous solution dissolved in 1% (w/w) acetic acid at pH 4 and 20 °C was 20 mPa·s. Acetic acid (100%, Rotipuran) used for adjustment of the pH of dissolved chitosan in water was purchased from Carl Roth GmbH+ Co. KG (Karlsruhe, Germany). Sodium alginate (viscosity 15–25 mPa·s 1% (w/w) in water at 25 °C) was purchased from Sigma-Aldrich (Saint Louis, MO, USA).

2.2. Preparation of Aqueous Coating Solutions

A chitosan coating solution with a solid content of 4% (w/w) was prepared by dissolving it in heated deionized water (70 °C), adding the chitosan powder in small amounts and stirring for 6 h at

400 rpm. Subsequently, acetic acid was added in small portions in order to achieve a pH of 4, measured constantly by a portable pH meter (inoLab pH 7110, WTW, Weilheim, Germany). This chitosan aqueous solution was stirred and heated for 4 h at 70 °C until a yellow solution was obtained and no visible particles were observed.

A sodium-alginate coating solution with a solid content of 4% (w/w) was prepared by dissolving sodium-alginate powder in deionized water at a neutral pH. The sodium-alginate powder was added to water in portions, stirred at 400 rpm, and the aqueous solution was heated for 6 h at 75 °C. After this time, sodium alginate was completely dissolved, resulting in a homogenous coating solution. Due to the heating and evaporation of water, the solid contents of both coating solutions slightly increased. Therefore, the solid content was remeasured using a moisture analyzer (HR73, Mettler Toledo, Columbus, OH, USA), and adjusted to 4% (w/w) with deionized water. Finally, the coating solutions with the adjusted and desired solid contents were cooled to room temperature. The viscosities (Brookfield II+, at 50 rpm, n = 3) of 4% (w/w) chitosan and 4% (w/w) alginate coating solutions, measured at room temperature, were 2911 mPa·s ± 57 and 1448 mPa·s ± 20, respectively.

2.3. Paper-Substrate Characterization

Two different commercial paper grades were used in the coating trials. The first was a paper (PF) made from 100% primary-fiber furnish (mixture of hardwood and softwood), mass-sized using 100% active liquid alkenyl succinic anhydride (ASA), and surface-sized using starch and a calender machine. The second substrate was a paperboard (SF) made from 100% secondary or recycled fibers with no surface treatment. Prior to coating, the basic properties of the substrates were measured, and are summarized in Table 1 (n = 15).

Table 1. Basic characterization of substrates used for barrier coating (n = 15).

Substrate	Furnish	Grammage (g·m^{-2})	Thickness (μm)	Sheet Density (g·cm^{-3})	Bendtsen Roughness (mL·min^{-1})
Primary fiber (PF) uncoated	100% Primary fiber	72.0 ± 0.5	96.0 ± 1.2	0.760 ± 0.01	150 ± 26
Secondary fiber (SF) uncoated	100% Secondary fiber	129.0 ± 0.6	197.0 ± 2.6	0.660 ± 0.01	1271 ± 211

2.4. Standardized Physical Paper Properties and Barrier Measurements

Prior to the measurements, the raw (uncoated) paper substrates and paper samples coated with the alginate and chitosan formulations were conditioned for 48 hours at 23 ± 1 °C and 50 ± 3% relative humidity (RH) [31]. Measurements for grammage, thickness, density, roughness, air permeability, water-vapor transmission rate, water absorptiveness, contact angle, and grease resistance were performed according to the standardized methods listed in Table 2.

2.5. Coating Trial with Laboratory Draw-Down Coater

A coating trial was performed using a laboratory draw-down coater from RK Printcoat Instruments Ltd. (Litlington, UK). A target coat weight of 6 g·m^{-2} (single-sided application), with a standard deviation of less than 10%, was achieved by applying two layers of barrier-coating solution. The coater speed was 4 m·min^{-1}, and the wet-film thickness for the first and second coating layers, defined by the wire-wounded rod used, was 40 μm for both coating solutions. Drying of the coated paper samples was performed with hot air at 150 °C for 60 s.

Table 2. Paper properties and standard methods used for the testing of uncoated and coated samples.

Property	Method	Standard
Grammage	Weighing	ISO 536 [32]
Thickness	Lehmann thickness tester	ISO 534 [33]
Density	Calculation	ISO 534 [33]
Air permeability	Bendtsen	ISO 5636-3 [34]
Bendtsen roughness	Bendtsen	ISO 8791-2 [35]
Contact angle	Fibrodat 1100	T 558 [36]
Grease resistance	KIT Test	T 559 [37]
Cobb	Frank-PTI Cobb tester	ISO 535 [38]
Water-vapor transmission rate	Gravimetric determination	T 448 om-09 [39]

2.6. Surface Evaluation of Uncoated and Coated Paper Substrates

The surface topography of uncoated raw paper, and chitosan- and alginate-coated paper was investigated using low-voltage scanning electron microscopy (LVSEM, Everhart-Thornley detector for the detection of secondary electrons; Zeiss Sigma 300, Oberkochen, Germany) [40]. The samples were cut (1 cm × 1 cm), then attached to SEM stubs using a double-sided conductive carbon tape, and imaging (magnification 500×) was performed at an acceleration voltage of 0.65 kV.

2.7. Migration Experiments

In this study, migration experiments were performed according to EU-Regulation No. 10/2011 [41]. As a food simulant, Tenax®, a poly (2,6-diphenyl-p-phenylene oxide) (Tenax® TA (refined), 60–80 mesh; SUPELCO, Bellefonte, PA, USA), was used for the simulation of dry foods such as rice, cereals, cocoa, coffee, and spices. The standard test conditions for long-term storage of these products for above and below six months at room temperature should be 60 °C for 10 days, but can be adapted using the Arrhenius equation. This was done, with conditions tested and set as 80 °C for two days. The Tenax® was applied in an amount of 4 g·dm^{-2}. The experiments were performed in triplicate in migration cells (MigraCell®; FABES Forschungs-GmbH, Munich, Germany) with a tested surface area of 0.32 dm^2. The cell was assembled according to the manufacturer's instructions, with the coated side facing the Tenax®, and placed in an oven for two days at 80 °C. Afterward, the Tenax® was drained into a glass vial with a screw cap, and 25 µL of an internal standard mix was added. The internal standard for migration experiments consisted of dodecane-d$_{26}$ (C$_{12}$D$_{26}$; EURISO-TOP SAS, Saint-Aubin, France), nonadecane-d$_{40}$ (C$_{19}$D$_{40}$; 98%; Cambridge Isotope Laboratories, Inc.; Tewksbury, MA, USA), benzophenone-d$_{10}$ (C$_{13}$D$_{10}$O; 99 at.%; Sigma-Aldrich Co., St. Louis, MO, USA), and bis(2-ethylhexyl)phthalate-d$_4$ and di-n-butyl phthalate-d$_4$ (both "analytical standard", purchased from Sigma-Aldrich Co., St. Louis, MO, USA). All were used at a concentration of 200 mg·L^{-1} in acetone (ROTISOLV® ≥99.9%, UV/IR-Grade; Carl Roth GmbH + Co. KG, Karlsruhe, Germany). The Tenax® was extracted three times with 10 mL of n-hexane (Picograde® for residue analysis; LGC Promochem GmbH; Wesel, Germany) and three minutes of vortexing. The extracts were combined through a folded filter in a 50-mL evaporation vial, and the solvent was evaporated to 0.5 mL in an automatic solvent evaporator (TurboVap® II; Biotage, Uppsala, Sweden). The extracts were then transferred into 1.5-mL glass vials with screw caps; the evaporation vials were rinsed with 0.5 mL of hexane, and this solvent added to the 1.5-mL vials. The extracts were stored in a refrigerator, and only a small amount was filled into a 1.5-mL glass vial with a micro insert and screw cap for measurements. The extracts were measured on a gas chromatograph with a flame ionization detector (GC–FID) to determine the overall migration. The separation was done using a Hewlett Packard 6890 Series GC System equipped with an Optima delta-6 capillary column (7.5 m × 100 µm × 0.10 µm, Macherey-Nagel, Germany). The oven was programmed to 60 °C (hold 1 min), then raised at 15 °C·min^{-1} to 300 °C (3 min). The carrier gas used was hydrogen with a linear velocity of 48 cm·s^{-1}. Aliquots of one microliter were injected with a split of 1:20. The injection-port temperature and detector temperature were set to 280 and 320 °C, respectively. Data evaluation was done using the "GC ChemStation" software, version B.04.03 (Agilent Technologies, Santa Clara, CA, USA).

For the analysis of MOSH and MOAH, online-coupled HPLC–GC–FID was used as described in [7,8]. Prior to analysis, a MOSH/MOAH internal standard mix was added in a concentration of 1.5–6 µg·mL^{-1}. The standard purchased by Restek Corporation (Bellefonte, PA, USA) contained the following substances in 1-mL ampoules in toluene: n-undecane (300 µg·mL^{-1}), n-tridecane (150 µg·mL^{-1}), bicyclohexyl (300 µg·mL^{-1}), cholestane (5-α-cholestane; 600 µg·mL^{-1}), 1-methylnaphthalene (300 µg·mL^{-1}), 2-methylnaphthalene (300 µg·mL^{-1}), n-pentylbenzene (300 µg·mL^{-1}), perylene (600 µg·mL^{-1}), and 1,3,5-tri-tert-butylbenzene (300 µg·mL^{-1}). For the calculation of retention indices and the determination of cutting fractions in HPLC–GC–FID, a "C_7–C_{40} saturated alkane standard" from SUPELCO (Bellefonte, PA, USA) was used. The concentration of the alkanes was 1000 µg·mL^{-1} in hexane, and was diluted to 1 µg·mL^{-1} with hexane prior to analysis.

The HPLC used was a Shimadzu LC-20AD (Shimadzu Corporation, Kyoto, Japan) equipped with an Allure Silica 5 µm column (250 mm × 2.1 mm). A gradient elution was used, starting with 100% n-hexane (flow 0.3 mL·min^{-1}), before being raised to 35% CH_2Cl_2 within 2 min (hold for 4.20 min). The column was then backflushed at 6.30 min with 100% CH_2Cl_2 (flow 0.5 mL·min^{-1}; hold for 9 min), and reconditioned to 100% n-hexane (flow 0.5 mL·min^{-1}; hold for 10 min). The flow was subsequently decreased to 0.3 mL·min^{-1} until the next injection. The UV-detector was equipped with a D_2-lamp set at 230 nm and a cell temperature of 40 °C. The GC was a Shimadzu GC 2010 dual-FID (Shimadzu Corporation, Kyoto, Japan), equipped with two guard columns, Restek MXT Siltek (10 m × 0.53 mm inner diameter (id)), and two analytical columns Restek MTX®-1 (15 m × 0.25 mm id × 0.1 µm d$_f$). The carrier gas used was hydrogen with an analysis pressure of 150 kPa, and an evaporation pressure of 87 kPa for MOSH and 85 kPa for MOAH. The oven was programmed to 60 °C (hold 6 min), and raised at 20 °C·min^{-1} to 100 °C (0 min) followed by 35 °C·min^{-1} to 370 °C (9.29 min). The LC–GC interface was controlled by a Chronect-LC–GC by Axel-Semrau (Sprockhövel, Germany); data evaluation was done using the LabSolutions software version 5.92. According to a proposed method published by the German Bundesinstitut für Risikobewertung (BfR), quantification was done by integration of the hump for various molecular weight regions. They propose the ranges of C_{16}–C_{25} and C_{25}–C_{35} for food-contact materials for dry non-fatty food and storage at room temperature [5].

2.8. Permeation Experiments

The used migration cell allowed a one-sided migration experiment (as described in Section 2.7), and a two-sided application for simultaneously testing the migration and permeation. When performing a two-sided test in the migration cell, three changes of the experimental set-up were made. Firstly, a piece of cellulose was placed at the bottom of the cell where the modeling substances for the permeation were spiked. Secondly, the metal plate in the middle of the cell was removed. Therefore, thirdly, the colorless silicone ring had to be replaced by a FEP (Fluorinated ethylene propylene)-coated red ring to prevent the contamination of the Tenax® with siloxanes. As modeling substances, deuterated n-alkanes of various chain lengths were chosen because they best simulated a possible migration of mineral oil hydrocarbons through the sample ($C_{14}D_{30}$, $C_{20}D_{42}$, and $C_{24}D_{50}$, 98%-at.%D, purchased from Cambridge Isotope Laboratories, Inc. (Tewksbury, MA, USA); $C_{16}D_{34}$, 99%-at.%D, purchased from abcr GmbH (Karlsruhe, Germany); and $C_{28}D_{58}$ 98%-at.%D, purchased from C/D/N/ Isotopes, Inc. (Pointe-Clair, QC, Canada)). The deuterated substances were used to prevent the interference of permeation and migration tests, because these n-alkanes were also present in the tested paper samples. To simulate aromatic permeability, a set of four aromatic compounds were selected (DL-Menthol, ≥95%; Eugenol, ReagentPlus®, 99%; Vanillin, ≥97%; and Acetovanillone, ≥98%; purchased from Sigma-Aldrich Co. (St. Louis, MO, USA)). One hundred microliters of a stock solution in acetone containing each of the mentioned substances in a concentration of 100 mg·L^{-1} were spiked into the bottom of the cell. The test conditions, extraction, and analysis stayed the same as described above.

The methods described were used to test the barrier efficiency of the two uncoated papers, and the papers coated with alginate and chitosan.

3. Results and Discussion

3.1. Physical Characterization of Coated Samples

The substrates made out of primary fibers (PF) and coated with chitosan or alginate were labeled as PF chitosan or PF alginate, respectively. For the secondary-fiber (SF) substrates, this principle resulted in the sample descriptions, SF chitosan and SF alginate. The average values of thickness, density, grammage, and coat weight, with their corresponding standard deviations are shown in Table 3.

Table 3. Thickness, density, grammage, and pick-up values of alginate- and chitosan-coated primary-fiber (PF) and secondary-fiber (SF) samples ($n = 15$).

Sample	Thickness (μm)	Density (g·cm^{-3})	Grammage (g·m^{-2})	Pick-Up (g·m^{-2})
PF chitosan	98.0 ± 1.0	0.780 ± 0.01	78.0 ± 0.7	6.0 ± 0.2
PF alginate	98.0 ± 1.4	0.800 ± 0.01	78.1 ± 0.6	6.1 ± 0.2
SF chitosan	199.0 ± 2.4	0.670 ± 0.01	135.1 ± 0.5	6.1 ± 0.2
SF alginate	198.0 ± 3.1	0.680 ± 0.01	135.0 ± 0.9	6.0 ± 0.3

3.2. Surface Evaluation, Film Formation, and Coating Quality

The topography of the coated and uncoated samples was assessed based on SEM images. The conventional technologies for the barrier coating of paper, such as extrusion or lamination, are based on the application of a specific polymer, and the formation of a distinct film is indispensable. Depending on the substrate and its specific physical properties, the amount of barrier coating sometimes exceeded 20 g·m^{-2} or 30 g·m^{-2}, in order to ensure good barrier efficiency [9]. In this respect, it was of interest whether bio-based coating materials must form a film on top of the paper surface in order to perform well as a barrier.

The SEM images of the uncoated raw substrates, SF and PF, showed the expected clear difference between the two different paper grades (Figure 1). Voids, and different sizes and alignments of fibers and pores between the fibers were visible, and the measured values for roughness and density (see Table 1) confirmed the difference in the structure of substrates produced from primary and secondary furnishes (Figure 1a,b).

Due to these differences between the paper substrates, it is obvious that chitosan and alginate interacted differently with the substrates, as illustrated by the SEM images. The PF substrate coated with chitosan had a completely covered surface, with no voids or pores visible, and a significant amount of the chitosan was also visible on the single fibers. In contrast, the SF chitosan sample showed that the fibers were not completely covered, and the coating appeared to have impregnated the material so that no clearly visible film was formed. A similar behavior was observed for the alginate coatings. Alginate also formed a film on the PF paper's surface, and covered the paper's surface completely, while it impregnated the SF paper. This can be explained by the higher roughness of the SF paper, resulting in impregnation rather than a full coverage of the paper's surface.

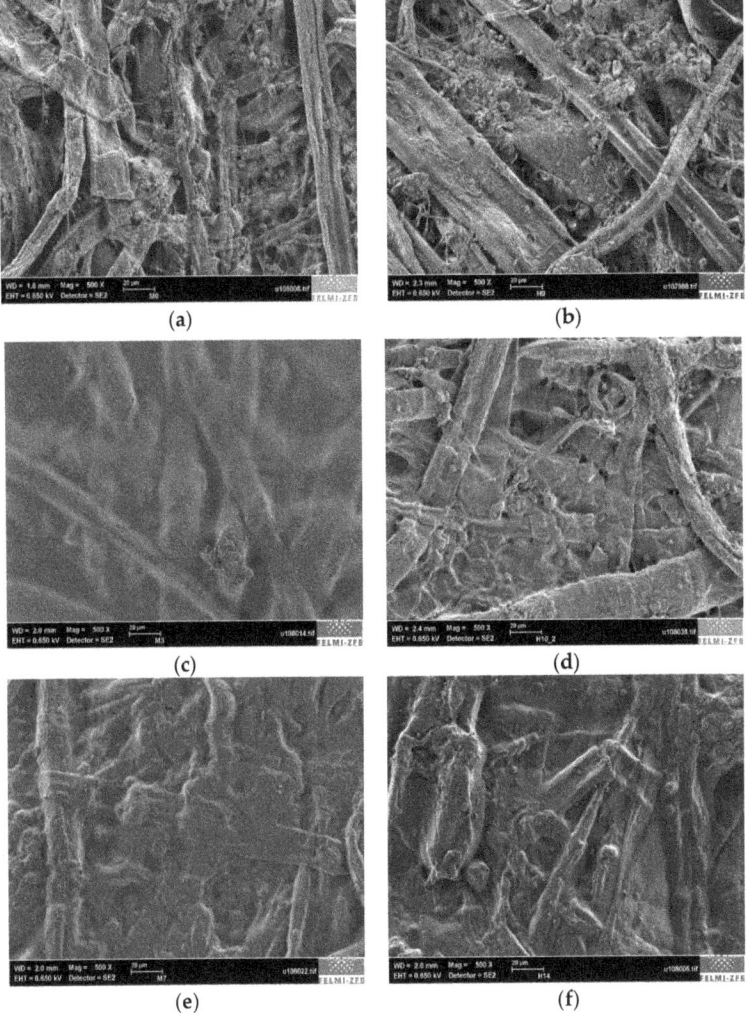

Figure 1. Scanning electron microscopy (SEM) images of uncoated and coated paper substrates at 500× magnification: (**a**) primary-fiber (PF) uncoated paper; (**b**) secondary-fiber (SF) uncoated paper; (**c**) chitosan-coated PF paper; (**d**) chitosan-coated SF paper; (**e**) alginate-coated PF paper; (**f**) alginate-coated SF paper.

3.3. Air Permeability and Grease Resistance (KIT Test)

Air permeability is a purely physical measurement, and gives the volume of ambient air able to pass through voids or pinholes through a paper substrate during a defined time period (one minute). The air permeability of the uncoated samples, SF and PF, was 809 mL·min^{-1} and 437 mL·min^{-1}, respectively (Figure 2).

After coating with alginate and chitosan, the measured air-permeability values for all samples were 0 mL·min^{-1}. Low air permeability also indicates that no pinholes or voids are present in the coated paper.

The KIT test is a common method for the evaluation of fat and grease resistance of paper. The method is primarily designed to evaluate fluorochemical-based coatings for grease barriers, but was successfully applied to bio-based barrier coatings as well [42]. KIT solutions are numbered from 1 to 12, with higher numbers indicating higher grease resistance, and vice versa. The grease resistance of a coated packaging material depends on its surface chemistry (hydrophilic or hydrophobic character), the barrier quality, density, present pores and voids, as well as thickness of the substrate and barrier. KIT solutions are organic, non-polar compounds (castor oil, toluene, and n-heptane) with low density (<1 g·cm^{-3}), able to penetrate easily through the porous structure of uncoated paperboard. In order to build a good barrier against grease, assessed with the KIT method, the barrier should, therefore, be hydrophilic rather than hydrophobic [43]. Alginate and chitosan manifest hydrophilic characteristics, and are able to close the voids and pores of the paper surface, thus meeting the initial criteria for a good grease barrier. Alginate applied on SF paper improved grease resistance to a medium level (KIT Number: 7.0 ± 1), which could already be of interest for some applications in the packaging industry. The PF substrate coated with alginate reached the maximum KIT number of 12.0 ± 0.5, and thus, is classified as an excellent barrier material against grease. Contrary to alginate, where the performance on SF and PF paper was significantly different, the chitosan barrier gave rather similar KIT values on both papers (6.0 ± 0.5 on PF, and 5.0 ± 0.5 on SF substrate). Although chitosan fully covered the fibers and closed the surface of the PF substrate, it did not reach such high KIT numbers when compared with alginate. The reason for this could be the distinctive hydrophilicity of the alginate, which is of course higher than that of chitosan, which may also have hydrophobic characteristics [44].

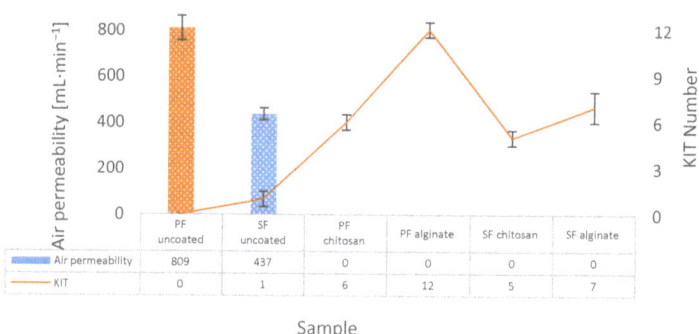

Figure 2. Air-permeability (n = 15) and grease-resistance values (KIT test, n = 9) for uncoated and coated primary-fiber (PF) and secondary-fiber (SF) paper.

3.4. Water-Vapor Transmission Rate (WVTR)

The water-vapor transmission rate is another important barrier property of packaging materials. Through the transmission of water vapor into a package, not only could the freshness of the packed food be affected, but the growth of microorganisms could also increase. In order to reduce and improve the water-vapor transmission rate (WVTR) of a fiber-based material, the coating barrier material should manifest resistance toward polar water vapor, and be able to close as many pores and voids as possible, preventing the interaction between the polar groups of cellulosic fibers and water vapor [45]. The WVTR was measured gravimetrically, and was expressed as an amount of water vapor in one gram able to pass through a material, usually within 24 h, in our study at 23 °C and 50% relative humidity. Raw untreated paper, as a hydrophilic and porous material, is known to be a poor barrier against water vapor. The WVTR values for uncoated PF and SF substrates were 690 g·m^{-2} × 24 h and 609 g·m^{-2} × 24 h, respectively (Figure 3). The coated samples showed significantly improved WVTR values, with chitosan coated on SF paper showing approximately 60% lower values when compared with uncoated paper, and an even better performance on PF substrate. The behavior of the alginate led to very similar WVTR values for both substrates, corresponding to a 35% reduction for alginate-coated SF and a 44% reduction for PF paper. Taking into account the different thicknesses of the samples, the water-vapor permeation coefficients (WVPCs) were calculated by multiplying the water-vapor transmission rate and the thickness of the sample (Figure 3). A significant reduction in the permeation coefficients was achieved with chitosan, where the WVPC values for both coated substrates were at least 50% lower than those of uncoated SF and PF paper. The same trend was observed and quantified for alginate-coated PF and SF samples (reductions of 35% and 42% for SF and PF, respectively). Both materials partially met the criteria stated above for the reduction in water-vapor transmission rate. After coating, the paper sheet was densified, and fibers were partially or totally covered with the coating material, resulting in the reduced interaction between cellulosic fibers and water vapor, and the reduced diffusion of water vapor. Thus, both materials, despite their hydrophilic characteristics, contributed to a reduction in water-vapor permeability.

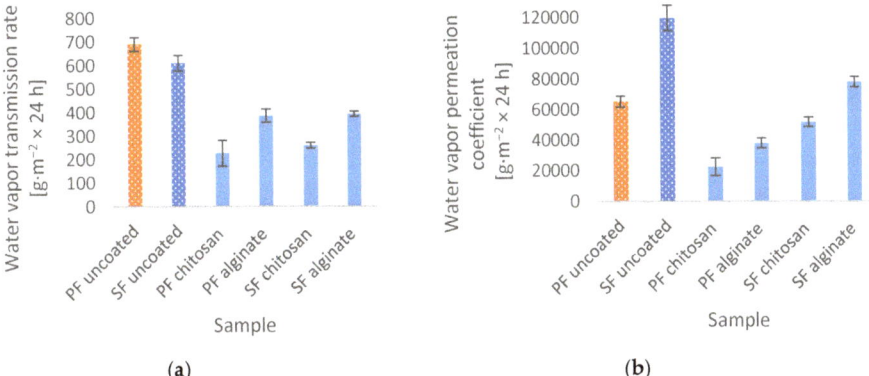

Figure 3. Water-vapor transmission rates (**a**) and water-vapor permeation coefficients (**b**) for uncoated and coated paper samples from primary fiber (PF) and secondary fiber (SF) (n = 6).

3.5. Wettability and Water Absorptiveness

The wettability of the uncoated and coated samples was assessed by a contact-angle (CA) measurement using deionized water (Figure 4). The water absorptiveness was characterized by performing Cobb measurements for 60 s, where the Cobb value was the amount of deionized water per area which could be absorbed by the substrate during the given period of time (Figure 5).

A contact angle below 90° is characteristic for hydrophilic surfaces. The uncoated PF substrate showed the highest CA, since this paper was already industrial-sized using ASA and starch. By applying chitosan and alginate onto the PF paper's surface, the initial contact angle decreased to ~80° (chitosan) and ~35° (alginate). On the other hand, the SF uncoated paper had very low CA, which was only measurable for eight seconds. The SF coated with chitosan exhibited a stable and higher CA (70° for 30 s) when compared with the uncoated SF paper. By coating the SF paper with alginate, the initial contact angle was lowered to 30°, but the time-dependent wettability was impacted, resulting in it being stable over the 30 s testing time.

Alginate-coated samples of both SF and PF were in a comparable range, when it came to surface hydrophobicity and water resistance.

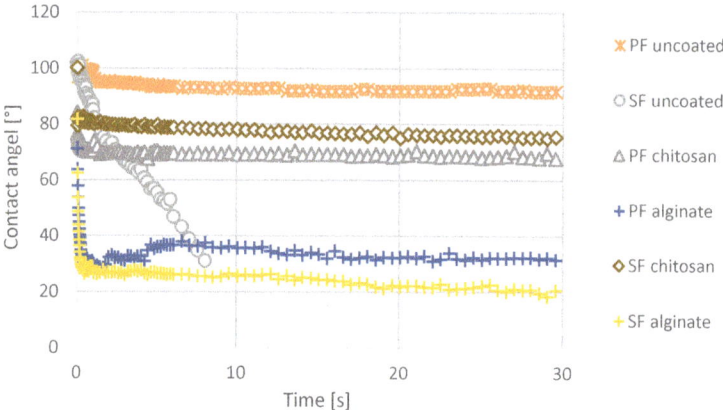

Figure 4. Contact angle of uncoated and coated samples measured with deionized water for 30 s (n = 6).

According to the contact-angle measurements, the PF uncoated substrate appeared to be a hydrophobic material (CA $\geq$ 90°). The Cobb 60 s value for the PF uncoated substrate was 25 g·m^{-2}. In contrast, the SF uncoated substrate reached saturation with water at 60 s, resulting in a higher water uptake (Cobb 60 s = 155 g·m^{-2}), and complete water penetration.

Chitosan- and alginate-coated PF samples were able to absorb at least 50% more water for 60 s when compared with the PF uncoated substrate. Consequently, the PF coated with chitosan or alginate became more hydrophilic, which coped very well with the CA measurements for PF samples.

On the other hand, the chitosan coating enormously affected the water absorptiveness of the SF paper, where a reduction of at least 80% was achieved. The Cobb value obtained with alginate-coated SF paper (149 g·m^{-2} for 60 s) was only slightly lower (<4%) when compared with uncoated SF paper, and no significant reduction was observed. SF paper is an unsized raw paper, which is considered to be very hydrophilic. By coating it with sodium alginate, which is also a hydrophilic material, the water uptake was not significantly reduced. On the other hand, the positively charged chitosan used for the coating of both substrates interacted very intensively with the negatively charged cellulosic fibers. The chitosan solution used for coating was only water-soluble in the presence of acetic acid at pH 4. Above this pH, chitosan was not water-soluble, and could be considered as "hydrophobic" [44]. Due to the fact that the SF paper was not treated with coating chemicals, and the fibers were fully available for positively charged chitosan, the interaction between the fibers and chitosan obviously took place. On that note, the pH could be shifted to the neutral or alkaline region, thus changing the paper's water absorptiveness, and making chitosan-coated SF paper water-repellent. The PF paper, which was mass- and surface-sized, manifested a very low water uptake, and interacted differently with alginate and chitosan when compared with the SF coated samples. The Cobb values of the PF coated samples were higher when compared with the uncoated PF paper. According to these measurements, different trends could be observed for alginate- and chitosan-coated samples. Irrespective of the paper substrate, alginate caused a hydrophilization effect, while the influence of chitosan on water uptake depended strongly on the paper substrate and its composition.

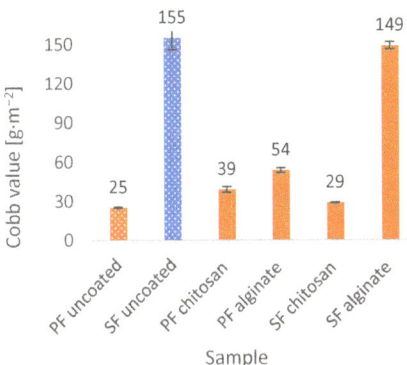

Figure 5. Cobb values for uncoated and coated samples measured for 60 s with deionized water (n = 6).

3.6. Migration Experiments

Since food-contact materials should not release any substances that cause unacceptable changes in the composition of the food, the overall migration needs to be kept as low as possible [41]. The two raw papers were of different qualities in this respect. A paper produced from secondary fiber is considered to be the worst case, especially in terms of contamination with mineral oil hydrocarbons, while a paper produced from clean primary fiber is preferable.

Table 4 shows the results of the migration tests performed in triplicate. Alginate and chitosan exhibited a good barrier performance for the SF paper. Setting the total migration of the uncoated SF paper to 100%, 63.8% ± 0.1% of the observed migration was accounted for as mineral oil, which, in turn, consisted of 57.8% ± 0.1% MOSH and 6.02% ± 0.16% MOAH. Using the alginate coating, the overall migration could be reduced to 16.3% ± 1.0%, of which 7.9% ± 0.25% were mineral oil hydrocarbons (MOH), consisting of 5.49% ± 0.18% MOSH and 2.41% ± 0.42% MOAH. The chitosan coating reduced the overall migration to 29.5% ± 1.6%, which consisted of 9.16% ± 0.3% MOH, divided into 8.43% ± 0.2% MOSH and 0.73% ± 0.34% MOAH.

The migration of the PF sample was naturally low, and coatings to reduce migration were not actually necessary. As expected, the values of the samples coated with alginate and chitosan were below the detection limit, and are, therefore, not given in Table 4. Unlike other barrier properties, which depended on the quality of the coating layer, densification, pores, voids and surface chemistry, it seems that MOSH and MOAH migration primarily depended on the change in surface chemistry rather than the other factors mentioned in our work. An explanation for such a low migration could be the hydrophilic and polar characteristics of these two materials, resulting in a high resistance toward organic non-polar compounds. As such, alginate performed better than chitosan most probably due to its higher polarity and slightly higher densification of the paper.

Table 4. Overall migration of mineral oil hydrocarbons (MOH), mineral oil saturated hydrocarbons and mineral oil aromatic hydrocarbons (MOSH/MOAH), in uncoated and coated paper samples from primary fiber (PF) and secondary fiber (SF) (%, n = 3).

Sample	Migration (%)	MOH (%)	MOSH (%)	MOAH (%)	Remnants * (%)
SF uncoated	100%	63.8 ± 0.10	57.80 ± 0.10	6.02 ± 0.16	36.2%
SF alginate	16.3 ± 1.0	7.90 ± 0.25	5.49 ± 0.18	2.41 ± 0.42	8.4%
SF chitosan	29.5 ± 1.6	9.16 ± 0.30	8.43 ± 0.20	0.73 ±0.34	20.3%

* Remnants consisted of substances with a retention time outside the range of C_{16}–C_{35}, and substances subtracted from the MOH (e.g., Diisopropylnaphthalene-DIPN).

3.7. Permeation Experiments

The use of deuterated n-alkanes allowed the performance of two-sided tests in the migration cells. This meant the determination of migration and permeation was possible in one experimental set-up without any interferences. This saved a lot of time and resources, and gave a quick and easy screening method for the barrier behavior of the natural polymers.

Gas-phase migration into dry food is limited by volatility. It was shown that it is relevant up to a chain length of C_{24}, and not detectable beyond a chain length of C_{28}, as substances with higher boiling points remain in the packaging material, and do not migrate [5]. According to the theory, the highest levels for permeation were found for d-C_{16} and d-C_{18}. An interesting observation was the behavior of the added aromatic active substances. Although all four compounds had a volatility and boiling points in the C_{12}–C_{16} range (Table 5), only menthol permeated through the papers (coated and uncoated). Apparently, the polar groups of the aromatic compounds interacted strongly with the polar groups of the paper.

Table 5. Boiling points of the aromatic compounds and n-alkanes in the same elution range.

Substance	Boiling Point (°C)
Menthol	212
C_{12}	216.2
C_{14}	253.5
Eugenol	254
Acetovanillon	265
Vanillin	285
C_{16}	286.8

From the two tested biopolymers, the alginate-coated samples showed better barrier properties against permeation than the chitosan-coated samples, as shown in Table 6. Under the given test conditions, the permeation rates of d-C_{14} to d-C_{20} were between 9.43 µg·dm^{-2}/day and 13.7 µg·dm^{-2}/day for the uncoated recycled fiber, and was reduced with the chitosan coating by 37%–50%, and was reduced with the alginate coating by 18%–50%. We observed similar permeation rates of the deuterated n-alkanes through the uncoated SF and PF papers, but lower permeation rates for primary fiber after the coating process, especially for the alginate coating. An explanation might be the difference between the two papers in terms of the three-dimensional structures, as well as the chemistry of the fiber surfaces, and a higher pressure on the SF barriers, due to the high load of possible migrants [46].

Table 6. Comparison of the permeation rates of deuterated n-alkanes of various chain lengths, and menthol through coated and uncoated paper samples from primary fiber (PF) and secondary fiber (SF) (µg·dm^{-2}/day; $n = 2$, data given individually).

ug/dm²/d	d-C_{14}		d-C_{16}		d-C_{20}		d-C_{24}		Menthol	
SF uncoated	12.85	9.43	13.7	10.58	12.88	11.12	3.8	3.4	8.77	6.38
SF alginate	3.26	3.43	5.45	5.6	2.34	2.37	0.49	0.44	3.39	3.59
SF chitosan	4.8	4.8	6.85	7.63	4.88	6.24	0.72	0.89	1.99	2.19
PF uncoated	12.68	12.51	13.54	13.39	14.63	14.05	3.23	3.23	10.76	10.16
PF alginate	1.34	0.81	1.87	0.92	0.7	0.35	0.3	0.25	3.39	3.19
PF chitosan	5.14	4.46	6.85	5.76	5.46	3.9	0.65	0.53	5.18	4.98

4. Conclusions

Biomaterials, such as alginate and chitosan, are biopolymers with higher degrees of complexity when compared with conventional synthetic surface-treatment chemicals. Therefore, the interactions between these two materials and the substrate are variable, and could bring about a comparative advantage for paper-based packaging producers when compared with synthetic barrier materials. Applying those two bio-based materials could improve the barrier properties of paperboard for food-packaging applications. Even with a pick-up weight of 6 g·m^{-2}, the permeability, migration, and transmission were significantly reduced. Depending on the paper substrate, specific barrier properties were differently affected, and could be selectively optimized and adjusted for the consumer's needs, thus giving the packaging producers certain flexibility for some specific applications. One of the most interesting findings resulted from the combination of SEM imaging, and the tests of migration and permeation. It was shown that a continuous surface layer of the biopolymeric materials was not necessary to substantially improve the barrier properties. This is an interesting aspect not only for future research, but also for coating and packaging technologists in the industry. Irrespective of the use of substrates made from primary or secondary fibers, medium-to-high grease resistance was accomplished. The water-vapor transmission rate was reduced by at least 35%. The water resistance or absorptiveness was clearly substrate-dependent, and optimal values were conditioned through the further utilization of packaging materials. The overall migration of organic volatile compounds was successfully reduced by 70% and 84% upon coating the SF substrate with chitosan or alginate, respectively. Migration for the PF sample was naturally low, and coatings were not actually necessary to reduce migration. The permeation of deuterated n-alkanes through both papers was reduced by up to 50%.

Summarizing all results, alginate and chitosan showed excellent barrier behavior.

Author Contributions: S.K. designed and performed the coating experiments, physical measurements, and barrier characterization. A.W. performed the migration and permeation experiments for the samples, including the determination of MOH levels using HPLC–GC–FID. A.Z. performed the SEM experiments. S.K., A.W., E.L. and W.B. analyzed the data, and wrote the paper.

Funding: This research was funded by the Austrian Research Promotion Agency (FFG), Austropapier–Vereinigung der Österreichischen Papierindustrie and Austrian Pulp, Paper and Packaging Industry (No. 855640).

Acknowledgments: The authors acknowledge the industrial partners, Mondi Group, Zellstoff Pöls AG, Delfort Group, W. Hamburger GmbH, Smurfit Kappa Nettingsdorf, the Austrian Research Promotion Agency (FFG), and Austropapier–Vereinigung der Österreichischen Papierindustrie, for their technical and financial support.

Conflicts of Interest: The authors declare no conflict of interest.

References

1. Coltelli, M.-B.; Wild, F.; Bugnicourt, E.; Cinelli, P.; Lindner, M.; Schmid, M.; Weckel, V.; Müller, K.; Rodriguez, P.; Staebler, A. State of the Art in the Development and Properties of Protein-Based Films and Coatings and Their Applicability to Cellulose Based Products: An Extensive Review. *Coatings* **2016**, *6*, 1. [CrossRef]
2. Andersson, C. New ways to enhance the functionality of paperboard by surface treatment—A review. *Packag. Technol. Sci.* **2007**, *21*, 339–373. [CrossRef]
3. Johansson, C.; Bras, J.; Mondragon, I.; Nechita, P.; Plackett, D.; Simon, P.; Gregor Svetec, D.; Virtanen, S.; Giacinti Baschetti, M.; Breen, C. Renewable fibers and bio-based materials for packaging applications—A review of recent developments. *Bioresources* **2012**, *7*, 2506–2552. [CrossRef]
4. Vartiainen, J.; Laine, C.; Willberg-Keyriläinen, P.; Pitkänen, M.; Ohra-aho, T. Biobased mineral-oil barrier-coated food-packaging films. *J. Appl. Polym. Sci.* **2016**, *134*, 44586. [CrossRef]
5. *Messung von Mineralöl—Kohlenwasserstoffen in Lebensmitteln und Verpackungsmaterialien*; Bundesinstitut für Risikobewertung (BfR): Berlin, Germany; Berlin und Kantonales Labor Zürich: Zürich, Switzerland, 2012; p. 103.
6. EFSA Panel on Contaminants in the Food Chain (CONTAM). Scientific Opinion on Mineral Oil Hydrocarbons in Food. *EFSA J.* **2012**, *10*, 2704. [CrossRef]
7. Biedermann, M.; Grob, K. On-line coupled high performance liquid chromatography–gas chromatography for the analysis of contamination by mineral oil. Part 2: Migration from paperboard into dry foods: Interpretation of chromatograms. *J. Chromatogr. A* **2011**, *1255*, 76–99. [CrossRef] [PubMed]
8. Biedermann, M.; Grob, K. On-line coupled high performance liquid chromatography–gas chromatography for the analysis of contamination by mineral oil. Part 1: Method of analysis. *J. Chromatogr. A* **2011**, *1255*, 56–75. [CrossRef] [PubMed]

9. Giles, H.F.; Wagner, J.R.; Mount, E.M. *Extrusion: The Definitive Processing Guide and Handbook*, 2nd ed.; William Andrew: Norwich, NY, USA, 2014; pp. 209–554. ISBN 978-1-4377-3481-2.
10. Ryan, N.M.; Mcnally, G.M.; Welsh, J. The Use of Aqueous-based Emulsion Polymers as Moisture Barrier Coatings for Carton Boards. *Dev. Chem. Eng. Miner. Process.* **2018**, *12*, 141–148. [CrossRef]
11. Fiselier, K.; Grob, K. Barriers against the Migration of Mineral Oil from Paperboard Food Packaging: Experimental Determination of Breakthrough Periods. *Packag. Technol. Sci.* **2011**, *25*, 285–301. [CrossRef]
12. Bastarrachea, L.J.; Wong, D.E.; Roman, M.J.; Lin, Z.; Goddard, J.M. Active Packaging Coatings. *Coatings* **2015**, *5*, 771–791. [CrossRef]
13. Kaiser, K.; Schmid, M.; Schlummer, M. Recycling of Polymer-Based Multilayer Packaging: A Review. *Recycling* **2018**, *3*, 1. [CrossRef]
14. Nesic, A.R.; Seslija, S.I. The influence of nanofillers on physical–chemical properties of polysaccharide-based film intended for food packaging. In *Nanotechnology in the Agri-Food Industry*, 1st ed.; Grumezescu, A., Ed.; Elsevier: Amsterdam, Netherlands, 2016; Volume VII, pp. 637–697. ISBN 9780128043738.
15. Wong, D.W.S.; Gregorski, K.S.; Hudson, J.S.; Pavlath, A.E. Calcium Alginate Films: Thermal Properties and Permeability to Sorbate and Ascorbate. *J. Food Sci.* **1995**, *61*, 337–341. [CrossRef]
16. Da Silva, M.A.; Iamanaka, B.T.; Taniwaki, M.H.; Kieckbusch, T.G. Evaluation of the Antimicrobial Potential of Alginate and Alginate/Chitosan Films Containing Potassium Sorbate and Natamycin. *Packag. Technol. Sci.* **2012**, *26*, 479–492. [CrossRef]
17. Fernandes, S.C.M.; Freire, C.S.R.; Silvestre, A.J.D.; Desbrières, J.; Gandini, A.; Neto, C.P. Production of Coated Papers with Improved Properties by Using a Water-Soluble Chitosan Derivative. *Ind. Eng. Chem. Res.* **2009**, *49*, 6432–6438. [CrossRef]
18. Nordqvist, D.; Idermark, J.; Hedenqvist, M.S.; Gällstedt, M.; Ankerfors, M.; Lindström, T. Enhancement of the Wet Properties of Transparent Chitosan−Acetic-Acid−Salt Films Using Microfibrillated Cellulose. *Biomacromolecules* **2006**, *8*, 2398–2403. [CrossRef] [PubMed]
19. Reis, A.B.; Yoshida, C.M.; Reis, A.P.C.; Franco, T.T. Application of chitosan emulsion as a coating on Kraft paper. *Polym. Int.* **2010**, *60*, 963–969. [CrossRef]
20. Thakhiew, W.; Devahastin, S.; Soponronnarit, S. Effects of drying methods and plasticizer concentration on some physical and mechanical properties of edible chitosan films. *J. Food Eng.* **2009**, *99*, 216–224. [CrossRef]
21. Vrabič Brodnjak, U.; Jesih, A.; Gregor-Svetec, D. Chitosan Based Regenerated Cellulose Fibers Functionalized with Plasma and Ultrasound. *Coatings* **2018**, *8*, 133. [CrossRef]
22. Khwaldia, K.; Arab-Tehrany, E.; Desobry, S. Biopolymer Coatings on Paper Packaging Materials. *Compr. Rev. Food Sci. Food Saf.* **2009**, *9*, 82–91. [CrossRef]
23. Bordenave, N.; Kemmer, D.; Smolic, S.; Franz, R.; Girard, F.; Coma, V. Impact of Biodegradable Chitosan-Based Coating on Barrier Properties of Papers. *J. Renew. Mater.* **2013**, *2*, 123–133. [CrossRef]
24. Vartiainen, J.; Tuominen, M.; Nättinen, K. Bio-hybrid nanocomposite coatings from sonicated chitosan and nanoclay. *J. Appl. Polym. Sci.* **2009**, *6*, 3638–3647. [CrossRef]
25. Bordenave, N.; Grelier, S.; Coma, V. Hydrophobization and Antimicrobial Activity of Chitosan and Paper-Based Packaging Material. *Biomacromolecules* **2009**, *11*, 88–96. [CrossRef] [PubMed]
26. Tang, C.; Chen, N.; Zhang, Q.; Wang, K.; Fu, Q.; Zhang, X. Preparation and properties of chitosan nanocomposites with nanofillers of different dimensions. *Polym. Degrad. Stab.* **2008**, *94*, 124–131. [CrossRef]
27. Rhim, J.-W.; Lee, J.-H.; Hong, S.-I. Water resistance and mechanical properties of biopolymer (alginate and soy protein) coated paperboards. *LWT–Food Sci. Technol.* **2005**, *39*, 806–813. [CrossRef]
28. Gällstedt, M.; Brottman, A.; Hedenqvist, M.S. Packaging-related properties of protein- and chitosan-coated paper. *Packag. Technol. Sci.* **2004**, *18*, 161–170. [CrossRef]
29. Rastogi, V.; Samyn, P. Bio-Based Coatings for Paper Applications. *Coatings* **2015**, *5*, 887–930. [CrossRef]
30. Kjellgren, H.; Engström, G. Influence of base paper on the barrier properties of chitosan-coated papers. *Nord. Pulp Pap. Res. J.* **2005**, *21*, 685–689. [CrossRef]
31. *T 402 SP-08 Standard Conditioning and Testing Atmospheres for Paper, Board, Pulp Handsheets, and Related Products*; TAPPI Standards Department: Norcross, GA, USA, 2008.
32. *ISO 536:2012 Paper and Board—Determination of Grammage*; International Organization for Standardization: Geneva, Switzerland, 2017.

33. *EN ISO 534:2011 Paper and Board—Determination of Thickness, Density and Specific Volume*; European Committee for Standardization: Brussels, Belgium, 2011.
34. *ISO/DIS 5636-3 Paper and Board—Determination of Air Permeance—Part 3: Bendtsen Method*; International Organization for Standardization: Geneva, Switzerland, 2013.
35. *ISO/DIS 8791-2 Paper and Board—Determination of Roughness/Smoothness—Part 2: Bendtsen Method*; International Organization for Standardization: Geneva, Switzerland, 2012.
36. *T 558 OM-15 Surface wettability and Absorbency of Sheeted Materials Using an Automated Contact Angle Tester*; TAPPI Standards Department: Norcross, GA, USA, 2012.
37. *T 559 CM-12 Grease Resistance Test for Paper and Paperboard*; TAPPI Standards Department: Norcross, GA, USA, 2012.
38. *EN ISO 535:2014 Paper and Board—Determination of Water Absorptiveness—Cobb Method*; European Committee for Standardization: Brussels, Belgium, 2011.
39. *T 448 OM-09 Water Vapor Transmission Rate of Paper and Paperboard at 23 °C and 50% RH*; TAPPI Standards Department: Norcross, GA, USA, 2009.
40. Fischer, W.J.; Zankel, A.; Ganser, C.; Schmied, F.J.; Schroettner, H.; Hirn, U.; Teichert, C.; Bauer, W.; Schennach, R. Imaging of the formerly bonded area of individual fibre to fibre joints with SEM and AFM. *Cellulose* **2014**, *21*, 251–260. [CrossRef]
41. *Union Guidance on Regulation (Eu) No 10/2011 On Plastic Materials And Articles Intended to Come into Contact With Food as Regards Information in the Supply Chain*; European Commission Health and Consumers Directorate-General: Luxembourg, Luxembourg, 2011; p. 89.
42. Ham-Pichavant, F.; Sèbe, G.; Pardon, P.; Coma, V. Fat resistance properties of chitosan-based paper packaging for food applications. *Carbohydr. Polym.* **2004**, *61*, 259–265. [CrossRef]
43. Kjellgren, H. Influence of Paper Properties and Polymer Coatings on Barrier Properties of Greaseproof Paper. Ph.D. Thesis, Karlstad Univeristy, Karlstad, Sweden, 2007.
44. Zhang, W.; Xiao, H.; Qian, L. Enhanced water vapour barrier and grease resistance of paper bilayer-coated with chitosan and beeswax. *Carbohydr. Polym.* **2014**, *101*, 401–406. [CrossRef] [PubMed]
45. Nilsson, L.; Wilhelmsson, B.; Stenstrom, S. The diffusion of water vapour trough pulp and paper. *Dry. Technol.* **1993**, *11*, 1205–1225. [CrossRef]
46. Biedermann-Brem, S.; Biedermann, M.; Grob, K. Taped Barrier Test for Internal Bags Used in Boxes of Recycled Paperboard: The Role of the Paperboard and Its Consequence for the Test. *Packag. Technol. Sci.* **2017**, *30*, 75–89. [CrossRef]

© 2018 by the authors. Licensee MDPI, Basel, Switzerland. This article is an open access article distributed under the terms and conditions of the Creative Commons Attribution (CC BY) license (http://creativecommons.org/licenses/by/4.0/).

Article

Superhydrophobic Bio-Coating Made by Co-Continuous Electrospinning and Electrospraying on Polyethylene Terephthalate Films Proposed as Easy Emptying Transparent Food Packaging

Maria Pardo-Figuerez [1,2], Alex López-Córdoba [3,4], Sergio Torres-Giner [1] and José M. Lagaron [1,*]

1. Novel Materials and Nanotechnology Group, Institute of Agrochemistry and Food Technology (IATA), Spanish National Research Council (CSIC), Calle Catedrático Agustín Escardino Benlloch 7, 46980 Paterna, Valencia, Spain; mpardo@iata.csic.es (M.P.-F.); storresginer@iata.csic.es (S.T.-G.)
2. Bioinicia R&D Department, Bioinicia S.L. Calle Algepser 65, nave 3, 46980 Paterna, Valencia, Spain
3. Laboratorio de Polímeros y Materiales Compuestos (LP&MC), Instituto de Física de Buenos Aires (IFIBA-CONICET), Departamento de Física, Facultad de Ciencias Exactas y Naturales, Universidad de Buenos Aires, Buenos Aires C1428EGA, Argentina; alexlcordoba@gmail.com
4. Escuela de Administración de Empresas Agropecuarias, Facultad Seccional Duitama, Universidad Pedagógica y Tecnológica de Colombia, Carrera18 con Calle 22, Duitama 150461, Boyacá, Colombia
* Correspondence: lagaron@iata.csic.es; Tel.: +34-963-900-022

Received: 17 September 2018; Accepted: 11 October 2018; Published: 16 October 2018

Abstract: Interest in coated films with micro/nanofeatures has grown rapidly in recent years due to their enhanced functional performance and better durability under demanding contact conditions or aggressive environments. In the current work, it is reported a one-step co-continuous bilayer coating process to generate a multilayer film that rendered superhydrophobicity to a polyethylene terephthalate (PET) substrate. A continuous coating based on ultrathin polylactide (PLA) fibers was deposited onto PET films by means of electrospinning, which increased the water contact angle of the substrate. Sequentially, nanostructured silica (SiO_2) microparticles were electrosprayed onto the coated PET/PLA films to achieve superhydrophobic behavior. This multilayer was then treated at different annealing temperatures, that is, 150 °C, 160 °C, and 170 °C, in order to create interlayers' adhesion to each other and to the substrate. It was found that co-continuous deposition of PLA fibers and nanostructured SiO_2 microparticles onto PET films constituted a useful strategy to increase the surface hydrophobicity of the PET substrate, achieving an optimal apparent water contact angle of 170° and a sliding angle of 6°. Unfortunately, a reduction in background transparency was observed compared to the uncoated PET film, especially after electrospraying of the SiO_2 microparticles but the films were seen to have a good contact transparency. The materials developed show significant potential in easy emptying transparent food packaging applications.

Keywords: electrospinning; electrospraying; superhydrophobicity; polyethylene terephthalate (PET); polylactide (PLA)

1. Introduction

Coatings are defined as mixtures of film-forming materials containing solvents and other additives that, when applied to a surface and after curing/drying process, yield a solid, protective, decorative, and/or functional adherent thin layer [1]. Surface coatings include paints, drying oils and varnishes, synthetic clear coatings and other products. Coating on substrates offer several advantages, such as the fact that coated materials have better durability against demanding contact conditions or aggressive environments than the raw materials [2]. Moreover, coatings allow the enhancement of certain properties that the raw materials could not provide, such as antimicrobial, self-cleaning, superhydrophobic and antifouling effects, and so forth [3–5].

Over the past years, nanotechnology has emerged as a potential tool in the development of surface coatings [2], resulting in their use in fields such as electronics, medical, food, pharmacy and aerospace. These surface coatings can contain micro/nanoscale features that offer more optimal and processing properties than conventional coatings, such as higher opacity, better interaction between the coating and surface and higher durability of the coating [2]. Moreover, coatings containing topographical cues may impart hydrophobic and oleophobic properties, thus improving corrosion resistance and enhancing either insulative or conductive properties [4].

The generation of superhydrophobic coatings has been gaining a fair interest in the field of coating technology [6,7]. Superhydrophobic coatings are specifically referred to water repellent layers with a water contact angle >150° [8]. Such coatings can be prepared either by fabricating rough architectures on the surface or by chemically modifying rough surfaces containing low surface free energy materials [8,9]. These surfaces can often be accomplished by common techniques such as dip coating [10], lithography [11,12], chemical vapor deposition [13], among others [14]. However, the common processes to obtain such coatings are considered complex and their industrial applications are still limited. Therefore, new developments in coating technology are necessary in order to expand the accessibility of superhydrophobic coatings.

Electrohydrodynamic processing (EHDP), including electrospinning and electrospraying, is a very appealing technology that can be used to apply topographical structured (such as fibers or particles) layers on a wide variety of substrates [15]. These technologies utilize high electrostatic potentials to draw polymer solutions or polymer melts into fibers or particles [16,17]. Electrospinning/electrospraying processes present the advantage to work at atmospheric pressure and are, therefore, easily integrate into continuous production lines. Moreover, unlike vapor deposition, these novel technologies allow the application of a continuous layer with the required surface roughness necessary to exhibit superhydrophobic properties [4]. The structure and the physicochemical properties of the electrospun fibers and electrosprayed particles generated by EHDP can also be tailored according to the end application, for instance packaging [18,19]. Moreover, both techniques enable to process many synthetic and natural polymers either alone or blended with other polymers and/or additives (e.g., surfactants). Among them, polylactide (PLA) is one of the most attractive materials for coating fabrication by electrospinning/electrospraying because it is easily spinnable, biodegradable, and thermoplastic [4,20].

Likewise, functional coatings containing particles of titanium dioxide (TiO_2), silicon dioxide (SiO_2, also known as silica), carbon black (CB), iron oxide (Fe_2O_3), and zinc oxide (ZnO) have been developed [3]. Among them, SiO_2-based coatings have been of particular interest as such coatings can offer self-cleaning properties as well as antimicrobial functionality. The use of SiO_2 microparticles for coating deposition allows for a change on the surface roughness and surface energy, leading to a superhydrophobic behavior without the need to perform any chemical modification [7,9]. As these surfaces are also highly water-repellent, they can be used in mirrors, self-cleaning windows, frames, bricks, wall paint, tiles, flat glass, and so forth [21]. Recently, Lasprilla-Botero et al. [8] reported an interesting approach to the design of superhydrophobic films by means of electrospinning/electrospraying, depositing electrospun ultrathin poly(ε-caprolactone) (PCL) fibers followed by electrosprayed nanostructured silica (SiO_2) microparticles onto low-density polyethylene (LDPE) films. The resultant coated films showed a high surface hydrophobicity with an apparent contact angle of 157°. Moreover, they showed good adhesion between layers and improved thermal stability.

In the current work, a similar but a one-step continuous electrospinning/electrospraying process was applied to generate a superhydrophobic PLA/SiO_2 bilayer on polyethylene terephthalate (PET) substrates with the additional aim of achieving maximum transparency. The process consisted of a deposition of PLA fibers and SiO_2 microparticles sequentially onto a PET film by electrospinning and electrospraying, respectively, and a subsequent annealing to improve the adhesion between the layers. To the best of our knowledge, this is the first time that PET films are coated with a PLA/SiO$_2$ bilayer obtaining not only high superhydrophobicity but also a good adhesion between layers using a scalable one-step continuous process.

2. Materials and Methods

2.1. Materials

The substrate PET, with a thickness of 77.0 ± 0.6 µm, was kindly provided from Belectric OPV GmbH (OPVIUS-Organic Photovoltaic Solutions, Kitzingen, Germany). For the films deposition, PLA was an Ingeo™ Biopolymer 2003D with melting features determined by differential scanning calorimetry (DSC) from 145 to 160 °C as reported by the manufacturer, NatureWorks LLC (Minnetonka, MN, USA). The SiO$_2$ microparticles (HDK® H18, Pyrogenic Silica) were obtained from Wacker Chemie AG (München, Germany). Polyvinylpyrrolidone (PVP), with a molecular weight (M_W) of 40,000 g/mol, was provided by Sigma Aldrich S.A. (Madrid, Spain). Chloroform, acetone, and ethanol with purities >99.5% were supplied by Panreac Química S.A. (Barcelona, Spain).

2.2. Preparation of Multilayer Films

2.2.1. Solutions

The solution for electrospinning was prepared by dissolving 7.5 wt.% PLA in a chloroform/acetone 7:3 (vol./vol.) solution and further stirring until complete dissolution. For electrospraying, a 1.5 wt.% of SiO$_2$ microparticles were suspended in ethanol and ultrasonicated for 1–2 min prior to electrospraying. Concentrations were optimized in preliminary experiments based on a previous work [22].

2.2.2. Electrospinning and Electrospraying of PLA Fibers and SiO$_2$ Microparticles

The tailor-made pilot equipment utilized for the formation of the bilayers consisted of two single needle injectors, one for electrospinning (PLA) and one for electrospraying (SiO$_2$), which sequentially deposited fibers and microparticles onto the PET film using an AC controlled proprietary Fluidnatek® LE500 pilot line from Bioinicia S.L. (Valencia, Spain, see Figure 1). The roll containing the PET films was moving over the roll-to-roll system in continuous at a speed of 1 mm/s. Initially, for optimization purposes, the PET films were only coated with an electrospun layer of PLA fibers. To this end, the PET films were first placed on the continuous roll-to-roll collector and the PLA solution was then pumped in one of the injectors at a flow-rate of 20 mL/h through a needle with an inner diameter of 0.8 mm and a distance to the collector of 15 cm. The electrical voltages of the dual polarizers were set at 17 kV and −4 kV using a high-voltage supply unit in which the spinneret filled with the polymer solution was connected to the positive terminal of the power supply and the collector was to the negative electrode (ground). Optimal deposition was found when the collector was placed at a distance of 12 cm. The process was carried out in the air conditioning (AC) system—controlled environmental chamber set at 23 °C and 40% relative humidity (RH).

Once the PLA deposition was optimized (PET/PLA films), the second injector for electrospraying was incorporated into the system to obtain a sequential electrospinning of PLA fibers, followed by the electrospraying of SiO$_2$ microparticles (PET/PLA/SiO$_2$ films). The process was carried out in identical conditions that those applied during electrospinning with the exemption that the SiO$_2$ solution was pumped at different flow-rates, from 20 to 50 mL/h, through the needle. In addition, for achieving a homogenous electrospraying deposition, the tip-to-collector distance was reduced to 10 cm for the SiO$_2$ deposition. Once optimization of both jets was achieved, the roll-to roll system worked in a manner where non-coated PET films went through the injector area for both depositions, continuously moving until reaching the roll collector in which a roll containing coated-PET films was obtained.

Figure 1. (**A**) Co-continuous electrospinning/electrospraying proprietary setup for the coating of the polyethylene terephthalate (PET) film by polylactide (PLA) fibers and silica (SiO_2) microparticles including an amplified image of the injector. (**B**) The roll containing the non-coated PET moving continuously from the area of the injector (for the deposition of fibers and microparticles) to the end of the roll collector, where it was collected in a coil-like manner.

2.2.3. Thermal Post-Treatment

PET/PLA films and PET/PLA/SiO_2 films were annealed by placing the PET coated films between Teflon sheets in a hydraulic hot-press model-4122 from Carver, Inc. (Wabash, IN, USA) without applying pressure. Firstly, different temperatures (90 °C, 120 °C, 140 °C, 160 °C, and 170 °C) were applied to the PET/PLA samples in order to find a balance between hydrophobicity, layer adhesion, and film transparency. An annealing time of 15 s was applied to obtain homogenous sample morphology where the structure of the mat was no longer seen. Similarly, for the PET/PLA/SiO_2 samples, different temperatures (150–170 °C) were applied to optimize interlayer adhesion.

2.3. Film Characterization

2.3.1. Thickness

The thicknesses of the deposited coatings were measured using an Elektrophysik Minitest 735 from Lumaquin, S.A. (Barcelona, Spain). Four values were measured at various points of the samples after each deposition, that is, PET film, PET/PLA fibers, and PET/PLA fibers/SiO_2 microparticles. The average and the standard deviation values were determined.

2.3.2. Morphology

The morphology of the resulting coatings and multilayer structures were examined by field emission scanning electron microscopy (FESEM) using a Hitachi S-4800 FESEM from Hitachi High Technologies Corp (Tokyo, Japan). An electron beam acceleration of 10 kV and a working distance of 8–10 mm were applied. The materials were sputtered for 3 min with a thin layer of gold/palladium layer prior to FESEM observation. Image analysis was carried out using the Image J Launcher v 1.05 software.

2.3.3. Transparency

The absorbance of the films, related to transparency, was determined by UV-Vis spectrophotometry using a UV4000 spectrophotometer from Dinko Instruments (Barcelona, Spain) in a spectral range between 390 and 800 nm. The transparency (T) was estimated from the absorption at 600 nm ($A600$) and the thickness of the film in mm (L) following the equation $T = A600/L$ (mm) [23]. For evaluating contact transparency, samples were placed on A4 paper sheets containing letters and imaged.

2.3.4. Surface Wettability

The water contact angle (θ) was measured to estimate the wettability of the films with an optical tensiometer in a Video-Based Contact Angle Meter, Theta Lite TL 101 model (Biolin Scientific, Espoo, Finland) using the OneAttension v 3.1 software. A 5-µL droplet of distilled water was deposited on the material surface at room temperature and the image of the drop was recorded. Three different measurements were taken and averaged at different parts of the samples. The sliding angle was determined on a water droplet deposited over the surface whilst being tilted and analyzed with the online software Ergonomics Ruler of the Universitat Politècnica de València (UPV), Valencia, Spain (https://www.ergonautas.upv.es/herramientas/ruler/ruler.php).

3. Results and Discussion

3.1. Morphology of Electrospun PLA and Electrosprayed SiO$_2$ Materials

Table 1 includes the mean thickness values of the uncoated PET film and the multilayer films after the deposition of the PLA fibers mat and also the electrosprayed nanostructured SiO$_2$ microparticles. One can observe that the neat PET film presented a mean thickness of 77 µm while the coatings of electrospun PLA fibers and SiO$_2$ microparticles were approximately 4.2 and 4.6 µm, respectively. The thickness was aimed at being the lowest under the screened conditions that could be applied and that could also generate reproducible superhydrophobic and adhesion performance in the final materials.

Table 1. Thicknesses of the non-coated polyethylene terephthalate (PET) film and PET films coated by electrospun polylactide (PLA) fibers and electrosprayed silica (SiO$_2$) microparticles.

PET	PET/PLA	PET/PLA/SiO$_2$
77.0 ± 0.6 µm	81.2 ± 0.9 µm	85.8 ± 1.3 µm

Figure 2 shows the top view of the non-coated and coated PET films. In Figure 2A, one can observe the smooth and continuous surface of the neat PET film. After electrospinning, a mat composed of uniform and randomly oriented PLA fibers with a mean fiber diameter of 1.5 ± 0.7 µm was obtained (see Figure 2B). Similar morphologies and sizes have been reported in the literature for PLA fibers obtained under similar electrospinning conditions and solution properties [24]. When SiO$_2$ microparticles were electrosprayed on the PET/PLA layers, important changes on the surface morphology were observed. Some fibers were coated with SiO$_2$ nanoparticles (41 ± 6 nm) and also some particles agglomerates (4 ± 0.5 µm) were randomly deposited onto the electrospun PLA mat (see Figure 2C). It has been reported that ultrafine particles such as silica nanoparticles tend to agglomerate or aggregate due to the strong cohesive forces between primary particles caused by their high surface-to-volume ratio and the small distance between them [7,25,26].

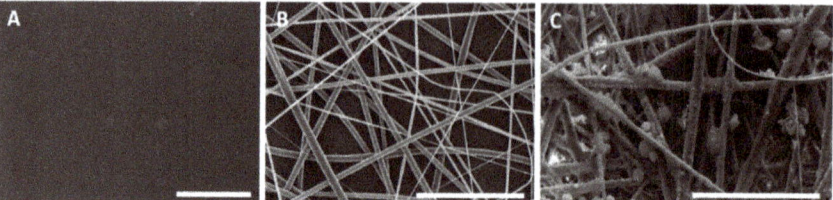

Figure 2. (**A**) Field emission scanning electron microscopy (FESEM) images of the as-received polyethylene terephthalate (PET) film; (**B**) Electrospun polylactide (PLA) fibers deposited onto the PET film; (**C**) Electrosprayed silica (SiO_2) microparticles onto the PET/PLA fibers. Scale bars of 50 μm.

3.2. Characterization of the Electrospun PLA Coated PET Films

The coated PET/PLA fibers were annealed at different temperatures and without pressure, in the range from 90 to 170 °C. Figure 3 shows the top view of the different PET films coated with PLA fibers and annealed at different temperatures. One can observe that, from 140 °C onwards, the PLA fiber morphology was lost due to a process of fibers coalescence. It has been early reported that the thermal post-processing of electrospun fibers well below their melting point leads to a packing of the material into a continuous film with virtually little or no porosity due to fibers coalescence [27,28].

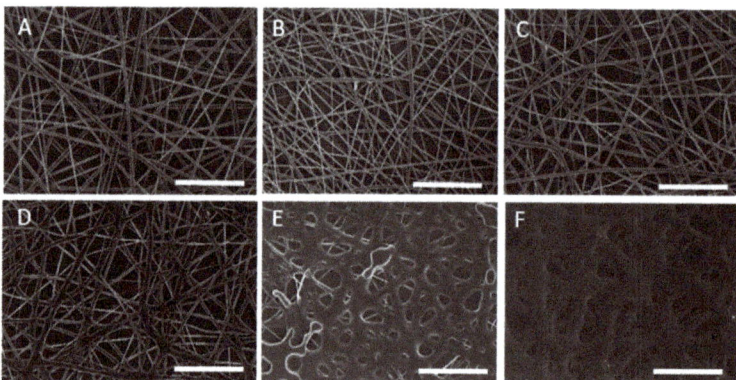

Figure 3. Field emission scanning electron microscope (FESEM) micrographs of the polyethylene terephthalate (PET) films coated with electrospun polylactide (PLA) fibers and annealed at different temperatures: (**A**) Without annealing; (**B**) 90 °C; (**C**) 120 °C; (**D**) 140 °C; (**E**) 160 °C; (**F**) 170 °C. The annealing time was 15 s in all samples and scale bars are 50 μm.

Figure 4 gathers relevant information about the transparency characteristics of the uncoated and coated PET films. Figure 4A shows quantitative measurements of transparency obtained by UV-Vis spectrophotometry in the wavelength range from ca. 400 nm to 800 nm. It can be observed that the uncoated PET film presented a high transparency, with an average light transmittance of 90% in the wavelength range screened. The PLA fibers deposition resulted in a high decrease in transparency to nearly 38% of light transmittance at 600 nm. A similar absorbance was observed for the bilayer structure annealed at 90 °C. The higher absorbance was seen to increase with increasing wavelength. It has been reported that nanofibrous films comprise large amount of air/fiber interfaces and therefore the incident light not only reflects and refracts many times at these interfaces but is absorbed within the mat, resulting in little light being transmitted through [29,30].

An increase in the light transmittance of the bilayer materials were observed when thermal treatments above 120 °C were applied. The sample annealed at 120 °C showed a slight increase in the light transmittance with respect to the non-heated sample. Interestingly, all bilayer structures annealed at higher temperatures, that is, 140 °C, 160 °C, and 170 °C, presented light transmittance values close to that of the uncoated PET film. Moreover, it was observed that the film transparency was progressively increased as the annealing temperature increased. This behavior was attributed to the fact that these annealing temperatures are closer or just above the first melting feature of the electrospun PLA fibers, previously reported at 154 °C by DSC [22] and in agreement with other previous study [31]. Therefore, it can be stated that the light transmittance of these bilayer films can be significantly enhanced by applying annealing temperatures of at least 140 °C. Similar observations were reported by Cherpinski et al. [28] when carrying out the post-processing optimization of electrospun submicron poly(3-hydroxybutyrate) (PHB) fibers to obtain continuous films.

Figure 4B includes the visual aspect of the film samples to evaluate their contact transparency. It can be observed that contact transparency of the films illustrated a high transparency for PET films and a clear decrease after the PLA fibers were deposited. Interestingly, more transparent films were observed after annealing, especially at temperatures higher than 140 °C. These observations are in expected coherence with the results of the light transmittance studies performed by UV-Vis spectrophotometry mentioned above.

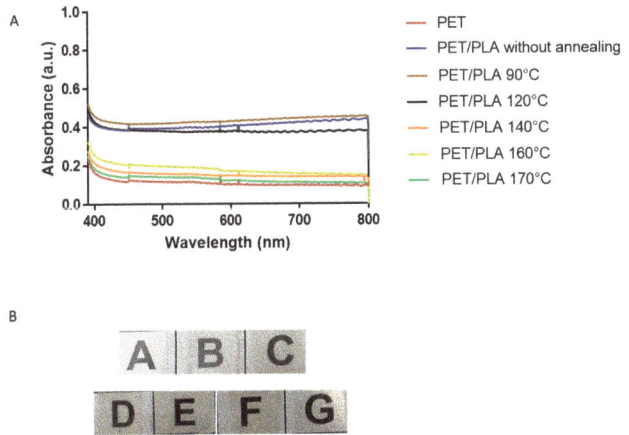

Figure 4. (**A**) Ultraviolet-visible (UV-Vis) transparency measures of the uncoated polyethylene terephthalate (PET) film and coated with electrospun polylactide (PLA) fibers at different annealing temperatures. (**B**) Contact transparency of the films of: A—Uncoated PET; B—PET/PLA without annealing; C—PET/PLA annealed at 90 °C; D—PET/PLA annealed at 120 °C; E—PET/PLA annealed at 140 °C; F—PET/PLA annealed at 170 °C; and G—PET/PLA annealed at 160 °C. The typical sample size of the films in the pictures is of ca. 2 × 1.5 cm^2.

Figure 5 provides the apparent water contact angle of the PET films coated with electrospun PLA fibers without annealing and annealed at different temperatures. One can observe that the PLA electrospun fibers generated a hydrophobic microdeposition with a value of 96.17° ± 10.79° that led to a slight increase of water contact angle when compared to the uncoated PET films (82.27° ± 0.83°). This higher contact angle indicates that the water droplet does not spread well on the substrate due to the intrinsic hydrophobicity of the electrospun PLA mats. Contact angles higher than 120° have been reported for PLA, PCL, and poly(lactic-co-glycolic acid) (PLGA) electrospun mats or coatings [8,32].

On the other hand, water contact angles of the PET/PLA films decreased as the temperature of the thermal post-treatment increased, obtaining values of 83.94° ± 3.54°, 80.64° ± 3.33°, 79.27° ± 9.08°, 75.3° ± 3.24°, and 73.42° ± 6.84° for the bilayer films treated at 90 °C, 120 °C, 140 °C, 160 °C, and 170 °C, respectively (see Figure 5). In particular, the PET/PLA samples annealed above 140 °C showed a reduction in the contact angles values higher than 20%, in comparison with the coated PET/PLA film without annealing. This effect can be related to the partial loss of the fibrilar roughness structure of the

PLA mat (see Figure 3) and, then, to a reduction of the overall porosity of the film surface, resulting in a decrease of the water contact angle. This behavior is in agreement with the work recently reported by Lasprilla-Botero et al. [8], where LDPE/PCL films were annealed at temperatures ranging from 55 to 90 °C and a loss of surface hydrophobicity was observed as a function of the annealing temperature.

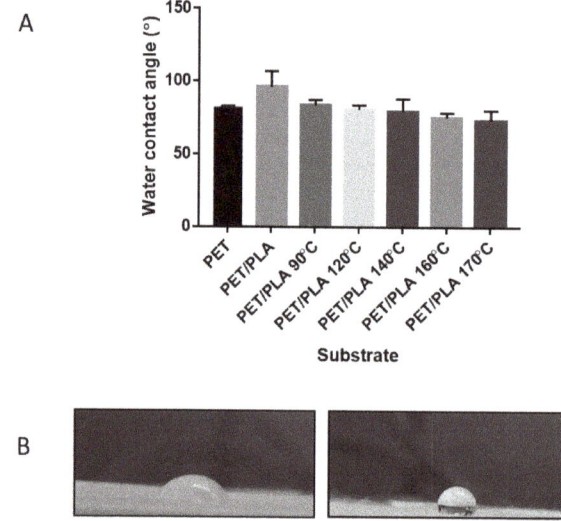

Figure 5. (**A**) Apparent water contact angle measurements of the uncoated polyethylene terephthalate (PET) film and coated with electrospun polylactide (PLA) fibers without annealing and annealed at 90 °C, 120 °C, 140°C, 160 °C, and 170 °C. The annealing time was 15 s in all samples. (**B**) Digital images of the contact angle performed on the PET films before (left) and after deposition (right) of the electrospun PLA fibers.

3.3. Characterization of the Multilayer PET Films

After optimization of the PLA deposition, achieving superhydrophobicity was attempted by depositing SiO_2 as a second layer on the PET/PLA bilayer structure. For this, different flow-rates were tested during electrospraying, that is, 20 mL/h, 30 mL/h, 40 mL/h, and 50 mL/h and the water contact angle was assessed prior to annealing. Figure 6 shows the values of the contact angle as a function of the flow-rate. One can observe that, after the SiO_2 deposition, superhydrophobicity (>150° on water contact angle) was achieved in all cases, regardless of the flow-rate used.

Since the all tested flow-rates resulted in superhydrophobicity of the multilayer, the lowest flow-rate, that is, 20 mL/h, was chosen for further applying the annealing. It provides higher reproducibility of the application of the SiO_2 microparticles and the optical properties would not be so impaired. Furthermore, superhydrophobicity was already achieved at this flow-rate so there would not be a need to use higher flow-rates, which would have implied a higher production cost in a potential upscaling.

In order to promote adhesion between layers, three different annealing temperatures were chosen, that is, 150 °C, 160 °C, and 170 °C, for the thermal post-treatment in view of the above results on the PET/PLA films. Top and cross-section views of the SEM images of the PET/PLA/SiO_2 films after applying these temperatures are gathered in Figure 7. All the annealing temperatures provoked strong changes in the surface of the multilayer films rendering a rougher topography. The PLA fibers were seen to adhere strongly to the substrate while the SiO_2 microparticles remained well-spread on the PET/PLA bedding substrate.

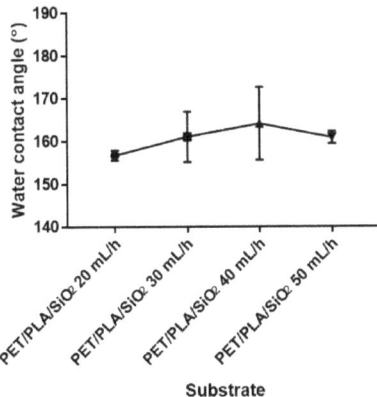

Figure 6. Apparent water contact angle of polyethylene terephthalate (PET)/polylactide (PLA)/silica (SiO$_2$) films at the electrospraying flow-rates of 20 mL/h, 30 mL/h, 40 mL/h, and 50 mL/h.

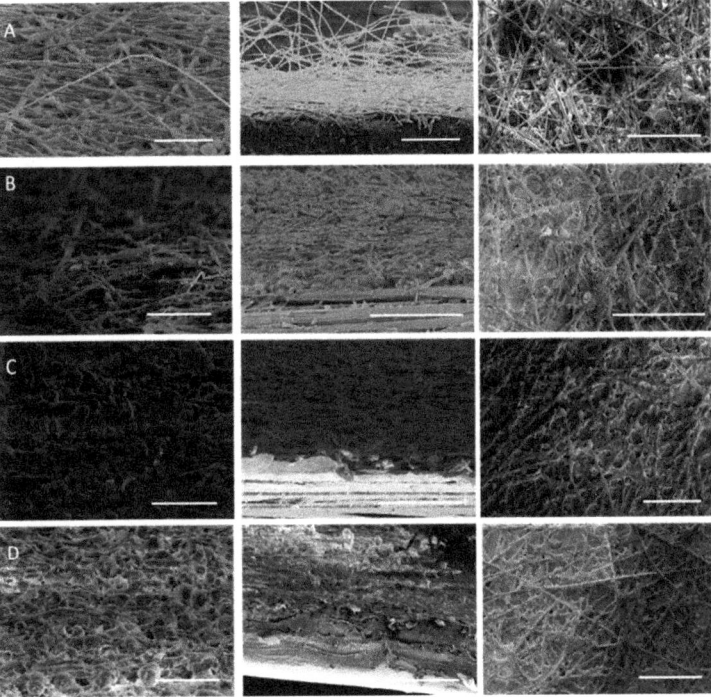

Figure 7. Field emission scanning electron microscope (FESEM) micrographs of the cross-section (left and middle columns) and top views (right column) of polyethylene terephthalate (PET)/polylactide (PLA)/silica (SiO$_2$) films: (**A**) Without annealing; (**B**) Annealed at 150 °C; (**C**) Annealed at 160 °C; (**D**) Annealed at 170 °C. The annealing time was 15 s in all samples and scale bars are 50 μm (left column) and 100 μm (middle and right columns).

As one can observe in Figure 8, the thermal treatments applied on the multilayers also resulted in an improvement of the contact transparency when compared with the untreated PET/PLA/SiO$_2$ film. The transparency values, which take into account the thickness differences among the samples, were, however, similar across the treatments. T values of 6.71, for the treatment at 150 °C, 6.50 for the treatment at 160 °C, and 5.65 for the highest temperature, that is, of 170 °C, were observed. The T value for the non-treated PET/PLA/SiO$_2$ was 7.35, while for the uncoated PET it was 1.32 and for the PET/PLA without annealing it was 4.42. In the case of the PET/PLA annealed at 150 °C and 160 °C, the T values were 2.15 and 2.50, respectively. A higher transparency was presented in the PET/PLA samples when comparing the T values with the transparency values of the PET/PLA/SiO$_2$ films. Likewise, the transmittance light values were lower as compared with the PET/PLA films (see Figure 4), due to the extra layer of sprayed SiO$_2$ microparticles that diffract the light.

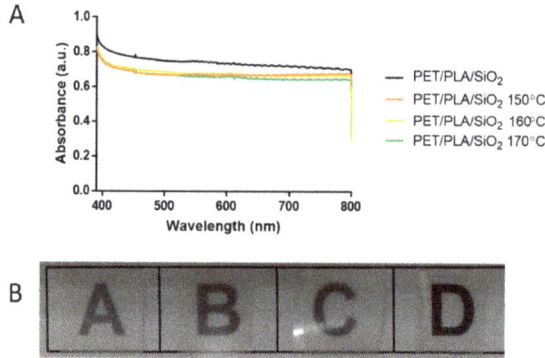

Figure 8. (**A**) Ultraviolet-visible (UV-Vis) transparency measurements of the polyethylene terephthalate (PET)/polylactide (PLA)/silica (SiO$_2$) films at different annealing temperatures. (**B**) Contact transparency of the PET/PLA/SiO$_2$ films: A—Without annealing; B—Annealed at 150 °C; C—Annealed at 160 °C; D—Annealed at 170 °C. The typical sample size of the films in the pictures is of ca. 2 × 1.5 cm^2.

Lastly, Figure 9 includes the results of the contact angle measurements of the PET/PLA/SiO$_2$ films. From this figure, it can be observed that all multilayer films rendered a superhydrophobic behavior with contact angle values in the 155°–170° range. The contact angle of the untreated multilayer films was 156.75° ± 1.16°. Su et al. [7] reported water contact angles around 165° for superhydrophobic structures fabricated via simultaneous electrospraying of SiO$_2$/dimethylacetamide (DMAc) colloids and electrospinning of polyvinylidene fluoride (PVDF)/DMAc solutions.

In this context, it was observed that the thermal post-treatment affected the surface superhydrophobicity of the multilayer systems. The film sample annealed at 150 °C showed a low reproducibility in the values of contact angles. It was probably due to the fact that at 150 °C, a partial loss of the fibrilar morphology occurred, provoking a non-homogeneous adherence between layers. When the film samples were treated with the highest temperature, that is, 170 °C, the PLA fibers were completely turned into a continuous film and therefore the contact angle was decreased up to 158.00° ± 4.37°. The film sample post-treated at 160 °C showed an increase in the contact angle values of around 8%, compared with the untreated multilayer film, obtaining a value of 170.63° ± 1.49°. Based on these results, it can be stated that thermal post-processing at 160 °C not only improved the adherence between layers but also provided the highest superhydrophobicity. The entrapped air and the roughness structures throughout the depth of the membrane was thought to lead to a continuous water-air-solid interface, resulting in a more hydrophobic surface [7]. In addition, this multilayer material promoted the desired hierarchical nanoparticle-agglomerates-fiber structure (see Figure 7) to increase the tortuosity of the water-air-solid interfaces, thus resulting in an easy sliding, with a sliding angle of 6°.

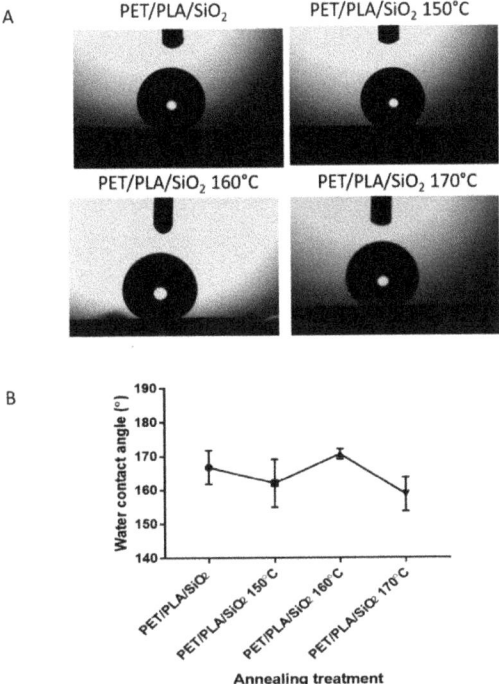

Figure 9. (**A**) Shape of the water droplets on the polyethylene terephthalate (PET)/polylactide (PLA)/silica (SiO$_2$) films; (**B**) Quantitative measurements of the apparent water contact angle of PET/PLA/SiO$_2$ films without annealing and annealed at 150 °C, 160 °C, and 170 °C.

4. Conclusions

A one-step co-continuous process that can increase the water repellency to PET substrates was herein developed. To this end, first, PLA ultrathin electrospun fibers were deposited onto the PET films by electrospinning and, thereafter, the coated PET/PLA films were post-treated at different annealing temperatures in the 90–170 °C range for 15 s. The control of the solution and electrospinning parameters allowed the creation of a homogeneous deposition of PLA fibers onto the PET films. The presence of the here-developed PLA coatings led to an increase of the surface hydrophobicity of the PET films but did not lead to superhydrophobic properties. For this reason, the co-deposition of SiO$_2$ microparticles by electrospraying was added to the process. Different electrospraying flow-rates of the SiO$_2$ microparticles, from 20 to 50 mL/h, generated multilayer films presenting superhydrophobicity. It was observed that the superhydrophobic behavior of the PET/PLA/SiO$_2$ multilayer films was somewhat improved when thermal post-treatments were applied up to 160 °C. In the optimal formulation, an apparent contact angle of 170° and a sliding angle of 6° were obtained.

Unfortunately, the very high contact and background transparency of PET could not be retained in the double coated samples but after annealing, the optical properties of the coated films with superhydrophobicity were optimal and showed good contact transparency. The generated PET films coated with PLA and SiO$_2$ reported in this work could be of potential use in easy emptying transparent packaging applications. Furthermore, the reported use of a scaled-up setup with a co-continuous process should facilitate its industrial implementation.

Author Contributions: Conceptualization was devised by J.M.L.; Methodology, Validation and Formal Analysis was carried out by M.P.-F., S.T.-G., A.L.-C. and J.M.L. Investigation, Resources, Data Curation, Writing Original Draft Preparation and Writing-Review & Editing was performed by M.P.-F., A.L.-C., S.T.-G. and J.M.L.; Supervision J.M.L.; Project Administration, J.M.L.

Funding: This study was funded by the EU H2020 OPTINANOPRO project (No. 686116).

Acknowledgments: S.T.-G. would like to thank the Ministry of Science, Innovation, and Universities (MICIU) for his Juan de la Cierva contract (IJCI-2016-29675).

Conflicts of Interest: The authors declare no conflict of interest.

References

1. Bierwagen, G.P. Surface coating. Available online: https://www.britannica.com/technology/surface-coating (accessed on 17 September 2018).
2. Maeztu, J.D.; Rivero, P.J.; Berlanga, C.; Bastidas, D.M.; Palacio, J.F.; Rodriguez, R. Effect of graphene oxide and fluorinated polymeric chains incorporated in a multilayered sol-gel nanocoating for the design of corrosion resistant and hydrophobic surfaces. *Appl. Surf. Sci.* **2017**, *419*, 138–149. [CrossRef]
3. Evstropiev, S.K.; Dukelskii, K.V.; Karavaeva, A.V.; Vasilyev, V.N.; Kolobkova, E.V.; Nikonorov, N.V.; Evstropyev, K.S. Transparent bactericidal ZnO nanocoatings. *J. Mater. Sci. Mater. Med.* **2017**, *28*, 102. [CrossRef] [PubMed]
4. Müller, K.; Bugnicourt, E.; Latorre, M.; Jorda, M.; Echegoyen Sanz, Y.; Lagaron, J.M.; Miesbauer, O.; Bianchin, A.; Hankin, S.; Bölz, U.; et al. Review on the processing and properties of polymer nanocomposites and nanocoatings and their applications in the packaging, automotive and solar energy fields. *Nanomaterials* **2017**, *7*, 74. [CrossRef] [PubMed]
5. Suyambulingam, G.R.T.; Jeyasubramanian, K.; Mariappan, V.K.; Veluswamy, P.; Ikeda, H.; Krishnamoorthy, K. Excellent floating and load bearing properties of superhydrophobic ZnO/copper stearate nanocoating. *Chem. Eng. J.* **2017**, *320*, 468–477. [CrossRef]
6. Barati Darband, G.; Aliofkhazraei, M.; Khorsand, S.; Sokhanvar, S.; Kaboli, A. Science and engineering of superhydrophobic surfaces: Review of corrosion resistance, chemical and mechanical stability. *Arab. J. Chem.* **2018**. [CrossRef]
7. Su, C.; Li, Y.; Dai, Y.; Gao, F.; Tang, K.; Cao, H. Fabrication of three-dimensional superhydrophobic membranes with high porosity via simultaneous electrospraying and electrospinning. *Mater. Lett.* **2016**, *170*, 67–71. [CrossRef]
8. Lasprilla-Botero, J.; Torres-Giner, S.; Pardo-Figuerez, M.; Álvarez-Láinez, M.; Lagaron, J.M. Superhydrophobic bilayer coating based on annealed electrospun ultrathin poly(ε-caprolactone) fibers and electrosprayed nanostructured silica microparticles for easy emptying packaging applications. *Coatings* **2018**, *8*, 173. [CrossRef]
9. Zhang, X.; Geng, T.; Guo, Y.; Zhang, Z.; Zhang, P. Facile fabrication of stable superhydrophobic SiO_2/polystyrene coating and separation of liquids with different surface tension. *Chem. Eng. J.* **2013**, *231*, 414–419. [CrossRef]
10. Nine, M.J.; Cole, M.A.; Johnson, L.; Tran, D.N.H.; Losic, D. Robust superhydrophobic graphene-based composite coatings with self-cleaning and corrosion barrier properties. *ACS Appl. Mater. Interfaces* **2015**, *7*, 28482–28493. [CrossRef] [PubMed]
11. Pozzato, A.; Zilio, S.D.; Fois, G.; Vendramin, D.; Mistura, G.; Belotti, M.; Chen, Y.; Natali, M. Superhydrophobic surfaces fabricated by nanoimprint lithography. *Microelectron. Eng.* **2006**, *83*, 884–888. [CrossRef]
12. Yang, X.; Liu, X.; Lu, Y.; Zhou, S.; Gao, M.; Song, J.; Xu, W. Controlling the adhesion of superhydrophobic surfaces using electrolyte jet machining techniques. *Sci. Rep.* **2016**, *6*, 23985. [CrossRef] [PubMed]
13. Hsieh, C.-T.; Chen, W.-Y.; Wu, F.-L. Fabrication and superhydrophobicity of fluorinated carbon fabrics with micro/nanoscaled two-tier roughness. *Carbon* **2008**, *46*, 1218–1224. [CrossRef]
14. Zhang, Z.; Wang, H.; Liang, Y.; Li, X.; Ren, L.; Cui, Z.; Luo, C. One-step fabrication of robust superhydrophobic and superoleophilic surfaces with self-cleaning and oil/water separation function. *Sci. Rep.* **2018**, *8*, 3869. [CrossRef] [PubMed]

15. Hu, C.; Liu, S.; Li, B.; Yang, H.; Fan, C.; Cui, W. Micro-/nanometer rough structure of a superhydrophobic biodegradable coating by electrospraying for initial anti-bioadhesion. *Adv. Healthc. Mater.* **2013**, *2*, 1314–1321. [CrossRef] [PubMed]
16. López-Córdoba, A.; Duca, C.; Cimadoro, J.; Goyanes, S. Electrospinning and electrospraying technologies and their potential applications in the food industry. In *Nanotechnology Applications in the Food Industry*; Rai, V.R., Bai, J.A., Eds.; CRC: Boca Raton, FL, USA, 2017.
17. Torres-Giner, S.; Pérez-Masiá, R.; Lagaron, J.M. A review on electrospun polymer nanostructures as advanced bioactive platforms. *Polym. Eng. Sci.* **2016**, *56*, 500–527. [CrossRef]
18. Torres-Giner, S.; Busolo, M.; Cherpinski, A.; Lagaron, J.M. Electrospinning in the packaging industry. In *Electrospinning: From Basic Research to Commercialization*; Kny, E., Ghosal, K., Thomas, S., Eds.; The Royal Society of Chemistry: London, UK, 2018; pp. 238–260.
19. Torres-Giner, S. Electrospun nanofibers for food packaging applications. In *Multifunctional and Nanoreinforced Polymers for Food Packaging*; Lagaron, J.-M., Ed.; Woodhead Publishing: Cambridge, UK, 2011; pp. 108–125.
20. López-Córdoba, A.; Lagarón, J.M.; Goyanes, S. Fabrication of electrospun and electrosprayed carriers for the delivery of bioactive food ingredients. In *Reference Module in Food Science*; Smithers, G.W., Ed.; Elsevier: Amsterdam, The Netherlands, 2018.
21. Umweltbundesamt Use of Nanomaterials in Coatings. Available online: https://www.umweltbundesamt.de/en/publikationen/use-of-nanomaterials-in-coatings (accessed on 17 September 2018).
22. Martínez-Sanz, M.; Lopez-Rubio, A.; Lagaron, J.M. Dispersing bacterial cellulose nanowhiskers in polylactides via electrohydrodynamic processing. *J. Polym. Environ.* **2014**, *22*, 27–40. [CrossRef]
23. Figueroa-Lopez, K.; Castro-Mayorga, J.; Andrade-Mahecha, M.; Cabedo, L.; Lagaron, J. Antibacterial and barrier properties of gelatin coated by electrospun polycaprolactone ultrathin fibers containing black pepper oleoresin of interest in active food biopackaging applications. *Nanomaterials* **2018**, *8*, 199. [CrossRef] [PubMed]
24. Casasola, R.; Thomas, N.L.; Trybala, A.; Georgiadou, S. Electrospun poly lactic acid (PLA) fibres: Effect of different solvent systems on fibre morphology and diameter. *Polymer* **2014**, *55*, 4728–4737. [CrossRef]
25. Syed, J.A.; Tang, S.; Meng, X. Super-hydrophobic multilayer coatings with layer number tuned swapping in surface wettability and redox catalytic anti-corrosion application. *Sci. Rep.* **2017**, *7*, 4403. [CrossRef] [PubMed]
26. Yao, W.; Guangsheng, G.; Fei, W.; Jun, W. Fluidization and agglomerate structure of SiO_2 nanoparticles. *Powder Technol.* **2002**, *124*, 152–159. [CrossRef]
27. Cherpinski, A.; Torres-Giner, S.; Cabedo, L.; Méndez Jose, A.; Lagaron, J.M. Multilayer structures based on annealed electrospun biopolymer coatings of interest in water and aroma barrier fiber-based food packaging applications. *J. Appl. Polym. Sci.* **2017**, *135*, 45501. [CrossRef]
28. Cherpinski, A.; Torres-Giner, S.; Cabedo, L.; Lagaron, J.M. Post-processing optimization of electrospun submicron poly (3-hydroxybutyrate) fibers to obtain continuous films of interest in food packaging applications. *Food Addit. Contam. Part A* **2017**, *34*, 1817–1830. [CrossRef] [PubMed]
29. Cherpinski, A.; Gozutok, M.; Sasmazel, H.; Torres-Giner, S.; Lagaron, J. Electrospun oxygen scavenging films of poly (3-hydroxybutyrate) containing palladium nanoparticles for active packaging applications. *Nanomaterials* **2018**, *8*, 469. [CrossRef] [PubMed]
30. Tang, C.; Liu, H. Cellulose nanofiber reinforced poly (vinyl alcohol) composite film with high visible light transmittance. *Compos. Part A Appl. Sci. Manuf.* **2008**, *39*, 1638–1643. [CrossRef]
31. Garlotta, D. A literature review of poly (lactic acid). *J. Polym. Environ.* **2001**, *9*, 63–84. [CrossRef]
32. Russo, V.; Tammaro, L.; Di Marcantonio, L.; Sorrentino, A.; Ancora, M.; Valbonetti, L.; Turriani, M.; Martelli, A.; Cammà, C.; Barboni, B. Amniotic epithelial stem cell biocompatibility for electrospun poly (lactide-co-glycolide), poly (ε-caprolactone), poly (lactic acid) scaffolds. *Mater. Sci. Eng. C* **2016**, *69*, 321–329. [CrossRef] [PubMed]

© 2018 by the authors. Licensee MDPI, Basel, Switzerland. This article is an open access article distributed under the terms and conditions of the Creative Commons Attribution (CC BY) license (http://creativecommons.org/licenses/by/4.0/).

Article

A Novel Way of Adhering PET onto Protein (Wheat Gluten) Plastics to Impart Water Resistance

Oisik Das [1,*], Thomas Aditya Loho [2], Antonio José Capezza [1,3], Ibrahim Lemrhari [1] and Mikael S. Hedenqvist [1,*]

1. Department of Fibre and Polymer Technology, Polymeric Materials Division, School of Engineering Sciences in Chemistry, Biotechnology and Health. KTH Royal Institute of Technology, Stockholm 10044, Sweden; ajcv@kth.se (A.J.C.); bramslem@hotmail.fr (I.L.)
2. Department of Chemical and Materials Engineering, The University of Auckland, Auckland 1142, New Zealand; thomas.loho@auckland.ac.nz
3. Department of Plant Breeding, Faculty of Landscape Planning, Horticulture and Crop Production Sciences, Swedish University of Agricultural Sciences, Alnarp 23053, Sweden
* Correspondence: oisik@kth.se or odas566@aucklanduni.ac.nz (O.D.); mikaelhe@kth.se (M.S.H.); Tel.: +46-790-469-886 (O.D.); +46-706-507-645 (M.S.H.)

Received: 9 October 2018; Accepted: 27 October 2018; Published: 31 October 2018

Abstract: This study presents an approach to protect wheat gluten (WG) plastic materials against water/moisture by adhering it with a polyethylene terephthalate (PET) film using a diamine (Jeffamine®) as a coupling agent and a compression molding operation. The laminations were applied using two different methods, one where the diamine was mixed with the WG powder and ground together before compression molding the mixture into plates with PET films on both sides. In the other method, the PET was pressed to an already compression molded WG, which had the diamine brushed on the surface of the material. Infrared spectroscopy and nanoindentation data indicated that the diamine did act as a coupling agent to create strong adhesion between the WG and the PET film. Both methods, as expected, yielded highly improved water vapor barrier properties compared to the neat WG. Additionally, these samples remained dimensionally intact. Some unintended side effects associated with the diamine can be alleviated through future optimization studies.

Keywords: PET; lamination; nanoindentation; water vapor barrier; interface

---

1. Introduction

The advent of the sustainability concept and the need for renewable materials has exposed wheat gluten (WG) as an attractive biopolymer [1]. The fact that WG is a by/co-product of the cereal processing industry, has good oxygen barrier properties in dry conditions, and is completely bio-based/biodegradable makes it an ideal candidate for propagating natural materials in the current polymer market [2]. WG possesses properties that make it suitable for replacing synthetic polymers in certain packaging, absorbent, and semi-structural applications [3]. WG can also be processed into films, foams, and solid 3D structures using extrusion, freeze-drying, compression, and injection molding [4]. Despite the aforementioned advantages of WG, there are certain inherent properties that adversely affect its applicability. WG is very susceptible to moisture and water [5], wherein the formed polymer can suffer from dimensional instability, inferior mechanical properties, microbial/fungal attack, and loss of barrier properties. If WG is intended to be used in e.g., food or medical packaging, it is critical in several cases that the polymer is resistant to moisture/water and has a low water vapor transmission rate. This would allow the moist food (e.g., cheese, cakes) to retain the required moisture and dry food (chips, fried, cookies) to remain crisp. Hence, the moisture resistance of the packaging material will enhance the shelf life of the stored food [6]. Moreover, if WG is planned to be applied as semi-structural elements, it is important that the polymer does not lose its structural integrity due to plasticization and degradation by microbes/fungus (which thrive on moisture). Therefore, to preserve the application potential of WG, it has become necessary to impart moisture and water resistance properties [7].

Several studies have been conducted where it was attempted to bestow WG and its films with the necessary moisture and water vapor resistance. Researchers have performed studies to chemically modify WG using cross-linkers to render them moisture resistant [8–10]. In other studies, different types of coatings (lipids, oils, and hydrophobic polymers) have been applied onto the WG substrate to create water resistant layers [8,11]. In a study by Cho et al. [12], WG films were laminated with poly(lactic acid) (PLA) to improve their water vapor barrier properties. The PLA coating was applied through the process of compression molding and the authors observed that the water vapor transmission rate (WVTR) was reduced irrespective of the content of glycerol in the WG film. However, the interlayer adhesion was deteriorated as a result of increased WG molding temperature (ca. 130 °C). Irissin-Mangata et al. [2] applied a UV-cured coating (hydrophobic cross-linked photopolymer) on WG films using a wire-wound applicator to render it moisture resistant. It was reported that the coating was able to decrease the water vapor transmission rate of the WG films. However, the study only visually analyzed the adhesion of the coating with the WG through scanning electron microscopy (SEM), and no quantitative measurement of the adhesive strength of the coated layer was presented. In an earlier study by the same authors [13], functionalized polyethylene films (ethylene/acrylic ester/maleic anhydride terpolymer and ethylene/glycidyl methacrylate copolymer) were compression molded onto a WG substrate. Although both coatings were able to reduce the water vapor transmission rate, only the terpolymer adhered to the WG film. The authors presented no quantitative measurements of the adhesion of the coating with the WG. Most probably, the adhesion of the terpolymer was due to the formation of covalent bonds, but the authors cited Van der Waal forces and hydrogen bonds as possible reasons.

Different types of hydrophobic lipids have also been applied (added to the film-forming solution) on WG films to provide water resistance [14]. It was observed that beeswax performed the best (amongst the tested lipids) to enhance the moisture barrier properties. However, the coating of the beeswax lipid reduced the transparency of the WG films while also exhibiting low puncture strength and a tendency to easily disintegrate in water. In another study by Gontard et al. [8], a thin layer of lipid was deposited (i.e., coated) onto WG films to create a water vapor barrier. Similar to the earlier study, beeswax and paraffin wax were reported to induce the highest improvement in the water vapor barrier properties. Nevertheless, the authors met with challenges since the hydrophilic WG under-layer expanded and fractured the brittle lipid layer when water reached this phase. Moreover, aside from being brittle, the lipid coating suffered from poor adhesion to the WG and oxidative instability. Fabra et al. [15] were able to reduce the water permeability of WG significantly (by ca. 88%) using a coating of annealed polyhydroxybutyrate electro spun fibers. The coated sample also exhibited a contact angle of 70°. Despite enhancing the barrier and mechanical properties, the color of the WG films turned brown and translucent. Elsewhere, Micard et al. [16] unsuccessfully endeavored to reduce the vapor permeability of WG films by chemical (formaldehyde vapors) and physical treatments (temperature and radiation). From the above-mentioned studies, it can be stated that most of the layers applied to WG lacked the proper adhesion (or its experimental determination thereof) while changing the appearance of the WG. This necessitates the exploration of other layering agents with reduced water transmission that can adhere strongly to WG with the aid of a coupling agent (preferably already used in treating WG polymers).

The central idea for the current study was obtained unintentionally. Jeffamine, a diamine, was tested for chemically cross-linking WG for the same purpose of rendering it more water resistant. While molding the WG sample (with the added diamine) using the hot press, polyethylene terephthalate (PET) films were used instead of the usual Teflon films. It was observed that post compression molding, the PET films where strongly attached to the WG sample, which indicated that the diamine acted as a coupling agent, reacting with both the PET layer and the WG polymer. The strongly adhered

PET layer would impart the necessary water resistance, ensuring the preservation of the high oxygen barrier properties of the WG component, similar to how high oxygen barrier properties are preserved for the ethylene vinyl alcohol (EVOH) layer when surrounded by polyethylene or polypropylene layers in food packaging. It has to be kept in mind though that the addition of PET to WG would reduce the sustainability aspect to some extent. However, bio-based PET is an upcoming material that is being increasingly researched [17] and holds potential to be used in the market soon. Therefore, this study lays the foundation for the application of PET, bio-based or otherwise, on WG plastics to impart water vapor barrier properties. The PET layered WG can be separated for recycling since WG is biodegradable and PET is not. Hence, for example, after the WG has degraded (facilitated by cutting the material into pieces to expose the WG for its decomposition), the PET can be recycled to form rPET (post separation from decomposed WG). More importantly, the research and development of bio-based plastics, like WG, should not be affected by the recycling landscape. Alaerts et al. [18] stated that if the bio-based plastics could be recycled on their own, the entire recycling landscape would be able to accommodate them eventually.

The overarching aim of this study was to apply a PET layer onto the WG polymer surface through two different methods and test their nanomechanical and water vapor resistance properties. Changes in the microscopic, chemical, and macro-mechanical properties as a result of the PET application were also evaluated accordingly. Since most of the previous studies did not attempt to definitively and quantitatively measure the adhesion properties of the coating with the WG, the current study employed the nanoindentation technique. Nanoindentation provided a new way to comprehend the nature of the adhesion of the PET layer to the WG component by determining the nanohardness and nanomodulus in the interfacial PET/WG region. In addition to the nanoindentation approach, the novelty of the work also lies in the method of using a diamine to laminate WG with a more water resistant layer (here PET).

2. Materials and Methods

2.1. Manufacturing of the Samples

The PET layer was applied on to the WG material (supplied by Lantmännen Reppe AB of Stockholm, Sweden with a gluten protein content of 77.7 wt %) using two different application methods. The first method involved compression molding (Fortijne Presses TP 400, Barendrecht, Netherlands) WG powder at 150 °C for 30 min at 290 kN force. Next, 15 wt % of Jeffamine (Jeffamine EDR 148 procured from Huntsman Corporation, Göteborg, Sweden) was brushed onto the molded WG surface and then compression molded again with the initially 250 µm thick PET films (Mylar® A, Synflex Insulation Systems, Blomberg, Germany) using the same pressing conditions. The amount of Jeffamine added was decided based on previous studies [19,20]. This sample was named "Brushed" since the Jeffamine was applied externally. In the other application method, 15 wt % Jeffamine was first added to the WG powder, then mixed through dry blending. This was followed by flash freezing in liquid N_2 and grinding (Retsch GmBH, 5657, Haan, Germany). The WG powder containing the Jeffamine was then compression molded along with the PET sheets using the same pressing conditions as mentioned previously. This sample was named "Ground". The neat WG sample (as the control) was also manufactured using the same process as for the "Ground" sample, but without the Jeffamine. The mold dimension was 100 × 100 × 4 mm^3, thereby yielding samples of the same sizes. The PET layer was attempted to be forcibly and manually removed to qualitatively gauge its adhesion. It was observed that the PET was completely stuck to the WG polymer. For the WVTR measurements, the mold was circular with a diameter of 70 mm and a thickness of 1 mm. The PET layers were applied to both sides of the samples.

2.2. Testing of the Samples:

The tensile strengths and moduli (chord moduli between 0.05% and 0.25% strain) of the laminated and control samples were determined on an Instron Universal Testing Machine (10 kN load cell) (Model No. 5566, Instron, Norwood, MA, USA), using specimens that were 4 mm (the WG part) thick, 12.7 mm wide, and 100 mm long. The samples were milled out from the produced plates. The gauge length was 50 mm and the crosshead speed used was 50 mm/min, following the ASTM D638 Protocol [21].

Three top-down samples and three cross-sectional samples were prepared for nanoindentation in this study. The top-down samples were cut into squares with a ca. 15 mm side length and the cross-sectional samples were cut into squares with a ca. 10 mm side length using a SiC cut-off wheel (Struers, Cleveland, OH, USA). The cross-section of the cross-sectional samples was ground with 240 and 600 grit-size SiC abrasive papers (Struers) and polished with a 6 µm diamond suspension (Struers). All samples were then cleaned with ethanol and fixed onto a mild steel plate using a cyanoacrylate adhesive. Figure 1 shows the orientation of the samples for the nanoindentation tests.

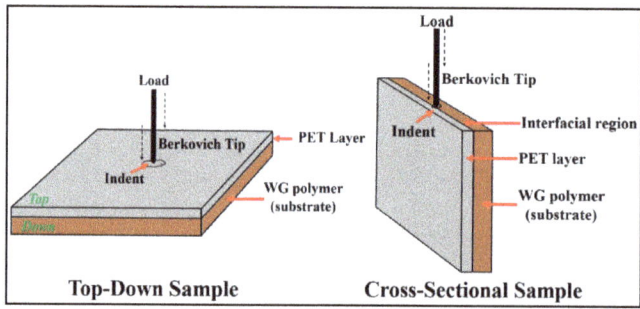

Figure 1. Sample orientation for nanoindentation.

The nanoindentation procedure was performed using a Hysitron TI950 TriboIndenter (Bruker, Billerica, MA, USA) at the University of Auckland, New Zealand. For both procedures, a 1 µm diamond conical tip, commonly used for the nanoindentation of non-metallic materials, was used. On the top-down samples, five sets of nine standard quasi-static indents (5 s loading, 2 s holding, and 5 s unloading) with increasing maximum loads from 1000 to 9000 µN were done at five random locations on the surface of the PET layer. On the cross-sectional samples, two sets of nine indents with a fixed maximum load of 5000 µN were done at the interfacial region of the PET layer and the WG substrate. The hardness and modulus were calculated from the load-displacement data and the details are available elsewhere [22].

The WVTR of the samples were determined using 25 cm^2 VF2201 permeability cups from TQC. The cups, with a volume of 20 mL water, were fitted with the circular samples (exposing 25 cm^2 of the film to water vapor) and then placed in a room with 50% relative humidity at 23 °C. The mass loss due to water vapor transmission was measured according to ASTM D1653 protocol [23]. The water vapor transmission rate was calculated based on the water loss data between the third and fourth day.

A goniometer (CAM 200, KSV Instruments Ltd., Espoo, Finland) was used wherein a droplet of distilled water was placed on the flat surface of the sample and its contact angles were measured.

Fourier transform infrared spectroscopy (FT-IR) of the samples was performed in ATR mode on a Perkin Elmer Spectrum 100 instrument, Waltham, MA, USA. The layered samples were placed over the crystal to comprehend the chemical changes in that region. For each sample, 64 scans were obtained and averaged from the 600 to 4000 cm^{-1} wavelength with a resolution of 4 cm^{-1}. The scanning electron microscopy (SEM) of the tensile fractured samples and of the interfacial (PET/WG) regions were conducted on a Hitachi TM 100 table top SEM, Tokyo, Japan (10 kV voltage, 6 mm working distance). All the tests mentioned in this manuscript were performed in replicates (tensile tests had five replicates, WVTR had two replicates, and the goniometer had ten replicates). The standard deviation measured was reported in the error bars of the figures and as numeric values in the table.

3. Results and Discussion

3.1. Mechanical Properties

3.1.1. Tensile Properties

The tensile properties of the laminated samples and the neat sample are summarized in Table 1. It can be observed that the neat WG had the highest tensile strength amongst the tested samples. Its tensile strength value of ca. 29 MPa is similar to that of the tensile strength of neat polypropylene [24], which is an important economic consideration. It is an essential first step in developing bio-based plastics that have the same tensile strength as a common and popular synthetic polymer (here polypropylene). The processing condition for the compression molding was decided after trial tests and a pressing temperature of 150 °C for 30 min was the most desirable in terms of good adhesion between PET and WG, and good cohesion of the WG matrix (enabled through sufficient melting during the molding cycle). It should be noted that a polyethylene film was also tested for adhesion to WG, but the effect observed with PET was absent. It was also observed that without pressure during the molding operation and without the diamine, the adhesion between PET and WG was absent. In addition, the particular processing condition enabled a heat-induced cross-linking (disulfide cross-linking) of the gluten network, which is known to enhance the mechanical strength of WG-based materials [25,26].

The addition of the diamine to WG caused some changes in the protein, which consequently reduced the tensile strength, ductility/strain at break and toughness. The effects were largest for the sample where the diamine had been mixed with the WG (Ground) and less for the sample with the diamine applied (Brushed) at the WG surface. It is known that at high temperatures, nitrogen-containing compounds such as diamine depolymerize polymers that possess amide groups [27]. The modulus, however, did not decrease in the presence of the diamine. An estimation of the laminate modulus employing the rule of mixtures for parallel geometry (using the neat WG modulus, Table 1 and PET modulus, 2.7 GPa) gave an unrealistically low value compared to those in Table 1. This revealed that the modulus of the WG component was higher in the presence of diamine due to the cross-linking [22,28]. During the tensile testing, no delamination of the PET layer was observed in any of the samples, which reaffirmed the fact that the bonding of PET to the WG polymer was strong. The results showed that in terms of mechanical properties, the brushing method was superior to the diamine/WG mixing (Ground) method. It should be noted though that both the Brushed and the Ground materials had sufficient mechanical properties for many applications (e.g., packaging, encapsulation of pest- or weed-control agents, toys, covers, frames for electrical devices, household appliances, and furniture) [13,29].

Table 1. Tensile properties.

Sample	Tensile Strength (MPa)	Tensile Modulus (GPa)	Strain at Break (%)	Toughness (J)
Neat WG	29.2 ± 2.1	1.2 ± 0.1	3.3 ± 2.3	1.36 ± 0.84
Ground	5.0 ± 1.6	1.3 ± 0.2	0.8 ± 0.4	0.04 ± 0.03
Brushed	17.3 ± 2.3	1.7 ± 0.3	2.3 ± 0.9	0.22 ± 0.04

3.1.2. Nanoindentation

The use of the nanoindentation technique has been previously proven to be effective in evaluating the interfacial (amongst other) properties of polymer composites [22,30,31]. Therefore, in this study, the same technique was applied to determine the nanomechanical properties of the PET laminated and neat WG samples (sample thickness of 4.2 mm). Figures 2 and 3 show the nanomechanical hardness and reduced modulus of the tested top-down samples, respectively. The hardness was essentially independent of the depth of the indentation and on the same level for both PET-layered WG samples, which was expected as the indentation was always in the PET layer. The indentation depth ranged from ca. 300 nm (for a 1000 µN load) to 2000 nm (for a 9000 µN load). This result suggests that the measured hardness was that of the PET with little to no substrate effect(s). For neat WG, there seemed to be a small but consistent increase in hardness inwards into the sample, however, based on standard deviation, the values were not significantly different.

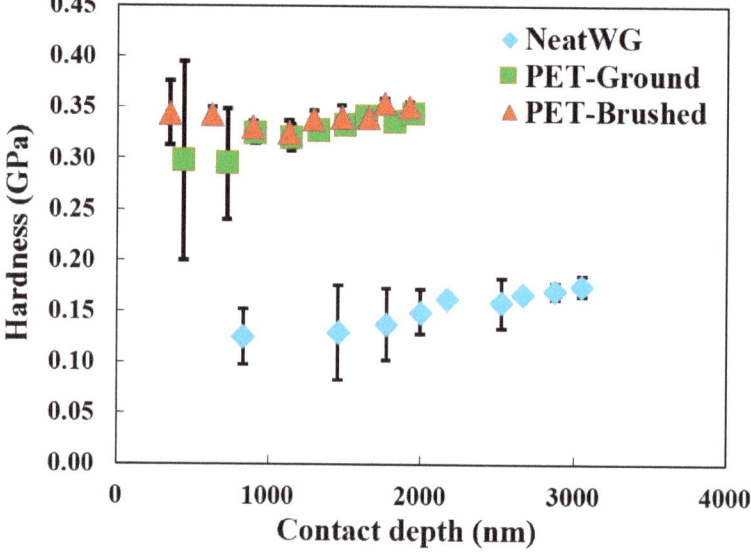

Figure 2. Nanoindentation hardness measured top-down.

From Figure 3, it can also be observed that the reduced modulus for neat WG increased with increasing depth, while the reduced modulus for the PET-layered samples remained depth-independent. Evidently, on a submicron to micron-range, the WG surface was less stiff than the interior, but still at a level comparable to that of the PET layer. From the nanoindentation hardness and reduced modulus values, it can be stated, that although the use of the diamine could have some adverse effects on the tensile strength and extensibility (see Section 3.1.1), the PET layers could increase the wear resistance of the samples, which is desirable (Figure 4). The wear resistance can be calculated as the ratio of the nanoindentation hardness to the modulus [32]. Figure 4 shows that the laminated samples had a significantly higher wear resistance when compared to that of the neat WG, and that the wear resistance of the laminated samples were the same.

The load vs. displacement curves of the samples tested top-down are presented in Figure 5. The unloading curves did not overlap the loading curves due to energy loss (damping) and a certain degree of plastic deformation. These effects were similar for the two PET-layered plates, but larger for the neat WG. Figure 6 shows the nanoindentation scan area of the top-down tested samples where the impression left after the conical-tip had indented the samples is clearly visible. This shows that the samples had deformed not only elastically, but also plastically, as also observed in Figure 5. Additionally, we noticed that the PET surfaces were flatter/smoother when compared to the WG material, implying a more elastic surface.

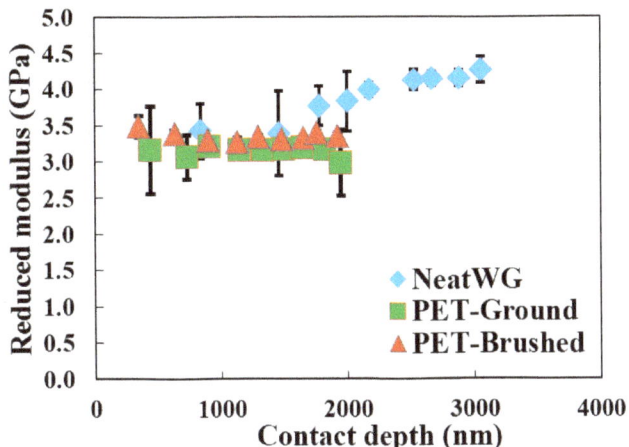

Figure 3. Reduced nanoindentation modulus measured top-down.

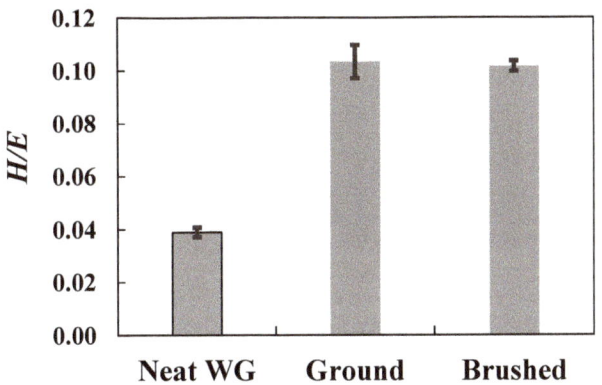

Figure 4. Wear resistance measured top-down.

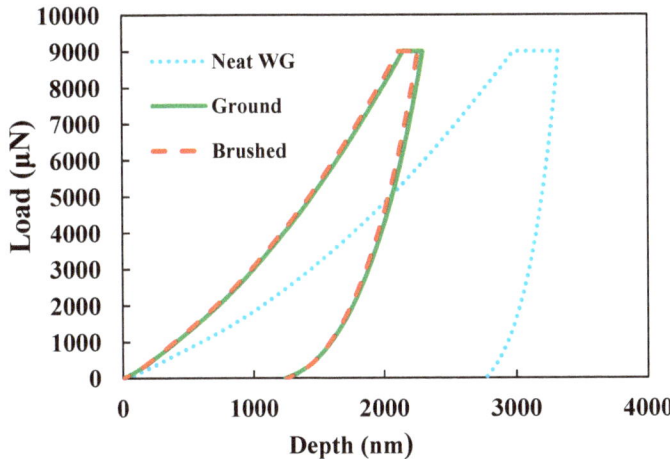

Figure 5. Load vs. displacement curves measured top-down.

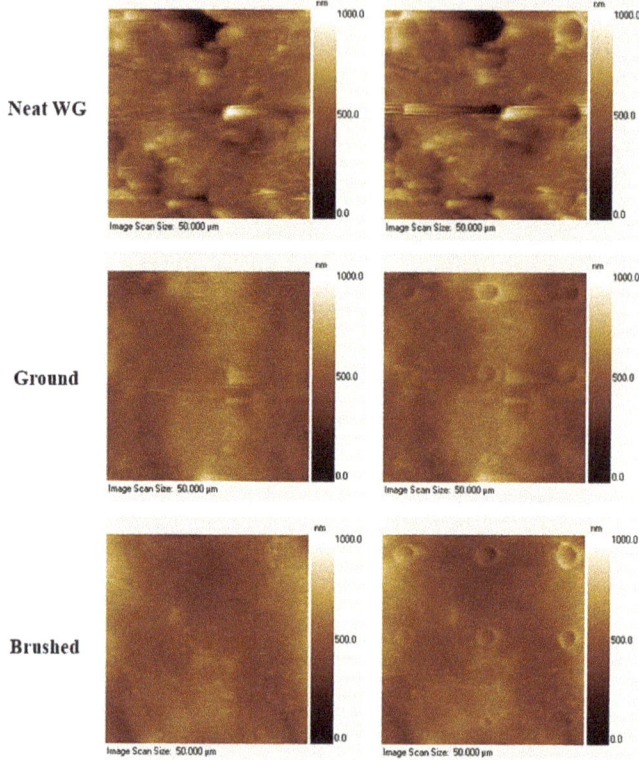

Figure 6. Nanoindentation regions (before and after indentation) measured top-down. The width of 50 μm in the images are the lateral dimension (x-y axes). The color bar to the right of each image represents the height of the sample surfaces.

To comprehend the mechanical properties of the interface between the PET and WG, the samples were also analyzed by nanoindentation in the cross-section. The nanomechanical properties of the interfacial region are critical to gauge the nature of the adhesion of the laminations. The nanoindentation regions of the cross-sectional specimens are displayed in Figure 7 (optical image). Figure 8 shows that the hardness of the area containing the interface of the Ground sample (0.18 GPa) was similar to that of the neat WG (0.19 GPa) and the PET (0.21 GPa). The area containing the interface of the Brushed sample had a significantly lower hardness (0.07 GPa) than the corresponding area of the Ground sample. The reduced modulus (Figure 9) was also significantly lower in the area containing the interface of the Brushed sample. Hence, the interface of the Brushed sample was both weaker and less stiff than in the Ground sample. This suggests that the adhesion between the PET layer and WG was stronger in the Ground sample when compared to the Brushed sample. This can be attributed to the method of PET application that created a higher cross-link density. In the Brushed sample, the PET was in direct contact with the diamine, which had an aminolytic effect, resulting in the loss of PET structural integrity (see Sections 3.4 and 3.5).

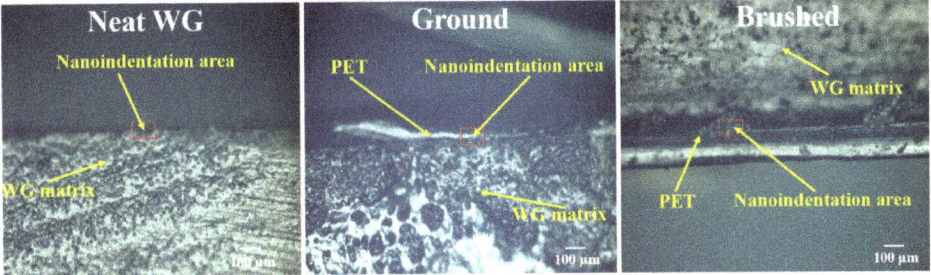

Figure 7. Nanoindentation regions of the cross-sectional samples.

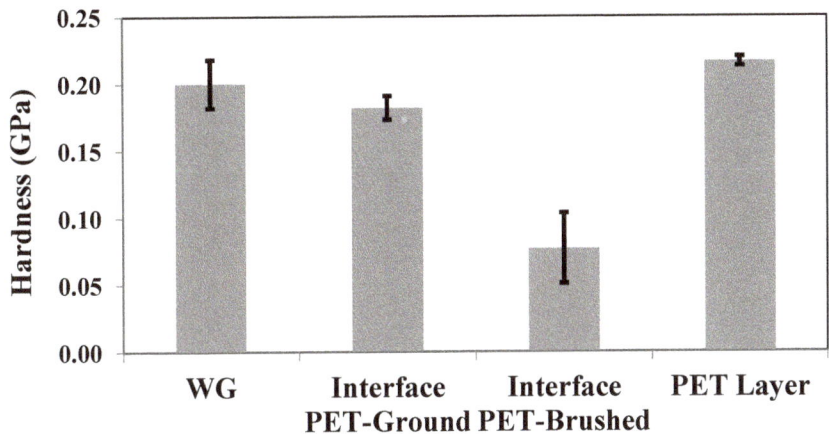

Figure 8. Nanoindentation hardness of the cross-sectional samples.

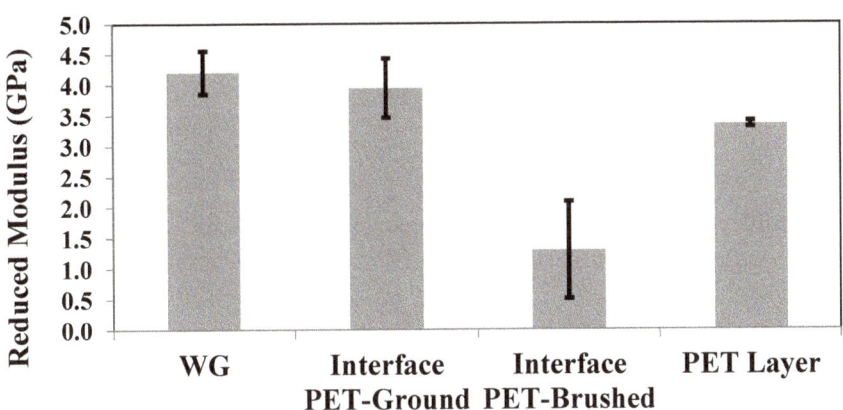

Figure 9. Nanoindentation modulus (reduced) of the cross-sectional samples.

3.2. Resistance Towards Water and Its Vapor

3.2.1. WVTR Tests

The water mass loss curves during the permeation measurement are given in Figure 10. The initial slow loss of water in the case of the WG film was due to an extensive uptake of water vapor in the film. Figure 11 shows how the initially flat film expanded and deformed during the test. The WVTR of the WG film (1 mm thick) was 79.1 ± 10.7 g·m^{-2}·day^{-1} (calculated between the third and fourth day). The WVTR of the neat PET film was, as expected, significantly lower: 11.2 ± 0.6 g·m^{-2}·day^{-1}. The Ground sample had a WVTR of 15.6 ± 4.2 g·m^{-2}·day^{-1} that was similar to the neat PET film value (within the standard deviation). The measured WVTR of the brushed sample was, however, higher (23.3 ± 4.9 g·m^{-2}·day^{-1}). The water vapor transmissions normalized to (WVTR values × total PET thickness) the actual PET thickness were 2800 ± 150, 3060 ± 820, and 4570 ± 960 g·µm/(m^2·day) for the neat PET, Ground, and Brushed samples, respectively. It was noted that there was a reduction in the total thickness (on both sides) of the PET (resulting thickness: (98 ± 5) × 2 = 196 µm) in the pressed material when compared to that of the neat PET (250 µm). The somewhat higher transmission rate of the Ground sample could be attributed to the change in the layer thickness. The changes in PET thickness were probably not the cause for the higher WVTR of the Brushed sample. It is more likely that the increase in WVTR was due to the migration of the diamine into the PET under the hot pressing. Not all diamine reacted with the PET (or the WG) and this low molar mass component may serve as an internal plasticizer of the PET component. In addition, the diamine increased the hydrophilicity of the PET layer. The migration was obviously higher for the Brushed samples as all the diamine was already available at the WG–PET interface, whereas it needed to migrate from the matrix to the PET in the Ground samples. Nevertheless, the Ground sample showed an almost 80% reduction in WVTR whereas the Brushed sample showed a reduction of 70% (compared to neat WG).

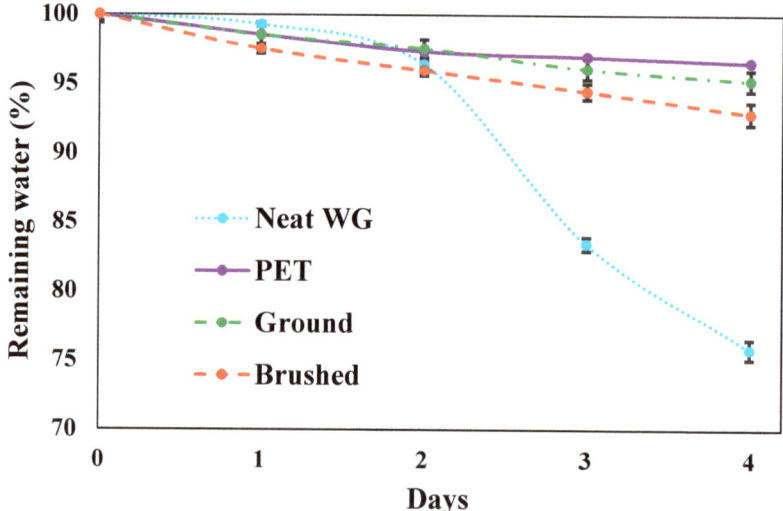

Figure 10. Water content inside the cup as a function of time.

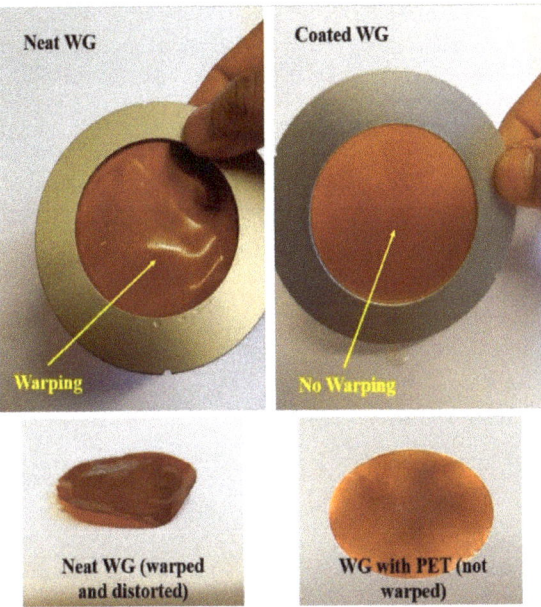

Figure 11. State of the neat and PET-layered WG film when exposed to saturated water vapor (100% RH) on the inside of the cup and 50% RH on the outside.

3.2.2. Contact Angle

The low contact angle at the outer PET-layered surface of the Brushed plate (35° ± 4°) verified that the polar and hygroscopic diamine had penetrated the entire PET layer (Figure 12). The PET film had a contact angle of 90° ± 1.2° and the more polar WG film had a value of 64° ± 1°. The value of the PET-layer on the Ground sample was 68° ± 2.5°. A possible explanation for this lower value compared to the neat (un-pressed) PET film may be because the PET film surface becomes slightly oxidized/hydrolyzed during the pressing operation, however, on a level small enough not to be observed by IR spectroscopy (see below).

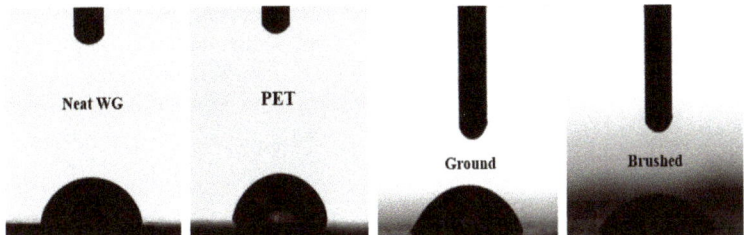

Figure 12. Water droplets on the samples during the goniometer test.

3.3. Infrared Spectroscopy

Figure 13 shows that the infrared (IR) spectrum of the Ground sample (measured top-down) was similar to that of the neat PET, showing that no diamine was present in the PET outer surface. The relative size of the PET ester peak (1712 cm^{-1}) to that of the reference peak (ca. 1410 cm^{-1}, phenylene-ring vibration) [33] was also the same (3.3) as that of un-pressed (neat) PET, indicating that the oxidation or hydrolysis of the PET layer was low enough not to be observed in the IR spectrum.

On the other hand, the spectrum of the PET layer of the Brushed sample showed a more pronounced absorbance in the 2800–3000 cm^{-1} region than that of neat PET due to the presence of the diamine. The 1712/1410 peak ratio was also lower (2.7) than those in the un-pressed PET and Ground samples, indicating a reaction between the diamine and PET.

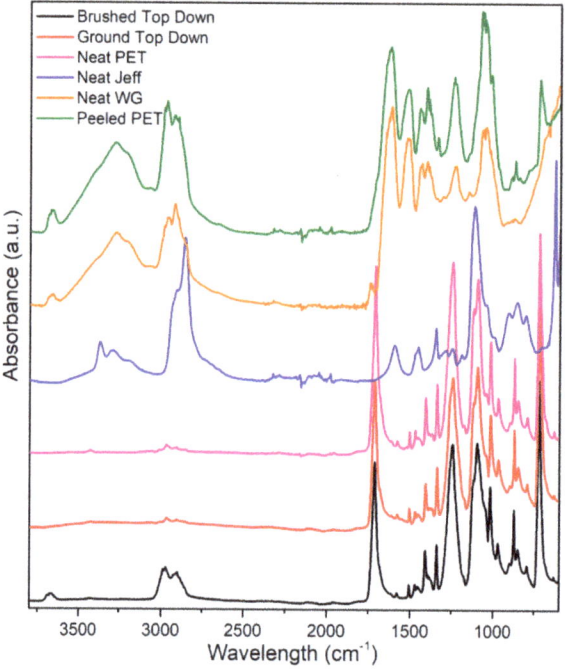

Figure 13. FT-IR spectra of the different neat components and the PET laminated WG.

In order to assess the strength of the bonding between the PET and WG, an attempt was made to peel off the PET layer from the WG and to run IR on the side of the layer that had faced the WG material. Figure 13 reveals that the inner surface (i.e., Peeled PET) showed clear features of WG (observed by the large amide I peak at 1700–1580 cm^{-1}). Additionally the NH$_2$/OH regions in the peeled PET was higher than that of the neat PET. This shows that during the peeling, the fracture went mainly through the WG material and not the PET/WG interface, which indicated a good bond between the PET and WG. Note also that this was the laminated WG with a weakest interface, i.e., the Brushed sample (weaker than that of the Ground sample).

3.4. Microscopy (SEM)

Figure 14 shows the cross-sections of the outer regions of the tensile fractured surfaces of the PET layered WG materials. Notable is the "intact" interface between the WG and the PET layer in the Ground sample. In the Brushed sample, the cross-section was rougher than that of the Ground sample, indicating a less brittle sample in accordance with the tensile properties given in Table 1. Contrary to the Ground sample, a damaged interface could be seen in the Brushed sample. This observation corroborates the results of the cross-sectional nanoindentation, contact angle, and WVTR tests. Due to the better interfacial bonding of the PET to the WG polymer in the Ground sample, it had higher interfacial hardness and better water vapor barrier properties. Moreover, the PET layer in the Ground sample seemed to retain its structural integrity during fracture whereas the same in the Brushed sample was lacking. The PET layer in the latter exhibited extensive cracks. This was the result of the direct contact of the PET layer with the diamine, which caused aminolysis of the PET (see Section 3.5). Consequently, the Brushed sample had inferior barrier properties when compared to the Ground sample.

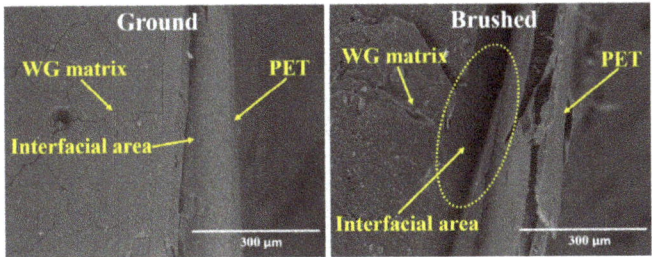

Figure 14. SEM micrographs of the tensile fractured surfaces.

3.5. Mechanisms of the Effects of the Diamine

Figure 15 illustrates the aminolysis of PET using a diamine [34], where the reaction led to a decrease in the PET molar mass. However, the amine terminated PET chain could react further with a second component (WG or another PET chain). Based on the IR data, it was observed that the ester groups of the PET reacted with the amine (decrease in the ester peak in the PET component of the Brushed sample). The good bond between WG and PET indicated that the un-reacted amine on the "aminated" PET reacted with WG at the interface. The presence of the diamine is a prerequisite for any bonding to occur at all between PET and WG. However, at high temperature, depolymerization involving amide containing polymers occurs in the presence of diamine [27]. Hence, the diamine reacts with the main chain protein amide causing a cleavage of the protein chain. Cross-linking reactions with the diamine as a coupling agent may also occur between the protein chains, however, the cleavage mechanism seems to dominate.

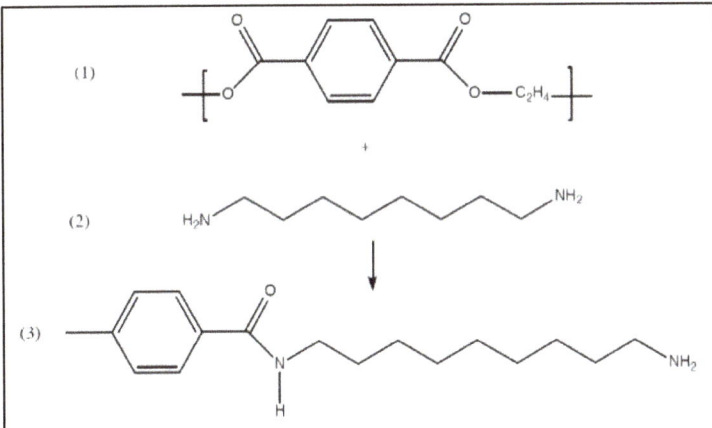

Figure 15. Aminolysis of PET.

4. Conclusions

Through the current study, an innovative step has been taken to apply a continuous, strong, and transparent layer aided by a coupling agent on WG plastics to impart water resistance. The diamine cross-linker (i.e., Jeffamine) that was used created a strong bond between the WG polymer and the

PET layer. The two methods used to laminate the protein plastic with PET had different advantages and drawbacks. In the Ground method, the strongest WG/PET interface was obtained but the protein was to some extent negatively affected (lower strength). The lower strength was due to the bulk WG material and not related to the adhesion with the PET layer. No delamination occurred between the WG and the PET layer in the Ground or the Brushed samples. In the Brushed method, the protein was less affected, but the interfacial strength and the hydrophobicity of the PET layer, containing migrated diamine, was lower, also leading to a somewhat poorer water vapor transmission rate. Nevertheless, both methods of lamination with PET led to significantly higher water vapor barrier properties than that of neat WG. The PET layered samples also remained undistorted during the water exposure. One of the biggest issues with WG is its dimensional instability under high humidity conditions. The PET layers, having a strong adherence to the WG, eliminated this issue. Still, there is room for improvement regarding the negative side-effects of the diamine and future work should include the optimization of the content of the added diamine as well as the processing conditions.

Author Contributions: Conceptualization, O.D. and M.S.H.; Methodology, O.D. and M.S.H.; Formal Analysis, O.D., T.A.L., and A.J.C.; Investigation, O.D. and I.L.; Data Curation, O.D., T.A.L., A.J.C., and M.S.H.; Writing–Original Draft Preparation, O.D.; Writing–Review & Editing, O.D., T.A.L., A.J.C., and M.S.H.; Supervision, O.D. and M.S.H.; Project Administration, M.S.H.; Funding Acquisition, M.S.H.

Funding: This research was funded by the Lantmännen Research Foundation (No. 2016H010).

Acknowledgments: The authors are grateful to the Lantmännen Research Foundation for providing the funding for this project. We also thank the University of Auckland, New Zealand for allowing us to use their nanoindentation instrument for this research.

Conflicts of Interest: The authors declare no conflict of interest.

References

1. Das, O.; Hedenqvist, M.S. Self-Reinforced Gluten Polymers: A Step towards a True Biocomposite. *Open Access Government*, May 2018; 178.
2. Irissin-Mangata, J.; Bauduin, G.; Boutevi, B. Bilayer films composed of wheat gluten film and UV-cured coating: Water vapor permeability and other functional properties. *Polym. Bull.* **2000**, *44*, 409–416. [CrossRef]
3. Capezza, A.J. *Novel Superabsorbent Materials Obtained from Plant Proteins—A Report*; Department of Plant Breeding, Swedish University of Agricultural Sciences: Alnarp, Sweden, 2017; pp. 1–54.
4. Cho, S.W.; Blomfeldt, T.O.; Halonen, H.; Gällstedt, M.; Hedenqvist, M.S. Wheat gluten-laminated paperboard with improved moisture barrier properties: A new concept using a plasticizer (Glycerol) containing a hydrophobic component (Oleic Acid). *Int. J. Polym. Sci.* **2012**, *2012*, 454359. [CrossRef]
5. Gennadios, A.; Weller, C.L.; Testin, R.F. Modification of physical and barrier properties of edible wheat gluten-based films. *Cereal Chem.* **1993**, *70*, 426–429.
6. Guilbert, S.; Seow, C.C.; Teng, T.T.; Quah, C.H. *Food Preservation and Moisture Control*; Elsevier Applied Science: London, UK, 1988; pp. 199–219.
7. Zhang, H.; Mittal, G. Biodegradable protein-based films from plant resources: A review. *Environ. Prog. Sustain. Energy* **2010**, *29*, 203–220. [CrossRef]
8. Gontard, N.; Marchessau, S.; Cuq, J.L.; Guilbert, S. Water vapour permeability of edible bilayer films of wheat gluten and lipids. *Int. J. Food Sci. Technol.* **1995**, *30*, 49–56. [CrossRef]
9. Hager, A.S.; Vallons, K.J.; Arendt, E.K. Influence of gallic acid and tannic acid on the mechanical and barrier properties of wheat gluten films. *J. Agric. Food Chem.* **2012**, *60*, 6157–6163. [CrossRef] [PubMed]
10. Hernández-Muñoz, P.; Villalobos, R.; Chiralt, A. Effect of cross-linking using aldehydes on properties of glutenin-rich films. *Food Hydrocoll.* **2004**, *18*, 403–411. [CrossRef]
11. Gennadios, A.; Weller, C.L.; Testin, R.F. Property modification of edible wheat, gluten-based films. *Trans. ASAE* **1993**, *36*, 465–470. [CrossRef]
12. Cho, S.W.; Gällstedt, M.; Hedenqvist, M.S. Properties of wheat gluten/poly (lactic acid) laminates. *J. Agric. Food Chem.* **2010**, *58*, 7344–7350. [CrossRef] [PubMed]
13. Irissin-Mangata, J.; Boutevin, B.; Bauduin, G. Bilayer films composed of wheat gluten and functionalized polyethylene: Permeability and other physical properties. *Polym. Bull.* **1999**, *43*, 441–448. [CrossRef]

14. Gontard, N.; Duchez, C.; Cuq, J.L.; Guilbert, S. Edible composite films of wheat gluten and lipids: Water vapour permeability and other physical properties. *Int. J. Food Sci. Technol.* **1994**, *29*, 39–50. [CrossRef]
15. Fabra, M.J.; Lopez-Rubio, A.; Lagaron, J.M. Effect of the film-processing conditions, relative humidity and ageing on wheat gluten films coated with electrospun polyhydryalkanoate. *Food Hydrocoll.* **2015**, *44*, 292–299. [CrossRef]
16. Micard, V.; Belamri, R.; Morel, M.H.; Guilbert, S. Properties of chemically and physically treated wheat gluten films. *J. Agric. Food Chem.* **2000**, *48*, 2948–2953. [CrossRef] [PubMed]
17. Iwata, T. Biodegradable and bio-based polymers: Future prospects of eco-friendly plastics. *Angew. Chem. Int. Ed.* **2015**, *54*, 3210–3215. [CrossRef] [PubMed]
18. Alaerts, L.; Augustinus, M.; Van Acker, K. Impact of Bio-Based Plastics on Current Recycling of Plastics. *Sustainability* **2018**, *10*, 1487. [CrossRef]
19. Wu, Q.; Andersson, R.L.; Holgate, T.; Johansson, E.; Gedde, U.W.; Olsson, R.T.; Hedenqvist, M.S. Highly porous flame-retardant and sustainable biofoams based on wheat gluten and in situ polymerized silica. *J. Mater. Chem. A* **2014**, *2*, 20996–21009. [CrossRef]
20. Wu, Q.; Yu, S.; Kollert, M.; Mtimet, M.; Roth, S.V.; Gedde, U.W.; Johansson, E.; Olsson, R.T.; Hedenqvist, M.S. Highly absorbing antimicrobial biofoams based on wheat gluten and its biohybrids. *ACS Sustain. Chem. Eng.* **2016**, *4*, 2395–2404. [CrossRef]
21. *ASTM D638-14 Standard Test Method for Tensile Properties of Plastics*; ASTM International: West Conshohocken, PA, USA, 2014.
22. Das, O.; Sarmah, A.K.; Bhattacharyya, D. Nanoindentation assisted analysis of biochar added biocomposites. *Compos. Part B Eng.* **2016**, *91*, 219–227. [CrossRef]
23. *ASTM D1653-13 Standard Test Methods for Water Vapor Transmission of Organic Coating Films*; ASTM International: West Conshohocken, PA, USA, 2013.
24. Das, O.; Bhattacharyya, D.; Hui, D.; Lau, K.T. Mechanical and flammability characterisations of biochar/polypropylene biocomposites. *Compos. Part B Eng.* **2016**, *106*, 120–128. [CrossRef]
25. Ali, Y.; Ghorpade, V.M.; Hanna, M.A. Properties of thermally-treated wheat gluten films. *Ind. Crop Prod.* **1997**, *6*, 177–184. [CrossRef]
26. Jansens, K. Gluten Cross-Linking and its Importance for the Mechanical Properties of Rigid Wheat Gluten Bioplastics. Ph.D. Thesis, KU Leuven, Leuven, Belgium, February 2013.
27. Hendrix, J.A.; Booij, M.; Frentzen, Y.H. Depolymerization of Polyamides. U.S. Patent 5,668,277, 16 September 1997.
28. Young's Modulus of Elasticity for Metals and Alloys. Available online: https://www.engineeringtoolbox.com/young-modulus-d_773.html (accessed on 14 September 2018).
29. Day, L.; Augustin, M.A.; Batey, I.L.; Wrigley, C.W. Wheat-gluten uses and industry needs. *Trend Food Sci. Technol.* **2006**, *17*, 82–90. [CrossRef]
30. Das, O.; Sarmah, A.K.; Bhattacharyya, D. Structure–mechanics property relationship of waste derived biochars. *Sci. Total Environ.* **2015**, *538*, 611–620. [CrossRef] [PubMed]
31. Das, O.; Bhattacharyya, D. Development of Polymeric Biocomposites: Particulate Incorporation, Interphase Generation and Evaluation by Nanoindentation. In *Interface/Interphase in Polymer Nanocomposites*; Netravali, A.N., Mittal, K.L., Eds.; John Wiley & Sons: Hoboken, NJ, USA, 2016; pp. 333–374.
32. Ni, W.; Cheng, Y.T.; Lukitsch, M.J.; Weiner, A.M.; Lev, L.C.; Grummon, D.S. Effects of the ratio of hardness to Young's modulus on the friction and wear behavior of bilayer coatings. *Appl. Phys. Lett.* **2004**, *85*, 4028–4030. [CrossRef]
33. Rusu, E.; Drobota, M.; Barboiu, V. Structural investigations of amines treated polyester thin films by FTIR-ATR spectroscopy. *J. Optoelectron. Adv. Mater.* **2008**, *10*, 377–381.
34. Noel, S.; Liberelle, B.; Robitaille, L.; De Crescenzo, G. Quantification of primary amine groups available for subsequent biofunctionalization of polymer surfaces. *Bioconj. Chem.* **2011**, *22*, 1690–1699. [CrossRef] [PubMed]

© 2018 by the authors. Licensee MDPI, Basel, Switzerland. This article is an open access article distributed under the terms and conditions of the Creative Commons Attribution (CC BY) license (http://creativecommons.org/licenses/by/4.0/).

Article

Raman Microscopy for Classification and Chemical Surface Mapping of Barrier Coatings on Paper with Oil-Filled Organic Nanoparticles

Pieter Samyn

Applied and Analytical Chemistry, Institute for Materials Research (IMO-IMOMEC), University of Hasselt, B-3590 Diepenbeek, Belgium; pieter.samyn@uhasselt.be; Tel.: +32-11-26-8594

Received: 11 February 2018; Accepted: 20 March 2018; Published: 24 April 2018

Abstract: The creation of functional papers requires a specific deposition of chemical moieties at the surface. In particular, water-repellent barrier coatings can be formed by the deposition of (poly(styrene-co-maleimide) nanoparticles filled with different vegetable oils. The analysis of coated paper surfaces by dispersive Raman spectroscopy allows for statistical classification of different coating types and chemical mapping of the lateral surface distribution of the coating components. The Raman spectra were used to quantify the amount of free oil and imide content. The partial least squares model with three principal components (PC) could differentiate between the type of oil (degree of saturation in PC-1), coating thickness (cellulose bands of paper substrate in PC-2), and organic coating phase (styrene, imide in PC-3). The chemical surface maps with average intensities indicate coating inhomogeneities for thin coatings located near the organic coating components, while the presence of free oil acts as a natural binder in between the organic phase and provides a more homogeneous coating. Depending on the type of oil, a higher amount of free oil coincides with lower imide content at the surface. The surface coverage of polyunsaturated oils overlaps relatively well with the areas of organic coating components, as the oil is largely encapsulated. The surface coverage for mono- and unsaturated oils is rather complementary to the organic phase as there are larger amounts of free oil. The latter is confirmed by single wavenumber maps and image processing constructing composite chemical surface maps.

Keywords: paper; coating; surface; Raman; microscopy; mapping

---

1. Introduction

The surface properties of paper are manipulated by the deposition of a coating layer containing an organic binder and different organic or inorganic fillers. In particular, the barrier properties and water repellence of cellulose substrates can be tuned by altering the chemical composition and topography of the surface. As cellulose materials are highly hydrophilic in nature, the protection against water is a primary requirement for controlling the barrier properties of packaging papers. The traditional methods of internal sizing and surface sizing provide a first barrier against water, but often cannot meet the requirements for modern packaging applications. Therefore, huge progress in nanoparticle deposition and functionalized additives was made in the last decade, as reviewed elsewhere [1]. The nanoparticles can be applied as local deposits on single cellulose fibers or as fillers in a paper coating to create various functionalities [2]. The metallic nanoparticles, e.g., silica [3] or titania [4], have been sprayed as a thin top-coating layer to improve the hydrophobicity. A simple, low-cost method involves the dispersion of silica nanoparticles in a silane/siloxane solution for the treatment of different paper grades by brush [5]. The fluorinated silica nanoparticles have also been deposited through one-step spray-coating, providing mechanical robustness together with self-cleaning and anti-icing [6]. Driven by environmental concerns, fluorine derivatives have been replaced by alternative materials with hydrophobic properties. Organic nanoparticles with surface-modified nanocellulose have been incorporated into coating formulations [7]. Other types of organics such as vegetable oils may also possess good potential to serve as a hydrophobic coating [8]. As an advantage of using nanoparticles instead of a polymer/oil mixture as protective barrier coating, the encapsulation of oils into an organic carrier material can provide better protection of the oil against oxidative degradation [9]. Moreover, the local presentation at the surface of functional chemical groups and hydrophobic moieties can be better controlled using nanoparticles. The high surface area of the nanoparticles allows us to increase the efficiency in presentation of the hydrophobic groups at the surface and the deposition of the nanoparticles in a top coating layer prevents the flow of the hydrophobic oil towards the bulk of the porous paper structure. In addition, nanoscale changes in the surface morphology contribute to the creation of a surface structure with hierarchical nano- to microscale roughness. However, the local deposition and lateral distribution of hydrophobic moieties at the surface are the most important parameter to create efficient barrier properties.

Raman spectroscopy is an appropriate analytical tool for paper analysis as it can be applied to water-based systems and water-sensitive materials such as cellulose fibers, where the water is strongly absorbing in the IR region but weak in Raman scattering. The confocal Raman imaging allows us to study the distribution of chemical compounds of paper coatings in the sub-micrometer range with adequate sensibility, spatial and depth resolution to differentiate between surface features and the substrate [10]. The local organization of a styrene–butadiene latex and calcium carbonate fillers in a paper coating has been observed, while information about the ink density in a printed top layer could be obtained simultaneously [11,12]. By appropriate selection of objectives and measuring times, the depth profiles with variations in latex content in z-direction of a coating were determined [13]. In particular, the migration of various pigments in the paper coating was followed [14]. Different paper coating fillers such as ground calcium carbonate (GCC) and precipitated calcium carbonate (PCC) could be differentiated and the spatial distribution of coatings containing a mixture of both components was monitored [15]. However, the technique was less effective at differentiating between minor coating components such as rheology modifiers and dispersants [16]. The other pigments such as calcium carbonate, talcum, gypsum, titanium dioxide (rutile/anatase crystalline forms), optical brighteners, and kaolinite could be detected [17]. The kaolinite has characteristic Raman bands, but they are often masked by fluorescence due to impurities in the sample [18]. Other disturbing effects for paper coatings include non-uniformities in geometry and morphology of the substrate, causing scattering effects. These effects can hinder the quantification of the coating components and calibration curves, baseline correction, or band intensity ratio calculations are needed to determine the relative amounts of each component. The more advanced systems with microcapsules in polymeric coatings have been more recently investigated by confocal Raman microscopy [19].

In this paper, the surface properties of a hydrophobic barrier coating on paper with poly(styrene-co-maleimide) nanoparticles and encapsulated vegetable oils (SMI/oil) are monitored as an example coating system by chemical mapping using Raman spectroscopy. The suitability of SMI/oil nanoparticles as a waterborne coating formulation has been proven by studying their rheological properties [20]: a comparison of pure SMI and SMI/oil dispersions (35 wt %) illustrates that the presence of oil induces viscoelastic behaviour and decreases the viscosity by an order of magnitude. The rotational rheometry shows good stability and reproducibility with no hysteresis for SMI/oil dispersions, in contrast to pure SMI dispersions. In particular, the present study focuses on an analytical method of describing the surface chemistry of the SMI/oil coatings. The understanding of in-plane lateral distribution (x, y) of the coating moieties (styrene, imide, oil) in relation to the type of selected oil type is important to gain better control over the surface properties and the presence of specific functional groups. First, Raman spectroscopy is used for statistical classification of coatings with different oil types by principal component analysis. Second, the distribution of different chemical moieties is visualized and shows various surface coverage depending on the degree of saturation and reactivity of the oil.

2. Materials and Methods

2.1. Materials for Paper Coating and Coating Application

An aqueous dispersion of hybrid organic nanoparticles was obtained by the imidization reaction of a given high-molecular-weight poly(styrene-co-maleic anhydride) or SMA copolymer (molecular weight M_w = 80.000 mol, 26 mol % maleic anhydride), in the presence of ammonium hydroxide (NH_4OH) and different types of vegetable oil. The resulting hybrid organic nanoparticles of poly(styrene-co-maleimide) or SMI with oil (SMI/oil) were obtained in combination with soy oil (SO), corn oil (CO), rapeseed oil (RO), sunflower oil (SfO), castor oil (CaO), and hydrogenated castor oil (HCO). More details about the reaction conditions and a full characterization of the aqueous nanoparticle dispersions as well as the dried nanoparticles can be found elsewhere [21]. Some characteristics of the oil qualities and SMI/oil nanoparticle dispersions are summarized in Table 1: the solid content (S.C.) was determined by infrared drying and weighing (LP16, Mettler Toledo, Greifensee, Switzerland); the viscosity was measured with a portable viscosity meter (Brookfield, DV-II Pro, Brookfield Engineering Laboratories, Middleboro, MA, USA), using a spindle n°5 at rotation speed of 100 rpm, and the particle sizes were measured by dynamic light scattering (Nanosizer ZS90, Malvern, Malvern, UK).

Table 1. Characteristics of the oils and SMI/oil nanoparticle dispersions for paper coatings.

Vegetable Oils			SMI/Oil Nanoparticle Dispersions				
Oil Type	Iodine Value g(I$_2$)/100 g	Coating Type	pH	S.C. (%)	Viscosity (cp)	z-Average Particle Size, Diameter (nm)	Poly-Dispersity
Soy	130	SMI/SO	5.48	49.8	146	149	0.163
Corn	119	SMI/CO	5.44	49.9	102	143	0.161
Rapeseed	98	SMI/RO	5.43	49.5	91	156	0.137
Sunflower	85	SMI/SfO	5.38	48.8	134	143	0.176
Castor	81	SMI/CaO	5.72	49.9	230	148	0.137
Hydrogenated Castor	5	SMI/HCO	5.54	49.4	116	132	0.152

The SMI/oil nanoparticle coatings were deposited onto a paper substrate with a laboratory-scale K303 Multi-coater (RK Print Coat Instruments Ltd., Royston, UK). Two different metering bars were used at a constant speed of 6 mm/s, resulting in a thin coating (dry weight 4.0 ± 0.2 g/m^2) and a thick coating (dry weight 6.0 ± 0.2 g/m^2). The base paper sheets (100 g/m^2, thickness 125 µm, Mondi Business Paper, Vienna, Austria) contained bleached long-fiber and short-fiber kraft pulp with internal sizing and calendering. All coated paper samples were immediately dried in a circulating hot-air oven for 2 min at 120 °C, and stored in a controlled environment (23 °C, 60% RH) until further use. The characterization was done on the as-deposited coatings without any further processing (no calendering) to intentionally keep the specific micro- to nanoscale coating morphology.

2.2. Paper Coating Characterization

The scanning electron microscopy (SEM) was performed on a Hitachi Tabletop TM3000 microscope (Manufacturer, Krefeld, Germany) at different magnifications (800×, 2000×) and 15 kV voltage. An ultra-high-resolution image was obtained by FEG-SEM analysis using a FEI Nova 600 NanoLab focused ion beam workstation (Manufacturer, Hillsboro, OR, USA).

The confocal micro-Raman microscopy was performed on a dispersive Perkin Elmer Raman Flex 400 equipment (Manufacturer, Rodgau, Germany) with a multichannel charge-coupled array detector (CCD). The measuring conditions were optimized to avoid fluorescence and provide a good signal-to-noise ratio, using a of diode NIR laser (785 nm) with a maximum power output of 100 mW at the head and selecting a 40 mW measured laser power output at the sample position. The laser light was coupled to an optical microscope (Olympus BX51, Hamburg, Germany) equipped with a motorized piezoelectric x, y micro-Raman stage. The chemical Raman maps were recorded over a surface area of 5 × 5 mm² using an objective lens of 20× (numerical aperture NA = 0.40) and a pinhole size of 50 µm. With a refractive index n = 1.5 for the samples, a lateral resolution of 2 µm and depth resolution of 5 µm was obtained. The Raman spectra were recorded at 100 × 100 points with 0.05 mm interdistance, 5 s exposure time, and six exposures per point (i.e., a total measuring time of about 14 h per map). The spectra were recorded at 3200–200 cm^{-1} with 4 cm^{-1} spectral resolution. The data were processed with Spectrum 10 analysis software (Perkin Elmer, Waltham, MA, USA) and Spectrum Image software (Version R1.7, Perkin Elmer, Waltham, MA, USA) to plot surface maps with average intensities, band ratio intensities, or single wavenumber intensities after normalization and baseline correction. Further processing of the Raman surface maps has been performed using the Image J processing software program (version 1.32j).

The statistical analysis of Raman spectra was done with Unscrambler 10.1 software. The input variables for the model were the baseline-corrected Raman spectra over the full wavenumber region. A calibration model for the principal component analysis (PCA) was developed by including 10 spectra per coating type (totally 60 spectra), recorded at random places over the coated surface area. The calibration model was verified by a cross-validation procedure, leaving out one of the calibration samples from the regression model and performing a new model for the remaining calibration samples. This validation method is considered when the number of samples is too small to have an independent training and validation set. The model was externally validated with an independent dataset (new coated paper samples), including five spectra per coating type that were recorded on independent surface areas.

3. Results

3.1. Paper Coating Morphology

The SEM images of thick and thin SMI/oil nanoparticle coatings on paper are shown in Figure 1 at magnifications of 800× and 2000×. A reference SEM image of the uncoated paper at the same magnifications can be found in the Supplementary Materials (see Figure S1 in Supplementary Materials).

The features of the base paper with cellulose fibers and internal paper fillers remain visible on the surface of thin coatings. The thin coating mainly fills the pores between the fibers and forms only a small layer on top of the fibers. The thick coatings almost fully cover the paper substrate with a more continuous layer. As pure nanoparticle dispersions are used here (no coating binder was used in order to study the intrinsic nanoparticle coating properties), the drying effects of the colloidal suspensions cause some coating defects, resulting in local cracks. The development of cracks in the drying of colloidal suspensions can be attributed to the role of packing and flocculation of the particles [22]. Therefore, the rheological properties of the SMI/oil nanoparticle types have been studied before [20], and the reduction in the surface tension (and consequent influences on capillary forces between nanoparticles) for the SMI/oil nanoparticles in comparison with the SMI nanoparticles is particularly beneficial to reduce the existence of cracks. Also, the smaller nanoparticle sizes of SMI/oil nanoparticles (20–50 nm) compared to pure SMI nanoparticles (100 nm) enhance the packing and reduce the tendency for cracking. The drying of the nanoparticle suspensions on porous substrates may induce stresses that can lead to a fracture of the thin film. However, the presence of oil may improve the coating homogeneity and partially functions as a natural binder. The dry thickness of thick coatings is about 10 ± 2 µm, and the dry thickness of thin coatings is about 7 ± 2 µm. Further detail of the coating structure (e.g., SMI/HCO) and presence of nanoparticles is illustrated in the FEG-SEM images in Figure 2.

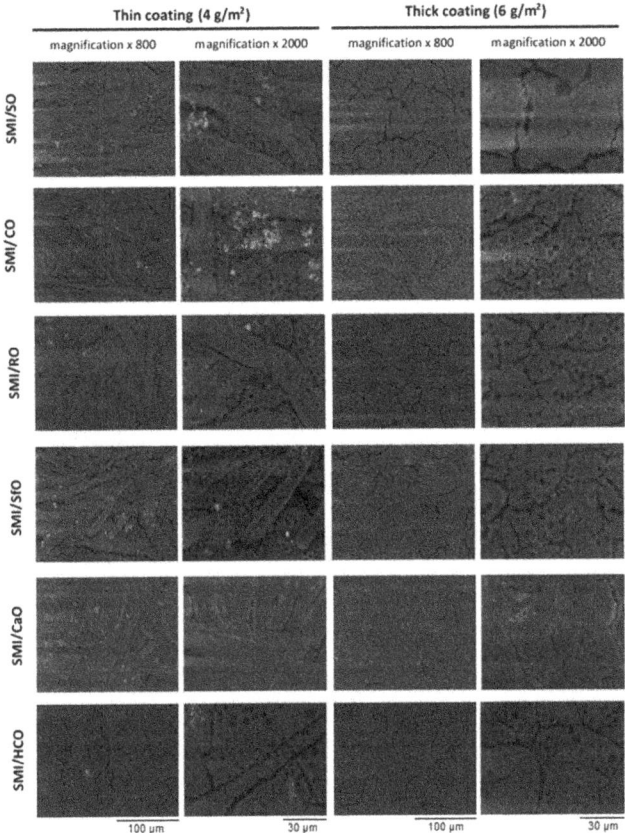

Figure 1. SEM microscopy of SMI/oil coatings on paper, including thin and thick coatings at different magnifications (indicated scale bars apply to full column).

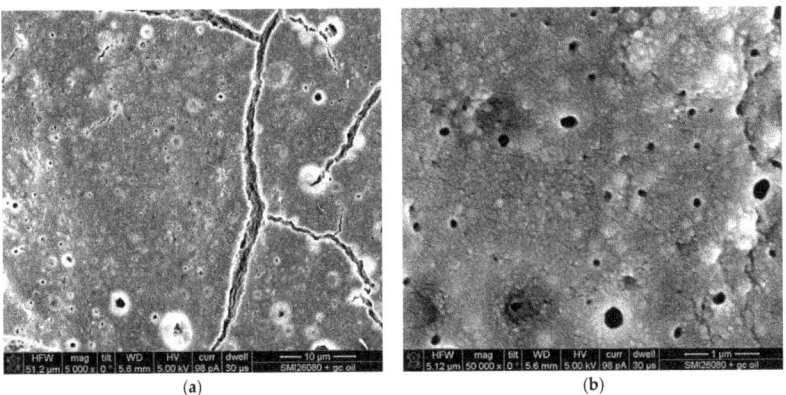

Figure 2. Detailed FEG-SEM microscopy of SMI/oil coating on paper (e.g., SMI/HCO): (**a**) magnification 5000×; (**b**) magnification 50000×.

3.2. Raman Spectroscopy and PCA Model

The fingerprint region of Raman spectra for different SMI/oil nanoparticle coatings on paper is illustrated in Figure 3 (see also Table 1 for band assignments in Supplementary Materials). For clarity of presentation, only the spectra of the thick coatings are represented, while the thin coatings give similar results within the fingerprint region (of course, with higher intensity of substrate-related cellulose bands).

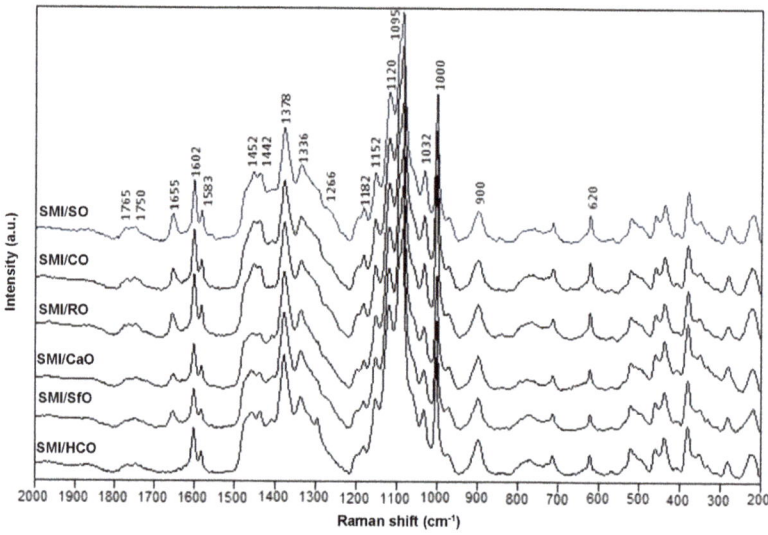

Figure 3. Raman spectra for SMI/oil coatings on paper.

The characteristic Raman bands for coating moieties (styrene, imide, and oil) and substrate (cellulose fibers) are clearly separated and allow for accurate characterization. The spectra have been normalized by integration of the region corresponding to the cellulose bands at 1170 to 1050 cm^{-1}. The C=O ester bands for imide (1765 cm^{-1}) and oil (1750 cm^{-1}) are clearly resolved as distinct peaks. The band at 1655 cm^{-1} represents cis conformation (C=C) of the fatty acids and naturally occurs in vegetable oils, in contrast with the trans-isomers at around 1680–1670 cm-$^{-1}$ in synthetic systems [23]. For fatty acids with the same chain length, the intensity at 1655 cm^{-1} increases in parallel with the degree of unsaturation [24], as follows: HCO (saturated fatty acids obtained after hydrogenation); SfO, CaO, RO (monounsaturated fatty acids); CO, SO (polyunsaturated fatty acids).

The imide content and amount of free oil can be quantified from the ratios of integrated Raman band areas of imide I (1765 cm^{-1}), oil (1655 cm^{-1}) and styrene (1602 cm^{-1}), after calibration with a fully imidized pure SMI coating (no oil), as calculated in Table 2. Based on the SMA grade with 26 mol % maleic anhydride, the calculated imide content should be compared to a theoretical maximum imide content (relative to styrene parts) of 35%. Depending on the oil type, the imide content is lower than the theoretical maximum, which can be explained by the interference of oil with the imidization reaction (reactive C=C oil) and consequently remaining ammonolyzed maleic anhydride. A progressively higher amount of unreacted or 'free' oil is present in the coating as the imide content lowers: the amount of free oil is low for polyunsaturated oil (SMI/SO, SMI/CO), intermediate for monounsaturated oils (SMI/RO, SMI/SfO, SMI/CaO), and extremely high for saturated oils (SMI/HCO).

Table 2. Quantification of SMI/oil coatings on paper based on Raman spectroscopy.

Coating Type	Imide Content/%	Free (Unreacted) Oil/%
SMI/SO	26.4	4
SMI/CO	23.2	5
SMI/RO	21.2	15
SMI/SfO	19.8	22
SMI/CaO	19.3	24
SMI/HCO	24.6	50

A model for the classification of the paper coatings based on principal component analysis (PCA) of the Raman spectra is presented in Figure 4. The analysis is used to define parameters that describe the variation in Raman spectra between the different coating types and illustrate major trends. The number of principal components (PC) included in the model is selected from a scree plot (Figure 4a). The variation in data points is most effectively described by a model with nearly zero residual variance and as few PCs as possible: the three principal components account for 97% of total variation (PC-1: 78%, PC-2: 12%, and PC-3: 7%), and the residual variance is very small for the higher-order components. The calibration model was evaluated by a leverage and residual analysis against PC-1, PC-2, and PC-3 to detect eventual outlier samples (Figure 4b). The samples with higher leverage are positioned further away from the mean value and have different characteristics: e.g., the leverage variation between the coating samples is somewhat larger for PC-1, while the leverage for PC-2 and PC-3 is almost the same. PC-1 consequently takes into account differences in oil type, discriminating between polyunsaturated, monounsaturated, and saturated oil. Only samples with high leverage and high residues would damage the model and should be removed from the model: as none of the samples simultaneously indicate high leverage and high residues, all samples fit into the model and each sample adds novel information to the model. The meaning of the different components in relation to the original Raman spectral bands can be recognized form the loading for individual PCs (Figure 4c): (i) the PC-1 (82% of the variation) includes specific characteristics related to the oil type (1655 cm^{-1}); (ii) the PC-2 (10% of the variation) includes characteristics of the substrate (cellulose: 1120–1095 cm^{-1}); and (iii) the PC-3 (4% of the variation) includes the characteristics of the organic phase (styrene: 1602, 1583, 1452, 1032, 1000, 620 cm^{-1}; imide: 1765, 1329 cm^{-1}). As such, single Raman bands make a major contribution to the individual PCs and can be used for statistical discrimination between the coatings. The two-dimensional score plots of individual PCs can be used for classification of the coated papers within the 95% confidence interval (Figure 4d). Test results of the graph include data for some representative samples from the calibration set (red points) and the validation set (blue points) with good agreement. From the score plot of PC-1 versus PC-2, the coated papers are classified according to the coating type in PC-1 and the coating thickness in PC-2. The discrimination of thick and thin coatings is related to the intensity of the Raman bands corresponding to the cellulose substrate, as reflected in PC-2. From the score plot PC-1 versus PC-3, the coated papers are classified according to the oil type, with polyunsaturated oils (SO, CO) having the highest PC values, monounsaturated oils (CaO, RO, SfO) having intermediate PC values, and the hydrogenated oils (HCO) having the lowest PC values. The latter classification is based on the relatively small variations in the organic composition of the coating (similar styrene content and slightly variable imide content), as reflected in PC-3.

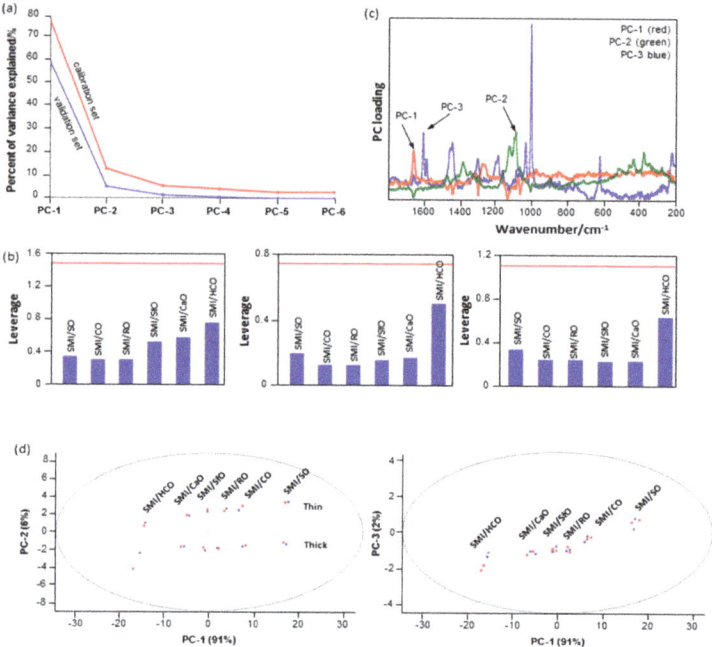

Figure 4. Statistical classification of SMI/oil paper coatings by statistical principal component analysis (PCA) of Raman spectra: (**a**) residual variance plots; (**b**) leverage plots; (**c**) PC loading plots; and (**d**) score plots with 95% confidence interval.

3.3. Raman Chemical Mapping (Average Intensities)

The chemical surface maps of coated papers (top view on 5 × 5 mm² surface area) with different SMI/oil compositions were recorded to study the surface coverage, homogeneity, and distribution of different chemical moieties within the coating layer.

The average intensity maps for thick and thin paper coatings are illustrated in Figure 5, with all maps having the same ordinate scale to make valid comparison between all SMI/oil coating types. The ordinate values are calculated as an average intensity value from the spectrum recorded at each (x, y) point. The surface maps consequently include data related to the intensities of the dominant Raman bands and illustrate the lateral homogeneity of the coating. As confirmed by the following more detailed analysis of single Raman bands, the ordinate intensities of average intensity maps can be related to the lateral distribution of organic coating moieties, with (i) high intensities (yellow to red) representing thick deposits of coating species and, (ii) low intensities (blue) corresponding to thin coating deposits and poor coverage. The surface areas with low average intensities consequently represent almost uncoated paper. For the thin coatings, inhomogeneities are recognized with striations parallel to the direction of the bar-coating. The average intensities are higher for thick coatings than for thin coatings as an indication for the surface coverage. The contrast in average intensities illustrate local heterogeneities: the coatings with high imide content and low amount of free oil are obviously most homogeneous, while the presence of a high amount of free oil is visualized as an island-like coating phase. The observations for average intensity maps can consequently be related to a combination of coating coverage and homogeneity, which depend on the imide content and amount of free oil. In order to confirm the significance of the average intensity maps and locations of oil, imide and styrene deposits, the features observed in the general surface maps are analyzed below in relation with specific Raman bands for the coating and paper substrate.

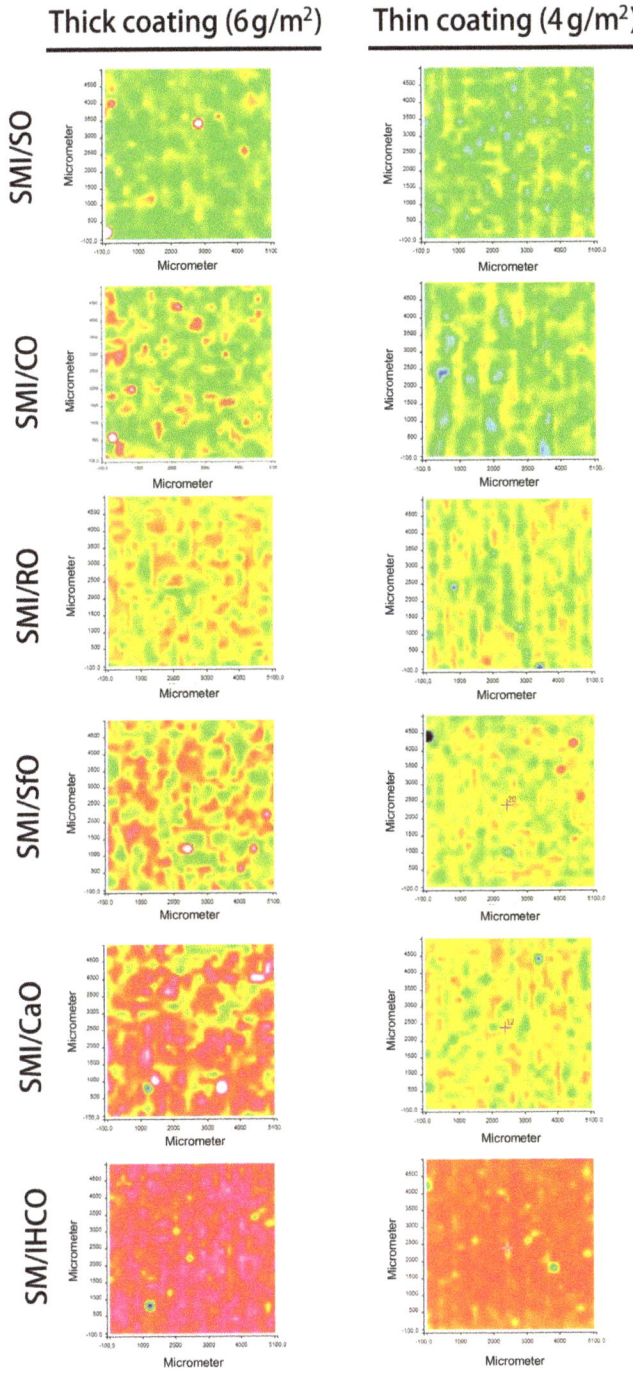

Figure 5. Raman maps (5 × 5 mm^2) with average intensities for SMI/oil paper coatings.

3.4. Raman Chemical Mapping (Band Ratios)

The intensity ratios of specific Raman bands related to the oil, styrene, imide, and cellulose components are further detailed to provide information on the local chemical composition and distribution of these elements over the paper coating.

The distribution of oil is evaluated from the intensity ratios of Raman bands for oil (1655 cm^{-1}), relative to styrene (1602 cm^{-1}) and cellulose (1095 cm^{-1}), as illustrated in Figure 6: the styrene band is chosen as a reference band for the distribution of oil within the coating, while the cellulose band is chosen as a reference for the distribution of coating over the paper substrate. The progressive increase in oil at the surface for different SMI/oil coatings agrees with previous calculations (see Table 2). The band intensity ratios and surface coverage of oil is comparable for thin and thick coatings when expressing the relative intensity of oil versus styrene within the coating itself: indeed, the same coating dispersions for deposition of thin and thick coatings were used. Contrarily, the differences in oil exposure between thin and thick coatings become clear when using the relative band intensity of oil versus cellulose. The surface patterns and locations covered with oil in the coating and those covered with oil are relative to the paper substrate overlap. Interestingly, the inhomogeneities for thin coatings observed in the average intensity maps are not reflected in the oil distribution over the surface. It is obvious that the inhomogeneities of the coating relative to the paper substrate are not expressed when taking the coating (styrene) as a reference. However, homogeneous distribution of the oil is also observed when taking the substrate (cellulose) as a reference. Therefore, the oil seems to be homogeneously spread over the surface and provides a continuous phase. The coating defects in thin coatings with striations can likely be attributed to the viscosity of the coating dispersions, where the presence of a free oil phase locally enhances the formation of a continuous coating due to better flow properties.

The distribution of organic coating moeities is evaluated from the intensity ratios of Raman bands for imide (1765 cm^{-1}), relative to styrene (1602 cm^{-1}) and cellulose (1095 cm^{-1}), as represented in Figure 7. The relative intensity of imide progressively decreases for different SMI/oil coatings, in agreement with previous calculations (see Table 2). The relative intensity of imide is comparable for thick and thin coatings when expressed against styrene, while the differences in imide between thick and thin coatings become clear relative to the paper substrate: the thick coatings are relatively homogeneous with even distribution of the imide, while thin coatings show striations at the same locations as observed in the average intensity maps. Therefore, the distribution of the organic imide phase plays a major role in the coating homogeneity. The imide phase is likely more rigid and viscous compared to the oil phase, with more risk of coating defects. There is good agreement in the surface maps of oil/cellulose versus imide/cellulose for the thin coatings, confirming that an amount of free oil is present between the imide phase and contributes to the formation of a homogeneous coating layer by 'filling' the non-imide zones as a binder. The complementarity of the oil and imide phases is visible for thin coatings with poly- and monounsaturated oils (e.g., SMI/CO, SMI/RO), where the calculated amounts of free oil are relatively small. The latter is less visible for thin coatings with polyunsaturated oils (e.g., SMI/SO) as there is only a limited amount of free oil, while it is partly visible for the coatings with unsaturated oil (e.g., SMI/HCO) as the oil coverage is much larger than the zones in between the imide phase.

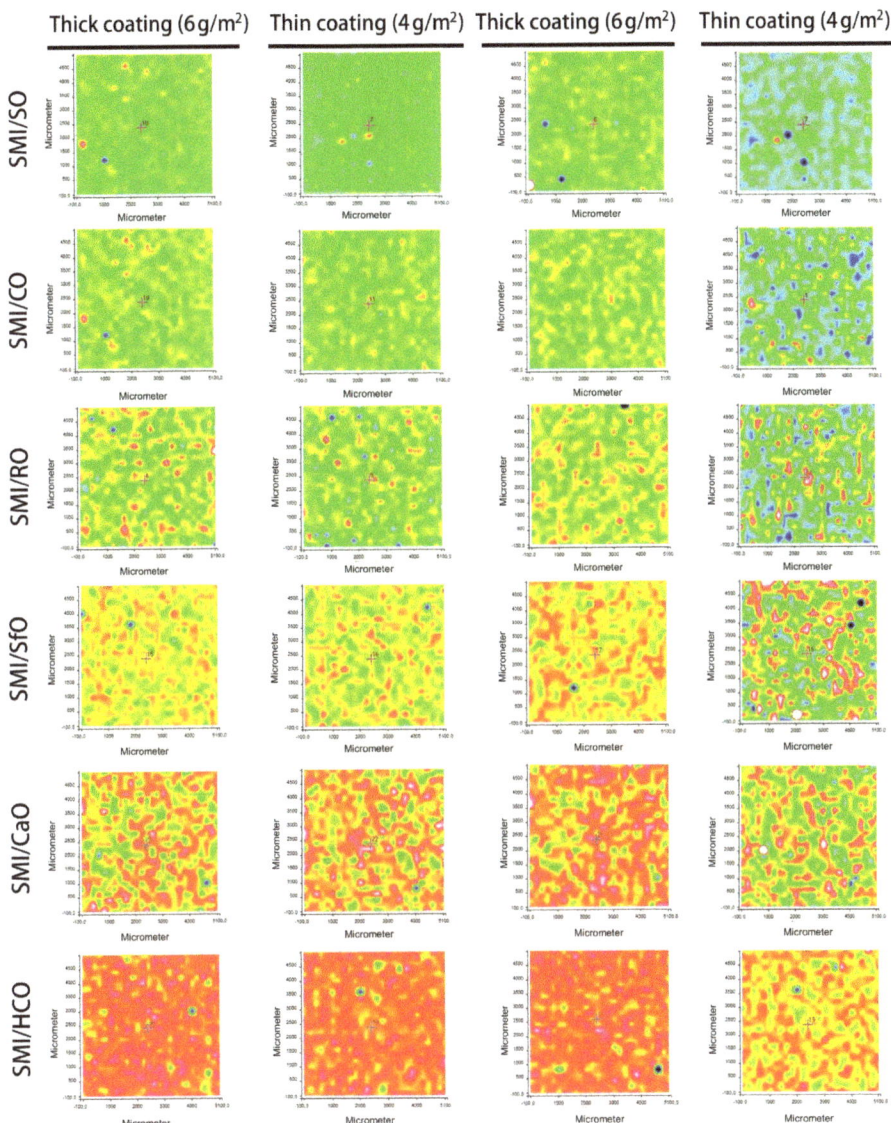

Figure 6. Raman maps (5 × 5 mm^2) with band intensity ratios illustrating exposure of oil at the surface for SMI/oil paper coatings (left two columns: oil/styrene; right two columns: oil/cellulose).

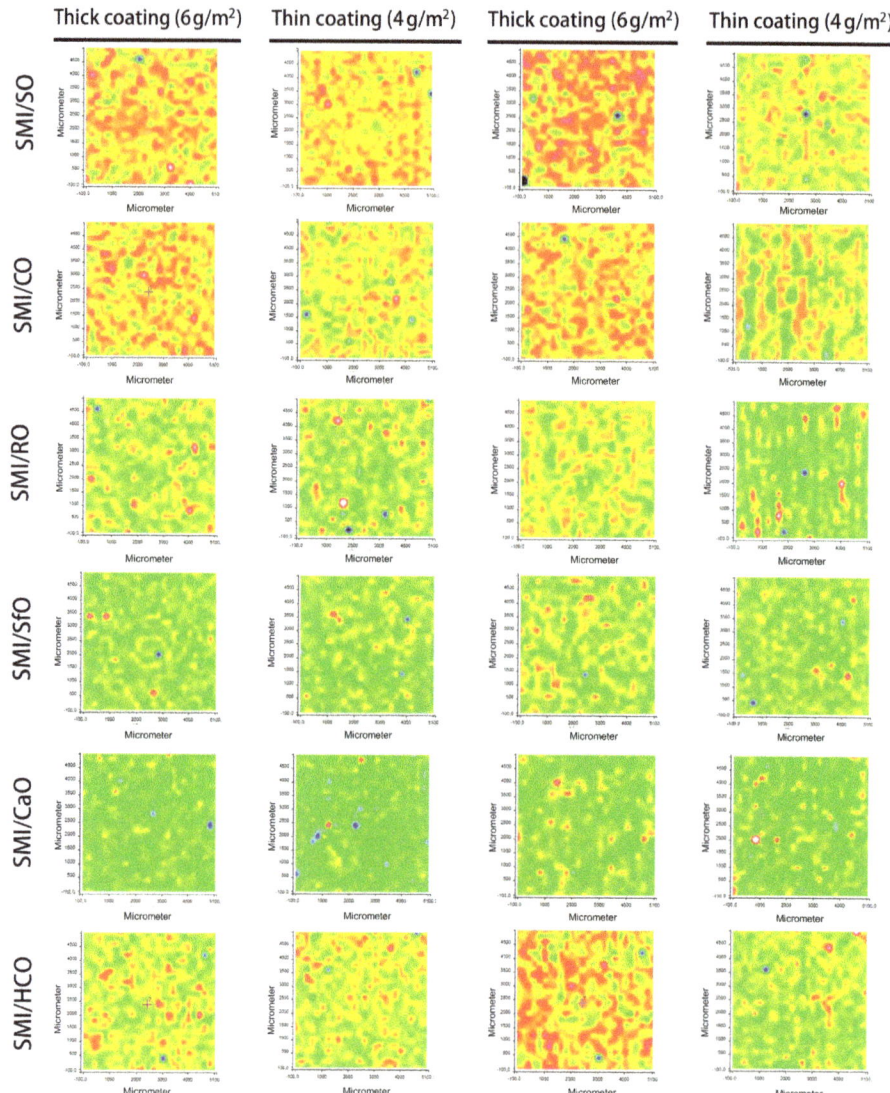

Figure 7. Raman maps (5 × 5 mm^2) with band intensity ratios illustrating imide moieties at the surface for SMI/oil paper coatings (left two columns: imide/styrene, right two columns: imide/cellulose).

The distribution of organic coating moieties from the intensity ratios of Raman bands for styrene (1602 cm^{-1} and 1000 cm^{-1}), relative to cellulose (1095 cm^{-1}), are shown in Figure 8. The maximum intensity of styrene bands is comparable for all SMI/oil coating types and does not strongly differ with the oil type, because all copolymers contain a fixed amount of 74 mol % styrene that remains inert during the synthesis of the SMI/oil nanoparticles. The differences in intensities of styrene therefore mainly correspond to variations in coating coverage. In particular, the effects of thin and thick coatings are visualized. The locations for coverage of styrene components represented by Raman bands at 1602 cm^{-1} and 1000 cm^{-1} are in agreement, but slight differences in intensity may be attributed to the specific orientation of the styrene parts. The styrene-related absorption bands are characteristic for either the aromatic C=C stretch (1602 cm^{-1}) or the aromatic C–H bending (1000 cm^{-1}). Their intensities depend on the bonding length of the π-conjugation between the vinyl group and benzene ring and change with local conformation of the aromatic styrene groups [25]. The styrene orientation in the SMI copolymer is induced by a specific molecular structure of the high-molecular-weight SMA precursor and induces nanoparticle formation through self-assembly [26]. Therefore, the variations in ratio of both styrene bands at 1602 and 1000 cm^{-1} can eventually be attributed to configuration or orientation effects: the influences of segment orientation in styrene bloc copolymers with polarized Raman scattering were detected [27]. In particular, the locations corresponding to the aromatic parts (1602 cm^{-1}) are more confined that the aliphatic parts of the styrene, as the aromatic styrene groups are likely more influenced by self-organization within the nanoparticles.

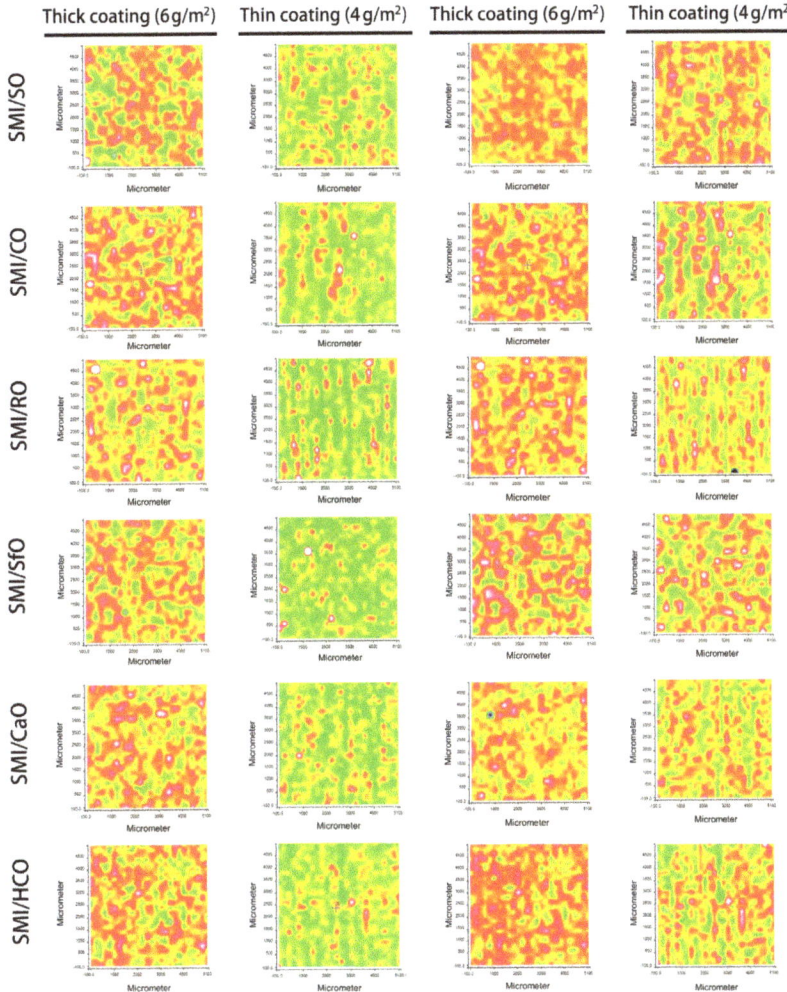

Figure 8. Raman maps (5 × 5 mm^2) with band intensity ratios illustrating styrene moieties at the surface for SMI/oil paper coatings (left two columns: 1602 cm^{-1}/cellulose, right two columns: 1000 cm^{-1}/cellulose).

In conclusion, the locations for styrene coverage strongly overlap with the previous locations for imide coverage (see Figure 6). The chemical mapping for the band ratios of organic coating moieties (imide and styrene) generally corresponds with the chemical mapping based on average intensities: the areas with high intensity in styrene (see Figure 7) overlap with the areas of high average intensity (see Figure 4). In case of coatings with polyunsaturated oils (SMI/SO, SMI/CO, SMI/SfO), the surface coverage with high intensities of band ratios for oil moieties (see Figure 5) overlap with the areas of styrene coverage (see Figure 7), as the oil is largely encapsulated in the organic parts. For coatings with monounsaturated oils (e.g., SMI/CaO), the surface areas with high intensities of oil moieties (see Figure 5) are complementary to the surface areas with high intensities of styrene moieties (see Figure 7), as there is a large amount of free oil in between the organic coating phase.

3.5. Raman Chemical Mapping (Single Bands)

The surface maps with intensities of single Raman bands provide direct chemical information on the surface composition of coated papers. As only a single band is considered, the spectra are first normalized (cellulose region 1170–1050 cm^{-1}) and baseline-corrected before selecting a specific band. An example of chemical surface maps for different single Raman bands of thick and thin coatings of SMI/SO and SMI/HCO is shown in Figure 9 (the maps for other coating types are given in Figures S2–S5 in Supplementary Materials). The selected Raman bands provide complementary information on the surface coverage of different chemical components.

The surface areas covered with high intensities of the styrene band (1000 cm^{-1}) are similar to those covered with high intensities of the imide band (1765 cm^{-1}). The same surface areas have also been covered by high intensities in previous surface maps with band intensity ratios of organic coating moieties. The surface maps of the cellulose bands (1378 cm^{-1} and 1095 cm^{-1}) are complementary to the holes observed in the surface maps with organic coating deposits, indicating that the paper is locally penetrating through the coating at those locations. Both cellulose bands cover the same surface areas and exposure of cellulose bands at the surface is more intense for thin coatings than for thick coatings. The intensity of the single band related to hydroxyl groups (3100 cm^{-1}) is highest at locations where cellulose fibers are exposed and lowest at locations covered by the coating. The exposure of cellulose hydroxyl groups at the surface is evidently more intense for thin coatings than for thick coatings. Therefore, interactions between the coating and the paper substrate are expected to happen through hydrogen bonding between hydroxyl groups at the cellulose fibers and the imidized SMI/oil nanoparticles. In a search for other complementary bands, the locations with low intensities in the 1765 cm^{-1} band (imide) correspond to locations with high intensities of the 1565 cm^{-1} band. The latter band can be assigned to the amide II band (NH bend + CN stretch) and represents amic acid groups as a remaining intermediate product after imidization (i.e., ammonolyzed maleic anhydride). The non-imidized parts obviously induce minor local defect spots and influence the formation of a fully homogeneous coating layer. These parts do not interact well with the paper substrate as they appear at similar surface areas with high intensities of hydroxyl bands. The surface areas with high intensities of the single oil band (1655 cm^{-1}) fully overlap with the imide zones for coatings with polyunsaturated oil (e.g., SMI/SO), corresponding to previous conclusions that oil is embedded in the organic phase and there is very little free oil. The overlap of oil coverage and imide locations is much less for coatings with monounsaturated oil (see Figures S3–5 in Supplementary Materials). In particular, the oil coverage largely exceeds the locations with imide deposits in the case of unsaturated oil types (e.g., SMI/HCO) due to the large amount of free oil.

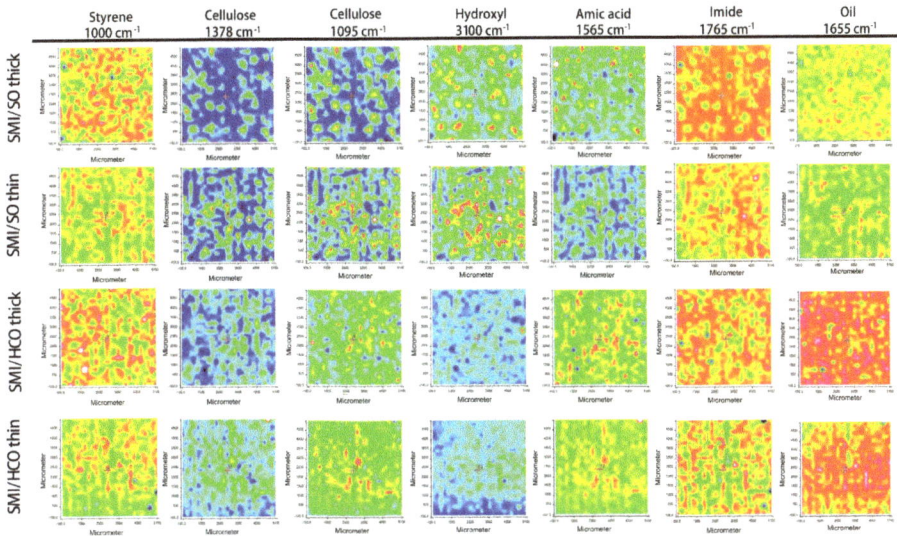

Figure 9. Raman maps (5 × 5 mm^2) with single wavenumbers representing different coating and substrate components at the surface for SMI/oil paper coatings.

The surface maps with simultaneous coverage of the different coating components have been constructed after image processing (Figure 10). The summation images with imide and oil coverage ("imide ADD oil") illustrate the degree of overlap and inclusion of oil within the organic coating phase, with following color code (see online for color figures: red = oil, black = imide). The quantification of thick and thin coatings is visible from the surface areas covered with the red phase. Almost full inclusion of the red surface areas within the black surface areas is recognized for SMI/SO, SMI/MO, and SMI/RO coatings, while the amount of red surface areas outside the black surface areas increases for SMI/SfO, SMI/CaO, and SMI/HCO coatings, which is in parallel with the localization of free oil. The difference in images between oil and imide ("oil SUBTRACT imide") directly illustrate the surface locations with free oil, following the color code (see online for color figures: blue = free oil). From the blue surface areas it can be observed that the amount of free oil increases in the formerly reported order of SMI/oil coating types, and is higher for thick coatings than for thin coatings. In conclusion, the processed images and original Raman surface maps allow us to accurately visualize the different components and local chemical composition at the coated paper surface, and systematically illustrate the effect of oil encapsulation depending on the degree of saturation of the vegetable oil.

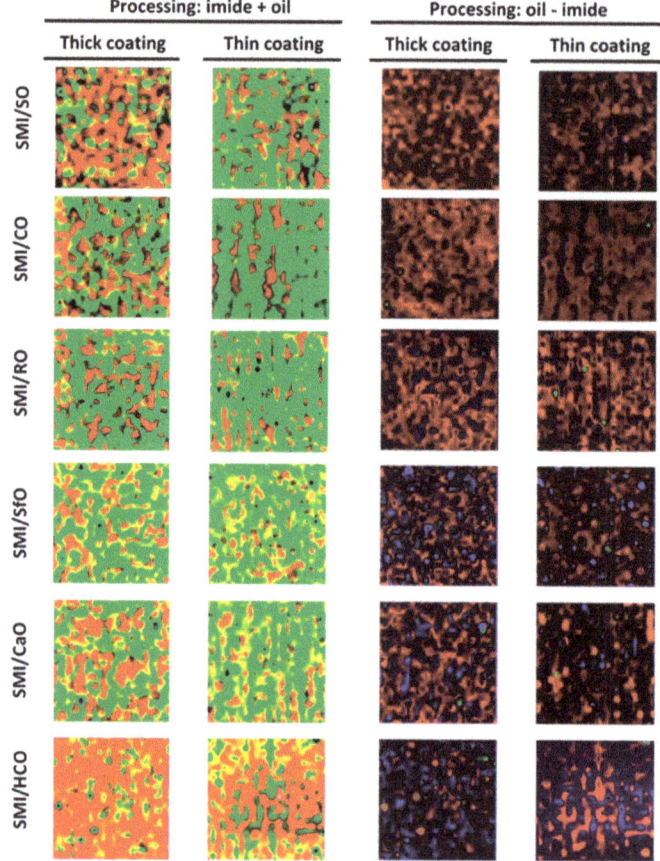

Figure 10. Processed images of single Raman wavenumbers for imide (1765 cm^{-1}) and oil (1655 cm^{-1}), illustrating lateral distribution of oil within the organic phase (+) and free oil (−).

4. Conclusions

Dispersive Raman spectroscopy and microscopy provide an efficient analytical tool for the characterization, discrimination, and quantification of thick and thin barrier coatings on paper with nanoparticle deposits of poly(styrene-co-maleimide) that are filled with six different types of vegetable oils, soy oil, corn oil, rapeseed oil, sunflower oil, castor oil, and hydrogenated castor oil. Depending on the degree of oil saturation (iodine value), the imide content gradually decreases and the amount of free oil gradually increases.

A statistical classification of the coatings can be made by the partial least squares model with three principal components (PC), as each of them include information about the different coating components, i.e., oil phase (PC-1), cellulose substrate (PC-2), and the organic coating components (PC-3). The overall classification of the paper coatings in score plots is based on the degree of saturation for the different oils and coating thicknesses.

The chemical surface maps with average intensities illustrate the coating coverage and homogeneity, where high amounts of free oil result in thick deposits with island-like locations. The thin coatings often result in the formation of striations parallel to the direction of coating, and can mainly be related to local deposits of the organic (imide) coating components. The presentation of free oil in different amounts depending on the coating type is complementary to the average intensity maps. The chemical surface maps with relative band ratios have been analyzed relative to the organic coating phase and the substrate. The distribution of styrene and imide over the surface agrees fairly well and largely corresponds with the chemical mapping based on average intensities. For coatings with polyunsaturated oils, the surface coverage with high intensities of oil overlaps with the areas of organic coating components, as the oil is largely encapsulated in the organic parts. For coatings with monounsaturated oils, the surface coverage with high intensities of band ratios for oil is complementary to the surface areas with organic coating components, as there is a large amount of free oil as a binder in between the organic coating phase.

In conclusion, the selection of specific oils incorporated in poly(styrene-co-maleic) anhydride allows us to create various distributions of chemical moieties at the surface. Depending on the type of oil chosen, the amount of imide, styrene, and free oil at the coating surface varies. As the resulting hydrophobic properties and barrier performance depend on both the chemical surface composition and topography, the coatings are expected to serve as a candidate for barrier coatings in future packaging applications.

Supplementary Materials: The following are available online at http://www.mdpi.com/2079-6412/8/5/154/s1, Figur S1: Reference SEM image of uncoated paper and the same magnifications of the coated paper samples; Figures S2–S5: Raman maps (5 × 5 mm^2) with single wavenumbers representing different coating and substrate components at the surface for SMI/oil paper coatings, Table S1: Raman shift assignment for SMI/oil coatings on paper.

Acknowledgments: The author acknowledges the support of Mr. Peter Mast (Ghent University—Department of Materials Science and Engineering, Belgium) with FEG-SEM. Samples were kindly supplied by Topchim N.V. (Wommelgem, Belgium).

Conflicts of Interest: The author declares no conflict of interest.

References

1. Samyn, P. Review paper: Wetting and hydrophobic modification of cellulose surfaces for paper applications. *J. Mater. Sci.* **2013**, *48*, 6455–6498. [CrossRef]
2. Samyn, P.; Barhoum, A.; Ohlund, T.; Dufresne, A. Review: Nanoparticles and nanostructured materials in papermaking. *J. Mater. Sci.* **2018**, *53*, 146–184. [CrossRef]
3. Ogihara, H.; Xie, J.; Okagaki, J.; Saji, T. Simple method for preparing superhydrophobic paper: Spray-deposited hydrophobic silica nanoparticle coatings exhibit high water-repellency and transparency. *Langmuir* **2012**, *88*, 4605–4608. [CrossRef] [PubMed]
4. Aromaa, M.; Arffman, A.; Suhonen, H.; Haapanen, J.; Keskinen, J.; Honkanen, M.; Nikkanen, J.P.; Levänen, E.; Messing, M.E.; Deppert, K.; et al. Atmospheric synthesis of superhydrophobic TiO$_2$ nanoparticle deposits in a single step using liquid flame spray. *J. Aerosol Sci.* **2012**, *52*, 57–68. [CrossRef]
5. Karapanagiotis, I.; Grosu, D.; Aslanidou, D.; Aifantis, K.E. Facile method to prepare superhydrophobic and water repellent cellulosic paper. *J. Nanomater.* **2015**, *2015*, 219013. [CrossRef]
6. Torun, I.; Onses, M.S. Robust superhydrophobicity on paper: Protection of spray-coated nanoparticles against mechanical wear by the microstructure of paper. *Surf. Coat. Technol.* **2017**, *319*, 301–308. [CrossRef]
7. Jonoobi, M.; Harun, J.; Mathew, A.P.; Hussein, M.Z.; Oksman, K. Preparation of cellulose nanofibers with hydrophobic surface characteristics. *Cellulose* **2010**, *17*, 299–307. [CrossRef]
8. Alam, M.; Akram, D.; Sharmin, E.; Zafar, F.; Ahmad, S. Vegetable oil based eco-friendly coating materials: A review article. *Arab. J. Chem.* **2014**, *7*, 469–479. [CrossRef]
9. Shen, Z.; Augustin, M.A.; Sanguansri, L.; Cheng, L.J. Oxidative stability of microencapsulated fish oil powders stabilized by blends of chitosan, modified starch, and glucose. *J. Agric. Food Chem.* **2010**, *58*, 4487–4493. [CrossRef] [PubMed]
10. Workman, J.J. Infrared and Raman spectroscopy in paper and pulp analysis. *Appl. Spectrosc. Rev.* **2001**, *36*, 139–168. [CrossRef]
11. Vyörykkä, J.; Juvonen, K.; Bousfield, D.; Vuorinen, T. Raman microscopy in lateral mapping of chemical and physical composition of paper coating. *Tappi J.* **2004**, *3*, 19–24.

12. Guyot, C.; Amram, B.; Ubrich, J.M. Raman-microscopic study of mottling of coated paper. *Wochenbl. Papierfabrikation* **1995**, *123*, 646.
13. Kenttä, E.; Juvonen, K.; Halttunen, M.; Vyörykkä, J. Spectroscopic methods for determination of latex content of coating layers. *Nordic Pulp Pap. Res. J.* **2000**, *15*, 579–585. [CrossRef]
14. Bitla, S.; Tripp, C.P.; Bousfield, D.W. Raman spectroscopic study of migration in paper coatings. *J. Pulp Pap. Sci.* **2003**, *29*, 382–385.
15. He, P.; Bitla, S.; Bousfield, D.; Tripp, C.P. Raman spectroscopic analysis of paper coatings. *Appl. Spectrosc.* **2002**, *56*, 1115–1121. [CrossRef]
16. Kugge, C.; Greaves, M.; Hands, K.; Scholes, F.H.; Vanderhoek, N.; Ward, J. Paper coating analysis by confocal Raman spectroscopy. *Appita J.* **2008**, *61*, 11–16.
17. Niemelä, P.; Hietala, E.; Ollanketo, J.; Tornberg, J.; Pirttinen, E.; Stenius, P. FT-Raman spectroscopy as a tool for analyzing the composition of recycled paper pulp. *Prog. Pap. Recycl.* **1999**, *8*, 15–24.
18. Frost, R.L. The structure of the kaolinite minerals: A FT-Raman study. *Clay Miner.* **1997**, *32*, 65–77. [CrossRef]
19. Lutz, A.; De Graeve, I.; Terryn, H. Non-destructive 3-dimensional mapping of microcapsules in polymeric coatings by confocal Raman spectroscopy. *Prog. Org. Coat.* **2015**, *88*, 32–38. [CrossRef]
20. Taheri, H.; Stanssens, D.; Samyn, P. Rheological behaviour of oil-filled polymer nanoparticles in aqueous dispersion. *Colloid Surf. A* **2016**, *499*, 31–45. [CrossRef]
21. Samyn, P.; Schoukens, G.; Stanssens, D.; Vonck, L.; Van den Abbeele, H. Incorporating different vegetable oils into an aqueous dispersion of hybrid organic nanoparticles. *J. Nanopart. Res.* **2012**, *14*, 1075–1099. [CrossRef]
22. Singh, K.B.; Bhosale, L.R.; Tirumkudulu, M.S. Cracking in drying colloidal films of flocculated dispersions. *Langmuir* **2009**, *25*, 4284–4287. [CrossRef] [PubMed]
23. Sadeghi-Jorabchi, H.; Wilson, R.H.; Belton, P.S.; Edwards-Webb, J.D.; Coxon, D.T. Quantitative analysis of oils and fats by Fourier transform Raman spectroscopy. *Spectrochim. Acta A* **1991**, *47*, 1449–1458. [CrossRef]
24. Johnson, G.L.; Machado, R.M.; Freidl, K.G.; Achenbach, M.L.; Clark, P.J.; Reidy, S.K. Evaluation of raman spectroscopy for determining cis and trans isomers in partially hydrogenated soybean oil. *Org. Proc. Res. Dev.* **2002**, *6*, 637–644. [CrossRef]
25. Choi, C.H.; Kertesz, M. Conformational information from vibrational spectra of styrene, trans-stilbene, and cis-stilbene. *J. Phys. Chem. A* **1997**, *101*, 3823–3831. [CrossRef]
26. Samyn, P.; Schoukens, G. Morphologies and thermal variability of patterned polymer films with poly(styrene-co-maleic anhydride). *Polymers* **2014**, *6*, 820–845. [CrossRef]
27. Archer, L.A.; Fuller, G.G. Segment orientation in a quiescent block copolymer melt studied by Raman scattering. *Macromolecules* **1994**, *27*, 4359–4363. [CrossRef]

© 2018 by the author. Licensee MDPI, Basel, Switzerland. This article is an open access article distributed under the terms and conditions of the Creative Commons Attribution (CC BY) license (http://creativecommons.org/licenses/by/4.0/).

Article

Efficiently Extracted Cellulose Nanocrystals and Starch Nanoparticles and Techno-Functional Properties of Films Made Thereof

Christoph Metzger [1], Solange Sanahuja [1], Lisa Behrends [1], Sven Sängerlaub [2,3], Martina Lindner [2,3] and Heiko Briesen [1,*]

[1] Chair of Process Systems Engineering, TUM School of Life Sciences Weihenstephan, Technical University of Munich, Gregor-Mendel-Str. 4, 85354 Freising, Germany; christoph.metzger@tum.de (C.M.); solange.sanahuja@tum.de (S.S.); lisa.behrends@tum.de (L.B.)

[2] Chair of Food Packaging Technology, Technical University of Munich, TUM School of Life Sciences Weihenstephan, Weihenstephaner Steig 22, 85354 Freising, Germany; sven.saengerlaub@ivv.fraunhofer.de (S.S.); martina.lindner@ivv.fraunhofer.de (M.L.)

[3] Fraunhofer Institute for Process Engineering and Packaging IVV, Giggenhauser Str. 35, 85354 Freising, Germany

* Correspondence: briesen@wzw.tum.de; Tel.: +49-8161-71-3272

Received: 2 March 2018; Accepted: 12 April 2018; Published: 14 April 2018

Abstract: Cellulose nanocrystals (CNC) and starch nanoparticles (SNP) have remarkable physical and mechanical characteristics. These properties particularly facilitate their application as high-performance components of bio-based packaging films as alternatives to fossil-based counterparts. This study demonstrates a time-efficient and resource-saving extraction process of CNC and SNP by sulfuric acid hydrolysis and neutralization. The yields of the hydrolyzed products were 41.4% (CNC) and 32.2% (SNP) after hydrolysis times of 3 h and 120 h, respectively. The nanoparticle dispersions were wet-coated onto poly(lactic acid) (PLA) and paper substrates and were incorporated into starch films. No purification or functionalization of the nanoparticles was performed prior to their application. Techno-functional properties such as the permeability of oxygen and water vapor were determined. The oxygen permeability of 5–9 cm^3 (STP) 100 µm m^{-2} d^{-1} bar^{-1} at 50% relative humidity and 23 °C on PLA makes the coatings suitable as oxygen barriers. The method used for the extraction of CNC and SNP contributes to the economic production of these nanomaterials. Further improvements, e.g., lower ion concentration and narrower particle size distribution, to achieve reproducible techno-functional properties are tangible.

Keywords: cellulose nanocrystals (CNC); starch nanoparticles (SNP); biopolymers; packaging; barrier films; nanomaterials; nanocomposites; bio-coatings; oxygen barrier; water vapor barrier

1. Introduction

Films and coatings made of cellulose nanocrystals (CNC) and starch nanoparticles (SNP) have considerable potential for application in sustainable and bio-based packaging materials [1–10]. Their techno-functional properties can supplement the limited gas barrier properties and the mechanical properties of renewable biopolymers such as poly(lactic acid) (PLA), paper and starch [11–15]. PLA is synthesized from fermented carbohydrates [16,17]. It is used in the packaging industry due to its thermal properties manifesting in good processing characteristics, its chemical and UV resistance, and its biodegradability [18–20]. However, the oxygen and water vapor permeability of PLA necessitates barrier enhancement for oxygen and water vapor sensitive packaging goods [21–23]. Pant et al. [24] reported an oxygen permeability of PLA of 153 (STP) 100 µm m^{-2} d^{-1} at 23 °C and 50% relative humidity and a water vapor transmission rate of 58 g (STP) 100 µm m^{-2} d^{-1} at 23 °C at a gradient in relative humidity of 85→0%. Similarly, the application of fiber-based packaging materials, such as paper, is restricted due to high sensitivity to moisture accompanied by poor barrier properties [25,26]. Starch can be converted into a continuous polymeric phase and is therefore processable using extrusion technology developed to produce fossil-based polymer packaging. The use of starch is nevertheless just as well limited due to its hydrophilic nature leading to moisture sensitivity that is compromising the dimensional stability and mechanical properties [27]. The techno-functional properties of these three materials can be enhanced by the introduction of CNC and SNP in the form of coatings or fillers in nanocomposites. Both CNC and SNP facilitate tailored physical and mechanical properties enabling the manufacturing of bio-based packaging materials to substitute or complement conventional, fossil-based polymers. Due to their presumably low environmental, health, and safety risks, packaging applications for fast moving consumer goods are conceivable [28].

However, the large-scale production of CNC and SNP is presently unattractive. Sulfuric acid hydrolysis is the commonly used method to extract CNC and SNP from suitable biopolymeric feed stocks [29,30]. Separating the hydrolyzed product from the acidic reaction solution is laborious and comprises a high consumption of resources and high material costs; usually applied strategies involve dilution with a large amount of water and quenching, sedimentation, and eventually centrifugation in combination with ultrafiltration or dialysis for purification [30]. In contrast, Müller et al. [31] suggested an efficient approach based on the neutralization of the acid. Flocculation of the nanoparticles is induced in the presence of cations at high ionic strengths. The salt concentration is subsequently reduced by centrifugation until the ion concentration level for peptization is reached. Hereby, a high yield of hydrolyzed product is achievable at an overall low net process time and low consumption of resources and materials.

Manufacturing limitations due to the nanoparticles' intrinsic physical properties further impede their application in coatings on substrates and as a filler in nanocomposites. These difficulties arise from the limited interfacial adhesion of the hydrophilic nanoparticles and the hydrophobic polymers, moisture absorption, and agglomeration issues. Several studies addressed the application of CNC and SNP in biopolymeric packaging materials [2,6–10,27,30,32–35]. Techno-functional properties such as gas permeability, mechanical properties, and thermal stability were adjusted by modifying the nanoparticles' surface chemistry. If further functionalization of the nanoparticles is desired, the use of non-green chemicals and additional time is required [36].

The aim of the present study was (i) to demonstrate a time-efficient and resource-saving post-processing method to produce CNC and SNP. The qualification of the nanomaterials as (ii) barrier coatings on flat substrates and (iii) as filler in nanocomposites was evaluated. The study tested several hypotheses (h). We tested whether (h1) stable dispersions of CNC and SNP are producible by the presented method and (h2) the nanoparticles lower the gas permeability of PLA and paper substrates and (h3) improve the water vapor permeability and the mechanical properties of solution-cast starch films. Good spreadability of the nanoparticles on the substrate materials and their miscibility in hydrophilic starch films are the necessary preconditions to achieve improved techno-functional properties.

2. Materials and Methods

2.1. Materials

2.1.1. Materials for Nanoparticle Processing

Cotton linters (water content 4 wt %) were purchased from Buch-Kunst-Papier (St. Ingbert Rentrisch, Germany). Corn starch (water content 15 wt %), sodium azide (99%), and sodium hydroxide (99%) were purchased from Carl Roth (Karlsruhe, Germany). Sulfuric acid (95%) was obtained from VWR (Ismaning, Germany). All chemicals were used as received. Ultrapure (type 1) water with a resistivity of 18.2 MΩ cm (Milli-Q Direct 8 system, Merck Chemicals, Schwalbach, Germany) was used for all experiments.

2.1.2. Substrate Materials

Poly(lactic acid) films (2002D, NatureWorks LLC, Minnetonka, MN, USA) with a thickness of 25 µm were provided by Fraunhofer IVV (Freising, Germany). Material properties of 2002D were determined by other groups: Murphy et al. [37] report a melting temperature T_m of 154 °C and a glass transition temperature T_g of 55 °C as well as a D-lactide content of 4% and a molecular weight M_W of 194,000 g mol^{-1}. Ge et al. [38] report values of T_m of 168.2 °C, T_g of 58.6 °C, a D-lactide content of 4%, a L-lactide content of 96% and a M_W of 121,400 g mol^{-1}. Mihai et al. [39] report semi-crystallinity of 2002D.

Paper sheets (Metalkote Evolution, Ahlstrom-Munksjö Group, Stockholm, Sweden) with a grammage of 65 g m^{-2} were also provided by Fraunhofer IVV (Freising, Germany). Untreated (bare) substrates were tested as reference materials.

2.1.3. Cast Films

Corn starch (water content 15 wt %) and glycerol (water content 14 wt %) were purchased from Carl Roth (Karlsruhe, Germany). Pure starch-glycerol films without nanofillers were tested as reference material.

2.1.4. Karl Fischer Titration

Formamide (99.5 wt %), iodine solution (Roti®hydroquant C5; 5 mg H_2O mL^{-1}, free of pyridine), methanol (Roti®hydroquant D; dry), and a water standard (Roti®hydroquant; 10 mg H_2O g^{-1}) were purchased from Carl Roth (Karlsruhe, Germany).

2.1.5. Pinhole Testing

Pinholes were determined with peanut oil containing Sudan Red III (Merck, Darmstadt, Germany) in a concentration of 1 part per thousand.

2.2. Nanoparticle Preparation

CNC were prepared by sulfuric acid hydrolysis and subsequent basic neutralization. The procedure was derived from the method described by Müller et al. [31]. 2 mol of sulfuric acid (64 wt %) was added to cut cotton linters in a mass ratio of 15:1 and transferred to a heated water bath at 50 °C. The raw cellulose was hydrolyzed for 3 h under vigorous stirring. Following hydrolysis, the reaction solution was diluted to 54 wt % H_2SO_4 to decrease the viscosity and then decanted to 3 mol of sodium hydroxide (7 mol kg^{-1}) in a cooled water bath. After homogenization for 15 min, the CNC were separated from the reaction solution by consecutive centrifugation steps until a pH of ~2 was reached. Excess ions were removed from the hydrolyzed product by decantation of the supernatant after each washing step and successive redispersion with H_2O. Eventually, the dispersion was ultrasonicated (8 kJ g^{-1} CNC) with a homogenizer (Sonoplus HD 3400 with the sonotrode VS 70 T, Bandelin, Berlin, Germany) and stored at 6 °C until further use.

The hydrolysis conditions to produce SNP was derived from the method demonstrated by Angellier et al. [40]. 1 mol of sulfuric acid (18 wt %) was added to corn starch in a mass ratio of 7:1 and then hydrolyzed for 120 h at 40 °C. Subsequently, the reaction mixture was decanted to 1.5 mol of sodium hydroxide (5 mol kg^{-1}) for neutralization. Similar to the extraction of CNC, SNP were separated from the salt solution by consecutive precipitation and redispersion until the pH stabilized at ~2.7. After ultrasonication, 0.01 g sodium azide L^{-1} was added as an antimicrobial agent before storing the product at 6 °C.

2.3. Coating of PLA and Paper

The coating strategy of the nanoparticle dispersions onto polymer substrates was empirically developed for CNC dispersions. Qualitative parameters were considered regarding the application of a nanoparticle dispersion onto a flat substrate and the subsequent drying process. PLA was corona-treated (Corona Station, Softal, Hamburg, Germany) at 400 W and 5 m min^{-1} to increase the surface energy with the aim to achieve good spreadability of the aqueous coatings on the substrate surface [41]. For paper this pretreatment was not necessary. The effect of viscosity of the coating medium was investigated for CNC dispersions with concentrations of $3 \leq c \leq 8$ wt %. It was found that with increasing concentration and hence increasing viscosity, contraction of the wet film could be fully avoided. The dynamic viscosity of a dispersion with 6 wt % of CNC was >2000 Pa s at a shear rate of 0.01 s^{-1} and decreased exponentially to 0.05 Pa s at 1000 s^{-1}.

The concentration of the nanoparticle dispersions was adjusted in a rotary evaporator (Rotavapor R-100, Büchi, Flawil, Switzerland) at 40 °C and 70 mbar to ~63 g kg^{-1} and ~73 g kg^{-1} of hydrolyzed product for cellulose and starch, respectively. The dispersions were applied onto the substrates with a semi-automatic coating unit (CUF5, Sumet Systems, Denklingen, Germany).

The interplay of blade velocities v of $5 \leq v \leq 60$ mm s^{-1} and wet film thicknesses d of $10 \leq d \leq 100$ μm was tested. At $v = 10$ mm s^{-1} and $d \approx 51$ μm under a normal load of 40 N, the coatings showed no macroscopic cracks or delamination on both PLA and paper. The same parameters led to conformable coatings using SNP dispersions on both substrates.

Drying was tested for temperatures of $40 \leq T \leq 60$ °C for PLA and $40 \leq T \leq 120$ °C for paper. The drying time was $1 \leq t \leq 10$ min. Drying at high temperatures and within narrow time spans led to contraction of the coating medium and eventually delamination of the coating layer on PLA. Furthermore, high temperature for extended time spans bears the risk of denaturation of plastic substrates and desulfation of the nanoparticles [42]. Eventually, a temperature of 40 °C for 10 min was chosen for PLA. Paper was dried at 70 °C for 3 min to avoid wrinkling [43]. To avoid possible pinholes in the coating layer on the paper substrate due to its surface roughness it was coated and dried two times with the same parameters.

2.4. Cast Film Preparation

The starch-glycerol and nanocomposite films were prepared by solution casting according to the method of Alves et al. [33]. Precursor solutions of the nanocomposite films with the total mass m_{total} had a water content c_{aq} of 97 wt %. Five different nanoparticle concentrations $0 \leq c_{NP} \leq 9$ wt % were added as well as glycerol as plasticizer with a concentration c_{gly} of 30 wt %, both relative to the weighed portion of starch m_{starch}. The weighed portions of chemicals were calculated according to Equation (1), considering the mass of water added m_{H_2O}, corrected for the water content of the other components.

$$m_{total} = \left(1 + c_{gly} + c_{NP}\right) m_{starch} + c_{aq} m_{H_2O} \tag{1}$$

Starch was added to preheated water with a temperature of 70 °C in a water bath and stirred for 1 h to allow plastification. Glycerol and nanoparticles were then added and stirring was continued for 20 min to allow plasticization and mixing of the composite. A dry film thickness of 50 μm was targeted. Before casting in polystyrene Petri dishes (Greiner Bio-One, distributed by VWR, Ismaning, Germany), the filmogenic solutions were homogenized in an ultrasonic bath for 3 min. Excess water was evaporated from the dishes overnight in a climatic chamber (ICH 110, Memmert, Schwabach, Germany) at 40 °C and a relative humidity (r.h.) of 47%. The films were then peeled off from the Petri dish and turned upside down for double-sided drying overnight.

2.5. Instrument Measurements

If applicable, all measurements were performed at least in triplicate and are presented with the 95% confidence interval of the mean. The uncertainty of quantities depending on multiple variables is given by the propagation of error. For the determination of the oxygen permeability, the mean value of two measurements is given with the minimum and the maximum value.

2.5.1. Particle Size and Viscosity

The hydrodynamic apparent particle size of the nanoparticle dispersions was measured by dynamic light scattering (DLS) using a Zetasizer Nano ZSP (Malvern Instruments, Worcestershire, UK). Aliquots were filtered with syringe filters with a hydrophilic PES membrane and a pore size of 1 µm (Chromafil PES, Macherey-Nagel, Düren, Germany). The harmonic intensity averaged particle diameter (z-average) and the polydispersity index (PdI) from the cumulants analysis were obtained for 0.025 wt % nanoparticle dispersions after equilibration for 30 min at 25 °C.

The volume-weighed particle size of residual microparticles was measured using a HELOS/KR laser diffraction particle size analyzer with a QUIXEL wet dispersion system (Sympatec, Clausthal-Zellerfeld, Germany) at 23 °C and an optical concentration of 10%. The balanced mean size $x_{1,3}$ and the *span* were evaluated according to Equations (2) and (3) as

$$x_{1,3} = \int_{x_{min}}^{x_{max}} x q_3(x) dx \text{ and} \tag{2}$$

$$span = \frac{x_{90} - x_{10}}{x_{50}}. \tag{3}$$

Here, x_{min} and x_{max} are the smallest and the largest particle size, x is the class midpoint and $q_3(x)$ is the volume-weighted particle size distribution. x_{10}, x_{50} and x_{90} are the particle sizes corresponding to 10%, 50% and 90% of the cumulative undersize distribution, respectively.

The viscosity of the concentrated CNC dispersion was determined using a Physica MCR 501 rheometer (Anton Paar, Graz, Austria) at 25 °C with a cone-plate geometry. Shear flow curves were measured in the range from 0.01 to 1000 s^{-1}.

2.5.2. Dry Mass, Ion Chromatography and Yield

The mass of nanoparticles m_{np} in the product was determined gravimetrically in combination with ion chromatography and calculated according to Equation (4):

$$m_{np} = m_{dry} - m_{wet} \sum_{i=0}^{n} \frac{\gamma_{IC,i} V_{IC}}{m_{IC}}. \tag{4}$$

About 2 g of wet product m_{wet} were freeze-dried (2–4 LSCplus, Christ, Osterode am Harz, Germany) and the dried product was weighed again to obtain m_{dry}. Ion chromatography (820 IC Separation Center, Metrohm, Herisau, Switzerland) was performed on aliquots with a mass m_{IC} of about 0.1 g, diluted in H$_2$O with a volume V_{IC} of 50 mL. The mass concentration $\gamma_{IC,i}$ of cations (Metrosep C 4—150/4.0 column) and anions (Metrosep A Supp 4—250/4.0 column) was obtained. The yield Y is the quotient of the initially provided mass of the feed stock m_{raw} and m_{np}.

2.5.3. Optical Properties of Dispersions, Coatings, and Cast Films

Optical microscopy was performed on a polarized light microscope (BX51-P, Olympus, Hamburg, Germany) equipped with a camera (XC50, Olympus, Hamburg, Germany) for the qualitative analysis of the coatings and the cast films.

Scanning electron microscopy (SEM) was carried out on a JEOL JSM-IT100 (Akishima, Japan) with a secondary electron detector and an acceleration voltage of 3–5 kV. The nanoparticle dispersions were freeze-dried beforehand and mounted on conductive carbon tape.

The optical absorbance A of dispersions and transparent films was measured using a Specord 50 Plus Spectrophotometer (Analytik Jena, Jena, Germany). The absorbance at a wavelength λ is proportional to the extinction coefficient $\varepsilon(\lambda)$, the concentration c and the optical path length d, according to the Beer-Lambert law (Equation (5)).

$$A = \varepsilon(\lambda) c d \tag{5}$$

2.5.4. Physical Properties

Water Content

The water content of the raw materials was determined by volumetric Karl Fischer titration (TitroLine KF, Xylem Analytics, Weilheim, Germany). The iodine titrant was determined with the water standard. Methanol and formamide were used as solvent in a ratio of 1.5:1. About 0.3 g of the sample was added to the solvent. The titration was started when the sample was completely dissolved. The solvent was exchanged after each measurement. The water content of the raw materials was considered regarding the calculation of the mass ratios for the preparation of nanoparticles and cast films.

Surface Tension

The surface tension of untreated and corona-treated PLA was evaluated with test ink pens (Arcotest, Mönsheim, Germany). The pens are filled with an ink of defined surface tension. The ink is applied with the pens to the substrate surface. If the line of ink does not separate into drops after at least 2 s, the surface energy of the substrate is the same or higher than the surface tension of the fluid. Then, the pen with a higher surface tension is applied until a separation of the line of fluid into drops is observed.

Pinhole Testing

The grease resistance of the coated films was measured using an internal method from Fraunhofer IVV [44]. A test area of 25 cm^2 of the film surface was covered with a fleece—for constant and sufficient covering—and saturated with a solution of colored peanut oil. No further weight was applied. After 24 h at 23 °C and 50% r.h., the fleece and oil residues were removed, and the stained area was characterized by digital image evaluation. At least four specimens were characterized. No pinholes are present, when no fatty spots on the back of the sample are detected.

Thickness

The film thickness was measured mechanically (Precision Thickness Gauge FT3, Rhopoint Instruments, East Sussex, UK) on 5 evenly distributed measuring points. The coating thickness $d_{laminate}$ was calculated from the thickness of a coated film and the same substrate without a coating $d_{substrate}$.

Oxygen and Water Vapor Permeability

Prior to the measurements, all samples were stored at 23 °C and 50% r.h. All tests were conducted at the Fraunhofer IVV, where these methods are accredited.

The reciprocal gas permeability of coated substrates $Q_{laminate}$ is the sum of the reciprocals of the gas permeability of the substrate $Q_{substrate}$ and the coating $Q_{coating}$ (Equation (6)). The oxygen permeability (OP) of the coating $OP_{coating}$ is calculated from the oxygen transmission rate (OTR) of the coating $OTR_{coating}$ and the respective coating thickness $d_{coating}$. The OTR was measured according to DIN 53380-3 [45] with an automatic high barrier oxygen transmission rate system (OX-TRAN 2/21, MOCON, Minneapolis, MN, USA). A humidity of 50% r.h. and a temperature of 23 °C was applied. With reference to the standard conditions (STP, 273 K; 1013 hPa), $OP_{coating}$ is given in cm^3 (STP) µm m^{-2} d^{-1} bar^{-1} according to Equation (7) [46]. The normalization to a thickness of 100 µm as $OP\ Q_{100}$ in cm^3 (STP) 100 µm m^{-2} d^{-1} bar^{-1} is reasonable.

$$\frac{1}{Q_{laminate}} = \frac{1}{Q_{substrate}} + \frac{1}{Q_{coating}} \quad (6)$$

$$OP_{coating} = OTR_{coating} \times d_{coating} \quad (7)$$

The water vapor transmission rate WVTR of coated polymer substrates and cast films was measured gravimetrically using the cup method described in the DIN 53122-1 [47]. Films with an exposed area A of 50.3 cm^2 for coated polymer films and 44.2 cm^2 for cast films with a thickness d were used. The bottom of the cup was filled with anhydrous silica gel. The initial weight of the sealed measuring cells was determined, and they were transferred to a desiccator containing a saturated KOH solution to maintain a humidity of 85% r.h. at 23 °C. The samples were weighed 5 times within a timeframe t of 144 h until the weight gain Δm over a time increment Δt was constant. Transmission rates of the coated films are calculated according to Equation (8). The transmission rate of the pure coating is determined by Equation (6). Films of different thicknesses can be compared via the water vapor permeability WVP in g (STP) µm m^{-2} d^{-1} as calculated by Equation (9) and after further normalization to a thickness of 100 µm (g (STP) 100 µm m^{-2} d^{-1}).

$$WVTR = \frac{\Delta m}{\Delta t A} \quad (8)$$

$$WVP = WVTR \times d \quad (9)$$

Mechanical Testing

Uniaxial tensile tests were carried out on a zwickiLine Z2.5 (Zwick, Ulm, Germany) testing machine as described in the DIN EN ISO 527-3 [48]. The samples were cut in dimensions of 100 mm × 20 mm. After a preload of 0.1 N, a constant extension rate of 25 mm min^{-1} was applied. The ultimate tensile strength σ_{UTS} and the elongation at break ε_f were read from the stress-strain curve σ vs. ε. Young's modulus E was evaluated according to Hooke's law from the linear-elastic relationship in the initial region of the stress-strain curve (Equation (10)):

$$E = \frac{\sigma}{\varepsilon} \quad (10)$$

2.6. Experiment Plan

All tests carried out for nanoparticle dispersions, coatings and cast films are summarized in Table 1.

Table 1. Overview of all measurement methods applied on nanoparticle dispersions, coatings and cast films.

Material	Dry Mass	DLS	Ion Concentration	Viscosity	Surface Tension	Pinhole Test	Thickness	OTR	WVTR	Mechanical Testing	SEM
Dispersions											
CNC	x	x	x	x							x
SNP	x	x	x								x
Coatings											
PLA					x		x	x	x		x
PLA-CNC							x	x	x		x
PLA-SNP							x	x	x		x
Paper							x				x
Paper-CNC						x	x				x
Paper-SNP						x	x				x
Cast Films											
Starch							x		x	x	x
Starch-CNC							x		x	x	x
Starch-SNP							x		x	x	x

3. Results

3.1. Properties of Dispersions

3.1.1. Product Concentration and Ion Content

The yield was measured gravimetrically. The gross yield of nanoparticles was corrected for the ion mass, determined by ion chromatography. Sulfuric acid hydrolysis followed by basic neutralization and successive centrifugation led to a gross yield of 41.4 ± 0.8 wt % of CNC from cotton linters. A gross yield of 32.2 ± 0.7 wt % of SNP was achieved from corn starch.

Redundant H_2O and excess ions were mostly removed during precipitation and redispersion. The relative nanoparticle concentrations in the hydrolyzed products were 53.6 ± 1.0 g kg^{-1} of product and 125.2 ± 2.6 g kg^{-1} of product for cellulose and starch, respectively. The concentration of Na$^+$ of 4.3 ± 0.1 mmol g^{-1} of CNC and 1.0 ± 0.3 mmol g^{-1} of SNP, and the concentration of SO_4^{2-} of 2.3 ± 0.1 mmol g^{-1} of CNC and 0.6 ± 0.0 mmol g^{-1} of SNP, respectively, indicate an excess of free sulfate ions in both hydrolyzed products. This finding is confirmed by the acidic pH of 2.07 and 2.71 after centrifugation for CNC and SNP, respectively. No other ions were detected, pointing at the purity of the raw materials. Thereby, more than 99.8% of the ionic residuals were removed during the washing step.

3.1.2. Particle Size

Both cellulose and starch feed stocks had a bimodal particle size distribution with a volume-weighted mean equivalent diameter of 50.1 ± 0.4 µm and 7.7 ± 0.0 µm, respectively, while the spans were 5.16 ± 0.1 µm and 1.1 ± 0.0 µm. After the sulfuric acid treatment, microscale hydrolyzed residues or agglomerates due to high salt concentrations with a diameter of 7.4 ± 0.1 µm and 4.3 ± 0.0 µm for cellulose and starch, respectively, were still present. The spans corresponding to the mean diameters were 2.7 ± 0.1 µm and 1.9 ± 0.0 µm, respectively.

The apparent mean size of the nanoparticles was 113.4 ± 0.6 nm for cellulose and 248.7 ± 15.7 nm for starch. Accordingly, the polydispersity indices were 0.225 ± 0.007 and 0.369 ± 0.015. Thus, both nanoscale and microscale particles coexist in the hydrolyzed product [49]. An overview of the dispersion properties and the nanoparticle sizes of CNC and SNP is given in Table 2.

Table 2. Properties of nanoparticle dispersions as prepared from cotton linters and corn starch.

Product Property	Cellulose Product	Starch Product
yield/wt %	41.4 ± 0.8	32.2 ± 0.7
hydrolyzed product g kg^{-1} product	53.6 ± 1.0	125.2 ± 2.6
Na$^+$ mmol g^{-1} polysaccharide	4.3 ± 0.1	1.0 ± 0.3
SO_4^{2-} mmol g^{-1} polysaccharide	2.3 ± 0.1	0.6 ± 0.0
apparent nanoparticle size by DLS/nm	113.4 ± 0.6	248.7 ± 15.7

3.1.3. Microscopy

Macro- and microscopically, no phase separation was detected in the concentrated nanoparticle dispersions after 16 weeks at 23 °C. The morphology of the raw materials is shown in Figure 1a,b. Cotton linters have a fibrillar structure, whereas corn starch has a granular shape. Sulfuric acid primarily degrades amorphous regions of the polysaccharides. Upon freeze-drying, agglomeration of nanoparticles promoted by the presence of residual salt occurs (Figure 1c,d). Additionally, microscale hydrolyzed residues with a high aspect ratio can be found in both products.

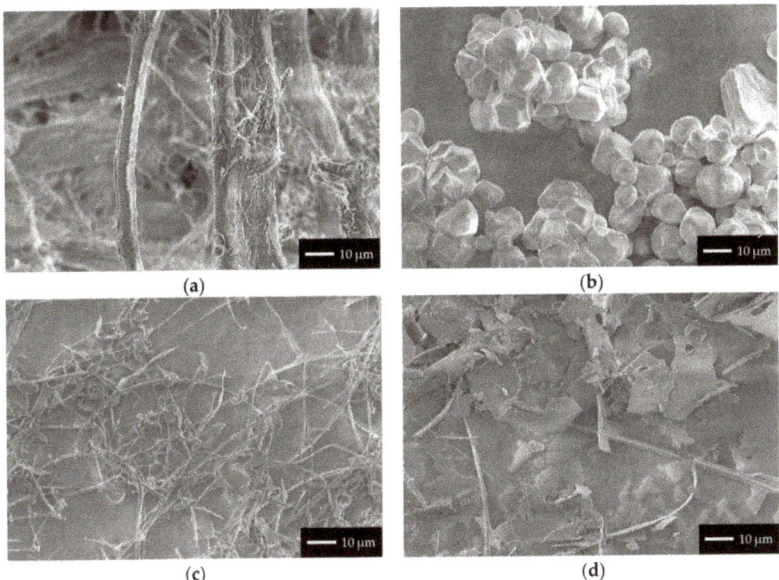

Figure 1. SEM images of (**a**) cotton linters, and (**b**) corn starch and the freeze-dried hydrolyzed products from (**c**) cotton linters, and (**d**) corn starch after hydrolysis and neutralization.

3.2. Properties of Nanoparticle Coatings

3.2.1. Surface Tension

Bare and untreated PLA had a surface tension of <34 mN m^{-1}, which is slightly below reported values (36–38 mN m^{-1} [50]). Corona-treating the substrate elevated the surface tension to 42 mN m^{-1}.

3.2.2. Dry Coating Thickness and Nanoparticle Loading

PLA and paper had thicknesses of 26.8 ± 0.1 µm and 62.5 ± 2.9 µm, respectively. Considering the concentration and the density of both the nanoparticles of 1.5 g cm^{-3} and Na$_2$SO$_4$ of 2.7 g cm^{-3}, the thickness of a single coated layer for a wet film thickness of 51 µm was estimated to be 2.5 µm for CNC and 2.7 µm for SNP. The actual coating thicknesses on PLA were 2.6 ± 0.8 µm for CNC and 5.9 ± 0.9 µm for SNP. Paper substrates were double-coated with both CNC and SNP, resulting in thicknesses of 9.5 ± 0.4 µm and 12.0 ± 1.0 µm, respectively.

The nanoparticle loading was calculated from the ratio of the mass of nanoparticles in the product m_{np} and the dry mass m_{dry} and was 75.6 ± 1.7 wt % in the CNC coating and 92.2 ± 2.6 wt % in the SNP coating.

3.2.3. Surface and Optical Properties

Coating PLA with CNC did not yield a uniform film (Figure 2a). Hydrolyzed residues and agglomerates were randomly distributed over the substrate surface. Hydrolyzed residues and agglomerates were found for SNP coatings as well, accompanied by fine fissures in the coating layer (Figure 2b).

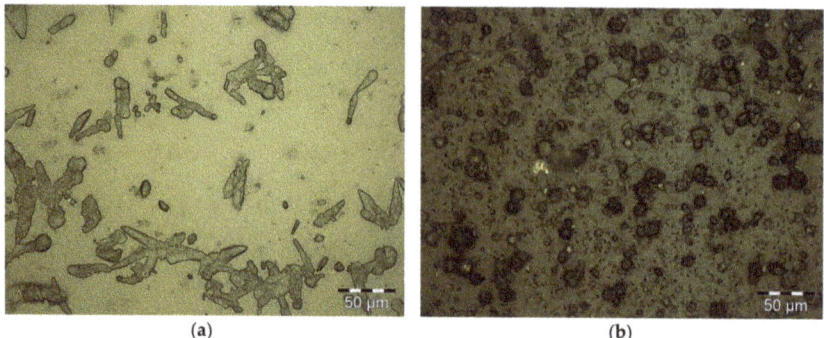

Figure 2. PLA coated with (**a**) CNC and (**b**) SNP in reflected bright-field microscopy.

The top side of the paper substrate was microscopically uniform (Figure 3a). The subjacent fibrous structure was visible via reflected light microscopy. While the top side appeared microscopically dense, the back side of the paper substrate showed pores with diameters in the micrometer range. Coating paper with CNC (Figure 3b) and SNP (Figure 3c) involved the deposition of hydrolyzed residues and agglomerates onto the surface, similar to the coatings on PLA. SEM imaging of the coated paper substrate shows the irregular surface topography caused by these residues (Figure 3e,f). Double-coating paper with SNP caused a more distinct topography.

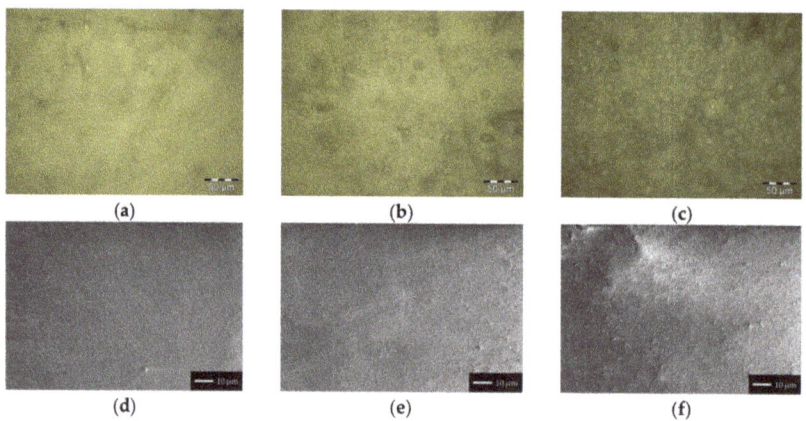

Figure 3. Paper (**a,d**) coated with CNC (**b,e**) or SNP (**c,f**) via reflected bright-field microscopy (**a–c**) and by secondary electron imaging via SEM (**d–f**).

The light transmittance of bare PLA of 91.4% ± 1.3% at a wavelength of 550 nm was reduced by the application of nanoparticle coatings. The CNC coating reduced the absolute light transmittance by 10.3% ± 1.6%, whereas SNP reduced the absolute light transmittance by 34.2% ± 1.7%.

3.2.4. Barrier Properties

Coating PLA with a single layer of CNC decreased the OP from 514.6 ± 3.8 cm^3 (STP) m^{-2} d^{-1} bar^{-1} to 129.7 ± 8.7 cm^3 (STP) m^{-2} d^{-1} bar^{-1} at 50% r.h. (74.8%) (Figure 4a). A decrease to an OP of 110.1 ± 14.2 cm^3 (STP) m^{-2} d^{-1} bar^{-1} (78.6%) was observed for a single coating layer of SNP. Considering the coating thickness, the OP Q_{100} of 4.7 ± 0.4 cm^3 (STP) 100 µm m^{-2} d^{-1} bar^{-1} and

8.5 ± 1.4 cm^3 (STP) 100 µm m^{-2} d^{-1} bar^{-1} for CNC and SNP resulted, emphasizing the noticeable barrier performance of CNC against oxygen compared to SNP (Figure 4b).

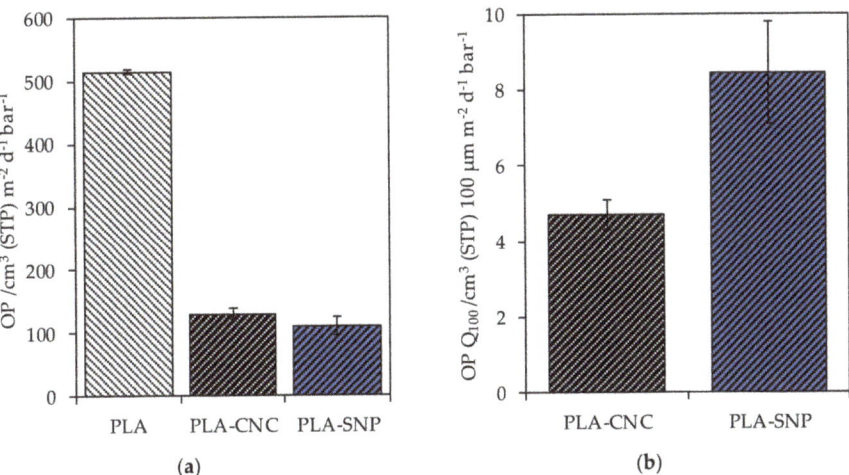

Figure 4. (a) The measured OTR of bare PLA and PLA substrate coated with CNC and SNP; (b) the normalized OTR to a layer thickness of 100 µm (*OTR* Q_{100}).

The nanoparticle coatings did not improve the water vapor barrier of coated PLA. The *WVP* Q_{100} of bare PLA substrate of 76.1 ± 3.1 g (STP) 100 µm m^{-2} d^{-1} remained almost constant for a coating with CNC (80.3 ± 4.6 g (STP) 100 µm m^{-2} d^{-1}; 85→0% r.h.) and SNP (81.4 ± 1.8 g (STP) 100 µm m^{-2} d^{-1}; 85→0% r.h.). The slight increase is explicable by water adsorption due to the hygroscopic character of both the coatings and the substrate in combination with the gravimetric measurement method.

Paper substrates double-coated with either CNC or SNP did not pass the pinhole test and were therefore excluded from the determination of barrier properties.

3.3. Properties of Cast Films

CNC and SNP were incorporated in hydrophilic starch matrices at different concentrations c_{filler} by solution casting. The target thickness was 50 µm. All experiments were repeated at least five times.

3.3.1. Microscopy and Optical Properties

Plastification at 70 °C for 1 h did not completely degrade the granular structure of corn starch. Swelling induced an increase of the grain size in the pure starch-glycerol film and ghost remnants were recognizable [51] (Figure 5a). Accordingly, the film surface displays the topography of the shells of the native starch granules (Figure 5d). Adding CNC (Figure 5b,e) or SNP (Figure 5c,f) in concentrations of $0 \leq c \leq 9$ wt % did not alter the microstructure of the film. Agglomerates or microscale residues from the hydrolyzed cellulose product were visible in both reflected bright-field microscopy and via SEM imaging. The visible accumulation suggests their segregation from the starch matrix during drying. Due to the similar appearance of the SNP and the starch matrix, no hydrolyzed starch residues were recognizable in these nanocomposites.

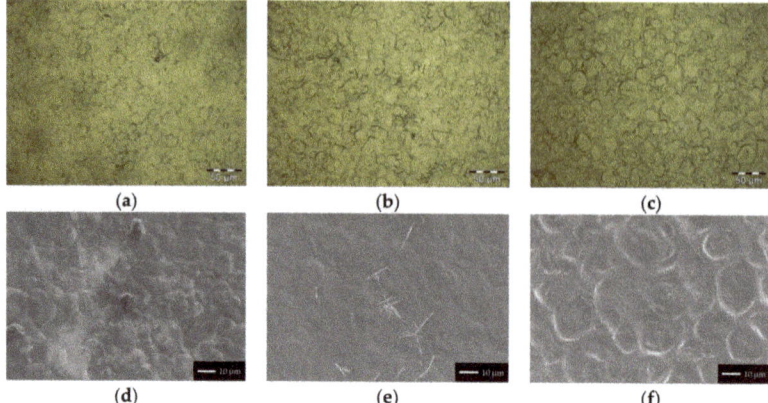

Figure 5. Pure starch-glycerol film (**a,d**), starch-CNC nanocomposite (**b,e**), and starch-SNP nanocomposite (**c,f**). The shown nanocomposites had a filler content of 3 wt %. (**a–c**) Reflected bright-field microscopy and (**d–f**) secondary electron imaging (SEM).

The light transmittance of a starch-glycerol film T_{starch} with a thickness d of 50 µm was 83.5 ± 2.1% (13.38 × 10^{-3} ± 2.3 × 10^{-3} µm^{-1}) at 550 nm. For better comparability, the light transmittance T was normalized with respect to d. The addition of CNC and SNP reduced the light transmittance with increasing nanoparticle concentration $c_{nanoparticles}$ (Figure 6a). This effect was more pronounced for the addition of SNP. Accordingly, the extinction coefficient ε decreased with increasing filler content and was overall higher for SNP nanocomposites (Figure 6b). The high uncertainties prevalent at low $c_{nanoparticles}$ arose from the strong relative weighting of variable film thicknesses.

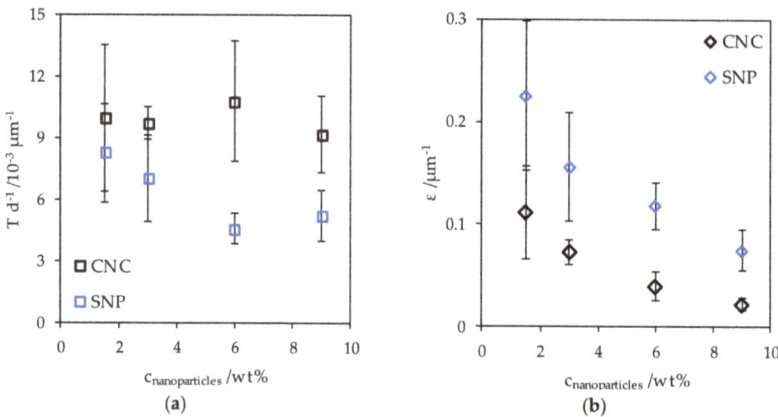

Figure 6. (**a**) The normalized transmittance of CNC and SNP nanocomposites and (**b**) the extinction coefficient at different nanoparticle concentrations.

3.3.2. Barrier Properties

A slight improvement of the WVP Q_{100} of cast starch-glycerol films was achieved by adding CNC or SNP (Table 3). However, all measured values lie within the 95% confidence interval of the WVP Q_{100} of 47.3 ± 20.6 g (STP) 100 µm m^{-2} d^{-1} (85→0% r.h.) of the pure starch-glycerol film.

Table 3. Normalized water vapor permeability of starch nanocomposites with different amounts of CNC or SNP.

Amount of Filler/wt %	WVP Q_{100}/g (STP) 100 µm m^{-2} d^{-1}	
	CNC	SNP
0.0	47.3 ± 20.6	–
1.5	40.3 ± 16.9	51.2 ± 15.7
3.0	43.6 ± 17.6	41.5 ± 14.5
6.0	45.4 ± 13.2	49.9 ± 12.4
9.0	43.7 ± 12.0	34.2 ± 16.8

3.3.3. Mechanical Properties

The ultimate tensile strength σ_{UTS} of the starch-glycerol films was 4.1 ± 2.1 MPa, Young's modulus E was 0.7 ± 0.4 MPa and the fracture strain ε_f was 23.7% ± 13.9%. The mechanical properties were changed by adding CNC and SNP, however, all measured values of nanocomposites lie within the 95% confidence interval of the values of starch-glycerol films (Figure 7). Although all samples were prepared with high diligence, the notch sensitivity of the organic composites caused the high uncertainties of the measured mechanical properties.

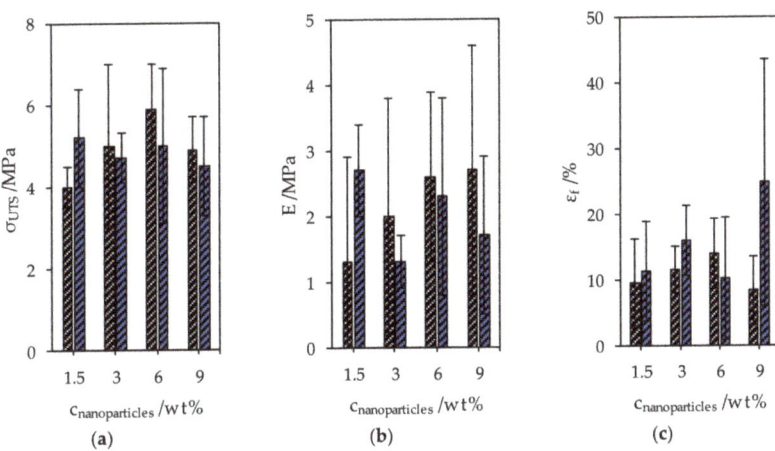

Figure 7. Mechanical properties of CNC and SNP nanocomposites. (a) The ultimate tensile strength, (b) Young's modulus, and (c) the elongation at break.

4. Discussion

4.1. Nanoparticle Dispersions

The extraction of nanoparticles from cotton linters and corn starch was performed by sulfuric acid hydrolysis followed by neutralization with sodium hydroxide. To reduce the amount of hydrolyzed cellulose residues in the CNC product, a comparably long hydrolysis time of 3 h was chosen. The achieved gross yield was still >40 wt %. Exemplarily, other studies addressing the extraction of CNC from cotton linters reported gross yields of 52.7 wt % after 45 min at 45 °C (64 wt % H_2SO_4, 1:17.5 g mL^{-1}) [52] and 54.4 wt % after 5 min at 45 °C (60 wt % H_2SO_4, 1:20 g mL^{-1}) [53]. Similar short hydrolysis times in combination with the here presented extraction method could facilitate a distinctly higher gross yield. To evaluate the degree of conversion of the raw cellulose to CNC and soluble residues, X-ray diffraction measurements could complement the process evaluation by giving information about the product crystallinity. Analogously, the gross yield of SNP from corn starch of 32.2 ± 0.7 wt % is comparably high (15% after 120 h at 40 °C and 25 wt % H_2SO_4 [54]).

The desired high ionic strength during washing enables flocculation of the nanoparticles. Therefore, a separation from the reaction solution is possible. The precipitation-redispersion mechanism enabled the removal of more than 99.8% of the ionic residues. Consequently, the nanoparticle dispersions showed no macroscopic phase separation over several weeks at 23 °C. Nevertheless, the presence of ions in the dispersions is expected to promote the formation of agglomerates [55,56] effecting larger apparent particle sizes. The actual particle size could be detected by atomic force microscopy and transmission electron microscopy.

A scale-up scenario regarding process time and the consumption of chemicals (Table 4) is derived from the applied process parameters and compared to a scaling approach documented by Reiner et al. [57]. Both approaches are normalized to a CNC product mass of 1 kg based on the respective yield. The rate-determining step of our approach is the hydrolysis time of 3 h. Further steps, comprising neutralization, washing, and homogenization, require only 1 h with the used equipment. Thus, the net process time is 4 h per batch. Reiner et al. used kraft pulp as feed stock and stopped the hydrolysis after 1.5 h by dilution and subsequent neutralization with NaOH. The unit operation times are given as 8 h for hydrolysis and neutralization, 24–48 h for gravity settling and initial purification and 24 h for filtration. The CNC produced by the overall faster process presented in this work has a higher residual ion content, but a markedly lower overall water consumption (75%) and facilitates a more compact reactor volume.

Table 4. Comparison of the masses of chemicals required to extract 1 kg of CNC from raw cellulose, based on the neutralization of sulfuric acid.

Chemical	Applied Parameters m/kg	Reiner et al. [57] m/kg
raw cellulose	2.4	2.0
sulfuric acid, 64 wt %	37.2	18.5
hypochlorite	–	0.003
sodium hydroxide	14.6	9.8
water	196.3	800.0

Analogously, 1 kg of SNP is produced from 3.1 kg corn starch converted with 26.3 kg H_2SO_4 (18 wt %). Neutralization and washing require 81.2 kg H_2O. Complete neutralization is achieved by adding 16.9 kg NaOH. Due to the protracted hydrolysis, the net process time is 121 h per batch.

4.2. Nanoparticle Coatings

PLA needed to be corona-treated prior to the application of aqueous nanoparticle dispersions onto the hydrophobic substrate by blade-coating of the coating medium without contraction. High nanoparticle concentrations and therefore increased viscosities further facilitated good spreadability of the nanoparticle dispersions on the substrates. In particular, the presence of ions induced the gel-like character of the CNC dispersion [58]. The dry coating thicknesses of both CNC and SNP coatings exceeded the targeted values on PLA. It was assumed that in both cases, non-dense layers formed on the substrate surface due to agglomeration and the presence of microparticles. Furthermore, the hydration of the salt residues as well as water absorption of the hygroscopic nanoparticles must be factored in. Same applies for the coatings on paper. The intrinsically less smooth surface of paper was assumed to additionally contribute to the deviating dry coating thicknesses. For both CNC and SNP coatings, hydrolyzed residues as well as agglomerates accumulated on the coated substrates and thereby reduced the optical transmittance of the films.

The observed effect of lower oxygen permeation of CNC and SNP coatings can be attributed to the size and the structural organization of the nanoparticles in the coating layer. The structural organization influences the diffusion path length of gas molecules in the film [59]. No improvement regarding the water vapor permeation was found. The intrinsic hydrophobicity of the nanoparticles in conjunction with the hygroscopic effect of the ionic residues is assumed to particularly impair the water vapor barrier properties. Microscopic cracks were found in starch coatings on PLA indicating embrittlement during solvent evaporation. Gentler drying conditions are not viewed as expedient. Instead, the addition of plasticizers may facilitate the prevention of cracks and lead to improved techno-functional properties [60].

Since the application of a double coating layer onto paper did not yield pinhole-free substrates, paper was excluded from further analyses. Alternatively to blade-coating, impregnating paper by dip-coating may lead to a pinhole-free substrate [5], however, accompanied by a higher expenditure of nanoparticles.

Results from other studies addressing the oxygen permeability of CNC coatings and cast films are shown in Table 5. A strong impact of the r.h., the substrate material and the nanoparticles themselves is recognizable. The $OP\ Q_{100}$ at 50% r.h. of coatings in the present study were in the same range as plasticized nanocellulose films. However, compared to other approaches, the $OP\ Q_{100}$ at 50% r.h. was up to 2 orders of magnitude higher. It was concluded that narrowing the particle size distributions of CNC and SNP by removing aggregates may be the decisive factor to further reduce the oxygen permeability of the nanoparticle coatings on polymer substrates.

Table 5. Comparison of oxygen permeabilities of coatings of CNC and SNP with results from other studies.

Barrier Film	r.h./%	T/°C	$OP\ Q_{100}/\text{cm}^3$ (STP) 100 μm m^{-2} d^{-1} bar^{-1}	Ref.
CNC; 2.6 μm on PLA substrate	50	23	4.7 ± 0.4	this study
SNP; 5.9 μm on PLA substrate	50	23	8.5 ± 1.4	
CNC, 1.5 μm on PET substrate OPP substrate OPA substrate	0	23	0.02 0.37 0.003	[2]
TEMPO-oxidized cellulose, 1.5 μm on PET substrate (50 μm) PLA substrate (25 μm) PET substrate (50 μm) PLA substrate (25 μm)	0 50 	23	0.0006–0.0046 0.0009–0.003 0.0225–0.09 0.38–0.96	[61]
Nanocellulose film	0 50	23	<0.01 0.3	
Nanocellulose film (plasticized)	0 50	23	0.03 17	[62]
Nanocellulose film (heat-treated)	0 50	23	<0.01 0.02	

4.3. Cast Films

Solvated and swollen starch granules formed the microstructure of the cast films. In CNC nanocomposites the agglomerates and hydrolyzed residues segregated from the starch-glycerol matrix. Similar behavior was assumed from the techno-functional properties of SNP nanocomposites. No improvement was determined regarding the WVP and the mechanical properties of CNC and SNP nanocomposites. Both types of nanoparticles may have a low percolation threshold in the starch-glycerol film. Thus, a negligible increase of the tortuosity and hindered force transmission in the polymer matrix resulted [59,63]. Agglomerates are considered as potential predetermined breaking points during applied mechanical stress. The reduced relative light transmittance and extinction coefficient with increasing CNC and SNP loading in the nanocomposites indicated increasing light scattering on nanoparticles and agglomerates.

Prolonged mixing at elevated temperature may have led to desulfation of the nanoparticles and promoted agglomeration [64]. Reducing mixing time and temperature in the polymer precursor as well as the addition of surfactants [32] and the adjustment of the surface energy by altering the ionic strength of the nanoparticle dispersions is considered beneficial.

5. Conclusions

The restricted efficiency of the extraction of CNC and SNP is one main limiting factor for the effective competition with already established additives for packaging materials [65]. While fossil-based polymers for packaging are still far ahead regarding economic competitiveness and techno-functional properties, such as barrier and mechanical properties, the potential of CNC and SNP is evident. In favor of the reduction of the complexity of multi-layer films and towards more ecofriendly and sustainable packaging solutions, CNC and SNP are promising candidates to enhance bio-based polymers with intrinsically attenuated barrier and mechanical properties. In this regard, an efficient post-processing method contributing to an overall economic extraction process was applied in this paper along with two examples of application. The use of non-functionalized CNC and SNP as coatings and nanofillers is attractive regarding an overall fast and green process.

From this study, we conclude:

(h1) The neutralization-based approach for the extraction of CNC and SNP from biopolymeric feed stocks in combination with the applied post-processing method is time-efficient and resource-saving. With hydrolysis times of 3 h for cellulose and 120 h for starch, gross yields of 41.4% and 32.2%, respectively, were achieved. The nanoparticle dispersions showed long-term stability.

(h2) Reduced oxygen permeabilities were achieved with coatings of CNC and SNP on PLA. The barrier improvement of paper failed due to pinholes in the substrate.

(h3) No improvement of the WVP and the mechanical properties of cast starch-glycerol films incorporating CNC or SNP were found. It was deduced from the results that agglomeration of the nanoparticles during the cast film preparation and segregation of the filler from the matrix during drying compromised the techno-functional properties.

Further purification of the hydrolyzed products regarding ionic residues and agglomerates is suggested to improve the techno-functional properties of CNC and SNP applied in coatings and cast nanocomposite films.

Acknowledgments: This work was supported by the German Research Foundation (DFG) and the Technical University of Munich (TUM) in the framework of the Open Access Publishing Program. We did not receive any specific grant from funding agencies in the public, commercial, or not-for-profit sectors. The authors thank Vesna Müller for helpful discussions in the early stage of the study as well as Petra Dietl, Zuzana Scheurer, and Markus Pummer for their experimental contributions. Moreover, we want to thank Michael Gebhardt (TUM, chair of zoology) for providing access to the scanning electron microscope.

Author Contributions: Christoph Metzger and Solange Sanahuja conceived and designed the experiments; Christoph Metzger and Lisa Behrends carried out the experiments; Christoph Metzger, Martina Lindner, Heiko Briesen, and Sven Sängerlaub analyzed the data and interpreted the results; Heiko Briesen, Sven Sängerlaub and Martina Lindner contributed reagents, materials and measurement equipment, and contributed to and edited the manuscript; Christoph Metzger wrote the manuscript.

Conflicts of Interest: The authors declare no conflict of interest. The funding sponsors had no role in the design of the study; in the collection, analyses, or interpretation of data; in the writing of the manuscript, and in the decision to publish the results.

References

1. LeCorre, D.; Dufresne, A.; Rueff, M.; Khelifi, B.; Bras, J. All starch nanocomposite coating for barrier material. *J. Appl. Polym. Sci.* **2014**, *131*, 39826. [CrossRef]
2. Li, F.; Biagioni, P.; Bollani, M.; Maccagnan, A.; Piergiovanni, L. Multi-functional coating of cellulose nanocrystals for flexible packaging applications. *Cellulose* **2013**, *20*, 2491–2504. [CrossRef]
3. Belbekhouche, S.; Bras, J.; Siqueira, G.; Chappey, C.; Lebrun, L.; Khelifi, B.; Marais, S.; Dufresne, A. Water sorption behavior and gas barrier properties of cellulose whiskers and microfibrils films. *Carbohydr. Polym.* **2011**, *83*, 1740–1748. [CrossRef]
4. Herrera, M.A.; Mathew, A.P.; Oksman, K. Gas permeability and selectivity of cellulose nanocrystals films (layers) deposited by spin coating. *Carbohydr. Polym.* **2014**, *112*, 494–501. [CrossRef] [PubMed]

5. Herrera, M.A.; Sirviö, J.A.; Mathew, A.P.; Oksman, K. Environmental friendly and sustainable gas barrier on porous materials. *Mater. Des.* **2016**, *93*, 19–25. [CrossRef]
6. Zheng, H.; Ai, F.; Chang, P.R.; Huang, J.; Dufresne, A. Structure and properties of starch nanocrystal-reinforced soy protein plastics. *Polym. Compos.* **2009**, *30*, 474–480. [CrossRef]
7. Angellier, H.; Molina-Boisseau, S.; Dufresne, A. Mechanical properties of waxy maize starch nanocrystal reinforced natural rubber. *Macromolecules* **2005**, *38*, 9161–9170. [CrossRef]
8. Pei, A.; Malho, J.-M.; Ruokolainen, J.; Zhou, Q.; Berglund, L.A. Strong nanocomposite reinforcement effects in polyurethane elastomer with low volume fraction of cellulose nanocrystals. *Macromolecules* **2011**, *44*, 4422–4427. [CrossRef]
9. Khan, A.; Khan, R.A.; Salmieri, S.; Le Tien, C.; Riedl, B.; Bouchard, J.; Chauve, G.; Tan, V.; Kamal, M.R.; Lacroix, M. Mechanical and barrier properties of nanocrystalline cellulose reinforced chitosan based nanocomposite films. *Carbohydr. Polym.* **2012**, *90*, 1601–1608. [CrossRef] [PubMed]
10. Fortunati, E.; Peltzer, M.; Armentano, I.; Torre, L.; Jimenez, A.; Kenny, J.M. Effects of modified cellulose nanocrystals on the barrier and migration properties of PLA nano-biocomposites. *Carbohydr. Polym.* **2012**, *90*, 948–956. [CrossRef] [PubMed]
11. Thompson, R.C.; Moore, C.J.; Vom Saal, F.S.; Swan, S.H. Plastics, the environment and human health: Current consensus and future trends. *Philos. Trans. R. Soc. B Biol. Sci.* **2009**, *364*, 2153–2166. [CrossRef] [PubMed]
12. Miller, S.A. Sustainable polymers: Opportunities for the next decade. *ACS Macro Lett.* **2013**, *2*, 550–554. [CrossRef]
13. Davis, G.; Song, J.H. Biodegradable packaging based on raw materials from crops and their impact on waste management. *Ind. Crops Prod.* **2006**, *23*, 147–161. [CrossRef]
14. Niaounakis, M. *Biopolymers: Processing and Products*; Elsevier Science: Burlington, NJ, USA, 2015.
15. Markarian, J. Biopolymers present new market opportunities for additives in packaging. *Plast. Addit. Compd.* **2008**, *10*, 22–25. [CrossRef]
16. Vink, E.T.H.; Rábago, K.R.; Glassner, D.A.; Gruber, P.R. Applications of life cycle assessment to NatureWorks™ polylactide (PLA) production. *Polym. Degrad. Stab.* **2003**, *80*, 403–419. [CrossRef]
17. Carothers, W.H.; Dorough, G.L.; van Natta, F.J. Studies of polymerization and ring formation—X. The reversible polymerization of six-membered cyclic esters. *J. Am. Chem. Soc.* **1932**, *54*, 761–772. [CrossRef]
18. Lunt, J. Large-scale production, properties and commercial applications of polylactic acid polymers. *Polym. Degrad. Stab.* **1998**, *59*, 145–152. [CrossRef]
19. Gruber, P.; O'Brien, M. Polylactides "NatureWorks® PLA". In *Biopolymers Online*; Steinbüchel, A., Ed.; Wiley-VCH: Weinheim, Germany, 2005.
20. Auras, R.; Harte, B.; Selke, S. An overview of polylactides as packaging materials. *Macromol. Biosci.* **2004**, *4*, 835–864. [CrossRef] [PubMed]
21. Auras, R.; Harte, B.; Selke, S. Effect of water on the oxygen barrier properties of poly(ethylene terephthalate) and polylactide films. *J. Appl. Polym. Sci.* **2004**, *92*, 1790–1803. [CrossRef]
22. Chaiwong, C.; Rachtanapun, P.; Wongchaiya, P.; Auras, R.; Boonyawan, D. Effect of plasma treatment on hydrophobicity and barrier property of polylactic acid. *Surf. Coat. Technol.* **2010**, *204*, 2933–2939. [CrossRef]
23. Siracusa, V.; Blanco, I.; Romani, S.; Tylewicz, U.; Rocculi, P.; Rosa, M.D. Poly(lactic acid)-modified Films for food packaging application: Physical, mechanical, and barrier behavior. *J. Appl. Polym. Sci.* **2012**, *125*, 390–401. [CrossRef]
24. Pant, A.F.; Sängerlaub, S.; Müller, K. Gallic acid as an oxygen scavenger in bio-based multilayer packaging films. *Materials* **2017**, *10*, 11. [CrossRef] [PubMed]
25. Pan, Y.; Xiao, H.; Song, Z. Hydrophobic modification of cellulose fibres by cationic-modified polyacrylate latex with core-shell structure. *Cellulose* **2013**, *20*, 485–494. [CrossRef]
26. Miller, K.S.; Krochta, J.M. Oxygen and aroma barrier properties of edible films: A review. *Trends Food Sci. Technol.* **1997**, *8*, 228–237. [CrossRef]
27. Rhim, J.-W.; Park, H.-M.; Ha, C.-S. Bio-nanocomposites for food packaging applications. *Prog. Polym. Sci.* **2013**, *38*, 1629–1652. [CrossRef]
28. Roman, M. Toxicity of cellulose nanocrystals—A review. *Ind. Biotechnol.* **2015**, *11*, 25–33. [CrossRef]
29. Rånby, B.G.; Banderet, A.; Sillén, L.G. Aqueous colloidal solutions of cellulose micelles. *Acta Chem. Scand.* **1949**, *3*, 649–650. [CrossRef]

30. Habibi, Y.; Lucia, L.A.; Rojas, O.J. Cellulose nanocrystals: Chemistry, self-assembly, and applications. *Chem. Rev.* **2010**, *110*, 3479–3500. [CrossRef] [PubMed]
31. Müller, V.; Briesen, H. Nanocrystalline Cellulose, Its Preparation and Uses of Such Nanocrystalline Cellulose. U.S. Patent 20170306056 A1, 26 October 2017.
32. Kim, J.; Montero, G.; Habibi, Y.; Hinestroza, J.P.; Genzer, J.; Argyropoulos, D.S.; Rojas, O.J. Dispersion of cellulose crystallites by nonionic surfactants in a hydrophobic polymer matrix. *Polym. Eng. Sci.* **2009**, *49*, 2054–2061. [CrossRef]
33. Alves, J.S.; dos Reis, K.C.; Menezes, E.G.T.; Pereira, F.V.; Pereira, J. Effect of cellulose nanocrystals and gelatin in corn starch plasticized films. *Carbohydr. Polym.* **2015**, *115*, 215–222. [CrossRef] [PubMed]
34. Li, F.; Mascheroni, E.; Piergiovanni, L. The potential of nanocellulose in the packaging field: A review. *Packag. Technol. Sci.* **2015**, *28*, 475–508. [CrossRef]
35. Liu, H.; Brinson, L.C. Reinforcing efficiency of nanoparticles: A simple comparison for polymer nanocomposites. *Compos. Sci. Technol.* **2008**, *68*, 1502–1512. [CrossRef]
36. Habibi, Y. Key advances in the chemical modification of nanocelluloses. *Chem. Soc. Rev.* **2014**, *43*, 1519–1542. [CrossRef] [PubMed]
37. Murphy, S.H.; Marsh, J.J.; Kelly, C.A.; Leeke, G.A.; Jenkins, M.J. CO_2 assisted blending of poly(lactic acid) and poly(ε-caprolactone). *Eur. Polym. J.* **2017**, *88*, 34–43. [CrossRef]
38. Ge, H.; Yang, F.; Hao, Y.; Wu, G.; Zhang, H.; Dong, L. Thermal, mechanical, and rheological properties of plasticized poly(L-lactic acid). *J. Appl. Polym. Sci.* **2013**, *127*, 2832–2839. [CrossRef]
39. Mihai, M.; Huneault, M.A.; Favis, B.D. crystallinity development in cellular poly(lactic acid) in the presence of supercritical carbon dioxide. *J. Appl. Polym. Sci.* **2009**, *113*, 2920–2932. [CrossRef]
40. Angellier, H.; Choisnard, L.; Molina-Boisseau, S.; Ozil, P.; Dufresne, A. Optimization of the preparation of aqueous suspensions of waxy maize starch nanocrystals using a response surface methodology. *Biomacromolecules* **2004**, *5*, 1545–1551. [CrossRef] [PubMed]
41. Lindner, M.; Rodler, N.; Jesdinszki, M.; Schmid, M.; Sängerlaub, S. Surface energy of corona treated PP, PE and PET films, its alteration as function of storage time and the effect of various corona dosages on their bond strength after lamination. *J. Appl. Polym. Sci.* **2018**, *135*, 45842. [CrossRef]
42. Beck, S.; Bouchard, J. Auto-catalyzed acidic desulfation of cellulose nanocrystals. *Nord. Pulp Pap. Res. J.* **2014**, *29*, 6–14. [CrossRef]
43. Lindner, M. Factors affecting the hygroexpansion of paper. *J. Mater. Sci.* **2018**, *53*, 1–26. [CrossRef]
44. Jost, V.; Kobsik, K.; Schmid, M.; Noller, K. Influence of plasticiser on the barrier, mechanical and grease resistance properties of alginate cast films. *Carbohydr. Polym.* **2014**, *110*, 309–319. [CrossRef] [PubMed]
45. DIN 53380-3 *Testing of Plastics—Determination of Gas Transmission Rate—Part 3: Oxygen-Specific Carrier Gas Method for Testing of Plastic Films and Plastics Mouldings*; German National Standard: Berlin, Germany, 1998.
46. Langowski, H.-C. Permeation of gases and condensable substances through monolayer and multilayer structures. In *Plastic Packaging: Interactions with Food and Pharmaceuticals*, 2nd ed.; Piringer, O.G., Baner, A.L., Eds.; Wiley-VCH: Weinheim, Germany, 2008.
47. DIN 53122-1 *Testing of Plastics and Elastomer Films, Paper, Board and Other Sheet Materials—Determination of Water Vapour Transmission—Part 1: Gravimetric Method*; German National Standard: Berlin, Germany, 2001.
48. DIN EN ISO 527-3 *Plastics—Determination of Tensile Properties—Part 3: Test Conditions for Films and Sheets*; German National Standard: Berlin, Germany, 2012.
49. LeCorre, D.; Bras, J.; Dufresne, A. Evidence of micro- and nanoscaled particles during starch nanocrystals preparation and their isolation. *Biomacromolecules* **2011**, *12*, 3039–3046. [CrossRef] [PubMed]
50. Zirkel, L. PLA for thermoforming. *Bioplast. Mag.* **2012**, *7*, 18–20.
51. Zhang, B.; Dhital, S.; Flanagan, B.M.; Gidley, M.J. Mechanism for starch granule ghost formation deduced from structural and enzyme digestion properties. *J. Agric. Food Chem.* **2014**, *62*, 760–771. [CrossRef] [PubMed]
52. Mascheroni, E.; Rampazzo, R.; Ortenzi, M.A.; Piva, G.; Bonetti, S.; Piergiovanni, L. Comparison of cellulose nanocrystals obtained by sulfuric acid hydrolysis and ammonium persulfate, to be used as coating on flexible food-packaging materials. *Cellulose* **2016**, *23*, 779–793. [CrossRef]
53. Chang, C.-P.; Wang, I.-C.; Hung, K.-J.; Perng, Y.-S. Preparation and characterization of nanocrystalline cellulose by acid hydrolysis of cotton linter. *Taiwan J. For. Sci.* **2010**, *25*, 251–264.
54. LeCorre, D.; Bras, J.; Choisnard, L.; Dufresne, A. Optimization of the batch preparation of starch nanocrystals to reach daily time-scale. *Starch* **2012**, *64*, 489–496. [CrossRef]

55. Phan-Xuan, T.; Thuresson, A.; Skepö, M.; Labrador, A.; Bordes, R.; Matic, A. Aggregation behavior of aqueous cellulose nanocrystals: The effect of inorganic salts. *Cellulose* **2016**, *23*, 3653–3663. [CrossRef]
56. Cherhal, F.; Cousin, F.; Capron, I. Influence of charge density and ionic strength on the aggregation process of cellulose nanocrystals in aqueous suspension, as revealed by small-angle neutron scattering. *Langmuir* **2015**, *31*, 5596–5602. [CrossRef] [PubMed]
57. Reiner, R.S.; Rudie, A.W. Process scale-up of cellulose nanocrystal production to 25 kg per batch at the forest products laboratory. In *Production and Applications of Cellulose Nanomaterials*; Postek, M.T., Moon, R.J., Rudie, A.W., Bilodeau, M.A., Eds.; TAPPI Press: Peachtree Corners, GA, USA, 2013; pp. 21–24.
58. Xu, Y.; Atrens, A.D.; Stokes, J.R. Rheology and microstructure of aqueous suspensions of nanocrystalline cellulose rods. *J. Colloid Interface Sci.* **2017**, *496*, 130–140. [CrossRef] [PubMed]
59. Nielsen, L.E. Models for the permeability of filled polymer systems. *J. Macromol. Sci. Part A Chem.* **1967**, *1*, 929–942. [CrossRef]
60. Sanyang, M.L.; Sapuan, S.M.; Jawaid, M.; Ishak, M.R.; Sahari, J. Effect of plasticizer type and concentration on physical properties of biodegradable films based on sugar palm (arenga pinnata) starch for food packaging. *J. Food Sci. Technol.* **2016**, *53*, 326–336. [CrossRef] [PubMed]
61. Fukuzumi, H.; Saito, T.; Isogai, A. Influence of TEMPO-oxidized cellulose nanofibril length on film properties. *Carbohydr. Polym.* **2013**, *93*, 172–177. [CrossRef] [PubMed]
62. Vartiainen, J.; Kaljunen, T.; Kunnari, V.; Lahtinen, P.; Salminen, A.; Seppälä, J.; Tammelin, T. Nanocellulose films: Towards large scale and continuous production. In Proceedings of the 26th IAPRI Symposium on Packaging 2013, Espoo, Finland, 10–13 June 2013; Nissi, M.V., Ed.; VTT: Espoo, Finland, 2013; pp. 197–209.
63. Favier, V.; Dendievel, R.; Canova, G.; Cavaille, J.Y.; Gilormini, P. Simulation and modeling of three-dimensional percolating structures: Case of a latex matrix reinforced by a network of cellulose fibers. *Acta Mater.* **1997**, *45*, 1557–1565. [CrossRef]
64. Jiang, F.; Esker, A.R.; Roman, M. Acid-catalyzed and solvolytic desulfation of H_2SO_4-hydrolyzed cellulose nanocrystals. *Langmuir* **2010**, *26*, 17919–17925. [CrossRef] [PubMed]
65. Tang, X.Z.; Kumar, P.; Alavi, S.; Sandeep, K.P. Recent advances in biopolymers and biopolymer-based nanocomposites for food packaging materials. *Crit. Rev. Food Sci. Nutr.* **2012**, *52*, 426–442. [CrossRef] [PubMed]

© 2018 by the authors. Licensee MDPI, Basel, Switzerland. This article is an open access article distributed under the terms and conditions of the Creative Commons Attribution (CC BY) license (http://creativecommons.org/licenses/by/4.0/).

MDPI
St. Alban-Anlage 66
4052 Basel
Switzerland
Tel. +41 61 683 77 34
Fax +41 61 302 89 18
www.mdpi.com

Coatings Editorial Office
E-mail: coatings@mdpi.com
www.mdpi.com/journal/coatings

www.ingramcontent.com/pod-product-compliance
Lightning Source LLC
LaVergne TN
LVHW070226100526
838202LV00015B/2098